ADVANCES IN EXPLORATION GEOPHYSICS 2

ELASTIC WAVE FIELD EXTRAPOLATION

REDATUMING OF SINGLE- AND MULTI-COMPONENT SEISMIC DATA

ADVANCES IN EXPLORATION GEOPHYSICS 2

ELASTIC WAVE FIELD EXTRAPOLATION

REDATUMING OF SINGLE- AND MULTI-COMPONENT SEISMIC DATA

C.P.A. WAPENAAR

AND

A.J. BERKHOUT

*Laboratory of Seismics and Acoustics, Delft University of Technology,
P.O. Box 5046, 2600 GA Delft, The Netherlands*

ELSEVIER
Amsterdam — Oxford — New York — Tokyo 1989

ELSEVIER SCIENCE PUBLISHERS B.V.
Sara Burgerhartstraat 25
P.O. Box 211, 1000 AE Amsterdam, The Netherlands

Distributors for the United States and Canada:

ELSEVIER SCIENCE PUBLISHING COMPANY INC.
655, Avenue of the Americas
New York, NY 10010, U.S.A.

ISBN 0-444-88472-6

Printed in The Netherlands

To my mother,

Kees Wapenaar

PREFACE

The essence of the seismic method is given by propagation and reflection. The source wave field propagates down into the subsurface, reflects at the layer boundaries, and the reflected wave field propagates back to the surface. Hence, the seismic response we measure at the surface represents a mixture of propagation and reflection information. The major part of seismic processing is dedicated to the elimination of propagation effects from the seismic response, aiming to correctly position the true amplitude reflectivity. Propagation is determined by the trend of the subsurface (macro layering) and reflection is determined by the detail of the subsurface (fine layering). Consequently, for the elimination of propagation effects a macro model of the subsurface should be available.

Seismic redatuming is a popular concept, indicating the wave field extrapolation process that transforms seismic measurements from the *actual* data acquisition surface (old datum) to a *simulated* data acquisition surface (new datum) down in the subsurface. Often the new datum will present the top of a target zone. After redatuming the propagation effects (down and up) of the target overburden have been removed from the target reflections and, particularly for a structurally complicated overburden, the target response will be simplified significantly.

It may be stated that wave field extrapolation has become one of the most important tools in modern seismic processing. In principle, stratigraphic and lithologic inversion in a target zone should be preceded by multi-offset redatuming. And, last but not least, redatuming is the 'heart' of any migration technique.

In this book wave field extrapolation is extensively discussed from a theoretical point of view. The acoustic case and the elastic case are treated separately. The reader who is just interested in the acoustic case may study the odd numbered chapters only. However, for each subject an instructive comparison between the acoustic and elastic case is obtained by reading an odd numbered chapter together with the next even numbered chapter.

We have prepared the material in this book for geophysicists who are professionally involved with advanced wave theory concepts. In particular this book will be useful for research geophysicists who are designing or using wave theory based processing techniques for single- or multi-component seismic data from structurally complicated areas.

This book would never have been finished without the help of many of the members of the professional staff of the "Delft Laboratory of Seismics and Acoustics". We are very grateful to Mr. G.C. Haimé who generated all the elastic data examples with his finite difference modeling software. Also he produced the examples on elastic inverse wave field extrapolation in chapter VIII. The elastic decomposition (chapter XII) was carried out by Mr. P. Herrmann; Mr. D.J. Verschuur generated all the examples on acoustic and elastic multiple elimination. The examples on inverse wave field extrapolation in high contrast media (chapters IX and X) were generated by Mr. C.G.M. de Bruin. Dr. N.A. Kinneging produced the examples on 2-D and 3-D acoustic redatuming. After he graduated, Mr. H.L.H. Cox adopted his software and carried out the elastic redatuming (chapter XII). The examples on true amplitude acoustic inverse wave field extrapolation (chapter VII) were prepared by Dr. G.L. Peels and Mr. V. Budejicky; Mrs. W.E.A. Rietveld and R. Arts generated the examples in chapters I and II. Several of the people mentioned above, as well as Mr. G. Blacquière and Mr. A.D. Lemaire critically carried out the proofreading. The support of all these people is greatly appreciated.
Many thanks are also due to Mr. A.S.G. de Knegt of the drawing office who prepared the illustrations. Finally, we are very grateful to Hanneke who designed the lay-out and patiently typed the manuscript.

Dr. C.P.A. Wapenaar
Dr. A.J. Berkhout
Delft, August 1989

TABLE OF CONTENTS

PREFACE VII

INTRODUCTION 1

CHAPTER I. ACOUSTIC WAVES

1. Introduction 7

2. Acoustic wave equation 7

2.1. Conservation of mass 7

2.2. Conservation of momentum 8

2.3 Constitutive relation 9

2.4. Linearization of the basic equations 10

2.5 Acoustic two-way wave equation 12

3. Spherical wave solutions of the acoustic two-way wave equation 13

3.1 Monopole and dipole sources 13

3.2 The wave equation in spherical coordinates 15

3.3 Causal and anti-causal solutions 16

3.4 Monopole and dipole wave fields 18

4. Plane wave solutions of the acoustic two-way wave equation 18

4.1 Homogeneous plane waves 19

4.2 Inhomogeneous plane waves 24

4.3 Acoustic one-way wave equations for plane waves... 28

5. References 33

CHAPTER II ELASTIC WAVES

1. **Introduction** 35
2. **Elastic wave equation** 35
2.1 Stress and strain 36
2.2 Conservation of momentum 39
2.3 Constitutive relation 42
2.4 Elastic two-way wave equation 45
2.5 P- and S-wave potentials 47
3. **Spherical wave solutions of the elastic two-way wave equation** 49
3.1 P- and S-wave sources 50
3.2 Spherical P- and S-wave fields 51
4. **Plane wave solutions of the elastic two-way wave equation** 54
4.1 Homogeneous plane waves 54
4.2 Inhomogeneous plane waves 66
4.3 Elastic one-way wave equations for plane waves 70
5. **References** 72

CHAPTER III ACOUSTIC TWO-WAY AND ONE-WAY WAVE EQUATIONS

1. **Introduction** 75
2. **Acoustic wave equations for horizontally layered media** 75
2.1 Temporal and spatial Fourier transformations 76
2.2 Acoustic two-way wave equation in the wavenumber-frequency domain 78
2.3 Acoustic one-way wave equations in the wavenumber-frequency domain 87
2.4 Acoustic one-way wave fields at interfaces in the wavenumber-frequency domain 94
3. **Acoustic wave equations for arbitrarily inhomogeneous media** 98
3.1 Acoustic two-way wave equation in the space-frequency domain 99
3.2 Acoustic one-way wave equations in the space-frequency domain 101
3.3 Acoustic one-way wave fields at interfaces in the space-frequency domain 110
4. **References** 113

CHAPTER IV ELASTIC TWO-WAY AND ONE-WAY WAVE EQUATIONS

1. **Introduction** 115
2. **Elastic wave equations for horizontally layered media** 115
2.1 Elastic two-way wave equation in the wavenumber-frequency domain 116
2.2 Elastic one-way wave equations in the wavenumber-frequency domain 119
2.3 Elastic one-way wave fields at interfaces in the wavenumber-frequency domain 138
3. **Elastic wave equations for arbitrarily inhomogeneous media** 143
3.1 Elastic two-way wave equation in the space-frequency domain 144
3.2 Elastic one-way wave equations in the space-frequency domain 146
3.3 Elastic one-way wave fields at interfaces in the space-frequency domain 153
4. **References** 156

CHAPTER V ACOUSTIC FORWARD WAVE FIELD EXTRAPOLATION

1. **Introduction** 159
2. **Acoustic reciprocity theorems** 159
3. **Acoustic representation theorems** 166
3.1 Acoustic Green's functions 166
3.2 Acoustic Kirchhoff-Helmholtz integral 171
4. **Acoustic two-way and one-way Rayleigh integrals** 174
4.1 Boundary conditions for the acoustic Green's functions 175
4.2 Acoustic two-way Rayleigh integrals 176
4.3 Acoustic one-way Rayleigh integrals 180
5. **Acoustic forward wave field extrapolation operators** 185
5.1 Integral formulation of acoustic forward wave field extrapolation 185
5.2 Matrix formulation of acoustic forward wave field extrapolation 194
6. **References** 199

CHAPTER VI ELASTIC FORWARD WAVE FIELD EXTRAPOLATION

1. Introduction 201

2. Elastic reciprocity theorems 201

3. Elastic representation theorems 206

3.1 Elastic Green's functions 206

3.2 Elastic Kirchhoff-Helmholtz integral 210

3.3 Modified elastic Green's functions for P- and S-waves 213

3.4 Modified elastic Kirchhoff-Helmholtz integrals for P- and S-waves 224

4. Elastic two-way and one-way Rayleigh integrals 228

4.1 Boundary conditions for the elastic Green's functions 228

4.2 Elastic two-way Rayleigh integrals 230

4.3 Elastic one-way Rayleigh integrals 231

5. Elastic forward wave field extrapolation operators 236

5.1 Integral formulation of elastic forward wave field extrapolation 236

5.2 Matrix formulation of elastic forward wave field extrapolation 245

6. References 255

CHAPTER VII ACOUSTIC INVERSE WAVE FIELD EXTRAPOLATION IN LOW CONTRAST MEDIA

1. Introduction 257

2. Acoustic inverse wave field extrapolation in laterally invariant media 259

2.1 Homogeneous media 259

2.2 Vertically inhomogeneous media 271

3. Acoustic inverse wave field extrapolation in arbitrarily inhomogeneous media 277

3.1 Introduction 277

3.2 Acoustic Kirchhoff-Helmholtz integrals for inverse extrapolation 279

3.3 Error analysis 281

3.4 Acoustic one-way Rayleigh integrals for inverse extrapolation 291

CHAPTER VII (continued)

3.5 Matrix formulation of acoustic inverse wave field extrapolation 294
3.6 Examples of acoustic inverse wave field extrapolation 296
4. References 301

CHAPTER VIII ELASTIC INVERSE WAVE FIELD EXTRAPOLATION IN LOW CONTRAST MEDIA

1. Introduction 305
2. Elastic inverse wave field extrapolation in homogeneous isotropic media 305
3. Elastic inverse wave field extrapolation in arbitrarily inhomogeneous anisotropic media 310
3.1 Introduction 310
3.2 Elastic Kirchhoff-Helmholtz integrals for inverse extrapolation 312
3.3 Error analysis 315
3.4 Elastic one-way Rayleigh integrals for inverse extrapolation 319
3.5 Matrix formulation of elastic inverse wave field extrapolation 323
3.6 Examples of elastic inverse wave field extrapolation 328
4. References 334

CHAPTER IX ACOUSTIC INVERSE WAVE FIELD EXTRAPOLATION IN HIGH CONTRAST MEDIA

1. Introduction 335
2. Recursive acoustic inverse wave field extrapolation 337
2.1 Principle 337
2.2 Example 339
3. Iterative acoustic inverse wave field extrapolation 345
3.1 Principle 345
3.2 Examples 351

CHAPTER X ELASTIC INVERSE WAVE FIELD EXTRAPOLATION IN HIGH CONTRAST MEDIA

1. **Introduction** ... 357
2. **Recursive elastic inverse wave field extrapolation** ... 358
3. **Iterative elastic inverse wave field extrapolation** ... 361
3.1 Principle ... 361
3.2 Examples ... 364

CHAPTER XI ACOUSTIC REDATUMING OF SINGLE-COMPONENT SEISMIC DATA

1. **Introduction** ... 367
2. **Forward model for single-component seismic data** ... 368
3. **Surface related acoustic pre-processing** ... 380
3.1 Introduction ... 380
3.2 Acoustic decomposition ... 381
3.3 Acoustic multiple elimination ... 385
4. **Acoustic redatuming** ... 387
4.1 Introduction ... 387
4.2 Acoustic redatuming of a multi-shot record ... 388
4.3 Acoustic redatuming of single-shot records and stacking ... 390
4.4 Examples of acoustic redatuming ... 392
5. **References** ... 398

CHAPTER XII ELASTIC REDATUMING OF MULTI-COMPONENT SEISMIC DATA

1. **Introduction** ... 399
2. **Forward model for multi-component seismic data** ... 399
3. **Surface related elastic pre-processing** ... 408
3.1 Introduction ... 408
3.2 Elastic decomposition ... 411
3.3 Elastic multiple elimination ... 415
4. **Elastic redatuming** ... 418
4.1 Introduction ... 418
4.2 Principle of elastic redatuming ... 419
4.3 Example of elastic redatuming ... 421
5. **References** ... 423

APPENDIX A MATRIX NOTATION

1. Introduction 425
2. Matrix notation for wave fields 426
3. Matrix notation for operators 428
4. References 431

APPENDIX B INTERACTIONS OF ONE-WAY ACOUSTIC WAVE FIELDS

1. Introduction 433
2. Analysis for forward propagation 434
3. Analysis for backward propgation 442
4. References 447

APPENDIX C INTERACTIONS OF ONE-WAY ELASTIC WAVE FIELDS

1. Introduction 449
2. Analysis for forward propagation 451
3. Analysis for backward propagation 459
4. Reference 462

INDEX 463

INTRODUCTION

Looking at the current approaches to multi-dimensional seismic inversion, we may distinguish the following two classes (Figure 1):

- Methods based on iterative forward modeling,
- Methods based on downward wave field extrapolation.

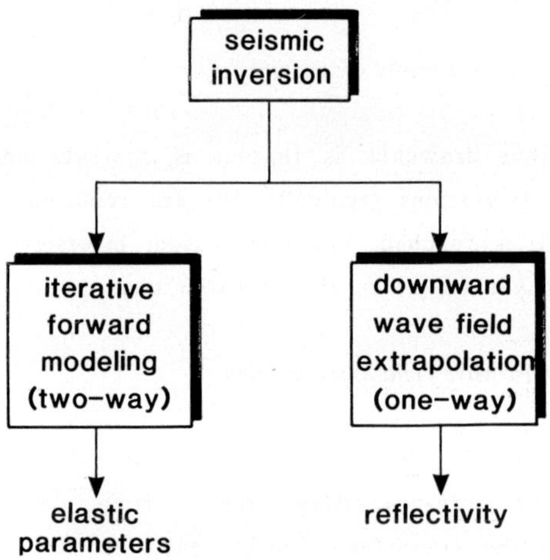

Figure 1: Two approaches to seismic inversion.

Inversion by iterative forward modeling

In this approach seismic data are simulated, given a description of the seismic source and an initial subsurface model. The simulated data are subtracted from the real data. Based on the data residue, the subsurface model is adjusted and the procedure is repeated iteratively. When the residue has reached a minimum value, the final subsurface model is assumed to give an accurate description of the real subsurface.

The most elegant aspect of this approach is that in principle any degree of accuracy can be reached by choosing the proper two-way wave equation[1)] (acoustic, elastic, Biot theory, 2-D, 3-D, etc.) in the forward modeling scheme (Figure 2).

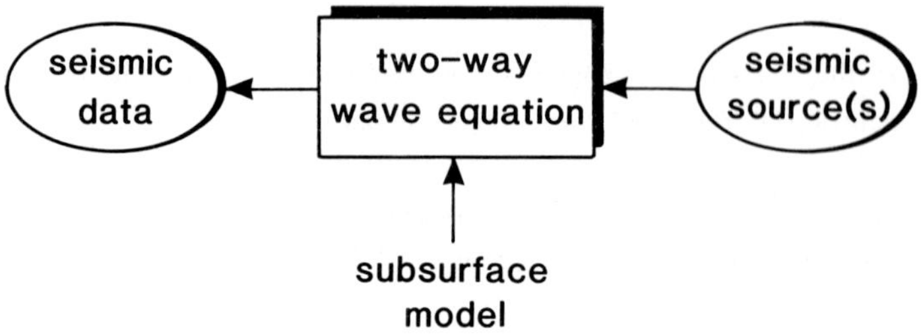

Figure 2: Two-way forward model of seismic data.

The most important drawback is that it is computationally very expensive. In practice many iterations (typically 50) are required before the minimum residue, if at all, is reached. For this reason, inversion by iterative forward modeling has not yet outgrown the research phase.

Inversion by downward wave field extrapolation

This is the approach that is currently used by the seismic industry (migration). After pre-processing the seismic waves are downward extrapolated into the subsurface and at each depth point the "reflectivity" is abstracted. This yields a reflectivity image of the subsurface, which is subsequently interpreted by geologists in the search for oil and gas.

In comparison with the previously described approach, migration is a very efficient solution to the seismic inversion problem. However, the theoretical foundation of the *currently* used schemes is questionable. The accuracy depends largely on the manner the wave fields are downward extrapolated from the earth's surface into the subsurface.

This book aims at giving a theoretically sound recipe for acoustic and elastic downward wave field extrapolation of, respectively, single- and

---

1) The two-way wave equation describes simultaneous downward *and* upward propagation.

multi-component seismic data from the earth's surface to a particular depth level in the subsurface (redatuming). A discussion of the post-processing of the redatumed data (imaging, amplitude versus angle analysis, lithologic inversion etc.) is beyond the scope of this book.

One-way forward model of seismic data

Downward wave field extrapolation is generally based on the one-way wave equations[1]. However, in seismic data acquisition two-way wave fields (including multiple reflections) are recorded. Hence, algorithms for downward wave field extrapolation can only be derived when we have a forward model available that properly combines the one-way aspects of wave propagation with the two-way aspects of seismic data acquisition.
In the one-way forward model depicted in Figure 3 we may distinguish three different model "layers".

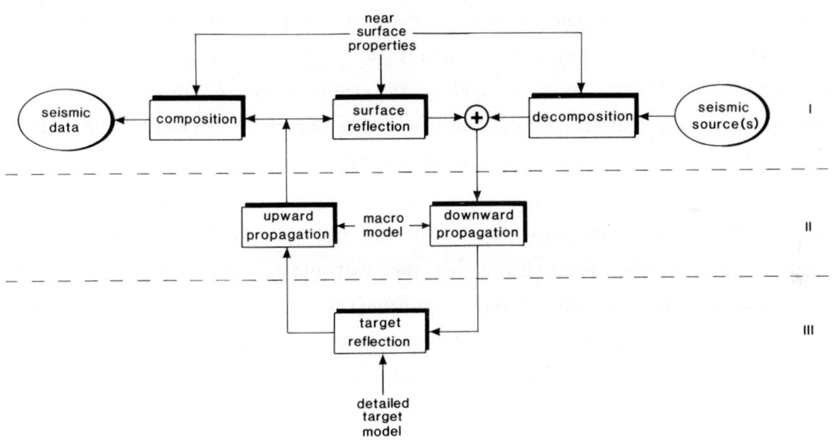

Figure 3: One-way forward model of seismic data.
(For simplicity only the response of a target zone is shown).

[1] A one-way wave equation describes either downward or upward wave propagation.

The first layer contains decomposition and composition operators that relate the two-way seismic data to the downward and upward propagating one-way wave fields. Also it contains a surface reflection operator that describes how the upgoing waves, arriving at the surface, are reflected back into the subsurface, thus giving rise to multiple reflections. The second layer contains operators that describe downward and upward wave propagation. Finally, the third model layer contains an operator that describes how the downgoing waves in the subsurface (for instance in a target zone) are transformed into upgoing waves.

Compare this one-way forward model with the two-way forward model of Figure 2. The two-way forward model is a "black box" that requires the seismic source and a subsurface model as input and that gives simulated seismic data as output. The one-way forward model, on the other hand, is composed of several "boxes", each of which has its own specific character. The boxes in the upper layer are fully determined by the near surface properties. The boxes in the second layer are fully determined by the macro properties (average velocities, main geologic boundaries, etc.) of the "overburden" (area between the surface and the target zone of interest). The box in the third layer is mainly determined by the detailed reflection properties of the target zone. Therefore the one-way forward model is an excellent starting point for "seismic inversion in steps".

Finally, it is interesting to note that the one-way forward model may be based either on the acoustic or on the full elastic wave equation. In the *acoustic* approximation, the earth's subsurface is assumed to be an ideal fluid which may support compressional waves only; the data acquisition is assumed to be carried out with single-component sources and receivers. Since the earth consists of solid rocks which may support both compressional (P) and shear (S) waves, the acoustic approximation is only partly valid. Particularly when the data are acquired with multi-component sources and receivers at large distances apart (large offsets), the acoustic approximation breaks down. In the *elastic* version of the one-way forward model, the decomposition and composition operators describe the relation between multi-component two-way seismic data and downward and upward propagating P- and S-waves. The propagation and reflection operators account for one-way P- and S-waves as well as for P-S and S-P conversions.

Outline

In chapters I and II a brief review of acoustic and elastic wave theory is presented. In chapters III and IV we discuss the relationship between two-way and one-way wave equations. Here we derive the important decomposition and composition operators as well as the reflection operators, both for the acoustic and for the elastic approach. In chapters V and VI we derive acoustic and elastic forward extrapolation operators for downgoing and upgoing waves. Note that chapters I to VI provide the basic tools for the one-way forward model of Figure 3.

Redatuming of single- and multi-component seismic data is essentially based on inverting the one-way forward model. Inverting the decomposition and composition operators is trivial: decomposition is inverse composition and vice versa. Inverting the extrapolation operators is certainly not trivial. Chapters VII and VIII deal with acoustic and elastic inverse wave field extrapolation in inhomogeneous (anisotropic) media with low contrasts. In chapters IX and X the theory is extended for media with high contrasts. Finally, in chapters XI and XII acoustic and elastic redatuming schemes are discussed. An interesting conclusion is that elastic redatuming of multi-component data, decomposed into P- and S-waves, is essentially equivalent to repeatedly acoustic redatuming of single-component data.

Summation convention

Throughout this book we make use of Einstein's summation convention for repeated indices.
Repeated Latin indices imply a summation from 1 to 3. E.g.,

$$\Delta c_{ijk\ell}\partial_\ell G_{k,m}\partial_j V_i$$

(equation (VI-23a)), stands for

$$\sum_{i=1}^{3}\sum_{j=1}^{3}\sum_{k=1}^{3}\sum_{\ell=1}^{3}\Delta c_{ijk\ell}\frac{\partial G_{k,m}}{\partial \ell}\frac{\partial V_i}{\partial j},$$

where i (or j, k, ℓ, m) = 1, 2, 3 stands for x, y, z, respectively. Of course the summation convention does not apply to repeated indices x, y or z. Exceptions throughout are the indices s (=scattered), t (=target) and c (=constrained). Other exceptions are mentioned when appropriate.
Repeated Greek indices imply a summation from 1 to 2. E.g.,

$$jk_\alpha \vec{\tilde{\tau}}_\alpha$$

(equation (IV-3a)), stands for

$$\sum_{\alpha=1}^{2} jk_\alpha \vec{\tilde{\tau}}_\alpha ,$$

where α=1, 2 stands for x, y, respectively. Exceptions throughout are the indices ϕ and ψ (denoting P- and S-waves, respectively).

I

ACOUSTIC WAVES

I.1. INTRODUCTION

Any mechanical disturbance in a solid, liquid or gaseous medium is accompanied with a force which tries to restore the equilibrium situation. This fundamental property of matter underlies wave propagation (i.e., propagation of the disturbance through the medium). The nature of the restoring forces depends largely on the state of the medium. Ideal fluids (i.e., ideal liquids or gasses) can only support forces of compressional nature, whereas solids can support compressional as well as shearing forces. Waves associated with compressional forces are called compressional waves (or P-waves); waves associated with shearing forces are called shear waves (or S-waves). In this book we will reserve the name *acoustic waves* for waves in fluids (P-waves only), whereas we will use the name *elastic waves* for waves in solids (P-waves and S-waves).

In this chapter we derive the acoustic wave equation for an inhomogeneous ideal fluid. Furthermore, we present spherical and plane wave solutions for the special case of a homogeneous fluid.

I.2. ACOUSTIC WAVE EQUATION

In this section we derive the basic equations which describe non-linear wave motion in an inhomogeneous ideal fluid. Next, we linearize these equations and derive the "acoustic two-way wave equation".

I.2.1. Conservation of mass

We consider a fluid in motion in which the particles move with a space ($\vec{r}$) and time (t) dependent velocity $\vec{v}(\vec{r},t)$[1]. The space and time dependent volume density of mass we denote by $\rho(\vec{r},t)$. In this fluid we consider a volume V enclosed by a surface S with outward pointing normal vector $\vec{n}$, see Figure I-1.

---

1) Here $\vec{r}$ is a short-hand notation for the Cartesian coordinates (x,y,z). Hence, $\vec{v}(\vec{r},t)$ stands for $\vec{v}(x,y,z,t)$. In this book we make use of both notations.

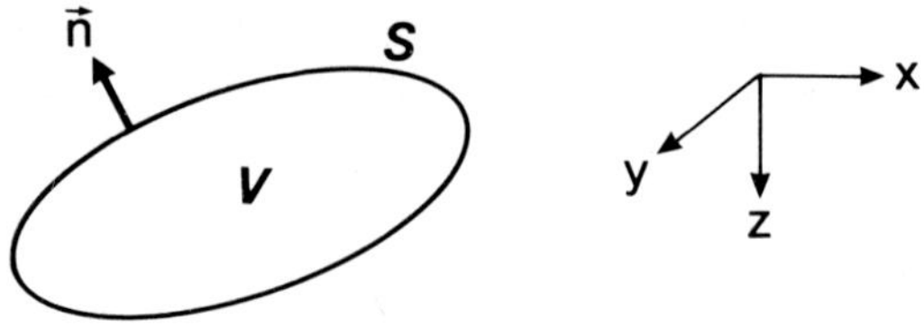

Figure I-1: Volume V, containing a fluid in motion.

The law of *conservation of mass* (Welty et al., 1976) states that the time rate of change of mass in V is equal to the incoming mass flux through S, increased with the time rate of mass injection, hence

$$\frac{\partial}{\partial t}\int_V \rho \, dV = -\oint_S \rho\vec{v}.\vec{n}\, dS + \frac{\partial}{\partial t}\int_V i_m \, dV, \tag{I-1}$$

where $i_m(\vec{r},t)$ represents a source distribution in terms of a volume density of mass injection. Applying the theorem of Gauss,

$$\oint_S \vec{a}.\vec{n}\, dS = \int_V \nabla.\vec{a}\, dV, \tag{I-2}$$

and taking volume V time-invariant, yields

$$\int_V \frac{\partial \rho}{\partial t} dV = -\int_V \nabla.(\rho\vec{v}) dV + \int_V \frac{\partial i_m}{\partial t} dV, \tag{I-3a}$$

or, since this equation holds for any volume V,

$$\frac{\partial \rho}{\partial t} + \nabla.(\rho\vec{v}) = \frac{\partial i_m}{\partial t} . \tag{I-3b}$$

This equation is known as the non-linear *equation of continuity.*

I.2.2. Conservation of momentum

Consider again the volume V, depicted in Figure I-1. The law of *conservation of momentum* (Welty et al., 1976) states that the rate of change of momentum of the particles in V is equal to the incoming momentum flux through S, increased with the resultant force acting on the particles in V (generalized Newton's law), hence

$$\frac{\partial}{\partial t}\int_V \rho\vec{v}\ dV = -\oint_S (\rho\vec{v})\vec{v}.\vec{n}\,dS + \vec{F}(V), \tag{I-4a}$$

where

$$\vec{F}(V) = -\oint_S p\vec{n}\,dS + \int_V \vec{f}\,dV. \tag{I-4b}$$

Here $p(\vec{r},t)$ denotes the pressure and $\vec{f}(\vec{r},t)$ is the volume density of external force. Note that shearing forces (due to viscosity) are not considered (we assumed an ideal fluid).

Applying the theorem of Gauss (I-2) to the three components of the vectorial integral equation (I-4) and combining the results, yields

$$\int_V \frac{\partial(\rho\vec{v})}{\partial t}\,dV = -\int_V [\vec{v}\nabla.(\rho\vec{v}) + (\rho\vec{v}.\nabla)\vec{v}]\,dV + \vec{F}(V), \tag{I-5a}$$

where

$$\vec{F}(V) = -\int_V \nabla p\ dV + \int_V \vec{f}\,dV, \tag{I-5b}$$

or, since this equation holds for any volume V,

$$\frac{\partial(\rho\vec{v})}{\partial t} + \vec{v}\nabla.(\rho\vec{v}) + (\rho\vec{v}.\nabla)\vec{v} + \nabla p = \vec{f}. \tag{I-5c}$$

This equation is known as the non-linear *equation of motion.*

I.2.3. Constitutive relation

The pressure p, the volume density of mass ρ and the temperature T of a fluid are mutually dependent. This is expressed by the *equation of state*, in its most general form given by

$$\chi(p,\rho,T) = 0, \tag{I-6a}$$

where χ is a non-linear function of p, ρ and T.
Assuming that compression and expansion occur adiabatically (i.e., assuming that heat exchanges can be neglected), the equation of state becomes (Zemansky, 1968)

$$p\rho^{-\kappa} = \text{constant}, \tag{I-6b}$$

with

$$\kappa = C_p/C_v, \tag{I-6c}$$

where C_p represents the specific heat at constant pressure and where C_v represents the specific heat at constant volume. Define

$$p(\vec{r},t) = p_0(\vec{r}) + \Delta p(\vec{r},t),$$

where p_0 represents the static pressure and Δp represents the pressure changes, caused by the acoustic wave field. Similarly, define

$$\rho(\vec{r},t) = \rho_0(\vec{r}) + \Delta\rho(\vec{r},t),$$

where ρ_0 represents the static mass density and $\Delta\rho$ represents the mass density changes, caused by the acoustic wave field. Thus the equation of state (I-6b) can be written as

$$p_0\left(1 + \frac{\Delta p}{p_0}\right)\rho_0^{-\kappa}\left(1 + \frac{\Delta\rho}{\rho_0}\right)^{-\kappa} = p_0\rho_0^{-\kappa}, \tag{I-7a}$$

or

$$\frac{\Delta\rho}{\rho_0} = \left(1 + \frac{\Delta p}{p_0}\right)^{\gamma} - 1; \quad \gamma = 1/\kappa, \tag{I-7b}$$

or

$$\frac{\Delta\rho}{\rho_0} = \gamma\frac{\Delta p}{p_0} + \frac{\gamma(\gamma-1)}{2}\left(\frac{\Delta p}{p_0}\right)^2 + \frac{\gamma(\gamma-1)(\gamma-2)}{6}\left(\frac{\Delta p}{p_0}\right)^3 + \ldots.. \tag{I-7c}$$

I.2.4. Linearization of the basic equations

In general, the particle velocity $\vec{v}(\vec{r},t)$ may represent an acoustic wave field, superposed on the steady flow of the fluid. In the following we assume that the flow velocity is zero, so that $\vec{v}(\vec{r},t)$ only represents the particle velocity associated to the acoustic wave field. For a wide range of applications it is justified to assume[1])

$$\left|\frac{\Delta p(\vec{r},t)}{p_0(\vec{r})}\right| << 1$$

and

$$\left|\frac{\Delta \rho(\vec{r},t)}{\rho_0(\vec{r})}\right| << 1.$$

With these assumptions we obtain from (I-3b) the linearized equation of continuity

$$\frac{1}{\rho_0}\frac{\partial(\Delta\rho)}{\partial t} + \nabla.\vec{v} = \frac{1}{\rho_0}\frac{\partial i_m}{\partial t} \hat{=} \frac{\partial i_v}{\partial t}, \tag{I-8a}$$

where we introduced the acoustically more realistic source distribution $i_v(\vec{r},t)$, which represents a volume density of volume injection (for example an airgun). Similarly, from (I-5c) we obtain the linearized equation of motion

$$\rho_0 \frac{\partial \vec{v}}{\partial t} + \nabla(\Delta p) = \vec{f}. \tag{I-8b}$$

Note that we assumed that the spatial variations of the static pressure are negligible in comparison with the spatial variations of the acoustic pressure,

$$|\nabla p_0(\vec{r})| << |\nabla(\Delta p(\vec{r},t))| .$$

Finally, from (I-7c) we obtain the linearized equation of state

$$\frac{\Delta\rho}{\rho_0} = \frac{\gamma}{p_0}\Delta p \hat{=} \frac{1}{K}\Delta p, \tag{I-8c}$$

where we introduced the adiabatic compression modulus $K(\vec{r})$.

---

1) For sound waves in air at normal atmospheric conditions, $|\Delta p/p_0|$ ranges from $2*10^{-10}$ to $6*10^{-5}$. For seismic waves in water, $|\Delta p/p_0|$ is at most 10^{-2} (except in the vicinity of the source).

From these equations $\Delta\rho$ can easily be eliminated by substituting (I-8c) into (I-8a). Now the linear acoustic wave field is fully represented by $\Delta p(\vec{r},t)$ and $\vec{v}(\vec{r},t)$, whereas the fluid is fully represented by $\rho_o(\vec{r})$ and $K(\vec{r})$. For notational convenience, in the remainder of this chapter we use the following substitutions:

$$\Delta p(\vec{r},t) \rightarrow p(\vec{r},t), \tag{I-9a}$$

and

$$\rho_o(\vec{r}) \rightarrow \rho(\vec{r}). \tag{I-9b}$$

Thus the linearized equation of continuity becomes

$$\frac{1}{K(\vec{r})}\frac{\partial p(\vec{r},t)}{\partial t} + \nabla.\vec{v}(\vec{r},t) = \frac{\partial i_v(\vec{r},t)}{\partial t} \tag{I-10a}$$

and the linearized equation of motion becomes

$$\rho(\vec{r})\frac{\partial \vec{v}(\vec{r},t)}{\partial t} + \nabla p(\vec{r},t) = \vec{f}(\vec{r},t). \tag{I-10b}$$

Note that (I-10b) implies

$$\nabla \times (\rho\vec{v}) = \vec{o}, \tag{I-10c}$$

in any region where $\vec{f}=\vec{o}$. Equation (I-10c) states that the mass flow vector $\rho\vec{v}$ in a linear acoustic wave field is curl-free. This is a fundamental property of acoustic wave motion. In a homogeneous medium equation (I-10c) simplifies to

$$\nabla \times \vec{v} = \vec{o}. \tag{I-10d}$$

I.2.5. Acoustic two-way wave equation

By eliminating the particle velocity $\vec{v}$ from the set of equations (I-10a) and (I-10b), we obtain the wave equation for the acoustic pressure p:

$$\rho\nabla.\left(\frac{1}{\rho}\nabla p\right) - \frac{\rho}{K}\frac{\partial^2 p}{\partial t^2} = -s, \tag{I-11a}$$

where

$$s = \rho \frac{\partial^2 i_v}{\partial t^2} - \rho \nabla . (\frac{1}{\rho} \vec{f}), \qquad \text{(I-11b)}$$

hence, $s(\vec{r},t)$ represents a source distribution in terms of the volume density of volume injection $i_v(\vec{r},t)$ and the volume density of force $\vec{f}(\vec{r},t)$. Wave equation (I-11) describes the propagation of linear acoustic waves in inhomogeneous fluids. In the following we call this equation the "acoustic two-way wave equation". This is opposed to the "acoustic one-way wave equations" which are derived in chapter III. One-way wave equations describe either "downgoing" or "upgoing" wave propagation and are in general not exact. The two-way wave equation (I-11), on the other hand, describes wave propagation in all directions and is exact (assuming linearity).

I.3. SPHERICAL WAVE SOLUTIONS OF THE ACOUSTIC TWO-WAY WAVE EQUATION

In this section we present the spherical wave solutions of the acoustic two-way wave equation for a point source of volume injection and for a point source of force in an unbounded homogeneous fluid. We discuss both causal and anti-causal solutions.

I.3.1. Monopole and dipole sources

We consider an unbounded homogeneous fluid and we define a *monopole* point source of volume injection, such that

$$\rho \frac{\partial^2 i_v(\vec{r},t)}{\partial t^2} \triangleq \delta(\vec{r})s(t), \qquad \text{(I-12a)}$$

where $s(t)$ is the source signature and where $\delta(\vec{r}) = \delta(x)\delta(y)\delta(z)$ represents a 3-D spatial delta function, which satisfies

$$\delta(\vec{r}) = 0 \qquad \text{for } \vec{r} \neq \vec{o} \qquad \text{(I-12b)}$$

and

$$\int_V \delta(\vec{r})dV = 1 \qquad \text{for any } V \text{ containing } \vec{r} = \vec{o}. \qquad \text{(I-12c)}$$

According to (I-11), the acoustic pressure $p_1(\vec{r},t)$ related to this source should satisfy

$$\nabla^2 p_1 - \frac{\rho}{K} \frac{\partial^2 p_1}{\partial t^2} = -\delta(\vec{r})s(t). \tag{I-12d}$$

We also define a point source of vertical force, such that

$$\vec{f}(\vec{r},t) = \begin{bmatrix} o \\ o \\ \delta(\vec{r})s(t) \end{bmatrix} . \tag{I-13a}$$

According to (I-11), the acoustic pressure $p_2(\vec{r},t)$ related to this source should satisfy

$$\nabla^2 p_2 - \frac{\rho}{K} \frac{\partial^2 p_2}{\partial t^2} = \frac{\partial \delta(\vec{r})}{\partial z} \; s(t). \tag{I-13b}$$

Note that the right-hand side of this equation could be written as

$$\lim_{\Delta z \to o} \frac{1}{\Delta z} \delta(x)\delta(y) \left[\delta\left(z+\frac{\Delta z}{2}\right) - \delta\left(z-\frac{\Delta z}{2}\right) \right] s(t),$$

hence, it represents a *dipole* source at $\vec{r}=\vec{o}$ with signature s(t). Differentiating both sides of equation (I-12d) with respect to z yields

$$\frac{\partial}{\partial z}\left(\nabla^2 p_1\right) - \frac{\partial}{\partial z}\left[\frac{\rho}{K} \frac{\partial^2 p_1}{\partial t^2}\right] = - \frac{\partial \delta(\vec{r})}{\partial z} s(t),$$

or, since ρ and K are constant

$$\nabla^2 \left[\frac{\partial p_1}{\partial z}\right] - \frac{\rho}{K} \frac{\partial^2}{\partial t^2} \left[\frac{\partial p_1}{\partial z}\right] = - \frac{\partial \delta(\vec{r})}{\partial z} s(t). \tag{I-13c}$$

Hence, assuming that $-\partial p_1/\partial z$ and p_2 satisfy the same initial conditions, we obtain

$$p_2(\vec{r},t) = - \frac{\partial}{\partial z} p_1(\vec{r},t). \tag{I-13d}$$

In the next sections we solve equation (I-12d) for the monopole wave field $p_1(\vec{r},t)$. The expressions for the dipole wave field $p_2(\vec{r},t)$ are then derived from $p_1(\vec{r},t)$ using (I-13d). Bear in mind that this procedure for deriving a dipole wave field cannot be followed in inhomogeneous media.

I.3.2. The wave equation in spherical coordinates

For $\vec{r} \neq \vec{o}$ wave equation (I-12d) reads

$$\nabla^2 p_1 - \frac{\rho}{K} \frac{\partial^2 p_1}{\partial t^2} = o. \tag{I-14a}$$

Around $\vec{r}=\vec{o}$ we consider a small volume V enclosed by a surface S with outward pointing normal vector $\vec{n}$. In analogy with sections I.2.1 and I.2.2 we rewrite wave equation (I-12d) around $\vec{r}=\vec{o}$ as

$$\int_V \nabla.\nabla p_1 dV - \int_V \frac{\rho}{K} \frac{\partial^2 p_1}{\partial t^2} dV = -\int_V \delta(\vec{r})s(t)dV, \tag{I-14b}$$

or, applying the theorem of Gauss (I-2) and using the definition of the delta function (I-12c),

$$\oint_S (\nabla p_1).\vec{n}\, dS - \frac{\rho}{K} \int_V \frac{\partial^2 p_1}{\partial t^2} dV = -s(t). \tag{I-14c}$$

We seek for a solution with spherical symmetry, i.e.,

$$p_1(\vec{r},t) = p_1(r,t),$$

where

$$r = |\vec{r}| = \sqrt{x^2+y^2+z^2} .$$

Thus equation (I-14a) may be transformed to

$$\frac{\partial^2 p_1}{\partial r^2} + \frac{2}{r} \frac{\partial p_1}{\partial r} - \frac{\rho}{K} \frac{\partial^2 p_1}{\partial t^2} = o \qquad \text{for } r \neq o. \tag{I-15a}$$

For the surface S in equation (I-14c) we choose a sphere around $\vec{r}=\vec{o}$ with radius $r=r_o$. Hence, equation (I-14c) may be transformed to

$$4\pi r_o^2 \left. \frac{\partial p_1}{\partial r} \right|_{r=r_o} - 4\pi \frac{\rho}{K} \int_o^{r_o} \frac{\partial^2 p_1}{\partial t^2} r^2 dr = -s(t). \tag{I-15b}$$

I.3.3. Causal and anti-causal solutions

The solution of (I-15a) reads

$$p_1^{\pm}(r,t) = \frac{u(\pm t - r/c)}{r} \qquad \text{for } r \neq o, \tag{I-16a}$$

with

$$c = \sqrt{K/\rho} \quad , \tag{I-16b}$$

where u may be any twice differentiable function. We will see below that c represents the propagation velocity. Substitution of (I-16a) into (I-15b) and taking the limit for $r_o \rightarrow o$, yields

$$-4\pi u(\pm t) = -s(t), \tag{I-16c}$$

hence, the solution of (I-15a) and (I-15b) together, reads

$$p_1^{\pm}(r,t) = \frac{1}{4\pi} \frac{s(t \mp r/c)}{r} \ . \tag{I-17a}$$

A wave front is defined as a surface on which

$$t \mp r/c = \text{constant}, \tag{I-17b}$$

hence, a wave front may be any sphere with midpoint $\vec{r} = \vec{o}$ and radius r, see Figure I-2. In Figure I-3a the solution $p_1^+(r,t)$ is shown as a function of r and t for a causal source function s(t), which means that s(t)=o for t<o. Note that the wave fronts propagate away from the source (at r=o) when time progresses, with *propagation velocity* c. Also note that the amplitude decreases with increasing distance r. We call $p_1^+(r,t)$ the causal or *forward propagating* solution of wave equation (I-15). In Figure I-3b the solution $p_1^-(r,t)$ is shown. Here the wave fronts propagate towards the source (at r=o) when time progresses. This solution is called the anti-causal or *backward propagating* solution of wave equation (I-15).

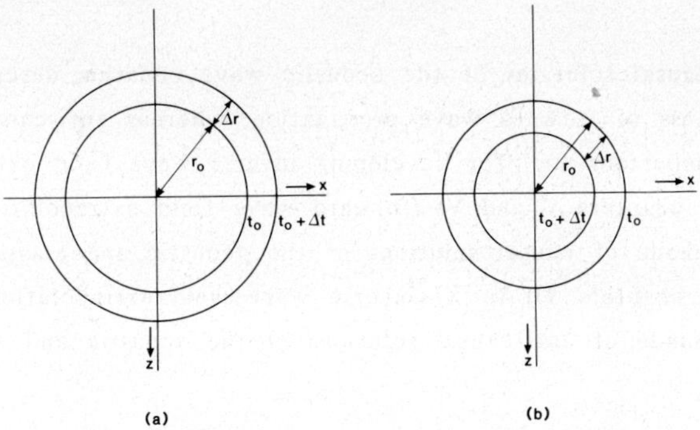

Figure I-2a: *Two-dimensional cross section (y=o) of a spherical* ***causal*** *wave front, defined by $t-r/c$=constant, at two times $t=t_o$ and $t=t_o+\Delta t$, respectively. By definition, the propagation velocity of this wave front is given by $\Delta r/\Delta t$. Since $t_o-r_o/c=t_o+\Delta t-(r_o+\Delta r)/c$, we obtain $\Delta r/\Delta t=c$, hence, c is the propagation velocity.*

Figure I-2b: *Two-dimensional cross section of a spherical* ***anti-causal*** *wave front, defined by $t+r/c$=constant, at two times $t=t_o$ and $t=t_o+\Delta t$, respectively.*

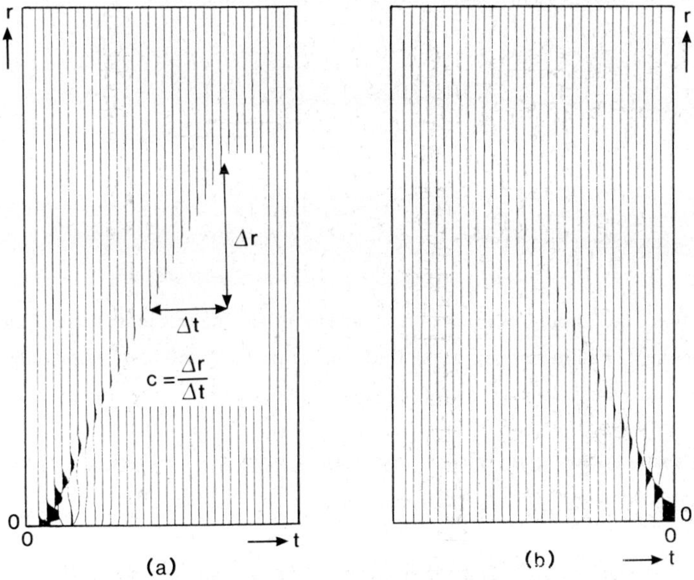

Figure I-3: *One-dimensional cross sections of spherical waves as a function of the distance r to the source, at subsequent travel times t.*

a. *Causal wave fronts $p_1^+(r,t)$, propagating away from the source.*

b. *Anti-causal wave fronts $p_1^-(r,t)$, propagating towards the source.*

In general, causal solutions of the acoustic wave equation describe the physical process of forward wave propagation, whereas anti-causal solutions provide an important tool for developing inverse wave field extrapolation operators. In chapters V and VI (forward wave field extrapolation) ample use will be made of causal solutions of the acoustic and elastic wave equations. In chapters VII to X (inverse wave field extrapolation) ample use will be made of anti-causal solutions of the acoustic and elastic wave equations.

I.3.4. Monopole and dipole wave fields

Consider expression (I-17a) for a causal monopole wave field in a homogeneous fluid,

$$p_1^+(r,t) = \frac{1}{4\pi} \frac{s(t-r/c)}{r} . \tag{I-18a}$$

Using equation (I-13d) we obtain the following expression for a dipole wave field in a homogeneous fluid, in spherical coordinates:

$$p_2^+(r,\alpha,t) = \frac{1}{4\pi} \cos\alpha \left[\frac{s(t-r/c)}{r^2} + \frac{1}{cr} \frac{\partial s(t-r/c)}{\partial t} \right] , \tag{I-18b}$$

with

$$\cos\alpha = \frac{z}{r} .$$

Note that the dipole wave field consists of a "near field term" and a "far field term". Furthermore, note that on a wave front the amplitude varies with $\cos\alpha$, where α represents the angle between $\vec{r}$ and the z-axis.

I.4 PLANE WAVE SOLUTIONS OF THE ACOUSTIC TWO-WAY WAVE EQUATION

In this section we present plane wave solutions of the acoustic two-way wave equation for a source-free, homogeneous fluid. We discuss both homogeneous ("propagating") and inhomogeneous ("evanescent") plane waves. Finally we introduce the concept "acoustic one-way wave equations".

I.4.1 Homogeneous plane waves

Consider a 3-D plane wave of the following form

$$p(\vec{r},t) = p_o(t-\vec{s}.\vec{r}), \tag{I-19a}$$

or

$$p(x,y,z,t) = p_o(t-s_x x-s_y y-s_z z). \tag{I-19b}$$

The tilt angle (dip) of $\vec{s}$[1] with the z-axis is α, with $o\leq\alpha\leq\pi$; the azimuth angle is β, with $o\leq\beta<2\pi$, (see Figure I-4), hence

$$\vec{s} = |\vec{s}| \begin{bmatrix} \sin\alpha \ \cos\beta \\ \sin\alpha \ \sin\beta \\ \cos\alpha \end{bmatrix}. \tag{I-19c}$$

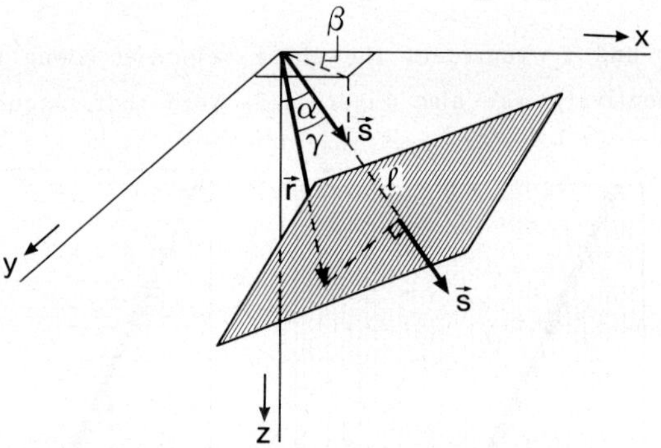

Figure I-4: *A homogeneous plane wave front, perpendicular to slowness vector $\vec{s}$, at time $t=t_o$. Note that $t_o-\vec{s}.\vec{r}=t_o-|\vec{s}|.|\vec{r}|\cos\gamma=t_o-|\vec{s}|\ell$=constant for any coordinate vector $\vec{r}=(x,y,z)$ of the wave front at t_o. The same wave front at $t_o+\Delta t$ satisfies $t_o+\Delta t-|\vec{s}|(\ell+\Delta\ell)=t_o-|\vec{s}|\ell$. Hence, the propagation velocity c is given by $c\triangleq\Delta\ell/\Delta t=1/|\vec{s}|$.*
The situation is shown for $s_x>o$, $s_y>o$ and $s_z>o$.

A wave front is defined as a surface on which

$$t-\vec{s}.\vec{r} = \text{constant}, \tag{I-20}$$

---

1) Vector $\vec{s}$ should not be confused with the source signature s in section I.3.

hence, a wave front may be any plane surface perpendicular to vector $\vec{s}$, see also Figure I-4. The wave fronts propagate in the direction of $\vec{s}$ with a propagation velocity c given by

$$c = \frac{1}{|\vec{s}|}. \tag{I-21}$$

Therefore, $\vec{s}$ is called the *slowness* vector. Its components s_x, s_y and s_z represent the phase slownesses along the x-, y- and z-axes, respectively, hence,

$$\vec{s} = \begin{bmatrix} s_x \\ s_y \\ s_z \end{bmatrix} \hat{=} \begin{bmatrix} c_x^{-1} \\ c_y^{-1} \\ c_z^{-1} \end{bmatrix} = \begin{bmatrix} c^{-1}\sin\alpha\cos\beta \\ c^{-1}\sin\alpha\sin\beta \\ c^{-1}\cos\alpha \end{bmatrix} , \tag{I-22}$$

where c_x, c_y and c_z represent the phase velocities along the x-, y- and z- axes, respectively, see also Figure I-5. Note that, according to (I-21),

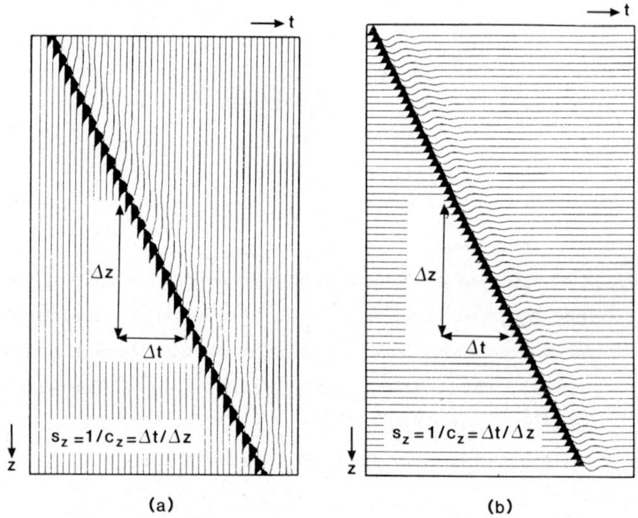

Figure I-5: *One dimensional cross-section along the z-axis (x=y=o) of the homogeneous plane wave $p_o(t-s_x x-s_y y-s_z z)$.*
The situation is shown for a positive phase slowness s_z.
a. Plane wave as a function of depth z at subsequent times t. Each trace represents a "snap-shot" of the wave field at fixed time.
b. Plane wave as a function of time t at subsequent depth levels z. Each trace represents a "registration" of the wave field at fixed depth.

$$\vec{s}\cdot\vec{s} = \frac{1}{c^2}, \tag{I-23a}$$

hence

$$s_x^2 + s_y^2 + s_z^2 = \frac{1}{c^2}, \tag{I-23b}$$

or

$$\frac{1}{c_x^2} + \frac{1}{c_y^2} + \frac{1}{c_z^2} = \frac{1}{c^2}. \tag{I-23c}$$

Assuming that c_x,c_y and c_z are *real* constants (positive or negative), we observe from (I-23c) that for homogeneous plane waves[1)], as defined by (I-19), any of the absolute phase velocities $|c_x|$, $|c_y|$ or $|c_z|$ is higher than or equal to the (real positive) propagation velocity c.

Upon substitution of the plane wave (I-19)

$$p(x,y,z,t) = p_o(t-s_x x-s_y y-s_z z) \tag{I-24a}$$

into the acoustic two-way wave equation (I-11) for a source-free, homogeneous fluid,

$$\nabla^2 p - \frac{\rho}{K}\frac{\partial^2 p}{\partial t^2} = 0, \tag{I-24b}$$

we obtain

$$s_x^2 + s_y^2 + s_z^2 = \frac{\rho}{K}, \tag{I-25a}$$

hence, considering (I-23b) the propagation velocity c is given by

$$c = \sqrt{K/\rho}\ . \tag{I-25b}$$

This result was already found in section I.3 for spherical wave solutions.

---

1) A plane wave is called homogeneous when its amplitude is constant on a wave front.

The particle velocity associated to the plane wave (I-24a) follows from the equation of motion (I-10b) for a source-free, homogeneous fluid,

$$\frac{\partial \vec{v}}{\partial t} = -\frac{1}{\rho} \nabla p, \tag{I-26a}$$

hence

$$\vec{v}(x,y,z,t) = \frac{1}{\rho} \vec{s}\, p_o(t-s_x x-s_y y-s_z z). \tag{I-26b}$$

Note that the wave motion occurs parallel to the propagation direction, hence, the plane wave is *longitudinal*. (Bear in mind that an acoustic wave is purely longitudinal only when the amplitudes are constant on the wave fronts).

Finally, we give the expression for *monochromatic* plane waves:

$$p(\vec{r},t) = p_o \cos[\phi(\vec{r},t)], \tag{I-27a}$$

where

$$\phi(\vec{r},t) = \omega_o(t-\vec{s}.\vec{r}) + \phi_o, \tag{I-27b}$$

or

$$\phi(x,y,z,t) = \omega_o(t-s_x x-s_y y-s_z z) + \phi_o. \tag{I-27c}$$

Here p_o denotes a constant amplitude factor and $\phi(x,y,z,t)$ denotes the space and time dependent phase. Furthermore, ω_o denotes the circular frequency and ϕ_o denotes the phase of the wave at the origin of the coordinate system at t=o. The time-periodicity T_o follows from

$$\phi(x,y,z,t+T_o) - \phi(x,y,z,t) = 2\pi, \tag{I-28a}$$

hence

$$T_o = \frac{2\pi}{\omega_o}. \tag{I-28b}$$

The wavelength λ along the propagation direction follows from

$$\phi(\vec{r} - \frac{\vec{s}}{|\vec{s}|}\lambda, t) - \phi(\vec{r}, t) = 2\pi, \tag{I-29a}$$

hence

$$\lambda = \frac{2\pi}{\omega_o}\frac{1}{|\vec{s}|} = \frac{2\pi}{\omega_o} c. \tag{I-29b}$$

The apparent wavelength λ_x along the x-axis follows from

$$\phi(x-\lambda_x, y, z, t) - \phi(x, y, z, t) = 2\pi, \tag{I-30a}$$

hence

$$\lambda_x = \frac{2\pi}{\omega_o}\frac{1}{s_x} = \frac{2\pi}{\omega_o} c_x = \frac{\lambda}{\sin\alpha\cos\beta}. \tag{I-30b}$$

Similarly, the apparent wavelengths λ_y and λ_z along the y- and z-axes, respectively, read

$$\lambda_y = \frac{2\pi}{\omega_o}\frac{1}{s_y} = \frac{2\pi}{\omega_o} c_y = \frac{\lambda}{\sin\alpha\sin\beta} \tag{I-30c}$$

and

$$\lambda_z = \frac{2\pi}{\omega_o}\frac{1}{s_z} = \frac{2\pi}{\omega_o} c_z = \frac{\lambda}{\cos\alpha}. \tag{I-30d}$$

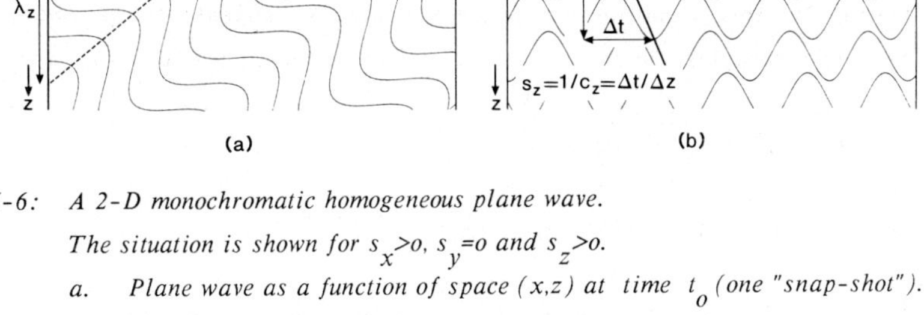

Figure I-6: *A 2-D monochromatic homogeneous plane wave.*
The situation is shown for $s_x>o$, $s_y=o$ and $s_z>o$.

a. *Plane wave as a function of space (x,z) at time t_o (one "snap-shot"). Note that $\lambda_x=\lambda/\sin\alpha>\lambda$ and $\lambda_z=\lambda/\cos\alpha>\lambda$.*

b. *1-D cross-section along the z-axis of the plane wave as a function of time t at subsequent depth levels z.*

Note that the apparent wavelengths λ_x, λ_y and λ_z are larger than or equal to the wavelength λ along the propagation direction. In Figure I-6 a two-dimensional monochromatic homogeneous plane wave (β=o $\rightarrow$ s_y=o) is visualized.

I.4.2. Inhomogeneous plane waves

Although plane waves do not often occur in reality, they play an important role in many seismic techniques, where arbitrary wave fields are decomposed into monochromatic plane waves. This decomposition is accomplished by applying multi-dimensional Fourier transformations (see section III.2.1).

In general, this decomposition yields a range of homogeneous as well as inhomogeneous plane waves[1)]. Homogeneous plane waves were introduced in the previous section, where we assumed that c_x, c_y and c_z are *real* (positive or negative) parameters. In this section we introduce inhomogeneous plane waves. Therefore we consider real phase velocities c_x and c_y, for which

$$\frac{1}{c_x^2} + \frac{1}{c_y^2} > \frac{1}{c^2}, \tag{I-31a}$$

or, equivalently,

$$s_x^2 + s_y^2 > \frac{1}{c^2}. \tag{I-31b}$$

Then, according to (I-23b),

$$s_z^2 = \frac{1}{c^2} - s_x^2 - s_y^2 < o, \tag{I-31c}$$

hence, for this situation the phase slowness s_z appears to be imaginary. To avoid complex functions, let us define a new parameter σ_z according to

$$\sigma_z^2 = s_x^2 + s_y^2 - \frac{1}{c^2} > o, \tag{I-32}$$

where σ_z is real (positive or negative).

1) A plane wave is called inhomogeneous when its amplitude is not constant on a wave front.

Consider the real monochromatic wave function

$$p(\vec{r},t) = p_o \cos[\phi(\vec{r},t)] e^{-\omega_o \sigma_z z}, \tag{I-33a}$$

where

$$\phi(\vec{r},t) = \omega_o(t-\vec{s}_o.\vec{r}) + \phi_o, \tag{I-33b}$$

with

$$\vec{s}_o = \text{Real}(\vec{s}) = \begin{bmatrix} s_x \\ s_y \\ o \end{bmatrix} = |\vec{s}_o| \begin{bmatrix} \cos\beta \\ \sin\beta \\ o \end{bmatrix}, \tag{I-33c}$$

hence

$$\phi(x,y,z,t) = \omega_o(t-s_x x-s_y y) + \phi_o. \tag{I-33d}$$

Note that the phase is independent of depth.
It can be verified that wave function (I-33) is a solution of the two-way wave equation (I-24b), with $\rho/K = 1/c^2$. A wave front is defined as a surface on which the phase

$$\phi(x,y,z,t) = \text{constant}, \tag{I-34}$$

hence, a wave front may be any plane surface perpendicular to vector $\vec{s}_o$, see also Figure I-7. The wave fronts propagate in the direction of $\vec{s}_o$ with a propagation velocity c_o given by

$$c_o = \frac{1}{|\vec{s}_o|} = \frac{1}{\sqrt{s_x^2 + s_y^2}}. \tag{I-35a}$$

Note that the propagation velocity of the inhomogeneous plane wave (I-33) is smaller than the homogeneous plane wave velocity:

$$c_o < c = \sqrt{\frac{K}{\rho}}. \tag{I-35b}$$

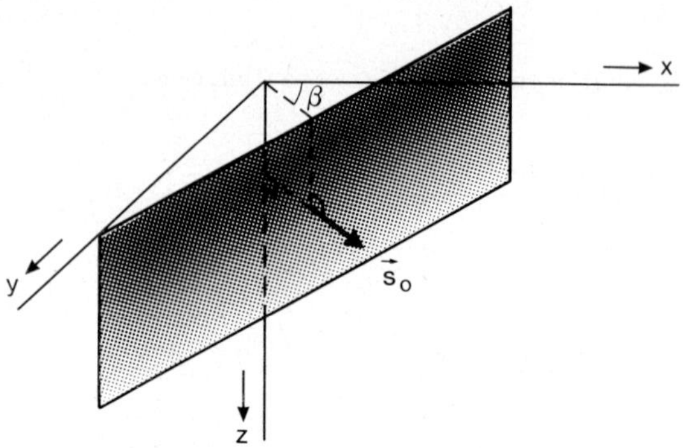

Figure I-7: *An inhomogeneous plane wave front, perpendicular to slowness vector $\vec{s}_o$, at time $t=t_o$. The velocity c_o in the direction of $\vec{s}_o$ is given by $c_o=1/|\vec{s}_o|$. The variable grey-level shows that the amplitude along the wave front decreases with depth.*
The situation is shown for positive s_x, s_y and σ_z, with $s_x^2 + s_y^2 = 1/c_o^2 > 1/c^2$.

According to (I-33), the amplitude along a wave front decreases or increases exponentially with depth for positive or negative σ_z, respectively. Opposed to the homogeneous plane wave functions (I-19) and (I-27), which have constant amplitudes on the wave fronts, plane wave function (I-33) is called an *inhomogeneous plane wave* (Brekhovskikh, 1980).

In analogy with (I-29), the wavelength λ_o along the propagation direction is related to the propagation velocity c_o according to

$$\lambda_o = \frac{2\pi}{\omega_o} \frac{1}{|\vec{s}_o|} = \frac{2\pi}{\omega_o} c_o < \lambda. \tag{I-36}$$

In analogy with (I-30), the apparent wavelengths λ_x, λ_y and λ_z are related to the phase slownesses s_x, s_y and s_z, respectively, according to

$$\lambda_x = \frac{2\pi}{\omega_o} \frac{1}{s_x} = \frac{2\pi}{\omega_o} c_x, \tag{I-37a}$$

$$\lambda_y = \frac{2\pi}{\omega_o} \frac{1}{s_y} = \frac{2\pi}{\omega_o} c_y, \tag{I-37b}$$

$$\lambda_z = \frac{2\pi}{\omega_o} \frac{1}{\mathrm{Real}(s_z)} = \infty. \tag{I-37c}$$

Note that

$$\frac{1}{\lambda_x^2} + \frac{1}{\lambda_y^2} = \frac{1}{\lambda_o^2} > \frac{1}{\lambda^2} . \qquad \text{(I-37d)}$$

In Figure I-8 a two-dimensional monochromatic inhomogeneous plane wave (β=o $\rightarrow$ s_y=o) is visualized.

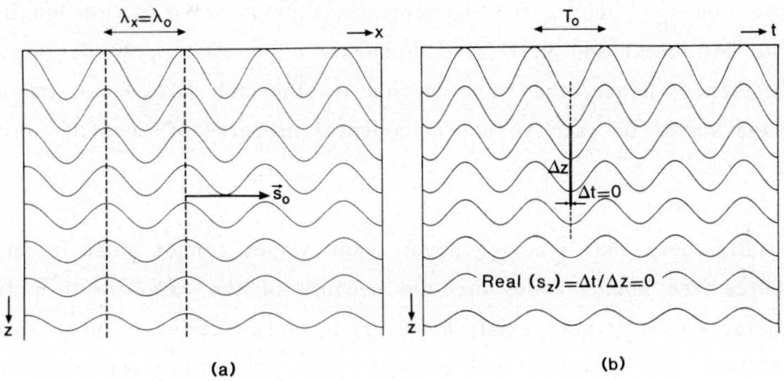

Figure I-8: *A 2-D monochromatic inhomogeneous plane wave.*
The situation is shown for $s_x=1/c_o>1/c$, s_y=o and σ_z>o; ω_o and c have the same values as in Figure I-6.

a. Inhomogeneous plane wave as a function of space (x,z) at time t_o (one "snap-shot"). Note that λ_o is smaller than λ in Figure I-6a.

b. 1-D cross-section along the z-axis of the inhomogeneous plane wave as a function of time t at subsequent depth levels z. Note that T_o has the same value as in Figure I-6b.

The particle velocity associated to the monochromatic inhomogeneous plane wave (I-33) follows from

$$\frac{\partial \vec{v}}{\partial t} = - \frac{1}{\rho} \nabla p, \qquad \text{(I-38a)}$$

hence

$$\vec{v}(\vec{r},t) = \frac{p_o}{\rho}\begin{bmatrix} s_x \cos[\phi(\vec{r},t)] \\ s_y \cos[\phi(\vec{r},t)] \\ \sigma_z \sin[\phi(\vec{r},t)] \end{bmatrix} e^{-\omega_o \sigma_z z} . \tag{I-38b}$$

Note that the wave motion does not occur parallel to the propagation direction $\vec{s}_o$, hence, the inhomogeneous plane wave is not longitudinal. The horizontal and vertical components of $\vec{v}$ have a 90 degrees phase difference, hence, the wave motion is elliptical. Of course, inhomogeneous plane waves do satisfy the fundamental property of acoustic wave motion $\nabla \times (\rho\vec{v}) = \vec{o}$.

Finally, note that inhomogeneous plane waves cannot exist in an unbounded source-free homogeneous medium because of the exponential increasing character. They may exist, however, near boundaries or near sources, both in fluids and in solids. For example, the well known Rayleigh-wave, which may propagate along the surface of a solid medium, is a superposition of inhomogeneous compressional and shear waves, see section II.4.2.

I.4.3. Acoustic one-way wave equations for plane waves

Consider the monochromatic homogeneous plane wave function (I-27)

$$p(x,y,z,t) = p_o \cos\left[\omega_o(t-s_x x-s_y y-s_z z) + \phi_o\right], \tag{I-39a}$$

with

$$s_x^2 + s_y^2 + s_z^2 = \frac{1}{c^2}, \tag{I-39b}$$

where c is the propagation velocity in the homogeneous fluid, according to

$$c = \sqrt{K/\rho} . \tag{I-39c}$$

The phase slownesses s_x, s_y and s_z are given by

$$s_x = c^{-1}\sin\alpha\ \cos\beta,$$
$$s_y = c^{-1}\sin\alpha\ \sin\beta, \qquad \text{(I-39d)}$$
$$s_z = c^{-1}\cos\alpha,$$

where α and β determine the tilt and azimuth angle, respectively, of the homogeneous plane wave, see Figure I-4. Note that s_x, s_y and s_z are real positive or negative constants. The sign of these constants is determined by the propagation direction. In seismics it is common practice to distinguish between downgoing waves which propagate in the positive z-direction (the z-axis is pointing downward) and upgoing waves which propagate in the negative z-direction. Therefore we slightly modify (I-39):

$$p^{\pm}(x,y,z,t) = p_o^{\pm} \cos\left[\omega_o(t - s_x x - s_y y \mp s_z z) + \phi_o\right], \qquad \text{(I-40a)}$$

with

$$s_z \triangleq + \sqrt{\frac{1}{c^2} - s_x^2 - s_y^2}\ . \qquad \text{(I-40b)}$$

Note that we defined s_z to be a positive constant. Now $p^+(x,y,z,t)$ defines a downgoing plane wave and $p^-(x,y,z,t)$ defines an upgoing plane wave[1], see Figure I-9.

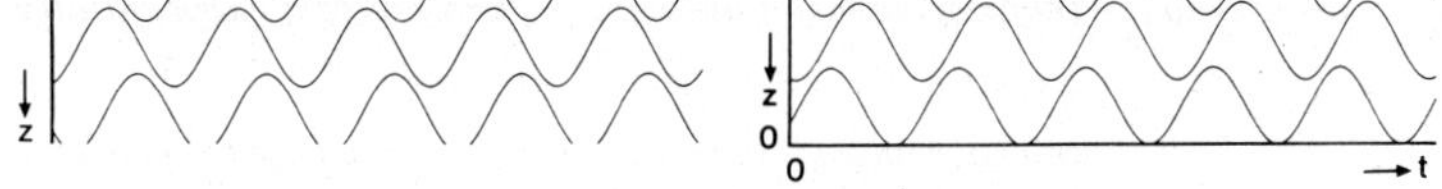

Figure I-9: 1-D cross-sections along the z-axis of homogeneous (or propagating) plane waves.

a. Downgoing homogeneous plane wave $p^+(x=0,y=0,z,t)$.

b. Upgoing homogeneous plane wave $p^-(x=0,y=0,z,t)$.

---

1) Actually plane waves are only purely downgoing or purely upgoing when $s_x = s_y = 0$. We will, however, use our definition for downgoing and upgoing plane waves also when $s_x \neq 0$ and/or $s_y \neq 0$.

In a similar way we slightly modify the monochromatic inhomogeneous plane wave function (I-33):

$$p^{\pm}(x,y,z,t) = p_o^{\pm} \cos\left[\omega_o(t-s_x x-s_y y) + \phi_o\right] e^{\mp \omega_o \sigma_z z}, \tag{I-41a}$$

with

$$\sigma_z = +\sqrt{s_x^2 + s_y^2 - \frac{1}{c^2}}\,. \tag{I-41b}$$

Note that we defined σ_z to be a positive constant.

Now $p^+(x,y,z,t)$ defines an inhomogeneous plane wave with exponentially decreasing amplitude in the *positive* z-direction and $p^-(x,y,z,t)$ defines an inhomogeneous plane wave with exponentially decreasing amplitude in the *negative* z-direction, see Figure I-10.

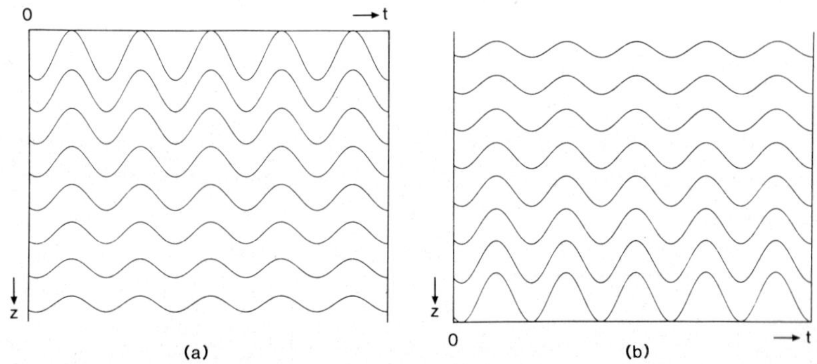

Figure I-10: 1-D cross-sections along the z-axis of inhomogeneous (or evanescent) plane waves.

a. Inhomogeneous plane wave $p^+(x=o,y=o,z,t)$; exponentially decreasing amplitude in the positive z-direction.

b. Inhomogeneous plane wave $p^-(x=o,y=o,z,t)$; exponentially decreasing amplitude in the negative z-direction.

The downgoing and upgoing homogeneous plane waves, defined by (I-40), are often referred to as *propagating* waves. On the other hand, the exponentially decreasing inhomogeneous plane waves, defined by (I-41), are often referred to as *evanescent*[1)] waves, (Berkhout, 1985).

1) Here the term "evanescent" only refers to the behaviour of the wave along the z-axis; along the x-axis and/or y-axis "evanescent" waves do propagate.

In both cases the tilt angle α and the azimuth angle β are related to the *real* phase slownesses s_x and s_y, according to

$$\sin\alpha = c\sqrt{s_x^2 + s_y^2}$$

and

$$\tan\beta = s_y/s_x,$$

respectively. Note that evanescent waves have a complex tilt angle α.

Monochromatic propagating and evanescent plane waves can be summarized by one expression when we introduce the complex notation

$$\hat{p}^{\pm}(x,y,z,t) = \hat{p}_o^{\pm} \exp\left[j\omega_o(t - s_x x - s_y y \mp s_z z)\right], \tag{I-42a}$$

with

$$s_z = +\sqrt{\frac{1}{c^2} - s_x^2 - s_y^2} \qquad \text{for } s_x^2 + s_y^2 \le \frac{1}{c^2}, \tag{I-42b}$$

$$s_z = -j\sqrt{s_x^2 + s_y^2 - \frac{1}{c^2}} \qquad \text{for } s_x^2 + s_y^2 > \frac{1}{c^2}, \tag{I-42c}$$

$$\hat{p}_o^{\pm} = p_o^{\pm} \exp(j\phi_o) \tag{I-42d}$$

and

$$j = +\sqrt{-1}\ . \tag{I-42e}$$

It can be easily verified that both the propagating waves (I-40) and the evanescent waves (I-41) are given by the real part of the complex wave function (I-42a):

$$p^{\pm}(x,y,z,t) = \text{Real}\left[\hat{p}^{\pm}(x,y,z,t)\right]. \tag{I-42f}$$

Note that

$$\frac{\partial\hat{p}^{+}(x,y,z,t)}{\partial z} = -j\omega_o s_z \hat{p}^{+}(x,y,z,t) \tag{I-43a}$$

and

$$\frac{\partial \hat{p}^-(x,y,z,t)}{\partial z} = +j\omega_o s_z \hat{p}^-(x,y,z,t). \tag{I-43b}$$

We call these the acoustic *one-way wave equations* for plane waves. Opposed to the acoustic two-way wave equation (I-11), which describes wave propagation in all directions, one-way wave equation (I-43a) describes only downgoing plane waves, whereas one-way wave equation (I-43b) describes only upgoing plane waves.

Note that the formal solution of (I-43) reads

$$\hat{p}^{\pm}(x,y,z,t) = \hat{w}^{\pm}(z)\hat{p}^{\pm}(x,y,o,t) \tag{I-44a}$$

where

$$\hat{w}^{\pm}(z) = \exp(\mp j\omega_o s_z z). \tag{I-44b}$$

Equation (I-44) describes "one-way wave field extrapolation" in its most simple form. The "boundary value" $\hat{p}^{\pm}(x,y,o,t)$ represents the downgoing or upgoing plane wave field at z=o,

$$\hat{p}^{\pm}(x,y,o,t) = \hat{p}_o^{\pm} \exp\left[j\omega_o(t-s_x x-s_y y)\right]; \tag{I-44c}$$

$\hat{p}^{\pm}(x,y,z,t)$ represents the extrapolated downgoing or upgoing wave field at depth level z. Finally, $\hat{w}^{\pm}(z)$ represents the one-way plane wave field extrapolation operator. Note that

$$|\hat{w}^{\pm}(z)| = 1 \quad \text{for } s_x^2 + s_y^2 \leq 1/c^2, \tag{I-44d}$$

hence, for propagating plane waves, operator $\hat{w}^{\pm}(z)$ describes a phase-shift only (the amplitude of a propagating plane wave is independent of depth, see also Figure I-9).

Also note that

$$\text{Arg}[\hat{w}^{\pm}(z)] = o \quad \text{for } s_x^2 + s_y^2 > 1/c^2, \tag{I-44e}$$

hence, for evanescent plane waves, operator $\hat{W}^{\pm}(z)$ describes an exponential amplitude effect only (the phase of an evanescent plane wave is independent of depth, see also Figure I-10).

In chapter III we generalize the one-way wave equations for arbitrary wave fields in horizontally layered and arbitrarily inhomogeneous acoustic media.

One-way wave equations provide an important tool for transforming two-way Kirchhoff-Helmholtz integrals into one-way Rayleigh integrals. In chapters V to X ample use will be made of acoustic and elastic one-way wave equations.

I.5. REFERENCES

Berkhout, A.J., 1985, Seismic migration: Imaging of acoustic energy by wave field extrapolation. A. Theoretical aspects: Elsevier Science Publ. Co., Inc.

Brekhovskikh, L.M., 1980, Waves in layered media: Academic Press.

De Hoop, A.T., 1988, Radiation and scattering of acoustic waves in fluids: Lecture notes, Delft University of Technology.

Welty, J.R., Wicks, C.E., and Wilson, R.E., 1976, Fundamentals of momentum, heat, and mass transfer: John Wiley and Sons.

Zemansky, M.W., 1968, Heat and thermodynamics: McGraw - Hill Kogakusha, Ltd.

II

ELASTIC WAVES

II.1 INTRODUCTION

Opposed to ideal fluids, which can only support compressional forces, solid media can support compressional as well as shearing forces. As a consequence, elastic waves in solids are in general more complicated than acoustic waves in ideal fluids. Elastic waves in arbitrarily inhomogeneous anisotropic solids consist of quasi-compressional waves (or qP-waves) and quasi-shear waves (or qS-waves), which are in general interrelated. Only in homogeneous[1)] isotropic[2)] solids, pure P-waves propagate completely independent from pure S-waves. The wave motion of pure P-waves is curl-free ($\nabla \times \vec{v}_p = \vec{o}$). This property also holds for acoustic waves in homogeneous fluids (equation (I-10d)). The wave motion of pure S-waves, on the other hand, is divergence-free ($\nabla . \vec{v}_s = o$).

In this chapter we derive the elastic wave equation for an inhomogeneous anisotropic solid. Furthermore, we present spherical P- and S-wave solutions for the special case of a homogeneous isotropic solid. Finally, we present plane wave solutions for homogeneous isotropic and anisotropic solids.

II.2. ELASTIC WAVE EQUATION

In this section we present a brief derivation of the basic equations which describe non-linear and linear wave motion in an inhomogeneous anisotropic solid. Next, we derive the "elastic two-way wave equation". Finally, we introduce potentials for P- and S-waves in homogeneous isotropic solids. For a more extensive derivation the reader is referred to Achenbach (1973), Pilant (1979) and Aki and Richards (1980).

---

1) A material is said to be *homogeneous* when its physical properties are independent of the position where they are measured.

2) A material is said to be *isotropic* when its physical properties are independent of the direction in which they are measured.

II.2.1. Stress and strain

a. Stress

Opposed to ideal fluids, solid media can support compressional as well as shearing forces. Consider an infinitesimal parallelepiped in a solid medium, as shown in Figure II-1.

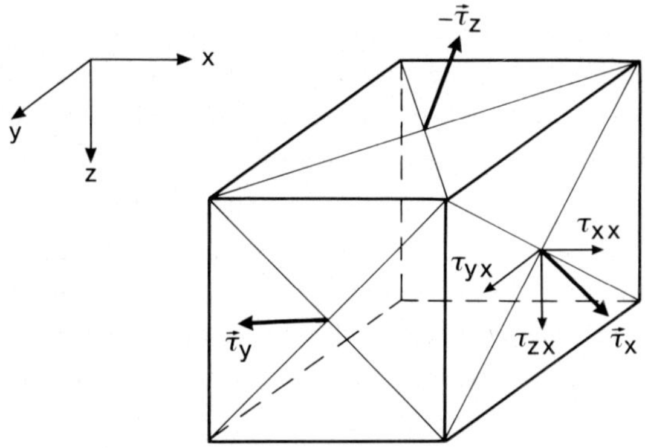

Figure II-1: Tractions acting on an infinitesimal parallelepiped in a solid medium

We assume that the parallelepiped is exposed to surface forces. The force per unit area, acting across the plane normal to the x-axis, is called the *traction* vector $\vec{\tau}_x$. Its components are the normal or *tensile stress* τ_{xx} and the tangential or *shearing stresses* τ_{yx} and τ_{zx}:

$$\text{traction} \rightarrow \vec{\tau}_x = \begin{bmatrix} \tau_{xx} \\ \tau_{yx} \\ \tau_{zx} \end{bmatrix} \begin{matrix} \leftarrow \text{tensile stress (tension)} \\ \left.\begin{matrix}\leftarrow \\ \leftarrow\end{matrix}\right] \text{shearing stresses (shear)} \end{matrix} \qquad \text{(II-1a)}$$

Similarly, the forces per unit area, acting across the planes normal to the y- and z-axes, are called the traction vectors $\vec{\tau}_y$ and $\vec{\tau}_z$, respectively, where

$$\vec{\tau}_y = \begin{bmatrix} \tau_{xy} \\ \tau_{yy} \\ \tau_{zy} \end{bmatrix} \quad \text{and} \quad \vec{\tau}_z = \begin{bmatrix} \tau_{xz} \\ \tau_{yz} \\ \tau_{zz} \end{bmatrix} . \qquad \text{(II-1b)}$$

The three traction vectors $\vec{\tau}_x$, $\vec{\tau}_y$ and $\vec{\tau}_z$ can be placed in the columns of a matrix, which is called the stress tensor τ:

$$\tau = \begin{bmatrix} \tau_{xx} & \tau_{xy} & \tau_{xz} \\ \tau_{yx} & \tau_{yy} & \tau_{yz} \\ \tau_{zx} & \tau_{zy} & \tau_{zz} \end{bmatrix} . \qquad \text{(II-2)}$$

Note that the diagonal elements of this tensor represent the tensile stresses; the off-diagonal elements represent the shearing stresses.
In the limiting case of ideal fluids (where shearing stresses cannot exist) the stress tensor simplifies to

$$\tau = \begin{bmatrix} -p & 0 & 0 \\ 0 & -p & 0 \\ 0 & 0 & -p \end{bmatrix} , \qquad \text{(II-3a)}$$

where p (the pressure) is the negative tensile stress:

$$p = -\tau_{xx} = -\tau_{yy} = -\tau_{zz} \qquad \text{(Pascal's law).} \qquad \text{(II-3b)}$$

b. Strain

We describe the deformation of an infinitesimal parallelepiped in terms of tensile strains and shearing strains. With reference to Figure II-2, we define the tensile strain e_{xx} by

$$e_{xx} \triangleq \frac{\delta X}{\Delta X} . \qquad \text{(II-4)}$$

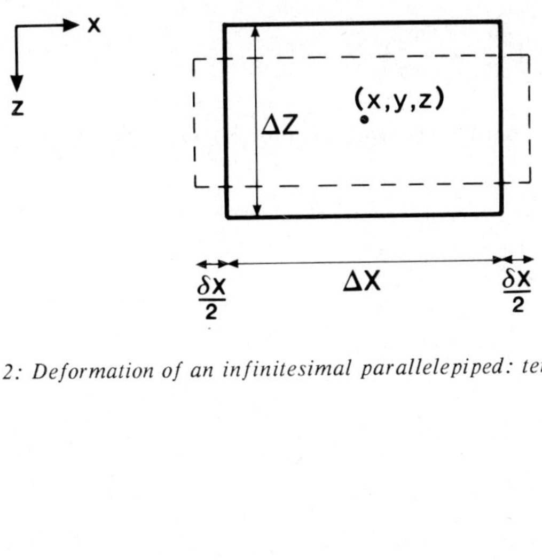

Figure II-2: Deformation of an infinitesimal parallelepiped: tensile strain.

If we define a displacement vector $\vec{u}$, with components u_x, u_y and u_z, then we may rewrite equation (II-4) as

$$e_{xx} = \lim_{\Delta X \to o} \frac{u_x\left(x + \frac{\Delta X}{2}\right) - u_x\left(x - \frac{\Delta X}{2}\right)}{\Delta X} = \frac{\partial u_x}{\partial x} . \qquad \text{(II-5a)}$$

In a similar way we define tensile strains e_{yy} and e_{zz}, according to

$$e_{yy} = \frac{\partial u_y}{\partial_y} \quad \text{and} \quad e_{zz} = \frac{\partial u_z}{\partial z} . \qquad \text{(II-5b)}$$

With reference to Figure II-3, we define the shearing strain e_{zx} by

$$e_{zx} = e_{xz} \triangleq \frac{1}{2}\left(\frac{\delta Z}{\Delta X} + \frac{\delta X}{\Delta Z}\right) . \qquad \text{(II-6)}$$

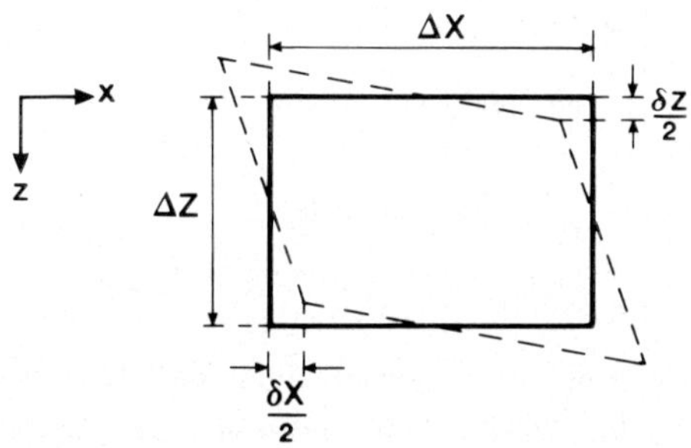

Figure II-3: Deformation of an infinitesimal parallelepiped: shearing strain.

Using again the components of the displacement vector $\vec{u}$, we may rewrite equation (II-6) as

$$e_{zx} = e_{xz} = \frac{1}{2}\left(\frac{\partial u_z}{\partial x} + \frac{\partial u_x}{\partial z}\right) . \qquad \text{(II-7a)}$$

In a similar way we define shearing strains $e_{yx}=e_{xy}$ and $e_{zy}=e_{yz}$, according to

$$e_{yx} = e_{xy} = \frac{1}{2}\left(\frac{\partial u_y}{\partial x} + \frac{\partial u_x}{\partial y}\right) \qquad \text{(II-7b)}$$

and

$$e_{zy} = e_{yz} = \frac{1}{2}\left(\frac{\partial u_z}{\partial y} + \frac{\partial u_y}{\partial z}\right) . \qquad \text{(II-7c)}$$

The tensile strains and shearing strains can be placed in a matrix, which is called the strain tensor e:

$$e = \begin{bmatrix} e_{xx} & e_{xy} & e_{xz} \\ e_{yx} & e_{yy} & e_{yz} \\ e_{zx} & e_{zy} & e_{zz} \end{bmatrix} . \qquad \text{(II-8)}$$

Note that the diagonal elements of this tensor represent the tensile strains; the off-diagonal elements represent the shearing strains.

Equations (II-5) and (II-7), which describe the elements of the strain tensor e, can be summarized by

$$e_{ij} = e_{ji} = \frac{1}{2}\left(\partial_j u_i + \partial_i u_j\right) , \qquad \text{(II-9)}$$

where i (or j) = 1, 2, 3 stands for x, y, z, respectively.

II.2.2. Conservation of momentum

Consider a solid medium with a space-dependent volume density of mass $\rho_o(\vec{r})$. In this medium we consider an elastic wave field, described by the space and time dependent particle velocity $\vec{v}(\vec{r},t)$. The mass density variations, associated to this wave field, we denote by $\Delta\rho(\vec{r},t)$, so that we may write for the total mass density

$$\rho(\vec{r},t) = \rho_o(\vec{r}) + \Delta\rho(\vec{r},t).$$

In the solid medium we consider a volume *V* enclosed by a surface *S* with outward pointing normal vector $\vec{n}$, see Figure II-4.

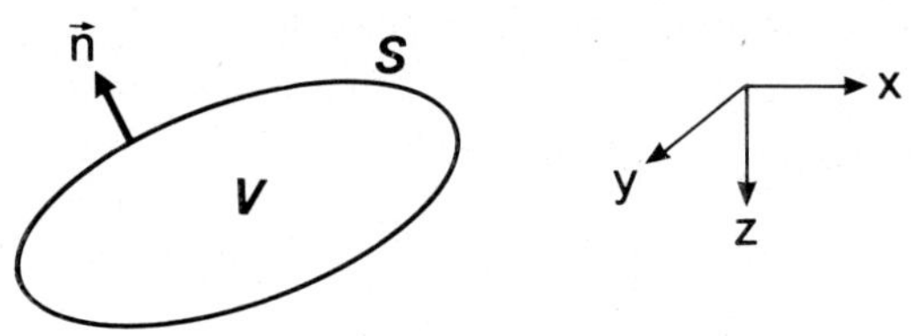

Figure II-4: Volume V in a solid medium.

The law of conservation of momentum states that the rate of change of momentum of the particles in V is equal to the incoming momentum flux through S, increased with the resultant force acting on the particles in V (generalized Newton's law), hence

$$\frac{\partial}{\partial t}\int_V \rho\vec{v}\,dV = -\oint_S (\rho\vec{v})\vec{v}.\vec{n}\,dS + \vec{F}(V), \qquad \text{(II-10a)}$$

where

$$\vec{F}(V) = \oint_S \vec{\tau}_n dS + \int_V \vec{f}\,dV. \qquad \text{(II-10b)}$$

Here $\vec{\tau}_n(\vec{r},t)$ denotes the traction on S and $\vec{f}(\vec{r},t)$ is the volume density of external force. Note that equation (II-10) is almost identical to equation (I-4); the only difference is that we replaced $-p\vec{n}$ (i.e., the traction normal to S) by $\vec{\tau}_n$, which may have any orientation (the sub-script n refers to the orientation of surface S). In the previous section we have defined traction vectors $\vec{\tau}_x$, $\vec{\tau}_y$ and $\vec{\tau}_z$, acting across planes normal to the x,y and z-axes, respectively. These vectors represent the first, second and third column, respectively, of the stress-tensor $\boldsymbol{\tau}$. They may therefore be denoted by

$$\vec{\tau}_x = \boldsymbol{\tau}\vec{i}_x, \quad \vec{\tau}_y = \boldsymbol{\tau}\vec{i}_y \quad \text{and} \quad \vec{\tau}_z = \boldsymbol{\tau}\vec{i}_z\,, \qquad \text{(II-11a)}$$

where $\vec{i}_x$, $\vec{i}_y$ and $\vec{i}_z$ denote unit vectors in the x, y and z-direction, respectively. The generalization of (II-11a) for a traction acting across an arbitrarily oriented surface with normal $\vec{n}$ is given by Cauchy's formula (Achenbach, 1973)

$$\vec{\tau}_n = \boldsymbol{\tau}\,\vec{n}. \qquad \text{(II-11b)}$$

Note that the i'th component of $\vec{\tau}_n$ is given by

$$\tau_{i,n} = \tau_{ix}n_x + \tau_{iy}n_y + \tau_{iz}n_z,$$

or

$$\tau_{i,n} = \tau_{ij} n_j \; .^{1)} \qquad \text{(II-11c)}$$

Applying the theorem of Gauss (I-2) to the three components of the vectorial integral equation (II-10) and combining the results, yields

$$\int_V \partial_t(\rho\vec{v})dV = -\int_V \left[\vec{v}\nabla.(\rho\vec{v}) + (\rho\vec{v}.\nabla)\vec{v}\right] dV + \vec{F}(V), \qquad \text{(II-12a)}$$

where

$$\vec{F}(V) = \int_V \partial_j\vec{\tau}_j dV + \int_V \vec{f} dV, \qquad \text{(II-12b)}$$

or, since this equation holds for any volume V

$$\partial_t(\rho\vec{v}) + \vec{v}\nabla.(\rho\vec{v}) + (\rho\vec{v}.\nabla)\vec{v} - \partial_j\vec{\tau}_j = \vec{f}. \qquad \text{(II-12c)}$$

This is the non-linear equation of motion for the elastic wave field in terms of $\vec{v}$, $\vec{\tau}_x$, $\vec{\tau}_y$, $\vec{\tau}_z$ and $\Delta\rho$ (bear in mind that $\rho=\rho_o+\Delta\rho$).
The linearized version reads

$$\rho_o(\vec{r})\partial_t\vec{v}(\vec{r},t) - \partial_j\vec{\tau}_j(\vec{r},t) = \vec{f}(\vec{r},t), \qquad \text{(II-13)}$$

or in scalar notation

$$\rho_o(\vec{r})\partial_t v_i(\vec{r},t) - \partial_j\tau_{ij}(\vec{r},t) = f_i(\vec{r},t). \qquad \text{(II-14)}$$

In the following we replace ρ_o by ρ for notational convenience. Equation (II-13) is the elastic equivalent of the acoustic linearized equation of motion (I-10b)

$$\rho(\vec{r})\partial_t\vec{v}(\vec{r},t) + \nabla p(\vec{r},t) = \vec{f}(\vec{r},t). \qquad \text{(II-15a)}$$

---

1) From here onwards we use the summation convention for repeated indices, see also the introduction.

The latter equation could be derived as a special case of the elastic equation of motion (II-13) by choosing

$$\tau_{ij}(\vec{r},t) = -p(\vec{r},t)\delta_{ij}, \tag{II-15b}$$

see also equation (II-3). In equation (II-15b) δ_{ij} is the Kronecker symbol:

$$\begin{aligned} \delta_{ij} &= 1 \quad \text{for } i=j, \\ \delta_{ij} &= 0 \quad \text{for } i\neq j. \end{aligned} \tag{II-15c}$$

Remark:

The linearized law of conservation of *angular* momentum states that the rate of change of angular momentum of the particles in V (Figure II-4) is equal to the resultant *moment* of forces acting on the particles in V. Achenbach (1973) and Aki and Richards (1980) show that this leads to the important symmetry property of the stress tensor:

$$\tau_{ij} = \tau_{ji},$$

see Figure II-5.

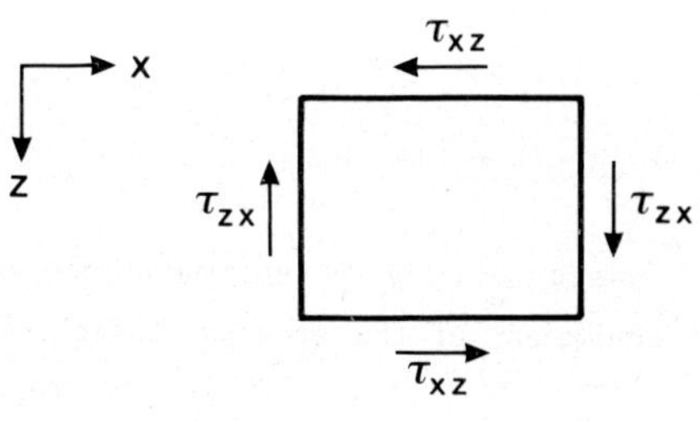

Figure II-5: Shearing stresses in the x,z-plane, acting on an infinitesimal parallelepiped: $\tau_{zx}=\tau_{xz}$.

II.2.3 Constitutive relation

The stress tensor $\tau(\vec{r},t)$ and the strain tensor $e(\vec{r},t)$ in a solid medium are mutually dependent, which is expressed by

$$\chi(\boldsymbol{\tau},\mathbf{e}) = 0, \tag{II-16a}$$

where χ is in general a non-linear function of the components of $\boldsymbol{\tau}$ and e. The stresses and strains associated to seismic waves are generally in the range where (II-16a) may be linearized. In its most general form this linear stress-strain relationship is described by

$$\tau_{ij}(\vec{r},t) = c_{ijk\ell}(\vec{r})e_{k\ell}(\vec{r},t), \tag{II-16b}$$

which is a generalization of Hooke's law. Here $c_{ijk\ell}(\vec{r})$ represents the 81 components of the stiffness tensor $c(\vec{r})$. The components of this tensor satisfy the following symmetry properties

$$c_{ijk\ell} = c_{jik\ell}, \tag{II-17a}$$

which follows from $\tau_{ij} = \tau_{ji}$ and

$$c_{ijk\ell} = c_{ij\ell k}, \tag{II-17b}$$

which follows from $e_{k\ell}=e_{\ell k}$. Assuming that the deformations occur adiabatically (i.e., assuming that heat exchanges can be neglected), it can also be shown that

$$c_{ijk\ell} = c_{k\ell ij}, \tag{II-17c}$$

(Aki and Richards, 1980). These symmetry properties reduce the number of independent stiffness coefficients to 21. In the presence of a source strain distribution $\mathbf{h}(\vec{r},t)$, equation (II-16b) should be modified, according to

$$\tau_{ij}(\vec{r},t) - c_{ijk\ell}(\vec{r})e_{k\ell}(\vec{r},t) = -\sigma_{ij}(\vec{r},t), \tag{II-18a}$$

with

$$\sigma_{ij}(\vec{r},t) = c_{ijk\ell}(\vec{r})h_{k\ell}(\vec{r},t), \tag{II-18b}$$

where $h_{k\ell}(\vec{r},t)$ represents the components of tensor $\mathbf{h}(\vec{r},t)$. Note that

$\sigma_{ij}(\vec{r},t)$ represents the stress distribution associated to the source.

Upon substitution of (II-9) into (II-18a) we obtain the linearized stress-displacement relation

$$\tau_{ij}(\vec{r},t) - \frac{1}{2} c_{ijk\ell}(\vec{r}) \left(\partial_\ell u_k(\vec{r},t) + \partial_k u_\ell(\vec{r},t)\right) = -\sigma_{ij}(\vec{r},t), \tag{II-19a}$$

or, using the symmetry property $c_{ijk\ell}(\vec{r}) = c_{ij\ell k}(\vec{r})$,

$$\tau_{ij}(\vec{r},t) - c_{ijk\ell}(\vec{r})\partial_\ell u_k(\vec{r},t) = -\sigma_{ij}(\vec{r},t). \tag{II-19b}$$

Finally, if we use the particle velocity vector $\vec{v}$, with

$$\vec{v}(\vec{r},t) = \partial_t \vec{u}(\vec{r},t), \tag{II-20}$$

we obtain the linearized stress-velocity relation

$$\partial_t \tau_{ij}(\vec{r},t) - c_{ijk\ell}(\vec{r})\partial_\ell v_k(\vec{r},t) = -\partial_t \sigma_{ij}(\vec{r},t), \tag{II-21}$$

where $v_k(\vec{r},t)$ for k=1,2,3 represents the components of vector $\vec{v}(\vec{r},t)$. Equation (II-21) is the elastic equivalent of the acoustic linearized equation of continuity (I-10a)

$$-\partial_t p(\vec{r},t) - K(\vec{r})\partial_k v_k(\vec{r},t) = -K(\vec{r})\partial_t i_v(\vec{r},t). \tag{II-22a}$$

Note that the latter equation can be derived as a special case of the elastic stress-velocity relation (II-21), by choosing

$$\tau_{ij}(\vec{r},t) = -p(\vec{r},t)\delta_{ij}, \tag{II-22b}$$

$$c_{ijk\ell}(\vec{r}) = K(\vec{r})\delta_{ij}\delta_{k\ell} \tag{II-22c}$$

and

$$\sigma_{ij}(\vec{r},t) = K(\vec{r})i_v(\vec{r},t)\delta_{ij}, \tag{II-22d}$$

with

$$i_v(\vec{r},t) = h_{kk}(\vec{r},t). \tag{II-22e}$$

It is often advantageous to write equation (II-21) in vector notation. Following Woodhouse (1974) we introduce stiffness-subtensors $C_{j\ell}(\vec{r})$ for j=1,2,3 and ℓ=1,2,3, which contain the stiffness coefficients $c_{ijk\ell}(\vec{r})$ according to

$$(C_{j\ell})_{ik} = c_{ijk\ell}, \tag{II-23a}$$

hence

$$\mathbf{C}_{j\ell} = \begin{bmatrix} c_{1j1\ell} & c_{1j2\ell} & c_{1j3\ell} \\ c_{2j1\ell} & c_{2j2\ell} & c_{2j3\ell} \\ c_{3j1\ell} & c_{3j2\ell} & c_{3j3\ell} \end{bmatrix} . \tag{II-23b}$$

With this notation, (II-21) can be rewritten as

$$\partial_t \vec{\tau}_j(\vec{r},t) - C_{j\ell}(\vec{r})\partial_\ell \vec{v}(\vec{r},t) = -\partial_t \vec{\sigma}_j(\vec{r},t). \tag{II-24}$$

II.2.4 Elastic two-way wave equation

By eliminating the stress tensor components τ_{ij} from the set of equations (II-14) and (II-21), we obtain the two-way wave equation for the particle velocity components v_i:

$$\partial_j(c_{ijk\ell}\partial_\ell v_k) - \rho\partial_t^2 v_i = -\partial_t(f_i - \partial_j \sigma_{ij}). \tag{II-25a}$$

Alternatively, by eliminating the traction $\vec{\tau}_j$ from the set of equations (II-13) and (II-24), we obtain the two-way wave equation for the particle velocity vector $\vec{v}$:

$$\partial_j(C_{j\ell}\partial_\ell \vec{v}) - \rho\partial_t^2 \vec{v} = -\partial_t(\vec{f} - \partial_j \vec{\sigma}_j). \tag{II-25b}$$

Two-way wave equation (II-25) describes the propagation of linear elastic waves in inhomogeneous anisotropic solids. For inhomogeneous *isotropic* solids the stiffness coefficients can be written as (Jeffreys and Jeffreys, 1972)

$$c_{ijk\ell}(\vec{r}) = \lambda(\vec{r})\delta_{ij}\delta_{k\ell} + \mu(\vec{r})\left[\delta_{ik}\delta_{j\ell}+\delta_{i\ell}\delta_{jk}\right], \tag{II-26a}$$

which reduces the number of independent coefficients to two. In (II-26a) $\lambda(\vec{r})$ and $\mu(\vec{r})$ represent the space-dependent Lamé coefficients. They are related to the *compression modulus* according to

$$K(\vec{r}) = \lambda(\vec{r}) + \frac{2}{3}\,\mu(\vec{r}). \tag{II-26b}$$

The Lamé coefficient $\mu(\vec{r})$ is also known as the *shear modulus*. Upon substitution of (II-26a) into (II-25a) we obtain

$$\partial_i(\lambda\partial_j v_j) + \partial_j\{\mu(\partial_j v_i+\partial_i v_j)\} - \rho\partial_t^2 v_i = -\partial_t(f_i-\partial_j\sigma_{ij}), \tag{II-27a}$$

where

$$\sigma_{ij} = \lambda\delta_{ij}h_{kk} + \mu(h_{ij}+h_{ji}), \tag{II-27b}$$

see also equation (II-18b).

For the same situation, equation (II-25b) can be rewritten as

$$\nabla\left[K_c\nabla.\vec{v}\right] - \nabla\times(\mu\nabla\times\vec{v}) - \rho\partial_t^2\vec{v}$$

$$+\ 2\left[(\nabla\mu.\nabla)\vec{v} - (\nabla\mu)\nabla.\vec{v} + (\nabla\mu)\times(\nabla\times\vec{v})\right] = -\partial_t(\vec{f}-\partial_j\vec{\sigma}_j), \tag{II-27c}$$

where we introduced the *constrained compression modulus*

$$K_c(\vec{r}) = \lambda(\vec{r}) + 2\mu(\vec{r}). \tag{II-27d}$$

Note that equation (II-27c) simplifies significantly when we consider a *homogeneous* isotropic solid

$$K_c \nabla(\nabla . \vec{v}) - \mu \nabla \times \nabla \times \vec{v} - \rho \partial_t^2 \vec{v} = -\partial_t(\vec{f} - \partial_j \vec{\sigma}_j). \tag{II-27e}$$

II.2.5 P- and S-wave potentials

In chapter I we have seen that acoustic wave motion in a homogeneous fluid is curl-free, which is denoted by

$$\nabla \times \vec{v} = \vec{o},$$

see equation (I-10d). This property does not apply in general for elastic waves in solids. For the special situation of elastic wave motion in a homogeneous isotropic solid we write

$$\vec{v}(\vec{r},t) \triangleq \vec{v}_p(\vec{r},t) + \vec{v}_s(\vec{r},t), \tag{II-28a}$$

where

$$\vec{v}_p(\vec{r},t) \triangleq \nabla \phi(\vec{r},t) \tag{II-28b}$$

and

$$\vec{v}_s(\vec{r},t) \triangleq \nabla \times \vec{\psi}(\vec{r},t), \tag{II-28c}$$

where ϕ and $\vec{\psi}$ are the Lamé potentials.
Note that, in analogy with acoustic wave motion, the particle velocity $\vec{v}_p$ is curl-free:

$$\nabla \times \vec{v}_p = \vec{o}. \tag{II-29a}$$

$\vec{v}_p$ is associated to *compressional waves* (or P-waves) and ϕ is the *P-wave potential*. The particle velocity $\vec{v}_s$ is divergence-free:

$$\nabla . \vec{v}_s = o. \tag{II-29b}$$

$\vec{v}_s$ is associated to *shear waves* (or S-waves) and $\vec{\psi}$ is the *S-wave potential.* We shall now derive independent wave equations for the P- and S-wave potentials ϕ and $\vec{\psi}$. For the moment we consider the source-free situation. Subsitution of (II-28) into two-way wave equation (II-27e) yields

$$K_c \nabla [\nabla^2 \phi - \frac{\rho}{K_c} \partial_t^2 \phi] + \mu \nabla \times [-\nabla \times \nabla \times \vec{\psi} - \frac{\rho}{\mu} \partial_t^2 \vec{\psi}] = \vec{o}.$$

The decomposition of this equation into independent equations for P- and S-waves is not unique. A convenient choice is

$$\nabla^2 \phi - \frac{\rho}{K_c} \partial_t^2 \phi = o \tag{II-30a}$$

and

$$-\nabla \times \nabla \times \vec{\psi} - \frac{\rho}{\mu} \partial_t^2 \vec{\psi} = \vec{o}. \tag{II-30b}$$

Note that this choice implies[1)]

$$\nabla \cdot \vec{\psi} = o. \tag{II-30c}$$

This property plays an important role in chapter VI, where we derive extrapolation operators for P- and S-wave potentials. Note that if we use the fundamental property

$$-\nabla \times \nabla \times \vec{\psi} + \nabla(\nabla.\vec{\psi}) = \nabla^2 \vec{\psi}, \tag{II-30d}$$

equation (II-30b) may be replaced by

$$\nabla^2 \vec{\psi} - \frac{\rho}{\mu} \partial_t^2 \vec{\psi} = \vec{o}; \quad \nabla.\vec{\psi} = o. \tag{II-30e}$$

In the remainder of this book we will use a slightly modified definition for P- and S-wave potentials, according to

---

1) Applying the divergence operator to both sides of equation (II-30b) yields $\partial_t^2 \nabla.\vec{\psi} = o$. For non-static wave fields equation (II-30c) follows immediately.

$$\partial_t \vec{v}(\vec{r},t) \triangleq -\frac{1}{\rho}[\nabla\phi(\vec{r},t) + \nabla \times \vec{\psi}(\vec{r},t)], \tag{II-31a}$$

with (for the source-free situation)

$$\nabla \cdot \vec{\psi}(\vec{r},t) \triangleq o. \tag{II-31b}$$

Note that for the limiting case of an ideal fluid the P-wave potential $\phi(\vec{r},t)$, as defined in (II-31a), is identical to the acoustic pressure $p(\vec{r},t)$, (compare (II-31a), for $\vec{\psi} = \vec{o}$, with (I-10b), for $\vec{f} = \vec{o}$).

Finally, note that according to (II-30) and (II-31a)

$$\nabla . \partial_t \vec{v} = -\frac{1}{\rho} \nabla^2 \phi = -\frac{1}{K_c} \partial_t^2 \phi,$$

or

$$\partial_t \phi(\vec{r},t) = -K_c \nabla . \vec{v}(\vec{r},t). \tag{II-31c}$$

Similarly,

$$\nabla \times \partial_t \vec{v} = -\frac{1}{\rho} \nabla \times \nabla \times \vec{\psi} = \frac{1}{\mu} \partial_t^2 \vec{\psi},$$

or

$$\partial_t \vec{\psi}(\vec{r},t) = \mu \nabla \times \vec{v}(\vec{r},t). \tag{II-31d}$$

In chapter VI, equations (II-31c) and (II-31d) will be used as alternative definitions of the P- and S-wave potentials ϕ and $\vec{\psi}$, respectively.

II.3 SPHERICAL WAVE SOLUTIONS OF THE ELASTIC TWO-WAY WAVE EQUATION

In this section we present the spherical wave solutions of the elastic two-way wave equation for a point source in an unbounded homogeneous isotropic solid. We discuss both P- and S-wave solutions.

II.3.1 P- and S- wave sources

We consider an unbounded homogeneous isotropic solid in which the elastic wave field satisfies equation (II-27e). In analogy with (II-31a), we also express the source distribution in the right-hand side of equation (II-27e) in terms of potentials. We define

$$-(\vec{f}-\partial_j\vec{\sigma}_j) \triangleq K\nabla i_v + \nabla\times\vec{m}. \tag{II-32}$$

If we choose $\vec{f} = \vec{o}$, this definition implies that the components σ_{ij} of the source tensor satisfy (see also equation (II-27b))

$$\sigma_{xx} = \sigma_{yy} = \sigma_{zz} = (\lambda + \tfrac{2}{3}\mu)h_{kk} \triangleq Ki_v, \tag{II-33a}$$

$$\sigma_{yz} = -\sigma_{zy} \triangleq m_x, \tag{II-33b}$$

$$-\sigma_{xz} = \sigma_{zx} \triangleq m_y \tag{II-33c}$$

and

$$\sigma_{xy} = -\sigma_{yx} \triangleq m_z, \tag{II-33d}$$

where m_x, m_y and m_z are the components of vector $\vec{m}$.

Note that $i_v(\vec{r},t)$ represents a source distribution in terms of a volume density of volume injection. We will see later on that this source function generates P-waves only. From equations (II-33b), (II-33c) and (II-33d), as well as from Figure II-6, we observe that, unlike in Figure II-5, the resultant moments on an infinitesimal parallelepiped do not vanish. Therefore $\vec{m}(\vec{r},t)$ may be interpreted as a source distribution in terms of a volume density of moment. We will see later on that this source function generates S-waves only.

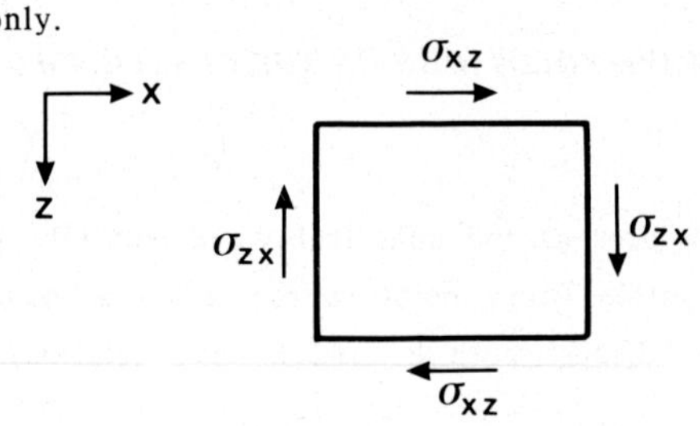

Figure II-6: Source shearing stresses in the x,z-plane, acting on an infinitesimal parallelepiped.
The situation is shown for a volume density of moment $m_y=\sigma_{zx}=-\sigma_{xz}$.

Upon substitution of (II-31a) and (II-32) into (II-27e), we obtain

$$K_c \nabla \left[\nabla^2 \phi - \frac{\rho}{K_c} \partial_t^2 \phi + \rho \frac{K}{K_c} \partial_t^2 i_v\right] + \mu \nabla \times \left[-\nabla \times \nabla \times \vec{\psi} - \frac{\rho}{\mu} \partial_t^2 \vec{\psi} + \frac{\rho}{\mu} \partial_t^2 \vec{m}\right] = \vec{o}. \qquad \text{(II-34)}$$

Again, the decomposition of this equation into independent equations for P- and S-waves is not unique. A convenient choice is

$$\nabla^2 \phi - \frac{\rho}{K_c} \partial_t^2 \phi = -\rho \frac{K}{K_c} \partial_t^2 i_v \qquad \text{(II-35a)}$$

and

$$-\nabla \times \nabla \times \vec{\psi} - \frac{\rho}{\mu} \partial_t^2 \vec{\psi} = -\frac{\rho}{\mu} \partial_t^2 \vec{m}. \qquad \text{(II-35b)}$$

Note that this choice implies

$$\nabla . \vec{\psi} = \nabla . \vec{m}. \qquad \text{(II-35c)}$$

Many authors choose $\nabla . \vec{m} = o$, thus imposing a dependency between the source components m_x, m_y and m_z. In section VI.3.3, however, we need *independent* sources for the different types of Green's S-waves and therefore we cannot make this restriction. Hence, using property (II-30d), equation (II-35b) may be replaced by

$$\nabla^2 \vec{\psi} - \frac{\rho}{\mu} \partial_t^2 \vec{\psi} = -\frac{\rho}{\mu} \partial_t^2 \vec{m} + \nabla(\nabla . \vec{m}); \ \nabla . \vec{\psi} = \nabla . \vec{m}. \qquad \text{(II-35d)}$$

In conclusion, for a source distribution in a homogeneous isotropic solid we have obtained two independent wave equations for P- and S-waves. Equation (II-35a) governs the P-wave potential $\phi(\vec{r},t)$, given a source distribution in terms of a volume density of volume injection $i_v(\vec{r},t)$. Equation (II-35d) governs the S-wave potential $\vec{\psi}(\vec{r},t)$, given a source distribution in terms of a volume density of moment $\vec{m}(\vec{r},t)$.

II.3.2 Spherical P- and S-wave fields

Consider wave equation (II-35a) for the P-wave potential $\phi(\vec{r},t)$ in a homogeneous isotropic solid:

$$\nabla^2 \phi - \frac{\rho}{K_c} \partial_t^2 \phi = - \rho \frac{K}{K_c} \partial_t^2 i_v . \qquad \text{(II-36)}$$

Note the resemblance with the acoustic two-way wave equation (I-11) for $\vec{f}=\vec{o}$:

$$\nabla^2 p - \frac{\rho}{K} \partial_t^2 p = - \rho \, \partial_t^2 i_v . \qquad \text{(II-37)}$$

Hence, for a point source of volume injection, defined by

$$i_v(\vec{r},t) \triangleq \delta(\vec{r})s(t), \qquad \text{(II-38a)}$$

the causal solution of (II-36) reads, in analogy with (I-17a),

$$\phi(r,t) = \rho \frac{K}{K_c} \frac{\partial_t^2 s(t-r/c_p)}{4\pi r} , \qquad \text{(II-38b)}$$

where

$$r = |\vec{r}| = \sqrt{x^2 + y^2 + z^2} \qquad \text{(II-38c)}$$

and

$$c_p = \sqrt{K_c/\rho} = \sqrt{(\lambda+2\mu)/\rho} . \qquad \text{(II-38d)}$$

Here c_p represents the P-wave propagation velocity.
The particle velocity associated to this P-wave potential reads, according to equation (II-31a),

$$\vec{v}(\vec{r},t) = - \frac{K}{K_c} \nabla \left(\frac{\partial_t s(t-r/c_p)}{4\pi r} \right) . \qquad \text{(II-38e)}$$

Consider wave equation (II-35d) for the S-wave potential $\vec{\psi}(\vec{r},t)$ in a homogeneous isotropic solid. The scalar formulation of this equation reads

$$\nabla^2 \psi_k - \frac{\rho}{\mu} \partial_t^2 \psi_k = - \frac{\rho}{\mu} \partial_t^2 m_k + \partial_k(\partial_\ell m_\ell) . \qquad \text{(II-39)}$$

For a point source of moment in the h-direction, defined by

$$m_k(\vec{r},t) \triangleq \delta_{kh}\delta(\vec{r})s(t), \tag{II-40a}$$

the causal solution of (II-39) reads, in analogy with (I-17a),

$$\psi_k(\vec{r},t) = [\delta_{kh}\frac{\rho}{\mu}\partial_t^2 - \partial_k\partial_h]\frac{s(t-r/c_s)}{4\pi r}, \tag{II-40b}$$

where

$$c_s = \sqrt{\mu/\rho}\,. \tag{II-40c}$$

Here c_s represents the S-wave propagation velocity. Note that for h=1, the source moment reads

$$\vec{m}(\vec{r},t) = \begin{bmatrix} \delta(\vec{r})s(t) \\ o \\ o \end{bmatrix}, \tag{II-40d}$$

the S-wave potential reads

$$\vec{\psi}(\vec{r},t) = \begin{bmatrix} \frac{\rho}{\mu}\partial_t^2 - \partial_x^2 \\ -\partial_y\partial_x \\ -\partial_z\partial_x \end{bmatrix} \frac{s(t-r/c_s)}{4\pi r} \tag{II-40e}$$

and, according to equation (II-31a), the particle velocity associated to this S-wave potential reads

$$\vec{v}(\vec{r},t) = \frac{1}{\mu}\begin{bmatrix} o \\ -\partial_z \\ \partial_y \end{bmatrix} \frac{\partial_t s(t-r/c_s)}{4\pi r}\,. \tag{II-40f}$$

Hence, for a source moment in the x-direction, the particle velocity is polarized in the plane perpendicular to the x-axis and therefore we speak of S_x-waves. Similarly, for a source moment in the y- or z-direction, the particle velocity is polarized in the plane perpendicular to the y- or z-axis and therefore we speak of S_y- or S_z-waves, respectively.

For a further discussion of spherical wave solutions we refer to section I.3.

II.4 PLANE WAVE SOLUTIONS OF THE ELASTIC TWO-WAY WAVE EQUATION

In this section we present plane wave solutions of the elastic two-way wave equation for a source-free homogeneous anisotropic solid. Next, for the isotropic situation we discuss both homogeneous ("propagating") and inhomogeneous ("evanescent") plane waves. Finally we introduce the concept "elastic one-way wave equations".

II.4.1 Homogeneous plane waves

In analogy with section I.4.1., consider a 3-D plane wave of the following form

$$\vec{v}(\vec{r},t) = \vec{v}_0 A(t-\vec{s}\cdot\vec{r}), \tag{II-41a}$$

or

$$\vec{v}(x,y,z,t) = \vec{v}_0 A(t-s_x x-s_y y-s_z z). \tag{II-41b}$$

Throughout this section we assume that the phase slownesses s_x, s_y and s_z are real [1)]. A wave front may be any plane surface perpendicular to the slowness vector $\vec{s}$, see also Figure I-4. The wave front velocity c in the direction of $\vec{s}$ is given by

$$c = \frac{1}{|\vec{s}|} . \tag{II-42}$$

Upon substitution of the plane wave (II-41) into the elastic two-way wave equation (II-25b) for a source-free homogeneous anisotropic solid,

$$\mathbf{C}_{j\ell}\partial_j\partial_\ell\vec{v} - \rho\,\partial_t^2\vec{v} = \vec{o}, \tag{II-43}$$

we obtain

$$\mathbf{K}\vec{v}_0 = \vec{o}, \tag{II-44a}$$

1) Vector $\vec{s}$ should not be confused with the source signature s in section II.3.

with

$$\mathbf{K} = s_j s_\ell C_{j\ell} - \rho \mathbf{I}, \tag{II-44b}$$

where **I** is the 3×3 identity matrix. This equation has non-trivial solutions only when

$$\det(\mathrm{K}) = 0. \tag{II-45}$$

In the following we will only consider a special form of anisotropy, namely *transverse isotropy* (see also Daley and Hron, 1977, and Van der Hijden, 1987, App. D). In the earth there are two main causes for transverse isotropy. Schoenberg (1983) shows that periodically stratified sedimentary rock may be replaced (in the long wavelength limit) by homogeneous rock with transverse isotropy. In this case the stiffness tensor exhibits rotational invariance in the horizontal x,y-plane, hence, the axis of symmetry is the vertical z-axis, see Figure II-7a.
Crampin (1984) shows that a solid medium with parallel fractures (e.g. a reservoir) may also be replaced by a homogeneous solid medium with transverse isotropy. E.g., for vertical fractures in the y,z-plane, the axis of symmetry is the horizontal x-axis, see Figure II-7b.

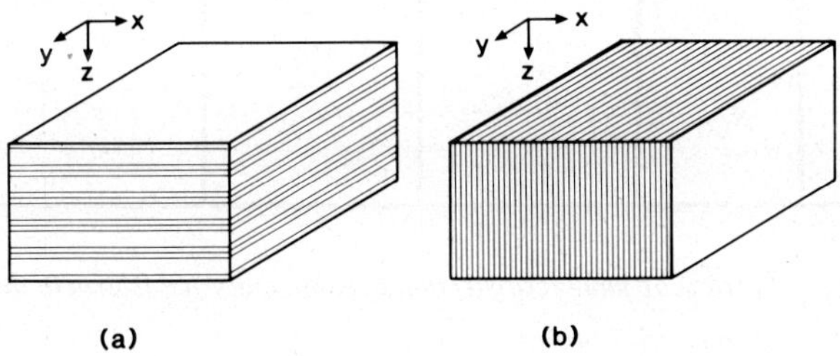

Figure II-7: Two typical situations of transverse isotropy
a. Periodically stratified sedimentary rock. The axis of symmetry is the z-axis.
b. Reservoir with vertical fractures in the y,z-plane. The axis of symmetry is the x-axis.

Any transverse isotropic medium can be described by five independent stiffness coefficients. Moreover, when the axis of symmetry coincides with one of the axes of the coordinate system (as in the two situations described above), then many of the stiffness coefficients are zero. Table II-1 gives an overview of the non-zero stiffness coefficients for isotropic and transverse isotropic solid media with symmetry axes in the x-, y- and z-direction, respectively.

Stiffness coefficients		Lamé coefficients	Modified Lamé coefficients		
Conventional notation	Voigt[1] notation	Pure isotropy	Transverse isotropy; symmetry axis: x-axis	y-axis	z-axis
c_{1111}	c_{11}	$\lambda + 2\mu$	$\lambda_{//} + 2\mu_{//}$	$\lambda_{\perp} + 2\mu_{\perp}$	$\lambda_{\perp} + 2\mu_{\perp}$
c_{2222}	c_{22}	$\lambda + 2\mu$	$\lambda_{\perp} + 2\mu_{\perp}$	$\lambda_{//} + 2\mu_{//}$	$\lambda_{\perp} + 2\mu_{\perp}$
c_{3333}	c_{33}	$\lambda + 2\mu$	$\lambda_{\perp} + 2\mu_{\perp}$	$\lambda_{\perp} + 2\mu_{\perp}$	$\lambda_{//} + 2\mu_{//}$
c_{1122}	c_{12}	λ	v	v	$\lambda_{\perp}$
c_{1133}	c_{13}	λ	v	$\lambda_{\perp}$	v
c_{2233}	c_{23}	λ	$\lambda_{\perp}$	v	v
c_{2323}	c_{44}	μ	$\mu_{\perp}$	$\mu_{//}$	$\mu_{//}$
c_{1313}	c_{55}	μ	$\mu_{//}$	$\mu_{\perp}$	$\mu_{//}$
c_{1212}	c_{66}	μ	$\mu_{//}$	$\mu_{//}$	$\mu_{\perp}$

Table II-1: Overview of non-zero stiffness coefficients for isotropic and transverse isotropic solid media.

In this table, the stiffness coefficients are expressed in terms of the Lamé coefficients λ and μ (for pure isotropy) and in terms of the modified Lamé coefficients $\lambda_{//}$, $\mu_{//}$, $\lambda_{\perp}$, $\mu_{\perp}$ and v (for transverse

1) Double sub-scripts in the conventional tensor notation are replaced by single sub-scripts in the Voigt notation, according to: 11→1, 22→2, 33→3, 23→4, 13→5, 12→6 (Nye, 1957).

isotropy). Note that for each of the three symmetry axes the same stiffness coefficients are non-zero. Let us now solve equation (II-45) for the isotropic situation and for the situation of transverse isotropy. Bearing in mind that the components of $\mathbf{C}_{j\ell}$ are given by $(\mathbf{C}_{j\ell})_{ik} = c_{ijk\ell}$, we may write for **K**, as defined by (II-44b),

$$\mathbf{K} = \begin{bmatrix} \left(s_x^2 c_{11} + s_y^2 c_{66} + s_z^2 c_{55} - \rho\right) & s_x s_y \left(c_{12} + c_{66}\right) & s_x s_z \left(c_{13} + c_{55}\right) \\ s_x s_y \left(c_{12} + c_{66}\right) & \left(s_x^2 c_{66} + s_y^2 c_{22} + s_z^2 c_{44} - \rho\right) & s_y s_z \left(c_{23} + c_{44}\right) \\ s_x s_z \left(c_{13} + c_{55}\right) & s_y s_z \left(c_{23} + c_{44}\right) & \left(s_x^2 c_{55} + s_y^2 c_{44} + s_z^2 c_{33} - \rho\right) \end{bmatrix}, \quad \text{(II-46)}$$

where we used the Voigt notation for the stiffness coefficients (see Table II-1). For the pure isotropic situation we find

$$\det(\mathbf{K}) =$$

$$\left\{ \left(s_x^2+s_y^2+s_z^2\right)\left(\lambda+2\mu\right) - \rho \right\} \left\{ \left(s_x^2+s_y^2+s_z^2\right)\mu - \rho \right\} \left\{ \left(s_x^2+s_y^2+s_z^2\right)\mu - \rho \right\}, \quad \text{(II-47)}$$

where we replaced the stiffness coefficients by the Lamé parameters (see Table II-1). Hence, for this situation the solution of $\det(\mathbf{K})=0$ is given by three concentric spheres in the 3-D slowness domain

$$s_x^2+s_y^2+s_z^2 = \rho/(\lambda+2\mu) \qquad \text{(see Figure II-8a)}, \qquad \text{(II-48a)}$$

$$s_x^2+s_y^2+s_z^2 = \rho/\mu \qquad \text{(see Figure II-8b)}, \qquad \text{(II-48b)}$$

$$s_x^2+s_y^2+s_z^2 = \rho/\mu \qquad \text{(see Figure II-8c)}. \qquad \text{(II-48c)}$$

Note that the latter two solutions are identical. The surfaces in the 3-D slowness domain are generally known as *slowness surfaces*. Any point on a slowness surface corresponds to a slowness vector $\vec{s}$ with components s_x,

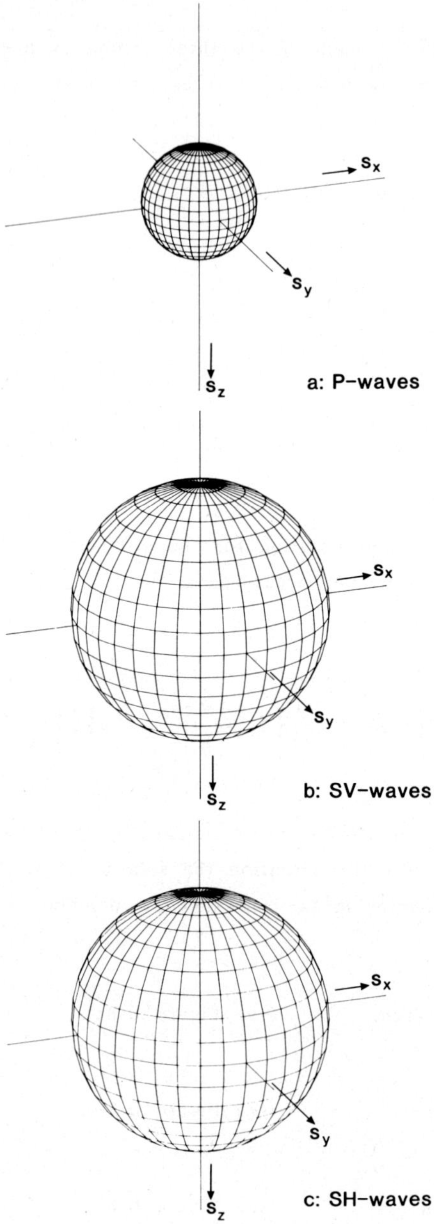

Figure II-8: Slowness surfaces for an isotropic medium. Note that Figures b and c are identical.

s_y and s_z. The reciprocal of the modulus of vector $\vec{s}$ represents the wave front velocity in the direction of $\vec{s}$. Hence, in this special situation of spherical slowness surfaces, the velocity is independent of the propagation direction. This is exactly what one would expect in an isotropic medium. Note that from (II-48a) we obtain a velocity c_p according to

$$c_p = 1\Big/\sqrt{s_x^2+s_y^2+s_z^2} = \sqrt{(\lambda+2\mu)/\rho} \;, \tag{II-49a}$$

whereas from (II-48b) and (II-48c) we obtain a velocity c_s according to

$$c_s = 1\Big/\sqrt{s_x^2+s_y^2+s_z^2} = \sqrt{\mu/\rho} \;. \tag{II-49b}$$

By substituting (II-49a) into (II-44a) for the isotropic situation, we find

$$\vec{v}_o = \kappa\vec{s}, \tag{II-50a}$$

where κ is an undefined parameter. This equation states that the particle motion in the plane wave occurs parallel to the propagation direction. Hence, c_p, as defined by (II-49a), represents the propagation velocity of longitudinal or compressional waves. The P-wave property $\nabla\times\vec{v}_p=\vec{o}$ is easily verified from (II-41b) and (II-50a).

By substituting (II-49b) into (II-44a) for the isotropic situation, we find

$$\vec{v}_o \,.\, \vec{s} = o. \tag{II-50b}$$

This equation states that the particle motion in the plane wave occurs perpendicular to the propagation direction. Hence, c_s, as defined by (II-49b), represents the propagation velocity of transversal or shear waves. The S-wave property $\nabla.\vec{v}_s=o$ is easily verified from (II-41b) and (II-50b).

Let us now consider the situation of transverse isotropy with a vertical symmetry axis (Figure II-7a). If we assume for the moment that s_y is equal to zero, then we find from equation (II-46) and Table II-1,

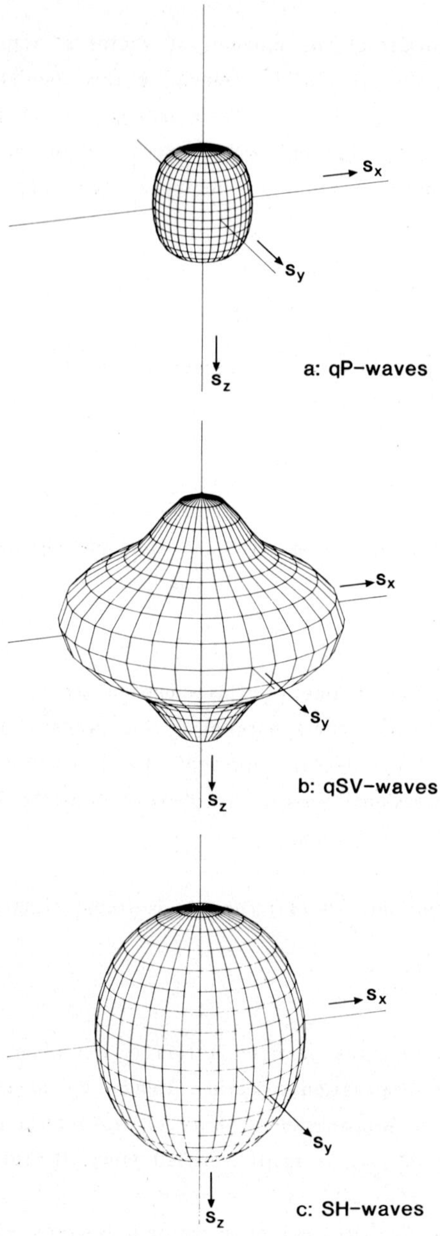

Figure II-9: Slowness surfaces for a transverse isotropic medium with a vertical symmetry axis (as in Figure II-7a).

$$\mathbf{K} = \begin{bmatrix} [s_x^2(\lambda_\perp+2\mu_\perp)+s_z^2\mu_{//}-\rho] & o & s_x s_z(\upsilon+\mu_{//}) \\ o & [s_x^2\mu_\perp+s_z^2\mu_{//}-\rho] & o \\ s_x s_z(\upsilon+\mu_{//}) & o & [s_x^2\mu_{//}+s_z^2(\lambda_{//}+2\mu_{//})-\rho] \end{bmatrix}, \tag{II-51a}$$

hence,

$$\det(\mathbf{K}) = [s_x^2\mu_\perp+s_z^2\mu_{//}-\rho]\Big\{[s_x^2(\lambda_\perp+2\mu_\perp) + s_z^2\mu_{//} - \rho]\times [s_x^2\mu_{//}+s_z^2(\lambda_{//}+2\mu_{//}) - \rho] - s_x^2 s_z^2(\upsilon+\mu_{//})^2\Big\}. \tag{II-51b}$$

If s_y is not equal to zero, then, on basis of the symmetry properties, we may conclude that in equation (II-51b) s_x^2 need be replaced by $s_x^2+s_y^2$, hence

$$\det(\mathbf{K}) = [s_r^2\mu_\perp+s_z^2\mu_{//}-\rho]\Big\{[s_r^2(\lambda_\perp+2\mu_\perp) + s_z^2\mu_{//} - \rho]\times [s_r^2\mu_{//}+s_z^2(\lambda_{//}+2\mu_{//}) - \rho] - s_r^2 s_z^2(\upsilon+\mu_{//})^2\Big\}, \tag{II-52a}$$

where

$$s_r^2 = s_x^2 + s_y^2. \tag{II-52b}$$

For this situation the solution of det(**K**)=o is given by the following surfaces in the 3-D slowness domain

$$s_z^2 = \rho\gamma_1 - \frac{\mu_\perp}{\mu_{//}}(1+\epsilon_1)s_r^2 - \sqrt{\rho^2\gamma_2^2 + \rho\epsilon_2 s_r^2 + \epsilon_3 s_r^4} \tag{II-53a}$$

(see Figure II-9a),

$$s_z^2 = \rho\gamma_1 - \frac{\mu_\perp}{\mu_{//}}(1+\epsilon_1)s_r^2 + \sqrt{\rho^2\gamma_2^2 + \rho\epsilon_2 s_r^2 + \epsilon_3 s_r^4} \tag{II-53b}$$

(see Figure II-9b) and

$$s_z^2 = \frac{\rho}{\mu_{//}} - \frac{\mu_\perp}{\mu_{//}} s_r^2 \tag{II-53c}$$

(see Figure II-9c), where

$$\gamma_1 = \frac{\lambda_{//} + 3\mu_{//}}{2(\lambda_{//}+2\mu_{//})\mu_{//}}, \tag{II-53d}$$

$$\gamma_2 = \frac{\lambda_{//} + \mu_{//}}{2(\lambda_{//}+2\mu_{//})\mu_{//}}, \tag{II-53e}$$

$$\epsilon_1 = \frac{(\lambda_\perp\lambda_{//}-\upsilon^2) + 2\mu_{//}(\lambda_\perp-\upsilon)}{2\mu_\perp(\lambda_{//}+2\mu_{//})}, \tag{II-53f}$$

$$\epsilon_2 = \frac{(\lambda_\perp+2\mu_\perp+\mu_{//})\mu_{//} - (\lambda_{//}+3\mu_{//})\mu_\perp(1+\epsilon_1)}{\mu_{//}^2(\lambda_{//}+2\mu_{//})} \tag{II-53g}$$

and

$$\epsilon_3 = \left[\frac{\mu_\perp(1+\epsilon_1)}{\mu_{//}}\right]^2 - \left[\frac{\lambda_\perp+2\mu_\perp}{\lambda_{//}+2\mu_{//}}\right]. \tag{II-53h}$$

From equation (II-53) and from Figure II-9 we observe that the slowness surfaces for the transverse isotropic situation are not spherical, hence, the velocity of the 3-D plane wave (II-41) depends on the tilt angle α of the slowness vector $\vec{s}$. Equation (II-53a) and Figure II-9a represent the slowness surface for quasi-longitudinal waves ($\vec{v}_o \approx \kappa\vec{s}$). For vertical wave propagation (s_r=o) we find $s_z^2=\rho/(\lambda_{//}+2\mu_{//})$. Hence, the velocity for qP-waves propagating parallel to the symmetry axis reads

$$c_{p,//} = \sqrt{\frac{\lambda_{//}+2\mu_{//}}{\rho}}. \tag{II-54a}$$

For horizontal wave propagation (s_z=o) we find $s_r^2=\rho/(\lambda_\perp+2\mu_\perp)$. Hence, the velocity for qP-waves propagating normal to the symmetry axis reads

$$c_{p,\perp} = \sqrt{\frac{\lambda_\perp+2\mu_\perp}{\rho}} \, . \qquad \text{(II-54b)}$$

Equation (II-53b) and Figure (II-9b) represent the slowness surface for quasi-transversal waves ($\vec{v}_o . \vec{s} \approx$ o). The polarization (i.e., the direction of vector $\vec{v}_o$) is in the vertical plane through $\vec{s}$ and the z-axis, therefore we speak of qSV-waves. For vertical wave propagation (s_r=o) we find $s_z^2=\rho/\mu_{//}$. Hence, the velocity for qSV-waves propagating parallel to the symmetry axis reads

$$c_{sv,//} = \sqrt{\frac{\mu_{//}}{\rho}} \, . \qquad \text{(II-55a)}$$

For horizontal wave propagation (s_z=o) we find $s_r^2=\rho/\mu_{//}$. Hence, the velocity for qSV-waves propagating normal to the symmetry axis also reads

$$c_{sv,\perp} = \sqrt{\frac{\mu_{//}}{\rho}} \, . \qquad \text{(II-55b)}$$

Equation (II-53c) and Figure (II-9c) represent the slowness surface for pure transversal waves ($\vec{v}_o \, . \, \vec{s}$ = o). The polarization is in the horizontal x,y-plane, therefore we speak of SH-waves. For vertical wave propagation (s_r=o) we find $s_z^2=\rho/\mu_{//}$. Hence, the velocity for SH-waves propagating parallel to the symmetry axis reads

$$c_{SH,//} = \sqrt{\frac{\mu_{//}}{\rho}} \, . \qquad \text{(II-56a)}$$

For horizontal wave propagation (s_z=o) we find $s_r^2=\rho/\mu_\perp$. Hence, the velocity for SH-waves propagating normal to the symmetry axis reads

$$c_{SH,\perp} = \sqrt{\frac{\mu_\perp}{\rho}} \, . \qquad \text{(II-56b)}$$

Note the interesting property

$$c_{sv,\perp} \neq c_{SH,\perp} . \qquad \text{(II-57)}$$

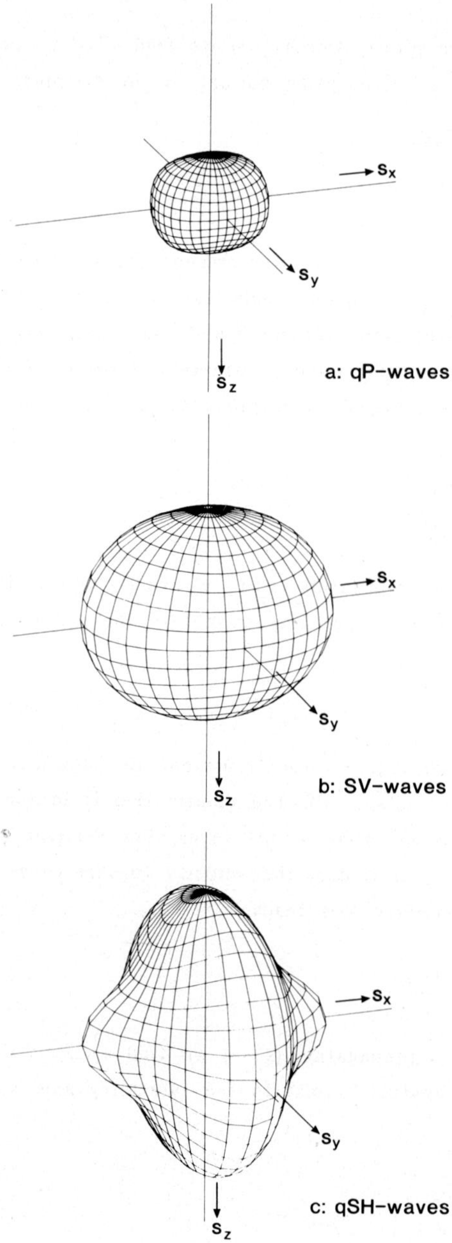

Figure II-10: Slowness surfaces for a transverse isotropic medium with a horizontal symmetry axis (as in Figure II-7b).

This equation implies that an arbitrarily polarized S-wave, which propagates normal to the symmetry axis of a transverse isotropic medium, splits up in vertically and horizontally polarized S-waves, propagating with different velocities (with $c_{SH,\perp} > c_{SV,\perp}$). This effect, which is not restricted to horizontal propagation, is commonly known as *shear wave birefringence.*

Finally, let us consider the situation of transverse isotropy with a horizontal symmetry axis (Figure II-7b). The expressions for the slowness surfaces can be obtained directly from equation (II-53) by making some simple substitutions (bear in mind that the modified Lamé parameters $\lambda_{//}$, $\mu_{//}$, $\lambda_{\perp}$, $\mu_{\perp}$ and υ are defined on basis of the orientation of the symmetry axis, see Table II-1). When the symmetry axis is in the x-direction, then the expressions for the slowness surfaces read

$$s_x^2 = \rho\gamma_1 - \frac{\mu_\perp}{\mu_{//}}\left(1+\epsilon_1\right)s_r^2 - \sqrt{\rho^2\gamma_2^2 + \rho\epsilon_2 s_r^2 + \epsilon_3 s_r^4} \qquad \text{(II-58a)}$$

(see Figure II-10a),

$$s_x^2 = \frac{\rho}{\mu_{//}} - \frac{\mu_\perp}{\mu_{//}} s_r^2 \qquad \text{(II-58b)}$$

(see Figure II-10b) and

$$s_x^2 = \rho\gamma_1 - \frac{\mu_\perp}{\mu_{//}}\left(1+\epsilon_1\right)s_r^2 + \sqrt{\rho^2\gamma_2^2 + \rho\epsilon_2 s_r^2 + \epsilon_3 s_r^4} \qquad \text{(II-58c)}$$

(see Figure II-10c), where

$$s_r^2 = s_y^2 + s_z^2 \qquad \text{(II-58d)}$$

and where γ_1, γ_2, ϵ_1, ϵ_2 and ϵ_3 are defined as in equation (II-53). Figure II-10 clearly shows that the slowness depends on the azimuth angle β (where $\tan\beta = s_y/s_x$, see Figure I-4). Therefore in this case we may also speak of *azimuthal anisotropy.*

II.4.2 Inhomogeneous plane waves

In this section we consider inhomogeneous plane waves in homogeneous isotropic media. First we discuss inhomogeneous plane P-waves. We consider real phase slownesses s_x and s_y for which

$$s_x^2 + s_y^2 > \frac{1}{c_p^2}, \tag{II-59a}$$

where c_p is the P-wave velocity, as defined by equation (II-49a). Note that

$$s_{z,p}^2 = \frac{1}{c_p^2} - s_x^2 - s_y^2 < o, \tag{II-59b}$$

hence, for this situation the phase-slowness $s_{z,p}$ appears to be imaginary. To avoid complex functions, we define a new parameter $\sigma_{z,p}$ according to[1])

$$\sigma_{z,p}^2 = s_x^2 + s_y^2 - \frac{1}{c_p^2} > o, \tag{II-59c}$$

where $\sigma_{z,p}$ is real (positive or negative). In analogy with section I.4.2, consider the monochromatic wave function

$$\vec{v}_p(\vec{r},t) = v_{p,o} \begin{bmatrix} s_x \cos[\phi(\vec{r},t)] \\ s_y \cos[\phi(\vec{r},t)] \\ \sigma_{z,p} \sin[\phi(\vec{r},t)] \end{bmatrix} e^{-\omega_o \sigma_{z,p} z}, \tag{II-60a}$$

where

$$\phi(\vec{r},t) = \omega_o(t - \vec{s}_o . \vec{r}) + \phi_o, \tag{II-60b}$$

with

$$\vec{s}_o = \text{Real}\ (\vec{s}) = \begin{bmatrix} s_x \\ s_y \\ o \end{bmatrix}. \tag{II-60c}$$

---

1) $\sigma_{z,p}$ should not be confused with the source stress σ_{ij} in section II.2.

Hence,

$$\phi(x,y,z,t) = \omega_o(t-s_x x-s_y y) + \phi_o. \tag{II-60d}$$

Note that the phase is independent of depth. Since

$$\nabla \times \vec{v}_p = \vec{o}, \tag{II-60e}$$

it can be easily verified that wave function (II-60) is a solution of the elastic two-way wave equation (II-27e), with $\rho/K_c=1/c_p^2$. A wave front is defined as a surface on which the phase $\phi(x,y,z,t)$ is constant. Hence, a wave front may be any plane surface perpendicular to vector $\vec{s}_o$, see also Figure I-7. The wave fronts propagate in the direction of $\vec{s}_o$ with a propagation velocity $c_{p,o}$ given by

$$c_{p,o} = \frac{1}{|\vec{s}_o|} = \frac{1}{\sqrt{s_x^2 + s_y^2}}. \tag{II-61a}$$

Note that

$$c_{p,o} < c_p. \tag{II-61b}$$

For a further discussion of this inhomogeneous plane P-wave we refer to section I.4.2.

Next we discuss inhomogeneous plane S-waves. We consider real phase slownesses s_x and s_y for which

$$s_x^2 + s_y^2 > \frac{1}{c_s^2}, \tag{II-62a}$$

where c_s is the S-wave velocity, as defined by equation (II-49b). Note that

$$s_{z,s}^2 = \frac{1}{c_s^2} - s_x^2 - s_y^2 < o. \tag{II-62b}$$

We define a real parameter $\sigma_{z,s}$ according to

$$\sigma_{z,s}^2 = s_x^2 + s_y^2 - \frac{1}{c_s^2} > o. \tag{II-62c}$$

Consider the monochromatic wave function

$$\vec{v}_s(\vec{r},t) = v_{s,o} \begin{bmatrix} \sigma_{z,s} s_x \cos[\phi(\vec{r},t)] \\ \sigma_{z,s} s_y \cos[\phi(\vec{r},t)] \\ (s_x^2+s_y^2) \sin[\phi(\vec{r},t)] \end{bmatrix} e^{-\omega_o \sigma_{z,s} z}, \tag{II-63a}$$

with $\phi(\vec{r},t)$ defined as in (II-60). Since

$$\nabla \cdot \vec{v}_s(\vec{r},t) = o, \tag{II-63b}$$

it can be easily verified that wave function (II-63) is a solution of the elastic two-way wave equation (II-27e), with $\rho/\mu=1/c_s^2$. Again, the wave fronts propagate in the direction of $\vec{s}_o$ with a propagation velocity $c_{s,o}$ given by

$$c_{s,o} = \frac{1}{|\vec{s}_o|} = \frac{1}{\sqrt{s_x^2 + s_y^2}}. \tag{II-64a}$$

Note that

$$c_{s,o} < c_s. \tag{II-64b}$$

Also note the interesting property that for fixed s_x and s_y (with $s_x^2+s_y^2 > 1/c_s^2 > 1/c_p^2$) we may write

$$c_{p,o} = c_{s,o}. \tag{III-65}$$

Hence, inhomogeneous plane P-waves and inhomogeneous plane S-waves may propagate with the same velocity.

Inhomogeneous plane waves cannot exist in an unbounded source-free homogeneous medium because of the exponential increasing character. They may exist, however, near boundaries or near sources. As an example, we consider the special situation of a homogeneous isotropic half-space z>o, bounded by a free surface z=o, on which the traction $\vec{\tau}_z$ vanishes. We search for a linear combination of inhomogeneous plane P- and S-waves

$$\vec{v}(\vec{r},t) = \vec{v}_p(\vec{r},t) + \vec{v}_s(\vec{r},t), \tag{II-66a}$$

with $\vec{v}_p(\vec{r},t)$ and $\vec{v}_s(\vec{r},t)$ defined by (II-60) and (II-63), respectively, such that the boundary condition

$$\vec{\tau}_z(\vec{r},t) = \vec{o} \qquad \text{at } z=o \tag{II-66b}$$

is satisfied for all x, y and t.
Substitution of (II-26a) into the stress-velocity relation (II-21) (for the source-free situation) yields for the components of the traction $\vec{\tau}_z$

$$\frac{\partial \tau_{xz}}{\partial t} = \mu \left(\frac{\partial v_x}{\partial z} + \frac{\partial v_z}{\partial x} \right), \tag{II-67a}$$

$$\frac{\partial \tau_{yz}}{\partial t} = \mu \left(\frac{\partial v_y}{\partial z} + \frac{\partial v_z}{\partial y} \right) \tag{II-67b}$$

and

$$\frac{\partial \tau_{zz}}{\partial t} = \lambda \nabla.\vec{v} + 2\mu \frac{\partial v_z}{\partial z}. \tag{II-67c}$$

Substitution of (II-66a) into (II-67) and solving for the boundary condition (II-66b), yields

$$\begin{bmatrix} -2\sigma_{z,p} & \frac{1}{c_s^2} - 2s_r^2 \\ \frac{1}{c_s^2} - 2s_r^2 & -2s_r^2\,\sigma_{z,s} \end{bmatrix} \begin{bmatrix} v_{p,o} \\ v_{s,o} \end{bmatrix} = \begin{bmatrix} o \\ o \end{bmatrix}, \tag{II-68a}$$

where

$$s_r^2 = s_x^2 + s_y^2. \tag{II-68b}$$

This equation has non-trivial solutions only when the matrix has a zero determinant:

$$4s_r^2 \sqrt{s_r^2 - 1/c_p^2}\ \sqrt{s_r^2 - 1/c_s^2} - \left(2s_r^2 - 1/c_s^2\right)^2 = 0. \tag{II-68c}$$

Hence, in a homogeneous isotropic half-space bounded by a free surface, a combination of inhomogeneous plane P- and S-waves may exist when the phase slowness $s_r = \sqrt{s_x^2 + s_y^2}$ satisfies equation (II-68c). The wave associated to the solution of (II-68) is called the *Rayleigh-wave*. Because its amplitude decays exponentially with depth [see equations (II-60a) and (II-63a)], the Rayleigh-wave belongs to the class of *surface-waves*.
The Rayleigh-wave velocity c_R is the reciprocal of the slowness s_r which satisfies (II-68c), hence, c_R is a solution of

$$4c_s^3 \sqrt{c_p^2 - c_R^2}\ \sqrt{c_s^2 - c_R^2} - c_p\left(2c_s^2 - c_R^2\right)^2 = 0. \tag{II-69}$$

The Rayleigh-wave velocity c_R is generally a few percent lower than the S-wave velocity c_s. For example, when $c_p = 1745$ m/s and $c_s = 800$ m/s, then equation (II-69) yields $c_R = 750$ m/s.

II.4.3 Elastic one-way wave equations for plane waves

In analogy with section I.4.3, it can be shown that (monochromatic) homogeneous and inhomogeneous plane P-waves in a homogeneous isotropic solid can be summarized by one expression when we introduce the complex notation

$$\vec{v}_p^{\pm}(x,y,z,t) = \text{Real}\left[\hat{\vec{v}}_p^{\pm}(x,y,z,t)\right], \tag{II-70a}$$

where

$$\hat{\vec{v}}_p^{\pm}(x,y,z,t) = \hat{\vec{v}}_{p,o}^{\pm} \exp\left[j\omega_o(t - s_x x - s_y y \mp s_{z,p} z)\right]. \tag{II-70b}$$

Here the vertical phase slowness $s_{z,p}$ is given by

$$s_{z,p} = +\sqrt{\frac{1}{c_p^2} - s_x^2 - s_y^2} \quad \text{for } s_x^2+s_y^2 \le \frac{1}{c_p^2} \tag{II-70c}$$

and

$$s_{z,p} = -j\sqrt{s_x^2 + s_y^2 - \frac{1}{c_p^2}} \quad \text{for } s_x^2+s_y^2 > \frac{1}{c_p^2} . \tag{II-70d}$$

The complex amplitude factor $\hat{\vec{v}}^{\pm}_{p,o}$ is given by

$$\hat{\vec{v}}^{\pm}_{p,o} = \hat{\kappa} \begin{bmatrix} s_x \\ s_y \\ \pm s_{z,p} \end{bmatrix} , \tag{II-70e}$$

where $\hat{\kappa}$ is an arbitrary complex parameter.
For homogeneous or propagating plane waves ($s_x^2+s_y^2 \le 1/c_p^2$) the super-scripts + and - refer to downgoing and upgoing waves, respectively (see also Figure I-9). For inhomogeneous or evanescent plane waves ($s_x^2+s_y^2>1/c_p^2$) the super-scripts + and - refer to the exponentially decreasing behaviour of the wave in the positive and negative z-direction, respectively (see also Figure I-10).

Similarly, (monochromatic) homogeneous and inhomogeneous plane S-waves in a homogeneous isotropic solid can be summarized by one expression

$$\vec{v}^{\pm}_s(x,y,z,t) = \text{Real}\left[\hat{\vec{v}}^{\pm}_s(x,y,z,t)\right], \tag{II-71a}$$

where

$$\hat{\vec{v}}^{\pm}_s(x,y,z,t) = \hat{\vec{v}}^{\pm}_{s,o} \exp\left[j\omega_o(t-s_x x-s_y y \mp s_{z,s} z)\right]. \tag{II-71b}$$

Here the vertical phase slowness $s_{z,s}$ is given by

$$s_{z,s} = +\sqrt{\frac{1}{c_s^2} - s_x^2 - s_y^2} \quad \text{for } s_x^2+s_y^2 \le \frac{1}{c_s^2} \tag{II-71c}$$

and

$$s_{z,s} = -j\sqrt{s_x^2 + s_y^2 - \frac{1}{c_s^2}} \quad \text{for } s_x^2+s_y^2 > \frac{1}{c_s^2}\,. \tag{II-71d}$$

The complex amplitude factor $\hat{\vec{v}}^{\pm}_{s,o}$ satisfies

$$\hat{\vec{v}}^{\pm}_{s,o} \cdot \begin{bmatrix} s_x \\ s_y \\ \pm s_{z,s} \end{bmatrix} = o. \tag{II-71e}$$

Note that the complex wave functions $\hat{\vec{v}}^{\pm}_p$ and $\hat{\vec{v}}^{\pm}_s$ satisfy the following elastic *one-way wave equations* for plane waves

$$\frac{\partial \hat{\vec{v}}^{\pm}_p(x,y,z,t)}{\partial z} = \mp j\omega_o s_{z,p} \hat{\vec{v}}^{\pm}_p(x,y,z,t) \tag{II-72a}$$

and

$$\frac{\partial \hat{\vec{v}}^{\pm}_s(x,y,z,t)}{\partial z} = \mp j\omega_o s_{z,s} \hat{\vec{v}}^{\pm}_s(x,y,z,t). \tag{II-72b}$$

In chapter IV we generalize the one-way wave equations for arbitrary wave fields in horizontally layered and arbitrarily inhomogeneous elastic media. One-way wave equations provide an important tool for transforming two-way Kirchhoff-Helmholtz integrals into one-way Rayleigh integrals. In chapters V to X ample use will be made of acoustic and elastic one-way wave equations.

II.5 REFERENCES

Achenbach, J.D., 1973, Wave propagation in elastic solids: North Holland Publ. Co.

Aki, K., and Richards, P.G., 1980, Quantitative seismology: W.H. Freeman and Co.

Crampin, S., 1984, Effective anisotropic elastic constants for wave propagation through cracked solids: Geophys. J.R. astr. Soc., *76*, 135 - 145.

Daley, P.F., and Hron, F., 1977, Reflection and transmission coefficients for transversely isotropic media: Bull., Seis. Soc. Am., *67*, 661 - 675.

Jeffreys, H., and Jeffreys, B.S., 1972, Methods of mathematical physics: Cambridge University Press.

Nye, J.F., 1957, Physical properties of crystals: Oxford University Press.

Pilant, W.L., 1979, Elastic waves in the earth: Elsevier Science Publ. Co., Inc.

Schoenberg, M., 1983, Reflection of elastic waves from periodically stratified media with interfacial slip: Geophysical Prospecting, *31*, 265 - 292.

Van der Hijden, J.H.M.T., 1987, Propagation of transient elastic waves in stratified anisotropic media: D.Sc. thesis, Delft University of Technology.

Woodhouse, J.H., 1974, Surface waves in a laterally varying layered structure: Geophys. J.R. astr. Soc., *37*, 461 - 490.

III

ACOUSTIC TWO-WAY AND ONE-WAY WAVE EQUATIONS

III.1 INTRODUCTION

It was mentioned already in section I.2.5 that we may distinguish between "two-way wave equations" and "one-way wave equations". The two-way wave equation is exact (under the linear assumption). Therefore it is valid for any angle of propagation of primary as well as multiply reflected waves. One-way wave equations, on the other hand, are generally not exact (for inhomogeneous media). They describe either "downgoing" or "upgoing" primary waves with moderate propagation angles. Despite of these shortcomings, the one-way wave equations play an important role in the practice of seismic processing. This is not surprising, because a seismic experiment can be described essentially in terms of downward propagation of the source wave and upward propagation of the reflected wave fields.

In this chapter we derive the acoustic two-way and one-way wave equations both in the wavenumber-frequency domain (for horizontally layered fluids) and in the space-frequency domain (for arbitrarily inhomogeneous fluids). We present formal solutions (in terms of Taylor series), but it is not our intention to elaborate the numerical aspects of these solutions. The main purpose of this chapter is to derive the mathematical relationship between two-way and one-way acoustic wave fields. The results will be used in chapters V and VII for transforming acoustic two-way Kirchhoff-Helmholtz integrals into acoustic one-way Rayleigh integrals and in chapter XI, where we discuss an acoustic processing scheme for single-component seismic data.

III.2 ACOUSTIC WAVE EQUATIONS FOR HORIZONTALLY LAYERED MEDIA

In this section we consider the special situation of wave propagation in horizontally layered acoustic media in which the medium parameters K and ρ are a function of depth only, hence

$$K = K(z)$$

and

$$\rho = \rho(z).$$

After a short introduction on temporal and spatial Fourier transformations, we derive the acoustic two-way wave equation in the wavenumber-frequency domain and present its exact solution. Next, we decompose the two-way wave equation into acoustic one-way wave equations for downgoing and upgoing waves and we present exact and approximate solutions. Finally, we derive reflection and transmission operators for one-way wave fields at interfaces.

III.2.1. Temporal and spatial Fourier transformations

The temporal Fourier transformation of a function h(x,y,z,t) from the space-time domain to the space-frequency domain we define as

$$H(x,y,z,\omega) = \int_{-\infty}^{\infty} h(x,y,z,t)e^{-j\omega t}dt \tag{III-1a}$$

and its inverse as

$$h(x,y,z,t) = \frac{1}{2\pi}\int_{-\infty}^{\infty} H(x,y,z,\omega)e^{j\omega t}d\omega, \tag{III-1b}$$

see Papoulis (1962) or Bracewell (1965).

In the following we will assume that h(x,y,z,t) is a real function, so that

$$H(x,y,z,-\omega) = H^{*}(x,y,z,\omega), \tag{III-2a}$$

where the asterisk (*) denotes complex conjugation. Apparently, negative frequency components do not provide information independent of the positive frequency components. Therefore throughout this book we will only consider positive frequencies, that is, we choose

$$\omega \geq 0 \tag{III-2b}$$

and we reformulate the inverse temporal Fourier transformation as

$$h(x,y,z,t) = \frac{1}{\pi}\,\mathrm{Real}\left[\int_{0}^{\infty} H(x,y,z,\omega)e^{j\omega t}d\omega\right]. \tag{III-3}$$

The double spatial Fourier transformation of the complex function H(x,y,z,ω) from the space-frequency domain to the wavenumber-frequency domain we define as

$$\tilde{H}(k_x,k_y,z,\omega) = \iint_{-\infty}^{\infty} H(x,y,z,\omega)e^{j(k_x x+k_y y)}dxdy \qquad \text{(III-4a)}$$

and its inverse as

$$H(x,y,z,\omega) = \frac{1}{4\pi^2}\iint_{-\infty}^{\infty} \tilde{H}(k_x,k_y,z,\omega)e^{-j(k_x x+k_y y)}dk_x dk_y, \qquad \text{(III-4b)}$$

see Goodman (1968), Dudgeon and Mersereau (1984) and Berkhout (1985).

Unless otherwise stated, we will use the following notation convention. Any function in the space-time domain (x,y,z,t) will be denoted by a lower case symbol. The corresponding function in the space-frequency domain (x,y,z,ω) will be denoted by the corresponding upper case symbol. The corresponding function in the wavenumber-frequency domain (k_x,k_y,z,ω) will be denoted by the same upper case symbol with a tilde (~).

Equations (III-1a) and (III-4a) can be combined, yielding the triple Fourier transformation from the space-time domain to the wavenumber-frequency domain, according to

$$\tilde{H}(k_x,k_y,z,\omega) = \iiint_{-\infty}^{\infty} h(x,y,z,t)e^{-j(\omega t-k_x x-k_y y)}dtdxdy. \qquad \text{(III-5a)}$$

Similarly, equations (III-3) and (III-4b) can be combined, yielding the inverse triple Fourier transformation from the wavenumber-frequency domain to the space-time domain, according to

$$h(x,y,z,t) =$$

$$\frac{1}{4\pi^3}\,\text{Real}\left[\int_0^{\infty}\left\{\iint_{-\infty}^{\infty} \tilde{H}(k_x,k_y,z,\omega)e^{j(\omega t-k_x x-k_y y)}dk_x dk_y\right\}d\omega\right]. \qquad \text{(III-5b)}$$

Compare the latter equation with the expression for monochromatic plane waves in a homogeneous medium

$$p(x,y,z,t) = \text{Real}\left[\hat{p}'_0(z)e^{j(\omega_0 t-\omega_0 s_x x-\omega_0 s_y y)}\right], \qquad \text{(III-6a)}$$

$$\text{with } \hat{p}'_0(z) = \hat{p}_0 e^{\mp j\omega_0 s_z z}, \qquad \text{(III-6b)}$$

(see equation (I-42)). Apparently, for fixed z equation (III-5b) describes the *synthesis* of a wave field h(x,y,z,t) from monochromatic plane waves, the complex amplitudes being given by $\tilde{H}(k_x,k_y,z,\omega)$ and the phase slownesses being given by

$$s_x = k_x/\omega \tag{III-7a}$$

and

$$s_y = k_y/\omega. \tag{III-7b}$$

Consequently, for fixed z equation (III-5a) describes the *decomposition* of an arbitrary wave field h(x,y,z,t) into monochromatic plane waves $\tilde{H}(k_x,k_y,z,\omega)$. Note that $\tilde{H}(k_x,k_y,z,\omega)$ represents *propagating* plane waves when

$$k_x^2 + k_y^2 \le \frac{\omega^2}{c^2}, \tag{III-8a}$$

(see equation (I-42b)), whereas it represents *evanescent* plane waves when

$$k_x^2 + k_y^2 > \frac{\omega^2}{c^2}, \tag{III-8b}$$

(see equation (I-42c)). The tilt angle α and the azimuth angle β of the plane waves are related to k_x, k_y and ω according to

$$\sin\alpha = c\sqrt{s_x^2 + s_y^2} = \frac{c}{\omega}\sqrt{k_x^2 + k_y^2} \tag{III-8c}$$

and

$$\tan\beta = s_y/s_x = k_y/k_x, \tag{III-8d}$$

see also equation (I-22).

III.2.2 Acoustic two-way wave equation in the wavenumber-frequency domain

In the space-time domain, the linearized equation of continuity (I-10a) reads (for horizontally layered media)

$$\nabla.\vec{v}(x,y,z,t) + \frac{1}{K(z)}\frac{\partial p(x,y,z,t)}{\partial t} = \frac{\partial i_v(x,y,z,t)}{\partial t}, \tag{III-9a}$$

whereas the linearized equation of motion (I-10b) reads

$$\nabla p(x,y,z,t) + \rho(z) \frac{\partial \vec{v}(x,y,z,t)}{\partial t} = \vec{f}(x,y,z,t). \tag{III-9b}$$

Differentiating both sides of equation (III-3) with respects to time yields

$$\frac{\partial h(x,y,z,t)}{\partial t} = \frac{1}{\pi} \text{Real} \left[\int_0^\infty j\omega H(x,y,z,\omega) e^{j\omega t} d\omega \right] . \tag{III-10}$$

Hence, differentiation with respect to time in the space-time domain corresponds to a multiplication by $j\omega$ in the space-frequency domain. Since the medium parameters K and ρ are time-invariant, equations (III-9a) and (III-9b) read in the space-frequency domain

$$\nabla . \vec{V}(x,y,z,\omega) + \frac{j\omega}{K(z)} P(x,y,z,\omega) = j\omega I_v(x,y,z,\omega) \tag{III-11a}$$

and

$$\nabla P(x,y,z,\omega) + j\omega\rho(z)\vec{V}(x,y,z,\omega) = \vec{F}(x,y,z,\omega), \tag{III-11b}$$

respectively.

Differentiating both sides of equation (III-4b) with respect to x yields

$$\frac{\partial H(x,y,z,\omega)}{\partial x} = \frac{1}{4\pi^2} \iint_{-\infty}^{\infty} -jk_x \tilde{H}(k_x,k_y,z,\omega) e^{-j(k_x x + k_y y)} dk_x dk_y. \tag{III-12a}$$

Similarly, differentiating both sides of equation (III-4b) with respect to y yields

$$\frac{\partial H(x,y,z,\omega)}{\partial y} = \frac{1}{4\pi^2} \iint_{-\infty}^{\infty} -jk_y \tilde{H}(k_x,k_y,z,\omega) e^{-j(k_x x + k_y y)} dk_x dk_y. \tag{III-12b}$$

Hence, a differentiation with respect to x or y in the space-frequency domain corresponds to a multiplication by $-jk_x$ or $-jk_y$, respectively, in the wavenumber-frequency domain.

Since the medium parameters K and ρ are laterally invariant, equations (III-11a) and (III-11b) read in the wavenumber-frequency domain

$$-jk_x\tilde{V}_x(k_x,k_y,z,\omega) - jk_y\tilde{V}_y(k_x,k_y,z,\omega) + \frac{\partial \tilde{V}_z(k_x,k_y,z,\omega)}{\partial z}$$

$$+ \frac{j\omega}{K(z)}\tilde{P}(k_x,k_y,z,\omega) = j\omega\tilde{I}_v(k_x,k_y,z,\omega) \tag{III-13a}$$

and

$$\begin{bmatrix} -jk_x\tilde{P}(k_x,k_y,z,\omega) \\ -jk_y\tilde{P}(k_x,k_y,z,\omega) \\ \dfrac{\partial \tilde{P}(k_x,k_y,z,\omega)}{\partial z} \end{bmatrix} + j\omega\rho(z)\begin{bmatrix} \tilde{V}_x(k_x,k_y,z,\omega) \\ \tilde{V}_y(k_x,k_y,z,\omega) \\ \tilde{V}_z(k_x,k_y,z,\omega) \end{bmatrix} = \begin{bmatrix} \tilde{F}_x(k_x,k_y,z,\omega) \\ \tilde{F}_y(k_x,k_y,z,\omega) \\ \tilde{F}_z(k_x,k_y,z,\omega) \end{bmatrix}, \tag{III-13b}$$

respectively.

Elimination of the particle velocity components $\tilde{V}_x$, $\tilde{V}_y$ and $\tilde{V}_z$ from equations (III-13a) and (III-13b) yields

$$\rho\frac{\partial}{\partial z}\left(\frac{1}{\rho}\frac{\partial \tilde{P}}{\partial z}\right) + (k^2 - k_x^2 - k_y^2)\tilde{P} = -\tilde{S}, \tag{III-14a}$$

where the source term $\tilde{S}(k_x,k_y,z,\omega)$ reads

$$\tilde{S} = -\omega^2\rho\tilde{I}_v + jk_x\tilde{F}_x + jk_y\tilde{F}_y - \rho\frac{\partial}{\partial z}\left(\frac{1}{\rho}\tilde{F}_z\right) \tag{III-14b}$$

and where

$$k^2(z) = \omega^2/c^2(z), \tag{III-14c}$$

with

$$c(z) = \sqrt{K(z)/\rho(z)} \tag{III-14d}$$

Equation (III-14) is the wavenumber-frequency domain equivalence of the acoustic two-way wave equation (I-11) in the space-time domain. Equation (III-14) is exact (under the linear assumption), but it has no straightforward solutions when the medium parameters c and ρ are arbitrary functions of depth (z). As an alternative, we derive a *first order* two-way wave equation for $\tilde{P}$ and $\tilde{V}_z$ by eliminating only $\tilde{V}_x$ and $\tilde{V}_y$ from equations

(III-13a) and (III-13b), yielding

$$\frac{\partial}{\partial z}\begin{bmatrix}\tilde{P}\\ \tilde{V}_z\end{bmatrix} = \begin{bmatrix}0 & -j\omega\rho\\ \frac{1}{j\omega\rho}(k^2-k_x^2-k_y^2) & 0\end{bmatrix}\begin{bmatrix}\tilde{P}\\ \tilde{V}_z\end{bmatrix}$$

(III-15)

$$+\begin{bmatrix}\tilde{F}_z\\ \frac{1}{j\omega\rho}\left(-\omega^2\rho\tilde{I}_v + jk_x\tilde{F}_x + jk_y\tilde{F}_y\right)\end{bmatrix}.$$

Many authors have used this matrix representation of the two-way wave equation (albeit most frequently for the source-free situation). For an overview the reader is referred to Ursin (1983).

In the following we use the more compact notation

$$\frac{\partial\vec{\tilde{Q}}}{\partial z} = \tilde{\mathbf{A}}_1\vec{\tilde{Q}} + \vec{\tilde{S}}, \tag{III-16a}$$

where the wave vector $\vec{\tilde{Q}}$ is defined as

$$\vec{\tilde{Q}}(k_x,k_y,z,\omega) = \begin{bmatrix}\tilde{P}(k_x,k_y,z,\omega)\\ \tilde{V}_z(k_x,k_y,z,\omega)\end{bmatrix}, \tag{III-16b}$$

the first order differential operator $\tilde{\mathbf{A}}_1$ is defined as

$$\tilde{\mathbf{A}}_1(k_x,k_y,z,\omega) = \begin{bmatrix}0 & -j\omega\rho(z)\\ \frac{k_z^2(k_x,k_y,z,\omega)}{j\omega\rho(z)} & 0\end{bmatrix}, \tag{III-16c}$$

the source vector $\vec{\tilde{S}}$ is defined as

$$\vec{\tilde{S}}(k_x,k_y,z,\omega) = \begin{bmatrix}\tilde{F}_z(k_x,k_y,z,\omega)\\ \frac{1}{j\omega\rho(z)}\left[-\omega^2\rho(z)\tilde{I}_v(k_x,k_y,z,\omega)+jk_x\tilde{F}_x(k_x,k_y,z,\omega)+jk_y\tilde{F}_y(k_x,k_y,z,\omega)\right]\end{bmatrix}$$

(III-16d)

and, finally,

$$k_z^2(k_x,k_y,z,\omega) = k^2(z) - k_x^2 - k_y^2 . \tag{III-16e}$$

We solve equation (III-16), assuming for the moment that the medium is source-free, hence, we solve

$$\frac{\partial \tilde{\vec{Q}}(z)}{\partial z} = \tilde{\mathbf{A}}_1(z)\tilde{\vec{Q}}(z). \tag{III-17}$$

For notational convenience we omitted the variables k_x, k_y and ω. The z-dependency, however, is essential in the following derivation.

We write the solution as a Taylor series expansion, according to

$$\tilde{\vec{Q}}(z) = \sum_{m=o}^{\infty} \frac{(z-z_o)^m}{m!} \left.\frac{\partial^m \tilde{\vec{Q}}(z)}{\partial z^m}\right|_{z_o} . \tag{III-18a}$$

Define the m'th order differential operator $\tilde{\mathbf{A}}_m$ by

$$\frac{\partial^m \tilde{\vec{Q}}(z)}{\partial z^m} = \tilde{\mathbf{A}}_m(z)\tilde{\vec{Q}}(z), \tag{III-18b}$$

then

$$\frac{\partial^{m+1} \tilde{\vec{Q}}(z)}{\partial z^{m+1}} = \frac{\partial \tilde{\mathbf{A}}_m(z)}{\partial z}\tilde{\vec{Q}}(z) + \tilde{\mathbf{A}}_m(z)\frac{\partial \tilde{\vec{Q}}(z)}{\partial z}, \tag{III-18c}$$

or, substituting two-way wave equation (III-17),

$$\frac{\partial^{m+1} \tilde{\vec{Q}}(z)}{\partial z^{m+1}} = \tilde{\mathbf{A}}_{m+1}(z)\tilde{\vec{Q}}(z), \tag{III-18d}$$

with

$$\tilde{\mathbf{A}}_{m+1}(z) = \frac{\partial \tilde{\mathbf{A}}_m(z)}{\partial z} + \tilde{\mathbf{A}}_m(z)\tilde{\mathbf{A}}_1(z) \tag{III-18e}$$

and

$$\tilde{\mathbf{A}}_o(z) = \mathbf{I}. \tag{III-18f}$$

Hence, solution (III-18a) may be rewritten as

$$\tilde{\vec{Q}}(z) = \tilde{U}(z,z_o)\tilde{\vec{Q}}(z_o), \tag{III-19a}$$

where

$$\tilde{U}(z,z_o) = \sum_{m=o}^{\infty} \frac{(z-z_o)^m}{m!} \tilde{\mathbf{A}}_m(z_o), \tag{III-19b}$$

with $\tilde{\mathbf{A}}_m(z_o)$ defined recursively by (III-18e) and (III-18f) for $z=z_o$.

For each k_x, k_y and ω, equation (III-19a) describes acoustic two-way wave field extrapolation of the wave vector $\tilde{\vec{Q}}(k_x,k_y,z_o,\omega)$ from depth level z_o, through a laterally invariant medium, to depth level z.

The two-way extrapolation operator[1)] $\tilde{U}$ satisfies the property

$$\tilde{U}(z_2,z_o) = \tilde{U}(z_2,z_1)\tilde{U}(z_1,z_o). \tag{III-20a}$$

Taking z_2 equal to z_o yields the property

$$\tilde{U}(z_o,z_o) = I = \tilde{U}(z_o,z_1)\tilde{U}(z_1,z_o) \tag{III-20b}$$

or

$$[\tilde{U}(z_1,z_o)]^{-1} = \tilde{U}(z_o,z_1). \tag{III-20c}$$

In deriving (III-19) it was assumed that the medium parameters c and ρ are continuously differentiable on the interval (z,z_o), see (III-16c) and (III-18e). Consider a piecewise continuously differentiable medium, as shown in Figure III-1, with "interfaces" at z_1, z_2 .... z_{i-1}, z_i ...... etc.
The boundary conditions state that the pressure and the vertical component of the particle velocity are continuous at the interfaces. Hence, for this situation we may apply (III-19a) recursively, according to

1) In the literature this operator is often named the "propagator" (Gilbert and Backus, 1966). We prefer the name "two-way extrapolation operator", opposed to the "one-way extrapolation operators" of Berkhout (1985).

$$\tilde{\vec{Q}}(z_1) = \tilde{U}(z_1,z_0)\tilde{\vec{Q}}(z_0), \tag{III-21a}$$

$$\tilde{\vec{Q}}(z_2) = \tilde{U}(z_2,z_1)\tilde{\vec{Q}}(z_1), \tag{III-21b}$$

$$\vdots$$

$$\tilde{\vec{Q}}(z_i) = \tilde{U}(z_i,z_{i-1})\tilde{\vec{Q}}(z_{i-1}), \tag{III-21c}$$

or

$$\tilde{\vec{Q}}(z_i) = \tilde{U}(z_i,z_0)\tilde{\vec{Q}}(z_0), \tag{III-22a}$$

with

$$\tilde{U}(z_i,z_0) = \tilde{U}(z_i,z_{i-1}) \;.....\; \tilde{U}(z_2,z_1)\tilde{U}(z_1,z_0). \tag{III-22b}$$

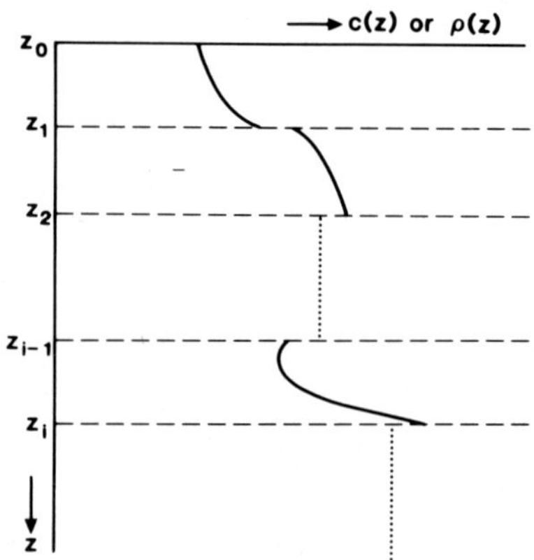

Figure III-1: The acoustic medium parameters c(z) and ρ(z) generally show discontinuities at layer interfaces (z_1, z_2 etc.).

For the special situation of a homogeneous medium equation (III-18e) simplifies to

$$\tilde{\mathbf{A}}_{m+1} = \tilde{\mathbf{A}}_m \tilde{\mathbf{A}}_1, \tag{III-23a}$$

or

$$\tilde{\mathbf{A}}_m = \tilde{\mathbf{A}}_1^{\,m} \,. \tag{III-23b}$$

Note that with this definition of $\tilde{\mathbf{A}}_m$ we may represent the Taylor series expansion (III-19b) symbolically by

$$\tilde{\mathbf{U}}(z,z_o) = \exp(\tilde{\mathbf{A}}_1 \Delta z), \tag{III-23c}$$

with

$$\Delta z = z - z_o. \tag{III-23d}$$

According to (III-16c) and (III-23b), $\tilde{\mathbf{A}}_m$ is given by

$$\tilde{\mathbf{A}}_m = (-1)^{m/2} \begin{bmatrix} k_z^m & 0 \\ 0 & k_z^m \end{bmatrix} \text{ for } \textit{even } m \tag{III-24a}$$

and

$$\tilde{\mathbf{A}}_m = (-1)^{(m-1)/2} \begin{bmatrix} 0 & -j\omega\rho k_z^{m-1} \\ \frac{1}{j\omega\rho} k_z^{m+1} & 0 \end{bmatrix} \text{ for } \textit{odd } m. \tag{III-24b}$$

Upon substitution into (III-19b) we obtain for propagating plane waves ($k_x^2+k_y^2 \leq k^2$)

$$\tilde{\mathbf{U}}(z,z_o) = \begin{bmatrix} \cos(k_z \Delta z) & -\frac{j\omega\rho}{k_z} \sin(k_z \Delta z) \\ \frac{k_z}{j\omega\rho} \sin(k_z \Delta z) & \cos(k_z \Delta z) \end{bmatrix}, \tag{III-25a}$$

with

$$k_z = + \sqrt{k^2 - k_x^2 - k_y^2}, \tag{III-25b}$$

whereas we obtain for evanescent plane waves ($k_x^2+k_y^2>k^2$)

$$\tilde{\mathbf{U}}(z,z_o) = \begin{bmatrix} \cosh(\ell_z \Delta z) & \frac{-j\omega\rho}{\ell_z} \sinh(\ell_z \Delta z) \\ \frac{-\ell_z}{j\omega\rho} \sinh(\ell_z \Delta z) & \cosh(\ell_z \Delta z) \end{bmatrix}, \tag{III-26a}$$

with

$$\ell_z = jk_z = + \sqrt{k_x^2 + k_y^2 - k^2} \quad . \tag{III-26b}$$

Let us now return to the situation where the medium is not homogeneous and not source-free. Hence, we solve equation (III-16a)

$$\frac{\partial \tilde{\vec{Q}}(z)}{\partial z} = \tilde{\mathbf{A}}_1(z)\tilde{\vec{Q}}(z) + \tilde{\vec{S}}(z), \tag{III-27a}$$

given a solution $\tilde{\mathbf{U}}(z,z_0)$ of equation (III-17) for the source-free situation, hence

$$\frac{\partial \tilde{\mathbf{U}}(z,z_0)}{\partial z} = \tilde{\mathbf{A}}_1(z)\tilde{\mathbf{U}}(z,z_0). \tag{III-27b}$$

We use the method of variation of parameters (Boyce and DiPrima, 1969) to find the solution of (III-27a). We seek a solution of the form

$$\tilde{\vec{Q}}(z) = \tilde{\mathbf{U}}(z,z_0)\left[\tilde{\vec{Q}}(z_0) + \tilde{\vec{Q}}_s(z)\right], \tag{III-28a}$$

with

$$\tilde{\vec{Q}}_s(z_0) = \vec{o}. \tag{III-28b}$$

Substitution of (III-28a) into two-way wave equation (III-27a), using property (III-27b), yields

$$\tilde{\mathbf{A}}_1(z)\tilde{\mathbf{U}}(z,z_0)\left[\tilde{\vec{Q}}(z_0) + \tilde{\vec{Q}}_s(z)\right] + \tilde{\mathbf{U}}(z,z_0)\frac{\partial \tilde{\vec{Q}}_s(z)}{\partial z} =$$

$$\tilde{\mathbf{A}}_1(z)\tilde{\mathbf{U}}(z,z_0)\left[\tilde{\vec{Q}}(z_0) + \tilde{\vec{Q}}_s(z)\right] + \tilde{\vec{S}}(z), \tag{III-29a}$$

or, using property (III-20c),

$$\frac{\partial \tilde{\vec{Q}}_s(z)}{\partial z} = \tilde{\mathbf{U}}(z_0,z)\tilde{\vec{S}}(z). \tag{III-29b}$$

The solution of (III-29b), with boundary condition (III-28b), reads

$$\vec{\tilde{Q}}_s(z) = \int_{z_0}^{z} \tilde{U}(z_0,z')\vec{\tilde{S}}(z')dz'. \tag{III-29c}$$

Substitution of this result into solution (III-28a), using property (III-20a), yields

$$\vec{\tilde{Q}}(z) = \tilde{U}(z,z_0)\vec{\tilde{Q}}(z_0) + \int_{z_0}^{z} \tilde{U}(z,z')\vec{\tilde{S}}(z')dz'. \tag{III-30}$$

This is the general solution of two-way wave equation (III-27a). It states that the two-way wave field $\vec{\tilde{Q}}$ at depth level z is found by extrapolating the two-way wave field $\vec{\tilde{Q}}$ from depth level z_0 to z and by adding the wave field at z related to all sources between z_0 and z.

III.2.3 Acoustic one-way wave equations in the wavenumber-frequency domain

Consider the acoustic two-way wave equation (III-16a) in the wavenumber-frequency domain,

$$\frac{\partial\vec{\tilde{Q}}(z)}{\partial z} = \tilde{A}_1(z)\vec{\tilde{Q}}(z) + \vec{\tilde{S}}(z), \tag{III-31a}$$

with

$$\vec{\tilde{Q}}(z) = \begin{bmatrix} \tilde{P}(z) \\ \tilde{V}_z(z) \end{bmatrix}, \tag{III-31b}$$

$$\tilde{\mathbf{A}}_1(z) = \begin{bmatrix} 0 & -j\omega\rho(z) \\ \dfrac{k_z^2(z)}{j\omega\rho(z)} & 0 \end{bmatrix}, \tag{III-31c}$$

$$\vec{\tilde{S}}(z) = \begin{bmatrix} \tilde{F}_z(z) \\ \dfrac{1}{j\omega\rho(z)}\left[-\omega^2\rho(z)\tilde{I}_v(z) + jk_x\tilde{F}_x(z) + jk_y\tilde{F}_y(z)\right] \end{bmatrix} \tag{III-31d}$$

and

$$k_z^2(z) = k^2(z) - k_x^2 - k_y^2 . \tag{III-31e}$$

The eigenvalue decomposition of matrix $\tilde{\mathbf{A}}_1(z)$ reads

$$\tilde{\mathbf{A}}_1(z) = \tilde{\mathbf{L}}(z)\tilde{\mathbf{\Lambda}}(z)\tilde{\mathbf{L}}^{-1}(z), \tag{III-32a}$$

with

$$\tilde{\mathbf{L}}(z) = \begin{bmatrix} 1 & 1 \\ \dfrac{k_z(z)}{\omega\rho(z)} & \dfrac{-k_z(z)}{\omega\rho(z)} \end{bmatrix} \tag{III-32b}$$

$$\tilde{\mathbf{\Lambda}}(z) = \begin{bmatrix} -jk_z(z) & 0 \\ 0 & jk_z(z) \end{bmatrix}, \tag{III-32c}$$

$$\tilde{\mathbf{L}}^{-1}(z) = \frac{1}{2}\begin{bmatrix} 1 & \dfrac{\omega\rho(z)}{k_z(z)} \\ 1 & \dfrac{-\omega\rho(z)}{k_z(z)} \end{bmatrix}, \tag{III-32d}$$

and

$$k_z(z) \triangleq +\sqrt{k^2(z) - k_x^2 - k_y^2} \quad \text{for } k_x^2 + k_y^2 \le k^2(z), \tag{III-32e}$$

$$k_z(z) \triangleq -j\sqrt{k_x^2 + k_y^2 - k^2(z)} \quad \text{for } k_x^2 + k_y^2 > k^2(z). \tag{III-32f}$$

Define a vector $\tilde{\vec{D}}(z)$ according to

$$\tilde{\vec{Q}}(z) = \tilde{\mathbf{L}}(z)\tilde{\vec{D}}(z), \tag{III-33a}$$

or, equivalently,

$$\tilde{\vec{D}}(z) = \tilde{\mathbf{L}}^{-1}(z)\tilde{\vec{Q}}(z), \tag{III-33b}$$

with

$$\tilde{\vec{D}}(z) = \begin{bmatrix} \tilde{P}^{+}(z) \\ \tilde{P}^{-}(z) \end{bmatrix} . \tag{III-33c}$$

It will be shown later on in this section that $\tilde{P}^{+}(z)$ and $\tilde{P}^{-}(z)$ represent downgoing and upgoing waves, respectively.

Upon substitution of (III-33a) into (III-31a) we obtain

$$\frac{\partial \tilde{\vec{D}}(z)}{\partial z} = \tilde{\mathbf{L}}^{-1}(z) \left[\left(\tilde{\mathbf{A}}_1(z)\tilde{\mathbf{L}}(z) - \frac{\partial \tilde{\mathbf{L}}(z)}{\partial z} \right) \tilde{\vec{D}}(z) + \tilde{\vec{S}}(z) \right] , \tag{III-34a}$$

or, with eigenvalue decomposition (III-32a),

$$\frac{\partial \tilde{\vec{D}}(z)}{\partial z} = \tilde{\mathbf{B}}_1(z)\tilde{\vec{D}}(z) + \tilde{\vec{S}}'(z), \tag{III-34b}$$

where

$$\tilde{\mathbf{B}}_1(z) = \tilde{\mathbf{\Lambda}}(z) - \tilde{\mathbf{L}}^{-1}(z) \frac{\partial \tilde{\mathbf{L}}(z)}{\partial z} \tag{III-34c}$$

and

$$\tilde{\vec{S}}'(z) = \tilde{\mathbf{L}}^{-1}(z)\tilde{\vec{S}}(z). \tag{III-34d}$$

Many authors have used this equation for downgoing and upgoing waves. For an overview the reader is referred to Ursin (1983).

The general solution of (III-34b) reads, in analogy with (III-30),

$$\tilde{\vec{D}}(z) = \tilde{\mathbf{W}}(z,z_0)\tilde{\vec{D}}(z_0) + \int_{z_0}^{z} \tilde{\mathbf{W}}(z,z')\tilde{\vec{S}}'(z')dz', \tag{III-35a}$$

where, in analogy with (III-19b),

$$\tilde{\mathbf{W}}(z,z_0) = \sum_{m=0}^{\infty} \frac{(z-z_0)^m}{m!} \tilde{\mathbf{B}}_m(z_0), \tag{III-35b}$$

with $\tilde{\mathbf{B}}_m(z_0)$ defined recursively by

$$\tilde{\mathbf{B}}_{m+1}(z_o) = \left.\frac{\partial\tilde{\mathbf{B}}_m(z)}{\partial z}\right|_{z_o} + \tilde{\mathbf{B}}_m(z_o)\tilde{\mathbf{B}}_1(z_o) \tag{III-35c}$$

and

$$\tilde{\mathbf{B}}_o(z_o) = \mathbf{I}. \tag{III-35d}$$

For the special situation of a homogeneous medium equation (III-35c) simplifies to

$$\tilde{\mathbf{B}}_{m+1} = \tilde{\mathbf{B}}_m\tilde{\mathbf{B}}_1, \tag{III-36a}$$

or

$$\tilde{\mathbf{B}}_m = \tilde{\mathbf{B}}_1^m = \tilde{\mathbf{\Lambda}}^m. \tag{III-36b}$$

Note that with this definition of $\tilde{\mathbf{B}}_m$ we may represent the Taylor series expansion (III-35b) symbolically by

$$\tilde{\mathbf{W}}(z,z_o) = \exp(\tilde{\mathbf{\Lambda}}\Delta z), \tag{III-37a}$$

with

$$\Delta z = z-z_o. \tag{III-37b}$$

According to (III-32c) and (III-36b), $\tilde{\mathbf{B}}_m$ is given by

$$\tilde{\mathbf{B}}_m = \begin{bmatrix} (-jk_z)^m & 0 \\ 0 & (jk_z)^m \end{bmatrix} . \tag{III-38}$$

Upon substitution into (III-35b) we obtain

$$\tilde{\mathbf{W}}(z,z_o) = \begin{bmatrix} \tilde{W}^+(z,z_o) & 0 \\ 0 & \tilde{W}^-(z,z_o) \end{bmatrix} , \tag{III-39a}$$

with

$$\tilde{W}^+(z,z_0) = \exp(-jk_z\Delta z) \tag{III-39b}$$

and

$$\tilde{W}^-(z,z_0) = \exp(jk_z\Delta z), \tag{III-39c}$$

or, substituting this result in (III-35a) for the source-free situation

$$\tilde{P}^+(z) = \tilde{W}^+(z,z_0)\tilde{P}^+(z_0) \tag{III-39d}$$

and

$$\tilde{P}^-(z) = \tilde{W}^-(z,z_0)\tilde{P}^-(z_0). \tag{II-39e}$$

Note the similarity with the downgoing and upgoing plane waves described by (I-44) in section I.4.3. Therefore, we may conclude that for each k_x, k_y and ω, $\tilde{P}^+(z)$ represents a downgoing plane wave, whereas $\tilde{P}^-(z)$ represents an upgoing plane wave.

We will now discuss the physical interpretation of equations (III-33) and (III-34).

In equation (III-33a), the wave vector $\vec{\tilde{D}}(z)$ contains downgoing and upgoing waves $\tilde{P}^+(z)$ and $\tilde{P}^-(z)$, respectively, whereas wave vector $\vec{\tilde{Q}}(z)$ contains the two-way wave field in terms of the total pressure $\tilde{P}(z)$ and the total particle velocity $\tilde{V}_z(z)$. Hence, in equation (III-33a) matrix $\tilde{L}(z)$ is a *composition* operator which composes the total wave field from its downgoing and upgoing constituents, according to

$$\tilde{P}(z) = \tilde{P}^+(z) + \tilde{P}^-(z) \tag{III-40a}$$

and

$$\tilde{V}_z(z) = \frac{k_z(z)}{\omega\rho(z)}\left[\tilde{P}^+(z) - \tilde{P}^-(z)\right]. \tag{III-40b}$$

Similarly, in equation (III-33b) matrix $\tilde{L}^{-1}(z)$ is a *decomposition* operator which decomposes the total wave field into downgoing and upgoing waves, according to

$$\tilde{P}^{+}(z) = \frac{1}{2}\left[\tilde{P}(z) + \frac{\omega\rho(z)}{k_z(z)}\tilde{V}_z(z)\right], \tag{III-41a}$$

and

$$\tilde{P}^{-}(z) = \frac{1}{2}\left[\tilde{P}(z) - \frac{\omega\rho(z)}{k_z(z)}\tilde{V}_z(z)\right]. \tag{III-41b}$$

Note that this decomposition breaks down for $k_z(z)=o$, that is, for waves which propagate in the horizontal direction ($\alpha=\pi/2$, see also equations (III-32e), (III-14c) and (III-8c)).

Equation (III-34) represents a coupled system of *one-way wave equations* for downgoing and upgoing waves, according to

$$\frac{\partial\tilde{P}^{+}(z)}{\partial z} = -jk_z(z)\tilde{P}^{+}(z) - \frac{1}{2}\frac{\rho(z)}{k_z(z)}\left[\frac{\partial}{\partial z}\left(\frac{k_z(z)}{\rho(z)}\right)\left(\tilde{P}^{+}(z)-\tilde{P}^{-}(z)\right)\right] + \tilde{S}^{+}(z) \tag{III-42a}$$

and

$$\frac{\partial\tilde{P}^{-}(z)}{\partial z} = +jk_z(z)\tilde{P}^{-}(z) - \frac{1}{2}\frac{\rho(z)}{k_z(z)}\left[\frac{\partial}{\partial z}\left(\frac{k_z(z)}{\rho(z)}\right)\left(\tilde{P}^{-}(z)-\tilde{P}^{+}(z)\right)\right] - \tilde{S}^{-}(z), \tag{III-42b}$$

where

$$\tilde{S}^{\pm}(z) = \frac{1}{2jk_z(z)}\left[-\omega^2\rho(z)\tilde{I}_v(z) + jk_x\tilde{F}_x(z) + jk_y\tilde{F}_y(z) \pm jk_z(z)\tilde{F}_z(z)\right]. \tag{III-42c}$$

Note that these equations are a generalization of the one-way wave equations for plane waves (I-43a) and (I-43b) that were derived in section I.4.3. $\tilde{S}^{+}(z)$ and $\tilde{S}^{-}(z)$ are the one-way representations of the source distribution. Apparently the downgoing and upgoing waves are coupled due to the vertical variations of the medium parameters, which is expressed by the term $\partial[k_z(z)/\rho(z)]/\partial z$.

It is common use to neglect $\tilde{P}^{-}(z)$ with respect to $\tilde{P}^{+}(z)$ in (III-42a) for downward propagation and to neglect $\tilde{P}^{+}(z)$ with respect to $\tilde{P}^{-}(z)$ in equation (III-42b) for upward propagation. This means that in both equations *multiple reflections* are neglected. Hence, *primary waves* fulfill the decoupled one-way wave equations

$$\frac{\partial \tilde{P}^{+}(z)}{\partial z} \approx \left[-jk_z(z) - \frac{1}{2} \frac{\rho(z)}{k_z(z)} \frac{\partial}{\partial z} \left(\frac{k_z(z)}{\rho(z)} \right) \right] \tilde{P}^{+}(z) + \tilde{S}^{+}(z) \qquad \text{(III-43a)}$$

and

$$\frac{\partial \tilde{P}^{-}(z)}{\partial z} \approx \left[+jk_z(z) - \frac{1}{2} \frac{\rho(z)}{k_z(z)} \frac{\partial}{\partial z} \left(\frac{k_z(z)}{\rho(z)} \right) \right] \tilde{P}^{-}(z) - \tilde{S}^{-}(z). \qquad \text{(III-43b)}$$

The general solutions of these equations read, in analogy with (III-30),

$$\tilde{P}^{+}(z) \approx \tilde{W}^{+}(z,z_0)\tilde{P}^{+}(z_0) + \int_{z_0}^{z} \tilde{W}^{+}(z,z')\tilde{S}^{+}(z')dz' \qquad \text{(III-44a)}$$

and

$$\tilde{P}^{-}(z) \approx \tilde{W}^{-}(z,z_0)\tilde{P}^{-}(z_0) - \int_{z_0}^{z} \tilde{W}^{-}(z,z')\tilde{S}^{-}(z')dz', \qquad \text{(III-44b)}$$

respectively, where $\tilde{W}^{+}(z,z_0)$ and $\tilde{W}^{-}(z,z_0)$ are solutions of (III-43a) and (III-43b), respectively, for the source-free situation. Hence,

$$\tilde{W}^{+}(z,z_0) \approx \left[\frac{k_z(z_0)}{\rho(z_0)} \right]^{\frac{1}{2}} \left[\frac{k_z(z)}{\rho(z)} \right]^{-\frac{1}{2}} \exp \int_{z_0}^{z} -jk_z(z')dz' \qquad \text{(III-44c)}$$

and

$$\tilde{W}^{-}(z,z_0) \approx \left[\frac{k_z(z_0)}{\rho(z_0)} \right]^{\frac{1}{2}} \left[\frac{k_z(z)}{\rho(z)} \right]^{-\frac{1}{2}} \exp \int_{z_0}^{z} +jk_z(z')dz'. \qquad \text{(III-44d)}$$

Expressions (III-44c) and (III-44d) are commonly known as the WKB-solutions, after Wentzel (1926), Kramers (1926) and Brillouin (1926). Note that these decoupled expressions for primary waves are not exact. In particular they break down for waves which propagate nearly horizontally, which occurs for $k_z(z) \rightarrow o$. Only in areas where the vertical variations of the medium parameters vanish, we obtain exact one-way wave equations directly from (III-42a) and (III-42b), according to

$$\frac{\partial \tilde{P}^{+}(z)}{\partial z} = -jk_z \tilde{P}^{+}(z) + \tilde{S}^{+}(z) \tag{III-45a}$$

and

$$\frac{\partial \tilde{P}^{-}(z)}{\partial z} = +jk_z \tilde{P}^{-}(z) - \tilde{S}^{-}(z). \tag{III-45b}$$

The general solutions are again given by (III-44a) and (III-44b), respectively (with ≈ replaced by =), with $\tilde{W}^{+}(z,z_0)$ and $\tilde{W}^{-}(z,z_0)$ given by (III-39b) and (III-39c), respectively.

III.2.4 Acoustic one-way wave fields at interfaces in the wavenumber-frequency domain

Both the exact solution (III-35) for the wave vector $\tilde{\vec{D}}(z)$, as well as the approximate WKB solution (III-44) for the decoupled waves $\tilde{P}^{+}(z)$ and $\tilde{P}^{-}(z)$ break down whenever the medium contains "interfaces" in the interval (z,z_0), see Figure III-1.

We derive the boundary conditions for an interface at $z=z_1$. Just above this interface the wave vector $\tilde{\vec{D}}(z)$ is related to the wave vector $\tilde{\vec{Q}}(z)$ according to

$$\lim_{\epsilon \downarrow 0} \tilde{\vec{Q}}(z_1-\epsilon) = \lim_{\epsilon \downarrow 0} \left[\tilde{\mathbf{L}}(z_1-\epsilon)\tilde{\vec{D}}(z_1-\epsilon)\right]. \tag{III-46a}$$

Similarly, just below the interface,

$$\lim_{\epsilon \downarrow 0} \tilde{\vec{Q}}(z_1+\epsilon) = \lim_{\epsilon \downarrow 0} \left[\tilde{\mathbf{L}}(z_1+\epsilon)\tilde{\vec{D}}(z_1+\epsilon)\right]. \tag{III-46b}$$

Since the wave vector $\tilde{\vec{Q}}(z)$ is continuous at z_1, we obtain the following boundary condition for the wave vector $\tilde{\vec{D}}(z)$ at z_1:

$$\lim_{\epsilon \downarrow 0} \left[\tilde{\mathbf{L}}(z_1-\epsilon)\tilde{\vec{D}}(z_1-\epsilon)\right] = \lim_{\epsilon \downarrow 0} \left[\tilde{\mathbf{L}}(z_1+\epsilon)\tilde{\vec{D}}(z_1+\epsilon)\right]. \tag{III-46c}$$

Let us now assume that z_1 represents an interface between two homogeneous half-spaces and derive the boundary conditions for the downgoing and upgoing plane waves $\tilde{P}^+(z)$ and $\tilde{P}^-(z)$, respectively. In the upper half-space $z<z_1$ the medium parameters read c_u and ρ_u; in the lower half-space $z>z_1$ the medium parameters read c_ℓ and ρ_ℓ. First consider the situation depicted in Figure III-2a. A plane wave $\tilde{P}_u^+(z)$ is incident to the interface from above, a plane wave $\tilde{P}_u^-(z)$ is reflected into the upper half-space and a plane wave $\tilde{P}_\ell^+(z)$ is transmitted into the lower half-space. Applying boundary condition (III-46c) to this situation yields

$$\tilde{\mathbf{L}}_u \begin{bmatrix} \tilde{P}_u^+(z_1) \\ \tilde{P}_u^-(z_1) \end{bmatrix} = \tilde{\mathbf{L}}_\ell \begin{bmatrix} \tilde{P}_\ell^+(z_1) \\ 0 \end{bmatrix}, \tag{III-47a}$$

where

$$\tilde{\mathbf{L}}_u = \begin{bmatrix} 1 & 1 \\ \dfrac{k_{z,u}}{\omega\rho_u} & \dfrac{-k_{z,u}}{\omega\rho_u} \end{bmatrix} \tag{III-47b}$$

and

$$\tilde{\mathbf{L}}_\ell = \begin{bmatrix} 1 & 1 \\ \dfrac{k_{z,\ell}}{\omega\rho_\ell} & \dfrac{-k_{z,\ell}}{\omega\rho_\ell} \end{bmatrix}, \tag{III-47c}$$

with

$$k_{z,u}^2 = \frac{\omega^2}{c_u^2} - \left(k_x^2 + k_y^2\right) \tag{III-47d}$$

and

$$k_{z,\ell}^2 = \frac{\omega^2}{c_\ell^2} - \left(k_x^2 + k_y^2\right). \tag{III-47e}$$

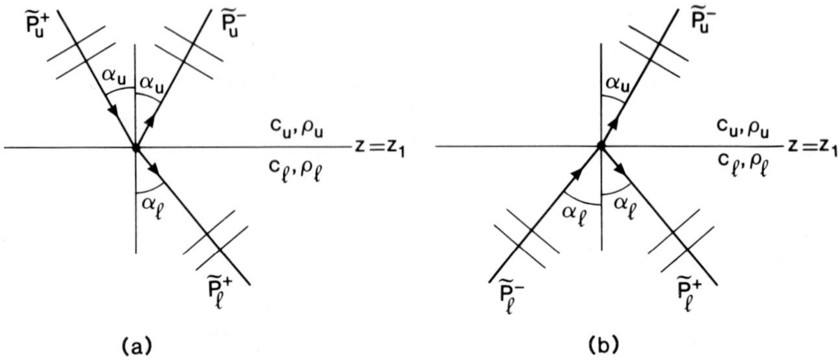

Figure III-2: Reflection and transmission of plane waves at an interface between two homogeneous acoustic half-spaces (the situation is shown for azimuth angles $\beta_u=\beta_\ell=o$).

a. Situation for an incident downgoing wave $\tilde{P}_u^+$.

b. Situation for an incident upgoing wave $\tilde{P}_\ell^-$.

Now define reflection and transmission operators $\tilde{R}^+(z_1)$ and $\tilde{T}^+(z_1)$, respectively, according to

$$\tilde{P}_u^-(z_1) = \tilde{R}^+(z_1)\tilde{P}_u^+(z_1) \tag{III-48a}$$

and

$$\tilde{P}_\ell^+(z_1) = \tilde{T}^+(z_1)\tilde{P}_u^+(z_1). \tag{III-48b}$$

Substitution into (III-47a) and dividing by $\tilde{P}_u^+(z_1)$, yields

$$\tilde{L}_u \begin{bmatrix} 1 \\ \tilde{R}^+(z_1) \end{bmatrix} = \tilde{L}_\ell \begin{bmatrix} \tilde{T}^+(z_1) \\ 0 \end{bmatrix}, \tag{III-49a}$$

or, writing all unknowns at the left-hand side

$$\begin{bmatrix} \left(\tilde{T}^+(z_1)\right)^{-1} \\ \tilde{R}^+(z_1)\left(\tilde{T}^+(z_1)\right)^{-1} \end{bmatrix} = \tilde{L}_u^{-1}\tilde{L}_\ell \binom{1}{0}, \tag{III-49b}$$

or, using (III-47b) and (III-47c),

$$\tilde{R}^+(z_1) = \frac{\rho_\ell k_{z,u} - \rho_u k_{z,\ell}}{\rho_\ell k_{z,u} + \rho_u k_{z,\ell}} \tag{III-50a}$$

and

$$\tilde{T}^+(z_1) = \frac{2\rho_\ell k_{z,u}}{\rho_\ell k_{z,u} + \rho_u k_{z,\ell}} = 1 + \tilde{R}^+(z_1). \tag{III-50b}$$

For the incident wave $\tilde{P}_u^+(z)$ and the reflected wave $\tilde{P}_u^-(z)$ we may associate a tilt angle α_u and an azimuth angle β_u to each k_x, k_y and ω value, according to

$$\sin\alpha_u = \frac{c_u}{\omega}\sqrt{k_x^2 + k_y^2} \tag{III-51a}$$

and

$$\tan\beta_u = k_y/k_x, \tag{III-51b}$$

see also equations (III-8c) and (III-8d). Hence, expressions (III-48a) and (III-48b) describe *angle-dependent* reflection and transmission, respectively. It can be easily seen from (III-47d) and (III-47e) that the reflection and transmission operators (III-50a) and (III-50b) depend on the tilt angle α_u only. For the transmitted wave $\tilde{P}_\ell^+(z)$ we may associate a tilt angle α_ℓ and an azimuth angle β_ℓ to each k_x, k_y and ω value, according to

$$\sin\alpha_\ell = \frac{c_\ell}{\omega}\sqrt{k_x^2 + k_y^2} \tag{III-52a}$$

and

$$\tan\beta_\ell = k_y/k_x. \tag{III-52b}$$

Note that

$$\sin\alpha_\ell = \frac{c_\ell}{c_u}\sin\alpha_u \tag{III-53a}$$

(Snell's law) and

$$\tan\beta_\ell = \tan\beta_u. \tag{III-53b}$$

If we define reflection and transmission operators $\tilde{R}^-$ and $\tilde{T}^-$ for the situation depicted in Figure III-2b, according to

$$\tilde{P}_\ell^+(z_1) = \tilde{R}^-(z_1)\tilde{P}_\ell^-(z_1) \tag{III-54a}$$

and

$$\tilde{P}_u^-(z_1) = \tilde{T}^-(z_1)\tilde{P}_\ell^-(z_1), \tag{III-54b}$$

then we obtain in a similar way as above

$$\begin{bmatrix} \tilde{R}^-(z_1)\left(\tilde{T}^-(z_1)\right)^{-1} \\ \left(\tilde{T}^-(z_1)\right)^{-1} \end{bmatrix} = \tilde{L}_\ell^{-1}\tilde{L}_u \begin{pmatrix} 0 \\ 1 \end{pmatrix}, \tag{III-55}$$

or

$$\tilde{R}^-(z_1) = \frac{\rho_u k_{z,\ell} - \rho_\ell k_{z,u}}{\rho_u k_{z,\ell} + \rho_\ell k_{z,u}} = -\tilde{R}^+(z_1) \tag{III-56a}$$

and

$$\tilde{T}^-(z_1) = \frac{2\rho_u k_{z,\ell}}{\rho_u k_{z,\ell} + \rho_\ell k_{z,u}} = 1 - \tilde{R}^+(z_1). \tag{III-56b}$$

III.3 ACOUSTIC WAVE EQUATIONS FOR ARBITRARILY INHOMOGENEOUS MEDIA

In this section we consider the situation of wave propagation in inhomogeneous acoustic media in which the medium parameters K and ρ are arbitrary functions of x,y and z, hence

$$K = K(x,y,z)$$

and

$$\rho = \rho(x,y,z).$$

First we derive the acoustic two-way wave equation in the space-frequency domain and present its solution in a formal operator notation. Next, we decompose the two-way wave equation into acoustic one-way wave equations for downgoing and upgoing waves and we present solutions in a formal operator notation as well as in a convenient matrix notation. Finally, we derive reflection and transmission operators for one-way wave fields at interfaces.

III.3.1 Acoustic two-way wave equation in the space-frequency domain

Consider the linearized equations of continuity and motion (I-10a) and (I-10b), in the space-frequency domain given by

$$\nabla.\vec{V}(x,y,z,\omega) + \frac{j\omega}{K(x,y,z)} P(x,y,z,\omega) = j\omega I_v(x,y,z,\omega) \tag{III-57a}$$

and

$$\nabla P(x,y,z,\omega) + j\omega\rho(x,y,z)\vec{V}(x,y,z,\omega) = \vec{F}(x,y,z,\omega), \tag{III-57b}$$

respectively.

We eliminate $\vec{V}$ from these equations, yielding, in analogy with (III-14), the following two-way wave equation

$$\rho\frac{\partial}{\partial z}\left(\frac{1}{\rho}\frac{\partial P}{\partial z}\right) + H_2 P = \omega^2\rho I_v + \rho\nabla.\left(\frac{1}{\rho}\vec{F}\right), \tag{III-58a}$$

where *operator* H_2 is defined as

$$H_2(x,y,z,\omega) = k^2(x,y,z,\omega) + \frac{\partial^2}{\partial x^2} + \frac{\partial^2}{\partial y^2}$$

$$- \frac{\partial \ln\rho(x,y,z)}{\partial x}\frac{\partial}{\partial x} - \frac{\partial \ln\rho(x,y,z)}{\partial y}\frac{\partial}{\partial y}, \tag{III-58b}$$

with

$$k^2(x,y,z,\omega) = \omega^2/c^2(x,y,z) \tag{III-58c}$$

and

$$c(x,y,z) = \sqrt{K(x,y,z)/\rho(x,y,z)}. \tag{III-58d}$$

Equation (III-58) is exact (under the linear assumption), but it has no straightforward solutions when the medium parameters c and ρ are arbitrary functions of x, y and z. As an alternative, we derive a first order two-way wave equation for P and V_z by eliminating only V_x and V_y from equations (III-57a) and (III-57b), yielding, in analogy with (III-16),

$$\frac{\partial \vec{Q}(x,y,z,\omega)}{\partial z} = \boldsymbol{A}_1(x,y,z,\omega)\vec{Q}(x,y,z,\omega) + \vec{S}(x,y,z,\omega), \tag{III-59a}$$

where the wave vector $\vec{Q}$ is defined as

$$\vec{Q}(x,y,z,\omega) = \begin{bmatrix} P(x,y,z,\omega) \\ V_z(x,y,z,\omega) \end{bmatrix}, \tag{III-59b}$$

the first order differential operator $\boldsymbol{A}_1$ is defined as

$$\boldsymbol{A}_1(x,y,z,\omega) = \begin{bmatrix} 0 & -j\omega\rho(x,y,z) \\ \frac{1}{j\omega\rho(x,y,z)} H_2(x,y,z,\omega) & 0 \end{bmatrix}, \tag{III-59c}$$

and where the source vector $\vec{S}$ is defined as

$$\vec{S}(x,y,z,\omega) = \begin{bmatrix} F_z(x,y,z,\omega) \\ j\omega I_v(x,y,z,\omega) - \frac{\partial}{\partial x}\left(\frac{F_x(x,y,z,\omega)}{j\omega\rho(x,y,z)}\right) - \frac{\partial}{\partial y}\left(\frac{F_y(x,y,z,\omega)}{j\omega\rho(x,y,z)}\right) \end{bmatrix}. \tag{III-59d}$$

The general solution of (III-59) reads, in analogy with (III-30),

$$\vec{Q}(z) = \boldsymbol{U}(z,z_o)\vec{Q}(z_o) + \int_{z_o}^{z} \boldsymbol{U}(z,z')\vec{S}(z')dz', \qquad \text{(III-60a)}$$

where the two-way wave field extrapolation operator $\boldsymbol{U}(z,z_o)$ is formally defined by

$$\boldsymbol{U}(z,z_o) = \sum_{m=o}^{\infty} \frac{(z-z_o)^m}{m!} \boldsymbol{A}_m(z_o), \qquad \text{(III-60b)}$$

with $\boldsymbol{A}_m(z_o)$ defined recursively by

$$\boldsymbol{A}_{m+1}(z_o) = \left.\frac{\partial \boldsymbol{A}_m(z)}{\partial z}\right|_{z_o} + \boldsymbol{A}_m(z_o)\boldsymbol{A}_1(z_o) \qquad \text{(III-60c)}$$

and

$$\boldsymbol{A}_o(z_o) = \mathbf{I}. \qquad \text{(III-60d)}$$

For notational convenience we omitted the variables x, y and ω. Relation (III-60) is the basis for numerical two-way wave field extrapolation algorithms which are valid for primary and multiply reflected waves in arbitrarily inhomogeneous acoustic media and which are accurate upto high tilt angles of propagation. A further discussion on the numerical aspects of acoustic two-way wave field extrapolation is beyond the scope of this book. The reader is referred to Kosloff and Baysal (1983) and Wapenaar and Berkhout (1986).

III.3.2 Acoustic one-way wave equations in the space-frequency domain

The derivation of the one-way wave equations for arbitrarily inhomogeneous media (including sources) is rather complicated, unless we make use of the matrix notation which is discussed in Appendix A. This matrix notation is a convenient alternative notation for generalized spatial convolution integrals. To show the principle first, we consider two-way wave equation (III-58a) for the source-free situation:

$$\rho \frac{\partial}{\partial z}\left(\frac{1}{\rho}\frac{\partial P(x,y,z,\omega)}{\partial z}\right) = -H_2(x,y,z,\omega)P(x,y,z,\omega) \;, \qquad \text{(III-61)}$$

with $H_2(x,y,z,\omega)$ defined by (III-58b).

We rewrite (III-61) as a generalized spatial convolution integral. Therefore we first write the lateral differentiations with respect to x and y as conventional spatial convolution integrals, according to

$$\frac{\partial^m P(x,y,z,\omega)}{\partial x^m} = \int_{-\infty}^{\infty} d_m(x-x')P(x',y,z,\omega)dx' \tag{III-62a}$$

and

$$\frac{\partial^m P(x,y,z,\omega)}{\partial y^m} = \int_{-\infty}^{\infty} d_m(y-y')P(x,y',z,\omega)dy'. \tag{III-62b}$$

Here $d_m(x)$ and $d_m(y)$ are *band-limited* lateral differentiation operators. They are defined as the inverse Fourier transform of operators $\tilde{d}_m(k_x)$ and $\tilde{d}_m(k_y)$, which represent properly chosen band-limited versions of $(-jk_x)^m$ and $(-jk_y)^m$, respectively (Berkhout, 1985, Appendix H). Hence

$$d_m(x) \triangleq \frac{1}{2\pi} \int_{-\infty}^{\infty} \tilde{d}_m(k_x)e^{-jk_x x} dk_x \tag{III-62c}$$

and

$$d_m(y) \triangleq \frac{1}{2\pi} \int_{-\infty}^{\infty} \tilde{d}_m(k_y)e^{-jk_y y} dk_y. \tag{III-62d}$$

Equation (III-62) is exact when $P(x,y,z,\omega)$ is a spatially band-limited function. With the definitions in (III-62), two-way wave equation (III-61) is rewritten as a generalized spatial convolution integral, according to

$$\rho\frac{\partial}{\partial z}\left(\frac{1}{\rho}\frac{\partial P(x,y,z;\omega)}{\partial z}\right) = -\iint_{-\infty}^{\infty} H_2(x,y,z;x',y',z'=z;\omega)P(x',y',z'=z;\omega)dx'dy', \tag{III-63a}$$

where

$$H_2(x,y,z;x',y',z'=z;\omega) = \tag{III-63b}$$

$$\frac{\omega^2}{c^2(x,y,z)}\delta(x-x')\delta(y-y') + d_2(x-x')\delta(y-y') + \delta(x-x')d_2(y-y')$$
$$- \frac{\partial \ln\rho(x,y,z)}{\partial x} d_1(x-x')\delta(y-y') - \frac{\partial \ln\rho(x,y,z)}{\partial y} \delta(x-x')d_1(y-y').$$

For the special situation of a homogeneous medium, H_2 is a function of x-x', y-y' and ω only, hence, for this situation (III-63) may be written as a conventional spatial convolution integral, according to

$$\frac{\partial^2 P(x,y,z,\omega)}{\partial z^2} = -\iint_{-\infty}^{\infty} H_2(x-x',y-y',\omega)P(x',y',z'=z,\omega)dx'dy', \tag{III-64a}$$

or, symbolically,

$$\frac{\partial^2 P(x,y,z,\omega)}{\partial z^2} = -H_2(x,y,\omega) * P(x,y,z,\omega), \tag{III-64b}$$

where

$$H_2(x,y,\omega) \triangleq \frac{\omega^2}{c^2}\delta(x)\delta(y) + d_2(x)\delta(y) + \delta(x)d_2(y). \tag{III-64c}$$

We return to the inhomogeneous situation. For notational convenience we define the coordinate vectors

$$\vec{r} = (x,y,z) \tag{III-65a}$$

and

$$\vec{r}\,' = (x',y',z'), \tag{III-65b}$$

so that the generalized spatial convolution integral (III-63) may be written as

$$\rho(\vec{r})\frac{\partial}{\partial z}\left(\frac{1}{\rho(\vec{r})}\frac{\partial P(\vec{r},\omega)}{\partial z}\right) = -\iint_{-\infty}^{\infty}\left[H_2(\vec{r},\vec{r}\,',\omega)P(\vec{r}\,',\omega)\right]_{z'=z}dx'dy'. \tag{III-66}$$

Let us now implicitly define an operator H_1 according to

$$H_2(\vec{r},\vec{r}\,'',\omega) = \iint_{-\infty}^{\infty} H_1(\vec{r},\vec{r}\,',\omega)H_1(\vec{r}\,',\vec{r}\,'',\omega)dx'dy', \tag{III-67a}$$

with

$$\vec{r}\,'' = (x'',y'',z''). \tag{III-67b}$$

In the seismic literature H_1 is referred to as the *square-root* operator.

For the situation that the medium parameters do not depend on the depth coordinate z, according to

$$c(x,y,z) = c(x,y,o) \tag{III-68a}$$

and

$$\rho(x,y,z) = \rho(x,y,o), \tag{III-68b}$$

we obtain the following first order differential equation:

$$\frac{\partial P(\vec{r},\omega)}{\partial z} = \pm j \iint_{-\infty}^{\infty} \left[H_1(\vec{r},\vec{r}',\omega)P(\vec{r}',\omega)\right]_{z'=z} dx'dy'. \tag{III-69}$$

This result can be verified by differentiating both sides with respect to z, yielding

$$\frac{\partial^2 P(\vec{r},\omega)}{\partial z^2} = \pm j \iint_{-\infty}^{\infty} \left[H_1(\vec{r},\vec{r}',\omega)\frac{\partial P(\vec{r}',\omega)}{\partial z}\right]_{z'=z} dx'dy', \tag{III-70a}$$

or, substituting (III-69),

$$\frac{\partial^2 P(\vec{r},\omega)}{\partial z^2} = -\iint_{-\infty}^{\infty}\left[H_1(\vec{r},\vec{r}',\omega)\iint_{-\infty}^{\infty}\left[H_1(\vec{r}',\vec{r}'',\omega)P(\vec{r}'',\omega)\right]_{z''=z'}dx''dy''\right]_{z'=z} dx'dy', \tag{III-70b}$$

or, changing the order of the integrations and substituting (III-67a),

$$\frac{\partial^2 P(\vec{r},\omega)}{\partial z^2} = -\iint_{-\infty}^{\infty} \left[H_2(\vec{r},\vec{r}'',\omega)P(\vec{r}'',\omega)\right]_{z''=z} dx''dy'', \tag{III-70c}$$

which is again two-way wave equation (III-66) for the situation described by (III-68). Note the similarity of (III-69) with the one-way wave equations (III-45a) and (III-45b) in the wavenumber-frequency domain, which read for the source free situation

$$\frac{\partial \tilde{P}^+(z)}{\partial z} = -jk_z\tilde{P}^+(z) \tag{III-71a}$$

and

$$\frac{\partial \tilde{P}^-(z)}{\partial z} = +jk_z\tilde{P}^-(z). \tag{III-71b}$$

Therefore we may conclude that (III-69) represents the one-way wave equations in the space-frequency domain for downgoing waves

$$\frac{\partial P^+(\vec{r},\omega)}{\partial z} = -j \int\int_{-\infty}^{\infty} \left[H_1(\vec{r},\vec{r}',\omega)P^+(\vec{r}',\omega)\right]_{z'=z} dx'dy' \qquad \text{(III-72a)}$$

and for upgoing waves

$$\frac{\partial P^-(\vec{r},\omega)}{\partial z} = +j \int\int_{-\infty}^{\infty} \left[H_1(\vec{r},\vec{r}',\omega)P^-(\vec{r}',\omega)\right]_{z'=z} dx'dy', \qquad \text{(III-72b)}$$

respectively.

With the matrix notation, discussed in Appendix A, we may also represent the two-way wave equation (III-70c) by

$$\frac{\partial^2 \vec{P}(z)}{\partial z^2} = -\mathbf{H}_2(z)\vec{P}(z) \qquad \text{(III-73a)}$$

and the one-way wave equations (III-72a) and (III-72b) by

$$\frac{\partial \vec{P}^+(z)}{\partial z} = -j\mathbf{H}_1(z)\vec{P}^+(z) \qquad \text{(III-73b)}$$

and

$$\frac{\partial \vec{P}^-(z)}{\partial z} = +j\mathbf{H}_1(z)\vec{P}^-(z), \qquad \text{(III-73c)}$$

where, in analogy with (III-67a),

$$\mathbf{H}_2(z) = \mathbf{H}_1(z)\mathbf{H}_1(z). \qquad \text{(III-73d)}$$

Here vectors $\vec{P}(z)$, $\vec{P}^+(z)$ and $\vec{P}^-(z)$ contain the discretized versions of the wave fields $P(\vec{r},\omega)$, $P^+(\vec{r},\omega)$ and $P^-(\vec{r},\omega)$, respectively, at depth level z (see Appendix A, section A.2). Furthermore, matrices $\mathbf{H}_1(z)$ and $\mathbf{H}_2(z)$ contain the discretized versions of the operators $H_1(\vec{r},\vec{r}',\omega)$ and $H_2(\vec{r},\vec{r}',\omega)$, respectively, at depth level z'=z (see Appendix A, section A.3). Note that $\mathbf{H}_2(z)=\mathbf{H}_2(o)$ and $\mathbf{H}_1(z)=\mathbf{H}_1(o)$, see also (III-68).
Next, we use this convenient matrix notation to derive the one-way wave equations for the situation where $c(x,y,z)$ and $\rho(x,y,z)$ are arbitrary

functions and where the medium contains sources.

Consider two-way wave equation (III-59a),

$$\frac{\partial \vec{Q}(\vec{r},\omega)}{\partial z} = \boldsymbol{A}_1(\vec{r},\omega)\vec{Q}(\vec{r},\omega) + \vec{S}(\vec{r},\omega), \tag{III-74}$$

with $\vec{Q}$, $\boldsymbol{A}_1$ and $\vec{S}$ defined in (III-59b), (III-59c) and (III-59d), respectively. With the matrix notation, discussed in Appendix A, we may rewrite this equation as

$$\frac{\partial \vec{Q}(z)}{\partial z} = \mathbf{A}_1(z)\vec{Q}(z) + \vec{S}(z), \tag{III-75a}$$

where

$$\vec{Q}(z) = \begin{bmatrix} \vec{P}(z) \\ \vec{V}_z(z) \end{bmatrix}, \tag{III-75b}$$

$$\mathbf{A}_1(z) = \begin{bmatrix} \mathbf{O} & -j\omega\mathbf{M}(z) \\ \frac{1}{j\omega}\mathbf{M}^{-1}(z)\mathbf{H}_2(z) & \mathbf{O} \end{bmatrix} \tag{III-75c}$$

and

$$\vec{S}(z) = \begin{bmatrix} \vec{F}_z(z) \\ j\omega\vec{I}_v(z) \end{bmatrix}, \tag{III-75d}$$

where we assumed for simplicity that $F_x = F_y = o$.
Note that $\mathbf{M}(z)$ is a diagonal matrix, containing the discretized version of the mass density $\rho(\vec{r})$ at depth level z (see Appendix A, section A.3); $\mathbf{O}$ is a matrix, containing zeroes only.

We can now follow the same approach as in section III.2.3. We decompose matrix $\mathbf{A}_1(z)$ according to

$$\mathbf{A}_1(z) = \mathbf{L}(z)\mathbf{\Lambda}(z)\mathbf{L}^{-1}(z), \tag{III-76a}$$

with

$$\mathbf{L}(z) = \begin{bmatrix} \mathbf{I} & \mathbf{I} \\ \frac{1}{\omega}\mathbf{M}^{-1}(z)\mathbf{H}_1(z) & \frac{-1}{\omega}\mathbf{M}^{-1}(z)\mathbf{H}_1(z) \end{bmatrix}, \tag{III-76b}$$

$$\mathbf{\Lambda}(z) = \begin{bmatrix} -j\mathbf{H}_1(z) & \mathbf{O} \\ \mathbf{O} & j\mathbf{H}_1(z) \end{bmatrix} \tag{III-76c}$$

and

$$\mathbf{L}^{-1}(z) = \frac{1}{2}\begin{bmatrix} \mathbf{I} & \omega\mathbf{H}_1^{-1}(z)\mathbf{M}(z) \\ \mathbf{I} & -\omega\mathbf{H}_1^{-1}(z)\mathbf{M}(z) \end{bmatrix}. \tag{III-76d}$$

We define a vector $\vec{D}(z)$ according to

$$\vec{Q}(z) = \mathbf{L}(z)\vec{D}(z), \tag{III-77a}$$

or, equivalently,

$$\vec{D}(z) = \mathbf{L}^{-1}(z)\vec{Q}(z), \tag{III-77b}$$

where $\vec{D}(z)$ represents the decomposed wave field in terms of the downgoing and upgoing wave vectors $\vec{P}^+(z)$ and $\vec{P}^-(z)$, according to

$$\vec{D}(z) = \begin{bmatrix} \vec{P}^+(z) \\ \vec{P}^-(z) \end{bmatrix}. \tag{III-77c}$$

The decomposition (III-77b) breaks down for waves which propagate in the horizontal direction (see also section III.2.3).

Upon substitution of (III-77a) into (III-75a), using property (III-76a), we obtain

$$\frac{\partial\vec{D}(z)}{\partial z} = \mathbf{B}_1(z)\vec{D}(z) + \vec{S}'(z), \tag{III-78a}$$

where

$$\mathbf{B}_1(z) = \mathbf{A}(z) - \mathbf{L}^{-1}(z)\,\frac{\partial \mathbf{L}(z)}{\partial z} \tag{III-78b}$$

and

$$\vec{S}'(z) = \mathbf{L}^{-1}(z)\vec{S}(z). \tag{III-78c}$$

This equation represents a coupled system of *one-way wave equations* for down-going and upgoing waves, according to

$$\frac{\partial \vec{P}^{+}(z)}{\partial z} = -j\mathbf{H}_1(z)\vec{P}^{+}(z)$$

$$- \frac{1}{2}\,\mathbf{H}_1^{-1}(z)\mathbf{M}(z)\left[\frac{\partial}{\partial z}\left(\mathbf{M}^{-1}(z)\mathbf{H}_1(z)\right)\left(\vec{P}^{+}(z)-\vec{P}^{-}(z)\right)\right] + \vec{S}^{+}(z) \tag{III-79a}$$

and

$$\frac{\partial \vec{P}^{-}(z)}{\partial z} = +j\mathbf{H}_1(z)\vec{P}^{-}(z)$$

$$- \frac{1}{2}\,\mathbf{H}_1^{-1}(z)\mathbf{M}(z)\left[\frac{\partial}{\partial z}\left(\mathbf{M}^{-1}(z)\mathbf{H}_1(z)\right)\left(\vec{P}^{-}(z)-\vec{P}^{+}(z)\right)\right] - \vec{S}^{-}(z), \tag{III-79b}$$

where

$$\vec{S}^{\pm}(z) = \frac{1}{2}\left[j\omega^2\mathbf{H}_1^{-1}(z)\mathbf{M}(z)\vec{\Gamma}_v(z) \pm \vec{F}_z(z)\right]. \tag{III-79c}$$

Here $\vec{S}^{+}(z)$ and $\vec{S}^{-}(z)$ are the one-way representations of the source distribution.

When we neglect $\vec{P}^{-}(z)$ with respect to $\vec{P}^{+}(z)$ in (III-79a) and $\vec{P}^{+}(z)$ with respect to $\vec{P}^{-}(z)$ in (III-79b), then we obtain decoupled one-way wave equations for *primary* downgoing and upgoing waves in arbitrarily inhomogeneous acoustic media, according to

$$\frac{\partial \vec{P}^{+}(z)}{\partial z} \approx -j\mathbf{H}_1^{+}(z)\vec{P}^{+}(z) + \vec{S}^{+}(z), \tag{III-80a}$$

and

$$\frac{\partial \vec{P}^{-}(z)}{\partial z} \approx +j\mathbf{H}_1^{-}(z)\vec{P}^{-}(z) - \vec{S}^{-}(z), \tag{III-80b}$$

where

$$j\mathbf{H}_1^{\pm}(z) = j\mathbf{H}_1(z) \pm \frac{1}{2}\mathbf{H}_1^{-1}(z)\mathbf{M}(z)\frac{\partial}{\partial z}\left(\mathbf{M}^{-1}(z)\mathbf{H}_1(z)\right). \tag{III-80c}$$

The general solution of these equations reads, in analogy with (III-44),

$$\vec{P}^{\pm}(z) \approx \mathbf{W}^{\pm}(z,z_0)\vec{P}^{\pm}(z_0) \pm \int_{z_0}^{z} \mathbf{W}^{\pm}(z,z')\vec{S}^{\pm}(z')dz', \tag{III-81a}$$

where the one-way wave field extrapolation matrices $\mathbf{W}^{\pm}(z,z_0)$ are defined by

$$\mathbf{W}^{\pm}(z,z_0) = \sum_{m=0}^{\infty} \frac{(z-z_0)^m}{m!}(\mp j)^m \mathbf{H}_m^{\pm}(z_0), \tag{III-81b}$$

with $\mathbf{H}_m^{\pm}(z_0)$ defined recursively by

$$\mathbf{H}_{m+1}^{\pm}(z_0) = \pm j \left.\frac{\partial \mathbf{H}_m^{\pm}(z)}{\partial z}\right|_{z_0} + \mathbf{H}_m^{\pm}(z_0)\mathbf{H}_1^{\pm}(z_0) \tag{III-81c}$$

and

$$\mathbf{H}_0^{\pm}(z_0) = \mathbf{I}. \tag{III-81d}$$

Finally, we rewrite the matrix products in (III-81a) and (III-79c) as generalized spatial convolution integrals, according to

$$P^{\pm}(\vec{r},\omega) \approx \iint_{-\infty}^{\infty} \left[W^{\pm}(\vec{r},\vec{r}',\omega)P^{\pm}(\vec{r}',\omega)\right]_{z'=z_0} dx'dy'$$
$$\pm\int_{z_0}^{z} \left[\iint_{-\infty}^{\infty} W^{\pm}(\vec{r},\vec{r}',\omega)S^{\pm}(\vec{r}',\omega)dx'dy'\right]dz', \tag{III-82a}$$

where

$$S^{\pm}(\vec{r},\omega) = \frac{1}{2} j\omega^2 \int\int_{-\infty}^{\infty} [H_1^{-1}(\vec{r},\vec{r}',\omega)\rho(\vec{r}')I_v(\vec{r}',\omega)]_{z'=z}dx'dy' \pm \frac{1}{2}F_z(\vec{r},\omega). \tag{III-82b}$$

Equation (III-82a) states that the one-way wave field at $\vec{r}$ is obtained from a surface integral over the one-way wave field at z_0 and a volume integral over the one-way sources between z_0 and z. These *one-way sources* are obtained, according to (III-82b), by spatially convolving the source function $I_v(\vec{r},\omega)$ with the *inverse square-root operator* and by adding the source function $\pm\frac{1}{2} F_z(\vec{r},\omega)$, (Wapenaar, 1989).

III.3.3. Acoustic one-way wave fields at interfaces in the space-frequency domain

Solution (III-81) breaks down whenever the medium contains "interfaces" in the interval (z,z_0).

We consider a horizontal interface at $z=z_1$ between two acoustic half-spaces in which the medium parameters vary laterally only. We use the sub-scripts u and ℓ to distinguish between the upper and lower half-space, respectively. In the upper half-space $z<z_1$ the medium parameters read $c_u(x,y)$ and $\rho_u(x,y)$; in the lower half-space $z>z_1$ the medium parameters read $c_\ell(x,y)$ and $\rho_\ell(x,y)$. First consider the situation depicted in Figure III-3a. A wave $P_u^+(x,y,z,\omega)$ is incident to the interface from above,

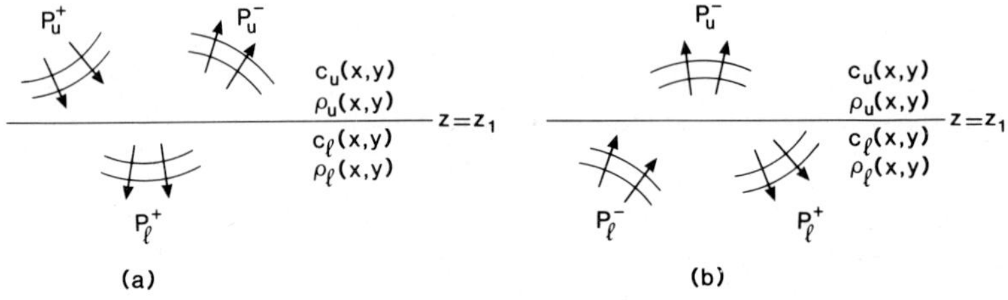

Figure III-3: Reflection and transmission at a horizontal interface.
a. Situation for an incident downgoing wave $P_u^+(x,y,z,\omega)$
b. Situation for an incident upgoing wave $P_\ell^-(x,y,z,\omega)$.

a wave $P^-_u(x,y,z,\omega)$ is reflected into the upper half-space and a wave $P^+_\ell(x,y,z,\omega)$ is transmitted into the lower half-space. In the following, the discretized versions of these wave fields are represented by vectors $\vec{P}^+_u(z)$, $\vec{P}^-_u(z)$ and $\vec{P}^+_\ell(z)$, respectively, (see Appendix A, section A.2).

In analogy with (III-48a) and (III-48b), we define reflection and transmission matrices $\mathbf{R}^+$ and $\mathbf{T}^+$ for the interface at z_1, according to

$$\vec{P}^-_u(z_1) = \mathbf{R}^+(z_1)\vec{P}^+_u(z_1) \tag{III-83a}$$

and

$$\vec{P}^+_\ell(z_1) = \mathbf{T}^+(z_1)\vec{P}^+_u(z_1), \tag{III-83b}$$

respectively.

In a similar way as described in section III.2.4, we may derive

$$\begin{bmatrix} \left(\mathbf{T}^+(z_1)\right)^{-1} \\ \\ \mathbf{R}^+(z_1)\left(\mathbf{T}^+(z_1)\right)^{-1} \end{bmatrix} = \mathbf{L}_u^{-1}\mathbf{L}_\ell \begin{bmatrix} \mathbf{I} \\ \mathbf{O} \end{bmatrix}, \tag{III-84}$$

or, using (III-76b) and (III-76d),

$$\mathbf{R}^+(z_1) = \left[\mathbf{I} - \mathbf{H}_{1,u}^{-1}\mathbf{M}_u\mathbf{M}_\ell^{-1}\mathbf{H}_{1,\ell}\right]\left[\mathbf{I} + \mathbf{H}_{1,u}^{-1}\mathbf{M}_u\mathbf{M}_\ell^{-1}\mathbf{H}_{1,\ell}\right]^{-1} \tag{III-85a}$$

and

$$\mathbf{T}^+(z_1) = 2\left[\mathbf{I} + \mathbf{H}_{1,u}^{-1}\mathbf{M}_u\mathbf{M}_\ell^{-1}\mathbf{H}_{1,\ell}\right]^{-1}. \tag{III-85b}$$

A special situation occurs when both half-spaces are homogeneous and have a density contrast only. Then

$$\mathbf{R}^+(z_1) = (\mathbf{M}_\ell - \mathbf{M}_u)(\mathbf{M}_\ell + \mathbf{M}_u)^{-1} \tag{III-86a}$$

and

$$\mathbf{T}^+(z_1) = 2\mathbf{M}_\ell(\mathbf{M}_\ell + \mathbf{M}_u)^{-1}. \tag{III-86b}$$

Note that for this special situation $\mathbf{R}^{+}(z_1)$ and $\mathbf{T}^{+}(z_1)$ are diagonal matrices.

For the situation depicted in Figure III-3b we define reflection and transmission matrices $\mathbf{R}^{-}$ and $\mathbf{T}^{-}$ for the interface at z_1, according to

$$\vec{P}_{\ell}^{+}(z_1) = \mathbf{R}^{-}(z_1)\vec{P}_{\ell}^{-}(z_1) \tag{III-87a}$$

and

$$\vec{P}_{u}^{-}(z_1) = \mathbf{T}^{-}(z_1)\vec{P}_{\ell}^{-}(z_1), \tag{III-87b}$$

respectively. In analogy with (III-84) we obtain

$$\begin{bmatrix} \mathbf{R}^{-}(z_1)\left(\mathbf{T}^{-}(z_1)\right)^{-1} \\ \left(\mathbf{T}^{-}(z_1)\right)^{-1} \end{bmatrix} = \mathbf{L}_{\ell}^{-1}\mathbf{L}_{u}\begin{bmatrix} \mathbf{O} \\ \mathbf{I} \end{bmatrix}, \tag{III-88}$$

or, using (III-76b) and (III-76d),

$$\mathbf{R}^{-}(z_1) = \left[\mathbf{I} - \mathbf{H}_{1,\ell}^{-1}\mathbf{M}_{\ell}\mathbf{M}_{u}^{-1}\mathbf{H}_{1,u}\right]\left[\mathbf{I} + \mathbf{H}_{1,\ell}^{-1}\mathbf{M}_{\ell}\mathbf{M}_{u}^{-1}\mathbf{H}_{1,u}\right]^{-1} \tag{III-89a}$$

and

$$\mathbf{T}^{-}(z_1) = 2\left[\mathbf{I} + \mathbf{H}_{1,\ell}^{-1}\mathbf{M}_{\ell}\mathbf{M}_{u}^{-1}\mathbf{H}_{1,u}\right]^{-1}. \tag{III-89b}$$

Again, for the situation of two homogeneous half-spaces having a density contrast only, we obtain diagonal matrices:

$$\mathbf{R}^{-}(z_1) = (\mathbf{M}_u - \mathbf{M}_{\ell})(\mathbf{M}_u + \mathbf{M}_{\ell})^{-1} \tag{III-90a}$$

and

$$\mathbf{T}^{-}(z_1) = 2\mathbf{M}_u(\mathbf{M}_u + \mathbf{M}_{\ell})^{-1}. \tag{III-90b}$$

The expressions for reflection and transmission, derived in this section, are strictly valid only for the situation of a horizontal interface between two acoustic half-spaces in which the medium parameters vary laterally only (Figure III-3). Because the expressions are formulated in the space-frequency domain they are in principle suited to handle more complex configurations. Of course the accuracy decreases with increasing complexity of the configuration.

III.4 REFERENCES

Berkhout, A.J., 1985, Seismic migration: Imaging of acoustic energy by wave field extrapolation. A. Theoretical aspects: Elsevier Science Publ. Co., Inc.

Boyce, W.E., and DiPrima, R.C., 1969, Elementary differential equations and boundary value problems: John Wiley & Sons, Inc.

Bracewell, R., 1965, The Fourier transform and its applications: McGraw-Hill.

Brillouin, L., 1926, Remarques sur la mécanique ondulatoire: J. Phys. Radium, *6*, 353 - 368.

Dudgeon, D.E., and Mersereau, R.M., 1984, Multi-dimensional digital signal processing: Prentice-Hall.

Gilbert, F., and Backus, G.E., 1966, Propagator matrices in elastic wave and vibration problems: Geophysics, *31*, 326 - 332.

Goodman, J.W., 1968, Introduction to Fourier optics: McGraw-Hill.

Kosloff, D.D., and Baysal, E., 1983, Migration with the full acoustic wave equation: Geophysics, *48*, 677 - 687.

Kramers, H.A., 1926, Wellenmechanik und halbzahlige Quantisierung: Z. Physik, *39* 828 - 840.

Papoulis, A., 1962, The Fourier integral and its applications: McGraw-Hill.

Ursin, B., 1983, Review of elastic and electromagnetic wave propagation in horizontally layered media: Geophysics, *48*, 1063 - 1081.

Wapenaar, C.P.A., 1989, Representation of seismic sources in the one-way wave equations (short note): submitted for publication in Geophysics.

Wapenaar, C.P.A., and Berkhout, A.J., 1986, Wave field extrapolation techniques for inhomogeneous media which include critical angle events. Part II. Methods using the two-way wave equation: Geophysical Prospecting, *34*, 147 - 179.

Wentzel, G., 1926, Eine Verallgemeinerung der Quantenbedingungen für die Zwecke der Wellenmechanik: Z. Physik, *38*, 518 - 529.

IV

ELASTIC TWO-WAY AND ONE-WAY WAVE EQUATIONS

IV.1. INTRODUCTION

In this chapter we derive the elastic two-way and one-way wave equations both in the wavenumber-frequency domain (for horizontally layered solids) and in the space-frequency domain (for arbitrarily inhomogeneous solids). We present formal solutions (in terms of Taylor series), but it is not our intention to elaborate the numerical aspects of these solutions. The main purpose of this chapter is to derive the mathematical relationship between two-way and one-way elastic wave fields. The results will be used in chapters VI and VIII for transforming elastic two-way Kirchhoff-Helmholtz integrals into elastic one-way Rayleigh integrals and in chapter XII, where we discuss an elastic processing scheme for multi-component seismic data.

IV.2 ELASTIC WAVE EQUATIONS FOR HORIZONTALLY LAYERED MEDIA

In this section we consider the special situation of wave propagation in horizontally layered elastic media in which the medium parameters $c_{ijk\ell}$ and ρ are a function of depth only, hence

$$c_{ijk\ell} = c_{ijk\ell}(z)$$

and

$$\rho = \rho(z).$$

We derive the elastic two-way wave equation in the wavenumber-frequency domain and present its exact solution. Next, we decompose the two-way wave equation into elastic one-way wave equations for downgoing and upgoing P- and S-waves and we present exact and approximate solutions. Finally, we derive reflection and transmission operators for one-way P- and S-wave fields at interfaces.

IV.2.1. Elastic two-way wave equation in the wavenumber-frequency domain

In the space-time domain, the linearized equation of motion (II-13) reads (for horizontally layered media)

$$\rho(z)\partial_t\vec{v}(x,y,z,t) - \partial_j\vec{\tau}_j(x,y,z,t) = \vec{f}(x,y,z,t), \tag{IV-1a}$$

whereas the linearized stress-velocity relation (II-24) reads

$$\partial_t\vec{\tau}_j(x,y,z,t) - \mathbf{C}_{j\ell}(z)\partial_\ell\vec{v}(x,y,z,t) = -\partial_t\vec{\sigma}_j(x,y,z,t). \tag{IV-1b}$$

A differentiation with respect to time in the space-time domain corresponds to a multiplication by $j\omega$ in the space-frequency domain. Since the medium parameters $\mathbf{C}_{j\ell}$ and ρ are time invariant, equation (IV-1a) and (IV-1b) read in the space-frequency domain

$$j\omega\rho(z)\vec{V}(x,y,z,\omega) - \partial_j\vec{\tau}_j(x,y,z,\omega) = \vec{F}(x,y,z,\omega)^{1)} \tag{IV-2a}$$

and

$$j\omega\vec{\tau}_j(x,y,z,\omega) - \mathbf{C}_{j\ell}(z)\partial_\ell\vec{V}(x,y,z,\omega) = -j\omega\vec{\sigma}_j(x,y,z,\omega), \tag{IV-2b}$$

respectively. A differentiation with respect to x or y in the space-frequency domain corresponds to a multiplication by $-jk_x$ or $-jk_y$, respectively, in the wavenumber-frequency domain. Since the medium parameters $\mathbf{C}_{j\ell}$ and ρ are laterally invariant, equations (IV-2a) and (IV-2b) read in the wavenumber-frequency domain

$$j\omega\rho(z)\tilde{\vec{V}}(k_x,k_y,z,\omega) + jk_\alpha\tilde{\vec{\tau}}_\alpha(k_x,k_y,z,\omega) - \frac{\partial\tilde{\vec{\tau}}_z(k_x,k_y,z,\omega)}{\partial z} = \tilde{\vec{F}}(k_x,k_y,z,\omega)^{2)} \tag{IV-3a}$$

and

---

1) Here and in the following equations the symbol j has two different meanings. When used as a factor, it denotes the imaginary unit $\sqrt{-1}$. Otherwise it is an index which may take the values 1, 2 or 3.

2) Greek indices may only take the values 1 or 2, see also the introduction.

$$j\omega\tilde{\vec{\tau}}_j(k_x,k_y,z,\omega) + jk_\beta \mathbf{C}_{j\beta}(z)\tilde{\vec{V}}(k_x,k_y,z,\omega) - \mathbf{C}_{j3}(z)\frac{\partial\tilde{\vec{V}}(k_x,k_y,z,\omega)}{\partial z} =$$

$$= -j\omega\tilde{\vec{\sigma}}_j(k_x,k_y,z,\omega), \tag{IV-3b}$$

respectively. In analogy with section III.2.2, we derive a *first-order* two-way wave equation for $\tilde{\vec{V}}$ and $\tilde{\vec{\tau}}_z$ by eliminating $\tilde{\vec{\tau}}_x$ and $\tilde{\vec{\tau}}_y$ from equations (IV-3a) and (IV-3b).

We express $\partial\tilde{\vec{V}}/\partial z$ and $\partial\tilde{\vec{\tau}}_z/\partial z$ in terms of operators acting on $\tilde{\vec{V}}$ and $\tilde{\vec{\tau}}_z$. From (IV-3b) we obtain

$$\frac{\partial\tilde{\vec{V}}}{\partial z} = \mathbf{C}_{33}^{-1}\left[jk_\beta \mathbf{C}_{3\beta}\tilde{\vec{V}} + j\omega\tilde{\vec{\tau}}_z + j\omega\tilde{\vec{\sigma}}_z\right], \tag{IV-4a}$$

whereas from (IV-3a) we obtain

$$\frac{\partial\tilde{\vec{\tau}}_z}{\partial z} = j\omega\rho\tilde{\vec{V}} + jk_\alpha\tilde{\vec{\tau}}_\alpha - \tilde{\vec{F}}, \tag{IV-4b}$$

or, upon substitution of (IV-3b),

$$\frac{\partial\tilde{\vec{\tau}}_z}{\partial z} = j\omega\rho\tilde{\vec{V}} + jk_\alpha\left[\frac{-1}{\omega}k_\beta \mathbf{C}_{\alpha\beta}\tilde{\vec{V}} + \frac{1}{j\omega}\mathbf{C}_{\alpha 3}\frac{\partial\tilde{\vec{V}}}{\partial z} - \tilde{\vec{\sigma}}_\alpha\right] - \tilde{\vec{F}}, \tag{IV-4c}$$

or, upon substitution of (IV-4a),

$$\begin{aligned}\frac{\partial\tilde{\vec{\tau}}_z}{\partial z} &= j\omega\rho\tilde{\vec{V}} + \frac{1}{j\omega}k_\alpha k_\beta\left(\mathbf{C}_{\alpha\beta} - \mathbf{C}_{\alpha 3}\mathbf{C}_{33}^{-1}\mathbf{C}_{3\beta}\right)\tilde{\vec{V}} \\ &\quad + jk_\alpha\left[\mathbf{C}_{\alpha 3}\mathbf{C}_{33}^{-1}(\tilde{\vec{\tau}}_z + \tilde{\vec{\sigma}}_z) - \tilde{\vec{\sigma}}_\alpha\right] - \tilde{\vec{F}}.\end{aligned} \tag{IV-4d}$$

Equations (IV-4a) and (IV-4d) can be combined in one equation, according to

$$\frac{\partial\tilde{\vec{Q}}}{\partial z} = \tilde{\mathbf{A}}_1\tilde{\vec{Q}} + \tilde{\vec{S}}, \tag{IV-5a}$$

where the wave vector $\tilde{\vec{Q}}$ is defined as

$$\tilde{\vec{Q}} = \begin{bmatrix} \tilde{\vec{V}} \\ \tilde{\vec{\tau}}_z \end{bmatrix}, \tag{IV-5b}$$

the first order differential operator $\tilde{\mathbf{A}}_1$ is defined as

$$\tilde{A}_1 = \begin{bmatrix} \tilde{\mathbf{a}}_{11} & \tilde{\mathbf{a}}_{12} \\ \tilde{\mathbf{a}}_{21} & \tilde{\mathbf{a}}_{22} \end{bmatrix}, \tag{IV-5c}$$

with

$$\tilde{a}_{11} = jk_\beta C_{33}^{-1} C_{3\beta} , \tag{IV-5d}$$

$$\tilde{a}_{12} = j\omega C_{33}^{-1}, \tag{IV-5e}$$

$$\tilde{a}_{21} = j\omega\rho I + \frac{1}{j\omega} k_\alpha k_\beta (C_{\alpha\beta} - C_{\alpha 3} C_{33}^{-1} C_{3\beta}) \tag{IV-5f}$$

and

$$\tilde{a}_{22} = jk_\alpha C_{\alpha 3} C_{33}^{-1} \tag{IV-5g}$$

and, finally, the source vector $\tilde{\vec{S}}$ is defined as

$$\tilde{\vec{S}} = \begin{bmatrix} \tilde{\vec{S}}_1 \\ \tilde{\vec{S}}_2 \end{bmatrix}, \tag{IV-5h}$$

with

$$\tilde{\vec{S}}_1 = j\omega C_{33}^{-1} \tilde{\vec{\sigma}}_z \tag{IV-5i}$$

and

$$\tilde{\vec{S}}_2 = jk_\alpha\left(C_{\alpha 3}C_{33}^{-1}\tilde{\vec{\sigma}}_z - \tilde{\vec{\sigma}}_\alpha\right) - \tilde{\vec{F}}. \tag{IV-5j}$$

The general solution of equation (IV-5a) reads, in analogy with equation (III-30),

$$\tilde{\vec{Q}}(z) = \tilde{U}(z,z_o)\tilde{\vec{Q}}(z_o) + \int_{z_o}^{z} \tilde{U}(z,z')\tilde{\vec{S}}(z')dz', \tag{IV-6a}$$

where the two-way extrapolation operator $\tilde{U}$ is defined as

$$\tilde{U}(z,z_o) = \sum_{m=o}^{\infty} \frac{(z-z_o)^m}{m!} \tilde{A}_m(z_o), \tag{IV-6b}$$

with

$$\tilde{A}_{m+1}(z_o) = \left.\frac{\partial \tilde{A}_m(z)}{\partial z}\right|_{z_o} + \tilde{A}_m(z_o)\tilde{A}_1(z_o) \tag{IV-6c}$$

and

$$\tilde{A}_o(z_o) = I. \tag{IV-6d}$$

Equation (IV-6a) states that the full elastic two-way wave field $\tilde{\vec{Q}}$ at depth level z is found by extrapolating the two-way wave field $\tilde{\vec{Q}}$ from depth level z_o to z and by adding the wave field at z related to all sources between z_o and z. For a further discussion we refer to section III.2.2.

IV.2.2 Elastic one-way wave equations in the wavenumber-frequency domain

Consider the elastic two-way wave equation (IV-5a) in the wavenumber-frequency domain,

$$\frac{\partial \tilde{\vec{Q}}(z)}{\partial z} = \tilde{A}_1(z)\tilde{\vec{Q}}(z) + \tilde{\vec{S}}(z). \tag{IV-7}$$

Define the eigenvalue decomposition of matrix $\tilde{\mathbf{A}}_1(z)$ as

$$\tilde{\mathbf{A}}_1(z) = \tilde{\mathbf{L}}(z)\tilde{\mathbf{\Lambda}}(z)\tilde{\mathbf{L}}^{-1}(z). \qquad \text{(IV-8a)}$$

Define a vector $\tilde{\vec{D}}(z)$ according to

$$\tilde{\vec{Q}}(z) = \tilde{\mathbf{L}}(z)\tilde{\vec{D}}(z), \qquad \text{(IV-8b)}$$

or, equivalently,

$$\tilde{\vec{D}}(z) = \tilde{\mathbf{L}}^{-1}(z)\tilde{\vec{Q}}(z), \qquad \text{(IV-8c)}$$

with

$$\tilde{\vec{D}}(z) = \begin{bmatrix} \tilde{\vec{D}}^{+}(z) \\ \\ \tilde{\vec{D}}^{-}(z) \end{bmatrix}. \qquad \text{(IV-8d)}$$

It will be shown later on in this section that the eigenvector matrix $\tilde{\mathbf{L}}$ can be organized in such a way that vector $\tilde{\vec{D}}^{+}(z)$ contains downgoing P- and S-waves whereas vector $\tilde{\vec{D}}^{-}(z)$ contains upgoing P- and S-waves. Hence, equation (IV-8b) describes *composition* of the elastic two-way wave field $\tilde{\vec{Q}}(z)$ from its downgoing and upgoing constituents $\tilde{\vec{D}}^{+}(z)$ and $\tilde{\vec{D}}^{-}(z)$. Similarly, equation (IV-8c) describes *decomposition* of the elastic two-way wave field into downgoing and upgoing P- and S-waves.
Upon substitution of (IV-8) into (IV-7) we obtain, in analogy with (III-34),

$$\frac{\partial \tilde{\vec{D}}(z)}{\partial z} = \tilde{\mathbf{B}}_1(z)\tilde{\vec{D}}(z) + \tilde{\vec{S}}'(z), \qquad \text{(IV-9a)}$$

where

$$\tilde{\mathbf{B}}_1(z) = \tilde{\mathbf{\Lambda}}(z) - \tilde{\mathbf{L}}^{-1}(z)\,\frac{\partial \tilde{\mathbf{L}}(z)}{\partial z} \qquad \text{(IV-9b)}$$

and

$$\tilde{\vec{S}}'(z) = \tilde{\mathbf{L}}^{-1}(z)\tilde{\vec{S}}(z). \qquad \text{(IV-9c)}$$

This equation represents a coupled system of elastic *one-way wave equations* for downgoing and upgoing P- and S-waves. The general solution of (IV-9) reads, in analogy with (III-35),

$$\tilde{\vec{D}}(z) = \tilde{\mathbf{W}}(z,z_o)\tilde{\vec{D}}(z_o) + \int_{z_o}^{z} \tilde{\mathbf{W}}(z,z')\tilde{\vec{S}}'(z')dz', \qquad \text{(IV-10a)}$$

where

$$\tilde{\mathbf{W}}(z,z_o) = \sum_{m=o}^{\infty} \frac{(z-z_o)^m}{m!} \tilde{\mathbf{B}}_m(z_o), \qquad \text{(IV-10b)}$$

with $\tilde{\mathbf{B}}_m(z_o)$ defined recursively by

$$\tilde{\mathbf{B}}_{m+1}(z_o) = \left. \frac{\partial \tilde{\mathbf{B}}_m(z)}{\partial z} \right|_{z_o} + \tilde{\mathbf{B}}_m(z_o)\tilde{\mathbf{B}}_1(z_o) \qquad \text{(IV-10c)}$$

and

$$\tilde{\mathbf{B}}_o(z_o) = \mathbf{I}. \qquad \text{(IV-10d)}$$

We shall now analyse equations (IV-8), (IV-9) and (IV-10) in more detail. Equation (IV-8a) represents an eigenvalue decomposition of the 6x6 matrix $\tilde{\mathbf{A}}_1(z)$. For the general anisotropic situation it is very difficult to perform this eigenvalue decomposition analytically. In the following we consider transverse isotropic media with symmetry axes in the x-, y- or z-direction (see also section II.4.1). Using equation (II-23) and Table II-1, we may now write for matrix $\tilde{\mathbf{A}}_1$, as defined in (IV-5),

$$\tilde{\mathbf{A}}_1 = \begin{bmatrix} \tilde{\mathbf{a}}_{11} & \tilde{\mathbf{a}}_{12} \\ \tilde{\mathbf{a}}_{21} & \tilde{\mathbf{a}}_{22} \end{bmatrix}, \qquad \text{(IV-11a)}$$

where

$$\tilde{a}_{11} = \begin{bmatrix} 0 & 0 & jk_x \\ 0 & 0 & jk_y \\ \frac{c_{13}}{c_{33}} jk_x & \frac{c_{23}}{c_{33}} jk_y & 0 \end{bmatrix}, \tag{IV-11b}$$

$$\tilde{a}_{12} = \begin{bmatrix} \frac{j\omega}{c_{55}} & 0 & 0 \\ 0 & \frac{j\omega}{c_{44}} & 0 \\ 0 & 0 & \frac{j\omega}{c_{33}} \end{bmatrix}, \tag{IV-11c}$$

$$\tilde{a}_{21} = \begin{bmatrix} [j\omega\rho + \frac{1}{j\omega}(\eta_1 k_x^2 + c_{66} k_y^2)] & \frac{1}{j\omega}\eta_3 k_x k_y & 0 \\ \frac{1}{j\omega}\eta_3 k_x k_y & [j\omega\rho + \frac{1}{j\omega}(c_{66} k_x^2 + \eta_2 k_y^2)] & 0 \\ 0 & 0 & j\omega\rho \end{bmatrix}, \tag{IV-11d}$$

and

$$\tilde{a}_{22} = \tilde{a}_{11}^T = \begin{bmatrix} 0 & 0 & \frac{c_{13}}{c_{33}} jk_x \\ 0 & 0 & \frac{c_{23}}{c_{33}} jk_y \\ jk_x & jk_y & 0 \end{bmatrix}, \tag{IV-11e}$$

with

$$\eta_1 = c_{11} - c_{13}^2/c_{33} , \tag{IV-11f}$$

$$\eta_2 = c_{22} - c_{23}^2/c_{33} \tag{IV-11g}$$

and

$$\eta_3 = c_{12} + c_{66} - c_{13}c_{23}/c_{33}. \tag{IV-11h}$$

We define a permutation of matrix $\tilde{\mathbf{A}}_1$ according to

$$\tilde{\mathbf{A}}_1^p = \begin{bmatrix} \mathbf{0} & \tilde{\mathbf{a}}_{12}^p \\ \tilde{\mathbf{a}}_{21}^p & \mathbf{0} \end{bmatrix}, \tag{IV-12a}$$

where

$$\tilde{\mathbf{a}}_{12}^p = \begin{bmatrix} \frac{j\omega}{c_{33}} & \frac{c_{13}}{c_{33}} jk_x & \frac{c_{23}}{c_{33}} jk_y \\ \frac{c_{13}}{c_{33}} jk_x & [j\omega\rho + \frac{1}{j\omega}(\eta_1 k_x^2 + c_{66}k_y^2)] & \frac{1}{j\omega}\eta_3 k_x k_y \\ \frac{c_{23}}{c_{33}} jk_y & \frac{1}{j\omega}\eta_3 k_x k_y & [j\omega\rho + \frac{1}{j\omega}(c_{66}k_x^2 + \eta_2 k_y^2) \end{bmatrix}, \tag{IV-12b}$$

$$\tilde{\mathbf{a}}_{21}^p = \begin{bmatrix} j\omega\rho & jk_x & jk_y \\ jk_x & j\omega/c_{55} & 0 \\ jk_y & 0 & j\omega/c_{44} \end{bmatrix} \tag{IV-12c}$$

and

$$\mathbf{0} = \begin{bmatrix} 0 & 0 & 0 \\ 0 & 0 & 0 \\ 0 & 0 & 0 \end{bmatrix}. \tag{IV-12d}$$

Note that $\tilde{\mathbf{A}}_1^p$ is related to $\tilde{\mathbf{A}}_1$ according to

$$\tilde{\mathbf{A}}_1^p = \mathbf{p}_1 \tilde{\mathbf{A}}_1 \mathbf{p}_1^{-1}, \tag{IV-13a}$$

where

$$\mathbf{p}_1 = \begin{bmatrix} 0&0&1&0&0&0 \\ 0&0&0&1&0&0 \\ 0&0&0&0&1&0 \\ 0&0&0&0&0&1 \\ 1&0&0&0&0&0 \\ 0&1&0&0&0&0 \end{bmatrix} \tag{IV-13b}$$

and

$$\mathbf{p}_1^{-1} = \mathbf{p}_1^T. \tag{IV-13c}$$

We define a permutation of matrix $\tilde{\mathbf{L}}$ according to

$$\tilde{\mathbf{L}}^p = \mathbf{p}_1 \tilde{\mathbf{L}}\, \mathbf{p}_2, \tag{IV-13d}$$

where

$$\mathbf{p}_2 = \begin{bmatrix} 1&0&0&0&0&0 \\ 0&1&0&0&0&0 \\ 0&0&1&0&0&0 \\ 0&0&0&-1&0&0 \\ 0&0&0&0&1&0 \\ 0&0&0&0&0&1 \end{bmatrix} . \tag{IV-13e}$$

With these definitions we may replace the eigenvalue decomposition (IV-8a) by[1)]

$$\tilde{\mathbf{A}}_1^p = \tilde{\mathbf{L}}^p \tilde{\mathbf{\Lambda}} (\tilde{\mathbf{L}}^p)^{-1}. \tag{IV-14}$$

Due to the simple structure of matrix $\tilde{\mathbf{A}}_1^p$, this eigenvalue problem is very well manageable. Define

$$\tilde{\mathbf{L}}^p = \begin{bmatrix} \tilde{\mathbf{L}}_1^p & \tilde{\mathbf{L}}_1^p \\ \tilde{\mathbf{L}}_2^p & -\tilde{\mathbf{L}}_2^p \end{bmatrix} \tag{IV-15a}$$

and

$$\tilde{\mathbf{\Lambda}} = \begin{bmatrix} \tilde{\mathbf{\Lambda}}_1 & \mathbf{0} \\ \mathbf{0} & -\tilde{\mathbf{\Lambda}}_1 \end{bmatrix} . \tag{IV-15b}$$

Then

---

1) This could also be accomplished when $\mathbf{p}_2$ in (IV-13d) were omitted. However, this would lead to a less elegant definition of $\vec{\tilde{D}}^-$ in (IV-30c).

$$(\tilde{\mathbf{L}}^p)^{-1} = \frac{1}{2}\begin{bmatrix} (\tilde{\mathbf{L}}_1^p)^{-1} & (\tilde{\mathbf{L}}_2^p)^{-1} \\ (\tilde{\mathbf{L}}_1^p)^{-1} & -(\tilde{\mathbf{L}}_2^p)^{-1} \end{bmatrix} \tag{IV-15c}$$

and

$$\tilde{\mathbf{A}}_1^p = \begin{bmatrix} 0 & \tilde{\mathbf{a}}_{12}^p \\ \tilde{\mathbf{a}}_{21}^p & 0 \end{bmatrix} = \begin{bmatrix} 0 & \tilde{\mathbf{L}}_1^p\tilde{\mathbf{\Lambda}}_1(\tilde{\mathbf{L}}_2^p)^{-1} \\ \tilde{\mathbf{L}}_2^p\tilde{\mathbf{\Lambda}}_1(\tilde{\mathbf{L}}_1^p)^{-1} & 0 \end{bmatrix}, \tag{IV-15d}$$

or

$$\tilde{\mathbf{a}}_{12}^p\,\tilde{\mathbf{a}}_{21}^p = \tilde{\mathbf{L}}_1^p\tilde{\mathbf{\Lambda}}_1^2(\tilde{\mathbf{L}}_1^p)^{-1} \tag{IV-16a}$$

and

$$\tilde{\mathbf{a}}_{21}^p\,\tilde{\mathbf{a}}_{12}^p = \tilde{\mathbf{L}}_2^p\tilde{\mathbf{\Lambda}}_1^2(\tilde{\mathbf{L}}_2^p)^{-1}. \tag{IV-16b}$$

Hence, for transverse isotropic media we have reduced the eigenvalue problem (IV-8a) for the 6 x 6 matrix $\tilde{\mathbf{A}}_1$ into two eigenvalue problems (IV-16a) and (IV-16b) for 3 x 3 matrices. The same procedure was followed by Ursin (1983) for pure isotropic media.

The eigenvalues ς (i.e., the elements of diagonal matrix $\tilde{\mathbf{\Lambda}}_1$) can be found by solving

$$\det[\tilde{\mathbf{a}}_{12}^p\,\tilde{\mathbf{a}}_{21}^p - \varsigma^2\mathbf{I}] = 0. \tag{IV-17a}$$

Subsequently, the eigenvectors $\vec{\ell}_1$ and $\vec{\ell}_2$ (i.e., the columns of the matrices $\tilde{\mathbf{L}}_1^p$ and $\tilde{\mathbf{L}}_2^p$) can be found by solving

$$(\tilde{\mathbf{a}}_{12}^p\,\tilde{\mathbf{a}}_{21}^p)\vec{\ell}_1 = \varsigma^2\vec{\ell}_1 \tag{IV-17b}$$

and

$$\left(\tilde{\mathbf{a}}_{21}^{p}\,\tilde{\mathbf{a}}_{12}^{p}\right)\vec{\ell}_2 = \zeta^2\,\vec{\ell}_2 \,, \tag{IV-17c}$$

respectively. First we consider the situation of pure isotropy. According to (IV-12b), (IV-12c) and Table II-1, we may write

$$\tilde{a}_{12}^{p} = \begin{bmatrix} \frac{j\omega}{\lambda+2\mu} & \left(\frac{\lambda}{\lambda+2\mu}\right) jk_x & \left(\frac{\lambda}{\lambda+2\mu}\right) jk_y \\ \left(\frac{\lambda}{\lambda+2\mu}\right) jk_x & \left[j\omega\rho+\frac{1}{j\omega}\left(\eta_1 k_x^2+\mu k_y^2\right)\right] & \frac{1}{j\omega}\eta_3 k_x k_y \\ \left(\frac{\lambda}{\lambda+2\mu}\right) jk_y & \frac{1}{j\omega}\eta_3 k_x k_y & \left[j\omega\rho+\frac{1}{j\omega}\left(\mu k_x^2+\eta_2 k_y^2\right)\right] \end{bmatrix}, \tag{IV-18a}$$

$$\tilde{a}_{21}^{p} = \begin{bmatrix} j\omega\rho & jk_x & jk_y \\ jk_x & j\omega/\mu & 0 \\ jk_y & 0 & j\omega/\mu \end{bmatrix}, \tag{IV-18b}$$

with

$$\eta_1 = \eta_2 = 4\mu\left(\frac{\lambda+\mu}{\lambda+2\mu}\right) \tag{IV-18c}$$

and

$$\eta_3 = \mu\left(\frac{3\lambda+2\mu}{\lambda+2\mu}\right) . \tag{IV-18d}$$

Substituting these expressions into (IV-17a) yields for the eigenvalues

$$\zeta_1^2 = -\left(\frac{\rho\omega^2}{\lambda+2\mu} - k_x^2 - k_y^2\right) \tag{IV-19a}$$

and

$$\zeta_2^2 = \zeta_3^2 = -\left(\frac{\rho\omega^2}{\mu} - k_x^2 - k_y^2\right). \tag{IV-19b}$$

Hence, we may write for the diagonal matrix $\tilde{\mathbf{\Lambda}}$, in analogy with (III-32c),

$$\tilde{\mathbf{\Lambda}} = \begin{bmatrix} \tilde{\mathbf{\Lambda}}_1 & 0 \\ 0 & -\tilde{\mathbf{\Lambda}}_1 \end{bmatrix}, \tag{IV-20a}$$

where

$$\tilde{\mathbf{\Lambda}}_1 = \begin{bmatrix} -jk_{z,p} & 0 & 0 \\ 0 & -jk_{z,s} & 0 \\ 0 & 0 & -jk_{z,s} \end{bmatrix}, \tag{IV-20b}$$

with

$$k_{z,p}^2 \triangleq -\varsigma_1^2 = k_p^2 - k_x^2 - k_y^2 , \tag{IV-20c}$$

$$k_{z,s}^2 \triangleq -\varsigma_2^2 = k_s^2 - k_x^2 - k_y^2 , \tag{IV-20d}$$

$$k_p^2 \triangleq \omega^2/c_p^2 = \rho\omega^2/(\lambda+2\mu) \tag{IV-20e}$$

and

$$k_s^2 \triangleq \omega^2/c_s^2 = \rho\omega^2/\mu, \tag{IV-20f}$$

where c_p and c_s are the propagation velocities for P- and S-waves, respectively.

Next, we solve (IV-17b) and (IV-17c) for the eigenvectors $\vec{\ell}_1$ and $\vec{\ell}_2$. Note that arbitrary scaling factors can be applied to these vectors. In the literature the scaling is generally chosen such that $(\tilde{\mathbf{L}}_1^p)^T = (\tilde{\mathbf{L}}_2^p)^{-1}$. However, we choose a different scaling for reasons which will be explained later on. Our eigenvector matrices read

$$\tilde{\mathbf{L}}_1^p = \frac{\mu}{\omega^2 \rho} \begin{bmatrix} \frac{\omega k_{z,p}}{\mu} & \frac{-\omega k_y}{\mu} & \frac{\omega k_x}{\mu} \\ -2k_x k_{z,p} & 2k_x k_y & \left(k_s^2 - 2k_x^2\right) \\ -2k_y k_{z,p} & -\left(k_s^2 - 2k_y^2\right) & -2k_x k_y \end{bmatrix} \tag{IV-21a}$$

and

$$\tilde{\mathbf{L}}_2^p = \frac{\mu}{\omega^2 \rho} \begin{bmatrix} -\left(k_s^2 - 2k_x^2 - 2k_y^2\right) & -2k_y k_{z,s} & -2k_x k_{z,s} \\ \frac{\omega k_x}{\mu} & \frac{-\omega k_x k_y}{\mu k_{z,s}} & \frac{-\omega\left(k_s^2 - k_x^2\right)}{\mu k_{z,s}} \\ \frac{\omega k_y}{\mu} & \frac{\omega\left(k_s^2 - k_y^2\right)}{\mu k_{z,s}} & \frac{\omega k_x k_y}{\mu k_{z,s}} \end{bmatrix} . \tag{IV-21b}$$

The inverse versions of these matrices read

$$\left(\tilde{\mathbf{L}}_1^p\right)^{-1} = \begin{bmatrix} \frac{\mu\left(k_s^2 - 2k_x^2 - 2k_y^2\right)}{\omega k_{z,p}} & \frac{-k_x}{k_{z,p}} & \frac{-k_y}{k_{z,p}} \\ \frac{-2\mu k_y}{\omega} & 0 & -1 \\ \frac{2\mu k_x}{\omega} & 1 & 0 \end{bmatrix} \tag{IV-22a}$$

and

$$\left(\tilde{\mathbf{L}}_2^p\right)^{-1} = \begin{bmatrix} -1 & \frac{2\mu k_x}{\omega} & \frac{2\mu k_y}{\omega} \\ \frac{k_y}{k_{z,s}} & \frac{-\mu k_x k_y}{\omega k_{z,s}} & \frac{\mu\left(k_s^2 - k_x^2 - 2k_y^2\right)}{\omega k_{z,s}} \\ \frac{-k_x}{k_{z,s}} & \frac{-\mu\left(k_s^2 - 2k_x^2 - k_y^2\right)}{\omega k_{z,s}} & \frac{\mu k_x k_y}{\omega k_{z,s}} \end{bmatrix} . \tag{IV-22b}$$

Now that we have solved the eigenvalue problem (IV-14) for the matrix $\tilde{A}_1^p$, we return to the original eigenvalue problem (IV-8) for the matrix $\tilde{\mathbf{A}}_1$. According to (IV-13d) we may write

$$\tilde{L} = p_1^{-1}\,\tilde{L}^p p_2^{-1} \tag{IV-23a}$$

and, consequently,

$$\tilde{L}^{-1} = p_2\left(\tilde{L}^p\right)^{-1} p_1. \tag{IV-23b}$$

Substituting (IV-13b), (IV-13e), (IV-15a), (IV-15c), (IV-21) and (IV-22) yields

$$\tilde{L} = \begin{bmatrix} \tilde{L}_1^+ & \tilde{L}_1^- \\ \tilde{L}_2^+ & \tilde{L}_2^- \end{bmatrix} \tag{IV-24a}$$

and

$$\tilde{L}^{-1} = \begin{bmatrix} \tilde{N}_1^+ & \tilde{N}_2^+ \\ \tilde{N}_1^- & \tilde{N}_2^- \end{bmatrix}, \tag{IV-24b}$$

where

$$\tilde{L}_1^+ = \frac{1}{\omega\rho}\begin{bmatrix} k_x & \mp\dfrac{k_x k_y}{k_{z,s}} & \mp\dfrac{(k_s^2-k_x^2)}{k_{z,s}} \\ k_y & \pm\dfrac{(k_s^2-k_y^2)}{k_{z,s}} & \pm\dfrac{k_x k_y}{k_{z,s}} \\ \pm k_{z,p} & -k_y & k_x \end{bmatrix}, \tag{IV-25a}$$

$$\tilde{L}_2^{\pm} = \frac{\mu}{\omega^2 \rho} \begin{bmatrix} \mp 2k_x k_{z,p} & 2k_x k_y & \left(k_s^2 - 2k_x^2\right) \\ \mp 2k_y k_{z,p} & -\left(k_s^2 - 2k_y^2\right) & -2k_x k_y \\ -\left(k_s^2 - 2k_x^2 - 2k_y^2\right) & \pm 2k_y k_{z,s} & \mp 2k_x k_{z,s} \end{bmatrix}, \tag{IV-25b}$$

$$\tilde{N}_1^{\pm} = \frac{\mu}{2\omega} \begin{bmatrix} 2k_x & 2k_y & \pm \dfrac{\left(k_s^2 - 2k_x^2 - 2k_y^2\right)}{k_{z,p}} \\ \mp \dfrac{k_x k_y}{k_{z,s}} & \pm \dfrac{\left(k_s^2 - k_x^2 - 2k_y^2\right)}{k_{z,s}} & -2k_y \\ \mp \dfrac{\left(k_s^2 - 2k_x^2 - k_y^2\right)}{k_{z,s}} & \pm \dfrac{k_x k_y}{k_{z,s}} & 2k_x \end{bmatrix} \tag{IV-26a}$$

and

$$\tilde{N}_2^{\pm} = \frac{1}{2} \begin{bmatrix} \mp \dfrac{k_x}{k_{z,p}} & \mp \dfrac{k_y}{k_{z,p}} & -1 \\ 0 & -1 & \pm \dfrac{k_y}{k_{z,s}} \\ 1 & 0 & \mp \dfrac{k_x}{k_{z,s}} \end{bmatrix}. \tag{IV-26b}$$

We shall now justify our unconventional scaling of the eigenvector matrices. Consider equation (II-31)

$$\partial_t \vec{v}(\vec{r},t) \hat{=} -\frac{1}{\rho} \left[\nabla\phi(\vec{r},t) + \nabla\times\vec{\psi}(\vec{r},t)\right], \tag{IV-27a}$$

with

$$\nabla.\vec{\psi}(\vec{r},t) \hat{=} o, \tag{IV-27b}$$

where ϕ and $\vec{\psi}$ are the potentials for P- and S-waves in homogeneous isotropic source-free solids. In the wavenumber-frequency domain these equations read

$$\tilde{\vec{V}} \triangleq \frac{-1}{j\omega\rho} \begin{bmatrix} -jk_x\tilde{\Phi}-jk_y\tilde{\Psi}_z-\partial_z\tilde{\Psi}_y \\ -jk_y\tilde{\Phi}+jk_x\tilde{\Psi}_z+\partial_z\tilde{\Psi}_x \\ \partial_z\tilde{\Phi}-jk_x\tilde{\Psi}_y+jk_y\tilde{\Psi}_x \end{bmatrix}, \tag{IV-28a}$$

with

$$-jk_x\tilde{\Psi}_x - jk_y\tilde{\Psi}_y + \partial_z\tilde{\Psi}_z \triangleq 0, \tag{IV-28b}$$

where $\tilde{\Psi}_x$, $\tilde{\Psi}_y$ and $\tilde{\Psi}_z$ are the components of vector $\tilde{\vec{\Psi}}$.

Define

$$\tilde{\Phi} = \tilde{\Phi}^+ + \tilde{\Phi}^- \tag{IV-29a}$$

and

$$\tilde{\vec{\Psi}} = \tilde{\vec{\Psi}}^+ + \tilde{\vec{\Psi}}^-, \tag{IV-29b}$$

where, in analogy with (II-72) or (III-45),

$$\frac{\partial\tilde{\Phi}^{\pm}}{\partial z} = \mp jk_{z,p}\tilde{\Phi}^{\pm} \tag{IV-29c}$$

and

$$\frac{\partial\tilde{\vec{\Psi}}^{\pm}}{\partial z} = \mp jk_{z,s}\tilde{\vec{\Psi}}^{\pm}. \tag{IV-29d}$$

Substituting (IV-29) into (IV-28) and eliminating $\tilde{\Psi}_z^{\pm}$ yields

$$\vec{\tilde{V}} = \tilde{L}_1^+ \, \vec{\tilde{D}}^+ + \tilde{L}_1^- \, \vec{\tilde{D}}^-, \tag{IV-30a}$$

where

$$\vec{\tilde{D}}^+ \triangleq \begin{bmatrix} \tilde{\Phi}^+ \\ \tilde{\Psi}_x^+ \\ \tilde{\Psi}_y^+ \end{bmatrix}, \qquad \vec{\tilde{D}}^- \triangleq \begin{bmatrix} \tilde{\Phi}^- \\ \tilde{\Psi}_x^- \\ \tilde{\Psi}_y^- \end{bmatrix} \tag{IV-30b,c}$$

and where $\tilde{L}_1^+$ and $\tilde{L}_1^-$ are defined by (IV-25a). Hence, our eigenvector matrices have been scaled in such a way that vectors $\vec{\tilde{D}}^+$ and $\vec{\tilde{D}}^-$ contain the *potentials* for downgoing and upgoing P- and S-waves, which were defined in section II.2.5. For this analysis we considered the homogeneous isotropic *source-free* situation. However, it can be shown that also for the situation with P- and S-wave sources, as defined in section II.3.1, vectors $\vec{\tilde{D}}^+$ and $\vec{\tilde{D}}^-$ contain the potentials for downgoing and upgoing P- and S-waves. A further discussion is beyond the scope of this book.

We summarize our results for the isotropic situation. *Composition* of the elastic two-way wave field from downgoing and upgoing P- and S-wave potentials reads, according to (IV-8b),

$$\vec{\tilde{Q}} = \tilde{L}\vec{\tilde{D}}, \tag{IV-31a}$$

or, upon substitution of (IV-5b), (IV-8d) and (IV-24a),

$$\vec{\tilde{V}} = \tilde{L}_1^+\vec{\tilde{D}}^+ + \tilde{L}_1^-\vec{\tilde{D}}^- \tag{IV-31b}$$

and

$$\vec{\tilde{\tau}}_z = \tilde{L}_2^+\vec{\tilde{D}}^+ + \tilde{L}_2^-\vec{\tilde{D}}^-, \tag{IV-31c}$$

with $\tilde{L}_1^{\pm}$ and $\tilde{L}_2^{\pm}$ defined by (IV-25a) and (IV-25b), respectively.

Decomposition of the elastic two-way wave field into downgoing and upgoing P- and S-wave potentials reads, according to (IV-8c),

$$\vec{\tilde{D}} = \tilde{L}^{-1}\vec{\tilde{Q}}, \tag{IV-32a}$$

or, upon substitution of (IV-5b), (IV-8d) and (IV-24b),

$$\vec{\tilde{D}}^{+} = \tilde{N}_1^{+}\vec{\tilde{V}} + \tilde{N}_2^{+}\vec{\tilde{\tau}}_z \tag{IV-32b}$$

and

$$\vec{\tilde{D}}^{-} = \tilde{N}_1^{-}\vec{\tilde{V}} + \tilde{N}_2^{-}\vec{\tilde{\tau}}_z\ , \tag{IV-32c}$$

with $\tilde{N}_1^{\pm}$ and $\tilde{N}_2^{\pm}$ defined by (IV-26a) and (IV-26b), respectively. Note that the decomposition operators are related to the composition operators, according to

$$\tilde{N}_1^{\pm} = \left(\tilde{L}_1^{\pm} - \tilde{L}_1^{\mp}(\tilde{L}_2^{\mp})^{-1}\tilde{L}_2^{\pm}\right)^{-1} \tag{IV-32d}$$

and

$$\tilde{N}_2^{\pm} = \left(\tilde{L}_2^{\pm} - \tilde{L}_2^{\mp}(\tilde{L}_1^{\mp})^{-1}\tilde{L}_1^{\pm}\right)^{-1}. \tag{IV-32e}$$

The decomposed wave fields satisfy, according to (IV-9a),

$$\frac{\partial\vec{\tilde{D}}}{\partial z} = \tilde{B}_1\vec{\tilde{D}} + \vec{\tilde{S}}'\ , \tag{IV-33a}$$

or, upon substitution of (IV-5h), (IV-8d), (IV-9b), (IV-9c), (IV-20a), (IV-24a) and (IV-24b),

$$\frac{\partial\vec{\tilde{D}}^{+}}{\partial z} = \tilde{\Lambda}_1\vec{\tilde{D}}^{+} - \tilde{N}_\alpha^{+}\left[\frac{\partial\tilde{L}_\alpha^{+}}{\partial z}\vec{\tilde{D}}^{+} + \frac{\partial\tilde{L}_\alpha^{-}}{\partial z}\vec{\tilde{D}}^{-}\right] + \vec{\tilde{S}}^{+} \tag{IV-33b}$$

and

$$\frac{\partial\vec{\tilde{D}}^{-}}{\partial z} = -\tilde{\Lambda}_1\vec{\tilde{D}}^{-} - \tilde{N}_\alpha^{-}\left[\frac{\partial\tilde{L}_\alpha^{+}}{\partial z}\vec{\tilde{D}}^{+} + \frac{\partial\tilde{L}_\alpha^{-}}{\partial z}\vec{\tilde{D}}^{-}\right] - \vec{\tilde{S}}^{-}, \tag{IV-33c}$$

where

$$\vec{\tilde{S}}^{\pm} = \pm\tilde{N}_\alpha^{\pm}\vec{\tilde{S}}_\alpha \tag{IV-33d}$$

and with $\tilde{\Lambda}_1$ defined by (IV-20b).

Equation (IV-33) represents a coupled system of *one-way wave equations* for downgoing and upgoing elastic waves $\tilde{\vec{D}}^+$ and $\tilde{\vec{D}}^-$, respectively. Note that these equations are a generalization of the elastic one-way wave equations for plane waves (II-72) that were derived in section II.4.3. $\tilde{\vec{S}}^+$ and $\tilde{\vec{S}}^-$ are the one-way representations of the source distribution. Apparently the downgoing and upgoing waves are coupled due to the vertical variations of the medium parameters, which is expressed by the terms $\partial\tilde{\mathbf{L}}_\alpha^-/\partial z$ in (IV-33b) and $\partial\tilde{\mathbf{L}}_\alpha^+/\partial z$ in (IV-33c).

It is common use to neglect $\tilde{\vec{D}}^-$ with respect to $\tilde{\vec{D}}^+$ in (IV-33b) for downward propagation and to neglect $\tilde{\vec{D}}^+$ with respect to $\tilde{\vec{D}}^-$ in (IV-33c) for upward propagation. This means that in both equations *multiple reflections* are neglected. Hence, primary elastic waves fulfill the decoupled one-way wave equations

$$\frac{\partial\tilde{\vec{D}}^+}{\partial z} \approx \left[\tilde{\mathbf{\Lambda}}_1 - \tilde{\mathbf{N}}_\alpha^+ \frac{\partial\tilde{\mathbf{L}}_\alpha^+}{\partial z}\right]\tilde{\vec{D}}^+ + \tilde{\vec{S}}^+ \tag{IV-34a}$$

and

$$\frac{\partial\tilde{\vec{D}}^-}{\partial z} \approx \left[-\tilde{\mathbf{\Lambda}}_1 - \tilde{\mathbf{N}}_\alpha^- \frac{\partial\tilde{\mathbf{L}}_\alpha^-}{\partial z}\right]\tilde{\vec{D}}^- - \tilde{\vec{S}}^- \quad . \tag{IV-34b}$$

Note that the terms $\tilde{\mathbf{N}}_\alpha^\pm \partial\tilde{\mathbf{L}}_\alpha^\pm/\partial z$ account for all amplitude effects (including *conversion*) during propagation of the primary waves.

For a source-free homogeneous region these equations simplify to

$$\frac{\partial\tilde{\vec{D}}^+}{\partial z} = \tilde{\mathbf{\Lambda}}_1\tilde{\vec{D}}^+ \tag{IV-35a}$$

and

$$\frac{\partial\tilde{\vec{D}}^-}{\partial z} = -\tilde{\mathbf{\Lambda}}_1\tilde{\vec{D}}^-, \tag{IV-35b}$$

or

$$\frac{\partial}{\partial z}\begin{bmatrix}\tilde{\Phi}^+ \\ \tilde{\Psi}_x^+ \\ \tilde{\Psi}_y^+\end{bmatrix} = \begin{bmatrix}-jk_{z,p} & 0 & 0 \\ 0 & -jk_{z,s} & 0 \\ 0 & 0 & -jk_{z,s}\end{bmatrix}\begin{bmatrix}\tilde{\Phi}^+ \\ \tilde{\Psi}_x^+ \\ \tilde{\Psi}_y^+\end{bmatrix} \tag{IV-35c}$$

and

$$\frac{\partial}{\partial z}\begin{bmatrix}\tilde{\Phi}^- \\ \tilde{\Psi}_x^- \\ \tilde{\Psi}_y^-\end{bmatrix} = \begin{bmatrix} +jk_{z,p} & 0 & 0 \\ 0 & +jk_{z,s} & 0 \\ 0 & 0 & +jk_{z,s}\end{bmatrix}\begin{bmatrix}\tilde{\Phi}^- \\ \tilde{\Psi}_x^- \\ \tilde{\Psi}_y^-\end{bmatrix}, \tag{IV-35d}$$

where $\tilde{\Phi}^+$ and $\tilde{\Phi}^-$ represent the potentials for downgoing and upgoing P-waves and where $\tilde{\Psi}_x^+$, $\tilde{\Psi}_y^+$ and $\tilde{\Psi}_x^-$, $\tilde{\Psi}_y^-$ represent the potentials for downgoing and upgoing S-waves. The solutions of these equations read[1)]

$$\begin{bmatrix}\tilde{\Phi}^+(z) \\ \tilde{\Psi}_x^+(z) \\ \tilde{\Psi}_y^+(z)\end{bmatrix} = \begin{bmatrix} \tilde{W}^+_{\phi,\phi}(z,z_o) & 0 & 0 \\ 0 & \tilde{W}^+_{\psi_x,\psi_x}(z,z_o) & 0 \\ 0 & 0 & \tilde{W}^+_{\psi_y,\psi_y}(z,z_o)\end{bmatrix}\begin{bmatrix}\tilde{\Phi}^+(z_o) \\ \tilde{\Psi}_x^+(z_o) \\ \tilde{\Psi}_y^+(z_o)\end{bmatrix} \tag{IV-36a}$$

and

$$\begin{bmatrix}\tilde{\Phi}^-(z) \\ \tilde{\Psi}_x^-(z) \\ \tilde{\Psi}_y^-(z)\end{bmatrix} = \begin{bmatrix} \tilde{W}^-_{\phi,\phi}(z,z_o) & 0 & 0 \\ 0 & \tilde{W}^-_{\psi_x,\psi_x}(z,z_o) & 0 \\ 0 & 0 & \tilde{W}^-_{\psi_y,\psi_y}(z,z_o)\end{bmatrix}\begin{bmatrix}\tilde{\Phi}^-(z_o) \\ \tilde{\Psi}_x^-(z_o) \\ \tilde{\Psi}_y^-(z_o)\end{bmatrix}, \tag{IV-36b}$$

where

$$\tilde{W}^+_{\phi,\phi}(z,z_o) = \exp(-jk_{z,p}\Delta z), \tag{IV-36c}$$

$$\tilde{W}^+_{\psi_x,\psi_x}(z,z_o) = \tilde{W}^+_{\psi_y,\psi_y}(z,z_o) = \exp(-jk_{z,s}\Delta z), \tag{IV-36d}$$

$$\tilde{W}^-_{\phi,\phi}(z,z_o) = \exp(jk_{z,p}\Delta z), \tag{IV-36e}$$

$$\tilde{W}^-_{\psi_x,\psi_x}(z,z_o) = \tilde{W}^-_{\psi_y,\psi_y}(z,z_o) = \exp(jk_{z,s}\Delta z) \tag{IV-36f}$$

and

---

1) The sub-scripts ϕ, ψ_x and ψ_y refer to the potentials $\tilde{\Phi}$, $\tilde{\Psi}_x$ and $\tilde{\Psi}_y$, respectively.

$$\Delta z = z - z_o . \tag{IV-36g}$$

We finalize this section by deriving the eigenvalues ζ for the situation of transverse isotropy with a vertical axis of symmetry. According to (IV-12b), (IV-12c) and Table II-1, we may write

$$\tilde{a}^p_{12} = \begin{bmatrix} \dfrac{j\omega}{\lambda_{//}+2\mu_{//}} & \left(\dfrac{\nu}{\lambda_{//}+2\mu_{//}}\right) jk_x & \left(\dfrac{\nu}{\lambda_{//}+2\mu_{//}}\right) jk_y \\ \left(\dfrac{\nu}{\lambda_{//}+2\mu_{//}}\right) jk_x & \left[j\omega\rho + \dfrac{1}{j\omega}(\eta_1 k_x^2 + \mu_\perp k_y^2)\right] & \dfrac{1}{j\omega}\eta_3 k_x k_y \\ \left(\dfrac{\nu}{\lambda_{//}+2\mu_{//}}\right) jk_y & \dfrac{1}{j\omega}\eta_3 k_x k_y & \left[j\omega\rho + \dfrac{1}{j\omega}(\mu_\perp k_x^2 + \eta_2 k_y^2)\right] \end{bmatrix}, \tag{IV-37a}$$

$$\tilde{a}^p_{21} = \begin{bmatrix} j\omega\rho & jk_x & jk_y \\ jk_x & j\omega/\mu_{//} & 0 \\ jk_y & 0 & j\omega/\mu_{//} \end{bmatrix}, \tag{IV-37b}$$

with

$$\eta_1 = \eta_2 = \lambda_\perp + 2\mu_\perp - \nu^2/(\lambda_{//}+2\mu_{//}) \tag{IV-37c}$$

and

$$\eta_3 = \lambda_\perp + \mu_\perp - \nu^2/(\lambda_{//}+2\mu_{//}). \tag{IV-37d}$$

Substituting these expressions into (IV-17a) yields for the eigenvalues

$$\zeta_1^2 = -\left[\omega^2\rho\gamma_1 - \frac{\mu_\perp}{\mu_{//}}(1+\epsilon_1)k_r^2 - \sqrt{(\omega^2\rho\gamma_2)^2 + \omega^2\rho\epsilon_2 k_r^2 + \epsilon_3 k_r^4}\right], \tag{IV-38a}$$

$$\zeta_2^2 = -\left[\omega^2\rho\gamma_1 - \frac{\mu_\perp}{\mu_{//}}(1+\epsilon_1)k_r^2 + \sqrt{(\omega^2\rho\gamma_2)^2 + \omega^2\rho\epsilon_2 k_r^2 + \epsilon_3 k_r^4}\right] \tag{IV-38b}$$

and

$$\zeta_3^2 = -\left[\frac{\omega^2\rho}{\mu_{//}} - \frac{\mu_\perp}{\mu_{//}} k_r^2\right] , \qquad \text{(IV-38c)}$$

where

$$k_r^2 = k_x^2 + k_y^2 \qquad \text{(IV-38d)}$$

and where γ_1, γ_2, ϵ_1, ϵ_2 and ϵ_3 are defined as in equation (II-53). Note that the expressions for ζ_1, ζ_2 and ζ_3 exhibit a high degree of similarity with the expressions for the 3-D slowness surfaces (II-53a), (II-53b) and (II-53c) for qP-waves, qSV-waves and SH-waves, respectively.

Finally, note that in a source-free homogeneous transverse isotropic medium (with a vertical axis of symmetry), the one-way wave equations read, in analogy with (IV-35),

$$\frac{\partial}{\partial z}\begin{bmatrix}\tilde{D}_1^+\\ \tilde{D}_2^+\\ \tilde{D}_3^+\end{bmatrix} = \begin{bmatrix}-jk_{z,P} & 0 & 0\\ 0 & -jk_{z,SV} & 0\\ 0 & 0 & -jk_{z,SH}\end{bmatrix}\begin{bmatrix}\tilde{D}_1^+\\ \tilde{D}_2^+\\ \tilde{D}_3^+\end{bmatrix} , \qquad \text{(IV-39a)}$$

and

$$\frac{\partial}{\partial z}\begin{bmatrix}\tilde{D}_1^-\\ \tilde{D}_2^-\\ \tilde{D}_3^-\end{bmatrix} = \begin{bmatrix}jk_{z,P} & 0 & 0\\ 0 & jk_{z,SV} & 0\\ 0 & 0 & jk_{z,SH}\end{bmatrix}\begin{bmatrix}\tilde{D}_1^-\\ \tilde{D}_2^-\\ \tilde{D}_3^-\end{bmatrix} , \qquad \text{(IV-39b)}$$

where

$$k_{z,P}^2 \triangleq -\zeta_1^2, \qquad \text{(IV-39c)}$$

$$k^2_{z,SV} \hat{=} -\zeta^2_2 \tag{IV-39d}$$

and

$$k^2_{z,SH} \hat{=} -\zeta^2_3 , \tag{IV-39e}$$

with ζ_1, ζ_2 and ζ_3 defined in (IV-38). In (IV-39a) the wave functions $\tilde{D}^+_1$, $\tilde{D}^+_2$ and $\tilde{D}^+_3$ represent *downgoing* qP-waves, qSV-waves and SH-waves, respectively. Similarly, in (IV-39b) the wave functions $\tilde{D}^-_1$, $\tilde{D}^-_2$ and $\tilde{D}^-_3$ represent *upgoing* qP-waves, qSV-waves and SH-waves, respectively.

IV.2.3 Elastic one-way wave fields at interfaces in the wavenumber-frequency domain

The general solution (IV-10) of the elastic one-way wave equations breaks down whenever the medium contains "interfaces" in the interval (z,z_0), see Figure III-1. We derive the boundary conditions for an interface at $z=z_1$. Just above this interface the wave vector $\tilde{\vec{D}}(z)$ is related to the wave vector $\tilde{\vec{Q}}(z)$ according to

$$\lim_{\epsilon\downarrow o} \tilde{\vec{Q}}(z_1-\epsilon) = \lim_{\epsilon\downarrow o} \left[\tilde{\mathbf{L}}(z_1-\epsilon)\tilde{\vec{D}}(z_1-\epsilon)\right]. \tag{IV-40a}$$

Similarly, just below the interface,

$$\lim_{\epsilon\downarrow o} \tilde{\vec{Q}}(z_1+\epsilon) = \lim_{\epsilon\downarrow o} \left[\tilde{\mathbf{L}}(z_1+\epsilon)\tilde{\vec{D}}(z_1+\epsilon)\right]. \tag{IV-40b}$$

The wave vector $\tilde{\vec{Q}}(z)$ contains the particle velocity vector $\tilde{\vec{V}}(z)$ and the traction vector $\tilde{\vec{\tau}}_z(z)$ which are continuous at z_1. The wave vector $\tilde{\vec{D}}(z)$ contains the downgoing P- and S-wave vector $\tilde{\vec{D}}^+(z)$ and the upgoing P- and S-wave vector $\tilde{\vec{D}}^-(z)$, which may be discontinuous at z_1. We obtain the following boundary condition for the wave vector $\tilde{\vec{D}}(z)$ at z_1:

$$\lim_{\epsilon\downarrow o} \left[\tilde{\mathbf{L}}(z_1-\epsilon)\tilde{\vec{D}}(z_1-\epsilon)\right] = \lim_{\epsilon\downarrow o} \left[\tilde{\mathbf{L}}(z_1+\epsilon)\tilde{\vec{D}}(z_1+\epsilon)\right]. \tag{IV-40c}$$

Let us now assume that z_1 represents an interface between two homogeneous isotropic half-spaces. In the following the sub-scripts u and ℓ

refer to the upper and lower half-space, respectively. We derive the boundary conditions for the downgoing and upgoing P- and S-waves. First consider the situation depicted in Figure IV-1a.

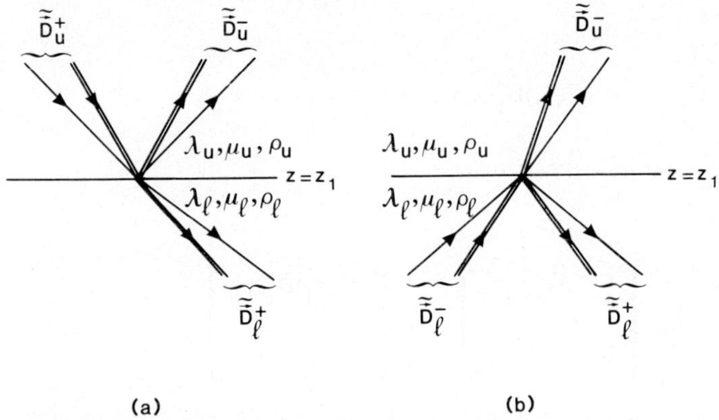

Figure IV-1: Reflection and transmission of plane waves at an interface between two homogeneous isotropic elastic half-spaces.

a. Situation for incident downgoing P- and S-waves
$\tilde{\vec{D}}_u^+ = \left(\tilde{\Phi}_u^+, \tilde{\Psi}_{x,u}^+, \tilde{\Psi}_{y,u}^+\right)^T.$

b. Situation for incident upgoing P- and S-waves
$\tilde{\vec{D}}_\ell^- = \left(\tilde{\Phi}_\ell^-, \tilde{\Psi}_{x,\ell}^-, \tilde{\Psi}_{y,\ell}^-\right)^T.$

Downgoing P- and S-waves $\tilde{\vec{D}}_u^+(z)$ are incident to the interface from above, upgoing P- and S-waves $\tilde{\vec{D}}_u^-(z)$ are reflected into the upper half-space and downgoing P- and S-waves $\tilde{\vec{D}}_\ell^+(z)$ are transmitted into the lower half-space. Applying boundary condition (IV-40c) to this situation yields

$$\tilde{L}_u \begin{bmatrix} \tilde{\vec{D}}_u^+(z_1) \\ \\ \tilde{\vec{D}}_u^-(z_1) \end{bmatrix} = \tilde{L}_\ell \begin{bmatrix} \tilde{\vec{D}}_\ell^+(z_1) \\ \\ \vec{0} \end{bmatrix}^{1)} , \qquad \text{(IV-41)}$$

with $\tilde{L}_u$ and $\tilde{L}_\ell$ defined by (IV-24a), (IV-25a) and (IV-25b). $\tilde{L}_u$ is related to the medium parameters of the upper half-space, $\tilde{L}_\ell$ is related to the medium parameters of the lower half-space. Now define reflection and

---

1) Of course, the summation convention does *not* apply here to the sub-scripts u (= upper) and ℓ (= lower).

transmission operators $\tilde{\mathbf{R}}^{+}(z_1)$ and $\tilde{\mathbf{T}}^{+}(z_1)$, respectively, according to

$$\vec{\tilde{D}}_u^{-}(z_1) = \tilde{R}^{+}(z_1)\vec{\tilde{D}}_u^{+}(z_1) \tag{IV-42a}$$

and

$$\vec{\tilde{D}}_\ell^{+}(z_1) = \tilde{T}^{+}(z_1)\vec{\tilde{D}}_u^{+}(z_1) \ . \tag{IV-42b}$$

Substitution into (IV-41) yields

$$\tilde{L}_u \begin{bmatrix} \vec{\tilde{D}}_u^{+}(z_1) \\ \tilde{R}^{+}(z_1)\vec{\tilde{D}}_u^{+}(z_1) \end{bmatrix} = \tilde{L}_\ell \begin{bmatrix} \tilde{T}^{+}(z_1)\vec{\tilde{D}}_u^{+}(z_1) \\ \vec{0} \end{bmatrix} , \tag{IV-43a}$$

or, since these equations should hold for any downgoing wave vector $\vec{\tilde{D}}_u^{+}(z_1)$.

$$\tilde{L}_u \begin{bmatrix} I \\ \tilde{R}^{+}(z_1) \end{bmatrix} = \tilde{L}_\ell \begin{bmatrix} \tilde{T}^{+}(z_1) \\ 0 \end{bmatrix} , \tag{IV-43b}$$

or, writing all unknowns at the left-hand side,

$$\begin{bmatrix} \left(\tilde{T}^{+}(z_1)\right)^{-1} \\ \tilde{R}^{+}(z_1)\left(\tilde{T}^{+}(z_1)\right)^{-1} \end{bmatrix} = \tilde{L}_u^{-1}\tilde{L}_\ell \begin{bmatrix} I \\ 0 \end{bmatrix} , \tag{IV-43c}$$

or, using (IV-24a) and (IV-24b),

$$\tilde{R}^{+}(z_1) = \left(\tilde{N}_{\alpha,u}^{-}\tilde{L}_{\alpha,\ell}^{+}\right)\left(\tilde{N}_{\alpha,u}^{+}\tilde{L}_{\alpha,\ell}^{+}\right)^{-1} \tag{IV-44a}$$

and

$$\tilde{\mathbf{T}}^{+}(z_1) = \left(\tilde{\mathbf{N}}^{+}_{\alpha,u}\tilde{\mathbf{L}}^{+}_{\alpha,\ell}\right)^{-1}, \tag{IV-44b}$$

with $\tilde{\mathbf{L}}^{+}_{\alpha,\ell}$ and $\tilde{\mathbf{N}}^{+}_{\alpha,u}$ for α=1, 2 defined by (IV-26).

If we define reflection and transmission operators $\tilde{\mathbf{R}}^{-}$ and $\tilde{\mathbf{T}}^{-}$ for the situation depicted in Figure IV-1b, according to

$$\tilde{\vec{\mathbf{D}}}^{+}_{\ell}(z_1) = \tilde{\mathbf{R}}^{-}(z_1)\tilde{\vec{\mathbf{D}}}^{-}_{\ell}(z_1) \tag{IV-45a}$$

and

$$\tilde{\vec{\mathbf{D}}}^{-}_{u}(z_1) = \tilde{\mathbf{T}}^{-}(z_1)\tilde{\vec{\mathbf{D}}}^{-}_{\ell}(z_1), \tag{IV-45b}$$

then we obtain in a similar way as above

$$\tilde{\mathbf{R}}^{-}(z_1) = \left(\tilde{\mathbf{N}}^{+}_{\alpha,\ell}\tilde{\mathbf{L}}^{-}_{\alpha,u}\right)\left(\tilde{\mathbf{N}}^{-}_{\alpha,\ell}\tilde{\mathbf{L}}^{-}_{\alpha,u}\right)^{-1} \tag{IV-46a}$$

and

$$\tilde{\mathbf{T}}^{-}(z_1) = \left(\tilde{\mathbf{N}}^{-}_{\alpha,\ell}\tilde{\mathbf{L}}^{-}_{\alpha,u}\right)^{-1}, \tag{IV-46b}$$

with $\tilde{\mathbf{L}}^{-}_{\alpha,u}$ and $\tilde{\mathbf{N}}^{+}_{\alpha,\ell}$ for α=1, 2 defined by (IV-26).

Finally, we consider the special situation of reflection at the free surface z_o of a homogeneous isotropic half-space. The situation is depicted in Figure IV-2. Upgoing P- and S-waves $\tilde{\vec{\mathbf{D}}}^{-}(z)$ are incident to the free surface from below, downgoing P- and S-waves $\tilde{\vec{\mathbf{D}}}^{+}(z)$ are reflected into the half-space.

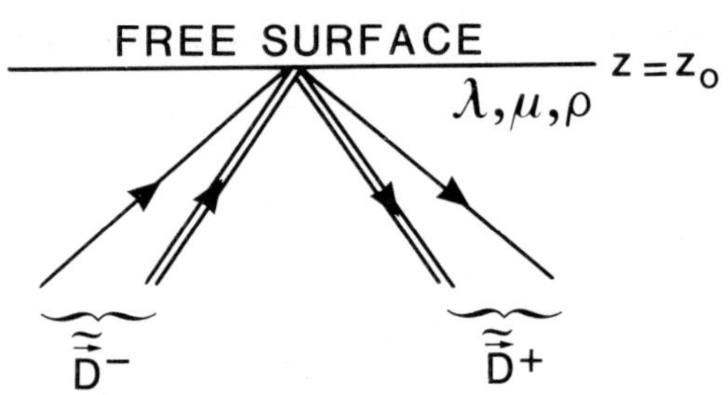

Figure IV-2: Reflection of plane waves at the free surface of a homogeneous isotropic elastic half-space.

For the wave fields at the free surface we write

$$\tilde{\vec{Q}}(z_o) = \tilde{L}\,\tilde{\vec{D}}(z_o), \tag{IV-47a}$$

or

$$\begin{bmatrix} \tilde{\vec{V}}(z_o) \\ \tilde{\vec{\tau}}_z(z_o)=\vec{o} \end{bmatrix} = \begin{bmatrix} \tilde{L}_1^+ & \tilde{L}_1^- \\ \tilde{L}_2^+ & \tilde{L}_2^- \end{bmatrix} \begin{bmatrix} \tilde{\vec{D}}^+(z_o) \\ \tilde{\vec{D}}^-(z_o) \end{bmatrix} . \tag{IV-47b}$$

Define a free surface reflection operator $\tilde{R}_{fr}^-(z_o)$, according to

$$\tilde{\vec{D}}^+(z_o) = \tilde{R}_{fr}^-(z_o)\tilde{\vec{D}}^-(z_o). \tag{IV-48}$$

Substitution into (IV-47b) yields

$$\vec{o} = \left(\tilde{L}_2^+\tilde{R}_{fr}^-(z_o) + \tilde{L}_2^-\right)\tilde{\vec{D}}^-(z_o), \tag{IV-49a}$$

or, since this equation should hold for any upgoing wave vector $\tilde{\vec{D}}^-(z_o)$,

$$\tilde{R}_{fr}^-(z_o) = -\left(\tilde{L}_2^+\right)^{-1}\tilde{L}_2^-. \tag{IV-49b}$$

It is interesting to note that $\tilde{R}_{fr}^-(z_o)$ is singular when

$$\det\left(\tilde{L}_2^+\right) = o, \tag{IV-50a}$$

or

$$-4k_r^2 k_{z,p} k_{z,s} - \left(k_s^2 - 2k_r^2\right)^2 = o, \tag{IV-50b}$$

with

$$k_r^2 = k_x^2 + k_y^2. \tag{IV-50c}$$

Hence, for wavenumbers k_r satisfying (IV-50b), "reflected" waves $\vec{\tilde{D}}^+$ may exist even in the absence of "incident" waves $\vec{\tilde{D}}^-$. Equation (IV-50b) has solutions only in the evanescent wavenumber area

$$k_r^2 > k_s^2 > k_p^2. \tag{IV-50d}$$

Hence, the "reflected" waves $\vec{\tilde{D}}^+$ for this situation are actually evanescent waves which propagate along the surface z_o and which decay exponentially with depth. This suggests that, for k_r satisfying (IV-50b), $\vec{\tilde{D}}^+$ represents the *Rayleigh-wave* which was introduced in section II.4.2. Indeed, if we associate a phase-velocity c_R to the wavenumber k_r, according to

$$c_R = \omega/k_r, \tag{IV-51a}$$

then (IV-50b) yields

$$4c_s^3 \sqrt{c_p^2 - c_R^2} \sqrt{c_s^2 - c_R^2} - c_p \left(2c_s^2 - c_R^2\right)^2 = o, \tag{IV-51b}$$

which is identical to equation (II-69) for the Rayleigh-wave velocity c_R.

IV.3 ELASTIC WAVE EQUATIONS FOR ARBITRARILY INHOMOGENEOUS MEDIA

In this section we consider the situation of wave propagation in inhomogeneous anisotropic elastic media in which the medium parameters $c_{ijk\ell}$ and ρ are arbitrary functions of x, y and z, respectively, hence

$$c_{ijk\ell} = c_{ijk\ell}(x,y,z)$$

and

$$\rho = \rho(x,y,z).$$

First we derive the elastic two-way wave equation in the space-frequency domain and present its solution in a formal operator notation. Next, we derive elastic one-way wave equations for downgoing and upgoing P- and S-waves and we present solutions in a convenient matrix notation. Finally, we derive reflection and transmission operators for one-way P- and S-wave fields at interfaces.

IV.3.1 Elastic two-way wave equation in the space-frequency domain

Consider the linearized equation of motion (II-13) and the linearized stress-velocity relation (II-24), in the space-frequency domain given by

$$j\omega\rho(x,y,z)\vec{V}(x,y,z,\omega) - \partial_j\vec{\tau}_j(x,y,z,\omega) = \vec{F}(x,y,z,\omega) \qquad \text{(IV-52a)}$$

and

$$j\omega\vec{\tau}_j(x,y,z,\omega) - C_{j\ell}(x,y,z)\partial_\ell\vec{V}(x,y,z,\omega) = -j\omega\vec{\sigma}_j(x,y,z,\omega), \qquad \text{(IV-52b)}$$

respectively. We derive a *first order* two-way wave equation for $\vec{V}$ and $\vec{\tau}_z$ by eliminating $\vec{\tau}_x$ and $\vec{\tau}_y$ from equations (IV-52a) and (IV-52b). In analogy with (IV-4a) and (IV-4d) we obtain

$$\frac{\partial\vec{V}}{\partial z} = C_{33}^{-1}\left[-C_{3\beta}\partial_\beta\vec{V} + j\omega\vec{\tau}_z + j\omega\vec{\sigma}_z\right] \qquad \text{(IV-53a)}$$

and

$$\frac{\partial\vec{\tau}_z}{\partial z} = j\omega\rho\vec{V} - \frac{1}{j\omega}\partial_\alpha\left[(C_{\alpha\beta} - C_{\alpha 3}C_{33}^{-1}C_{3\beta})\partial_\beta\vec{V}\right]$$

$$-\partial_\alpha\left[C_{\alpha 3}C_{33}^{-1}(\vec{\tau}_z + \vec{\sigma}_z) - \vec{\sigma}_\alpha\right] - \vec{F}. \qquad \text{(IV-53b)}$$

Equations (IV-53a) and (IV-53b) can be combined in one equation, according to

$$\frac{\partial\vec{Q}(x,y,z,\omega)}{\partial z} = \boldsymbol{A}_1(x,y,z,\omega)\vec{Q}(x,y,z,\omega) + \vec{S}(x,y,z,\omega), \qquad \text{(IV-54a)}$$

where the wave vector $\vec{Q}$ is defined as

$$\vec{Q}(x,y,z,\omega) = \begin{bmatrix} \vec{V}(x,y,z,\omega) \\ \\ \vec{\tau}_z(x,y,z,\omega) \end{bmatrix}, \qquad \text{(IV-54b)}$$

the first order differential operator $\boldsymbol{A}_1$ is defined as

$$\boldsymbol{A}_1(x,y,z,\omega) = \begin{bmatrix} \boldsymbol{a}_{11}(x,y,z,\omega) & \boldsymbol{a}_{12}(x,y,z,\omega) \\ \boldsymbol{a}_{21}(x,y,z,\omega) & \boldsymbol{a}_{22}(x,y,z,\omega) \end{bmatrix}, \tag{IV-54c}$$

where

$$\boldsymbol{a}_{11} = -C_{33}^{-1}C_{3\beta}\partial_\beta, \tag{IV-54d}$$

$$\boldsymbol{a}_{12} = j\omega C_{33}^{-1}, \tag{IV-54e}$$

$$\boldsymbol{a}_{21} = j\omega\rho I - \frac{1}{j\omega}\left[\partial_\alpha\left(C_{\alpha\beta}-C_{\alpha 3}C_{33}^{-1}C_{3\beta}\right)\partial_\beta + \left(C_{\alpha\beta}-C_{\alpha 3}C_{33}^{-1}C_{3\beta}\right)\partial_\alpha\partial_\beta\right], \tag{IV-54f}$$

$$\boldsymbol{a}_{22} = -\left[\partial_\alpha\left(C_{\alpha 3}C_{33}^{-1}\right) + C_{\alpha 3}C_{33}^{-1}\partial_\alpha\right] \tag{IV-54g}$$

and, finally, the source vector $\vec{\mathcal{S}}$ is defined as

$$\vec{\mathcal{S}}(x,y,z,\omega) = \begin{bmatrix} \vec{\mathcal{S}}_1(x,y,z,\omega) \\ \vec{\mathcal{S}}_2(x,y,z,\omega) \end{bmatrix}, \tag{IV-54h}$$

where

$$\vec{\mathcal{S}}_1 = j\omega C_{33}^{-1}\vec{\sigma}_z \tag{IV-54i}$$

and

$$\vec{\mathcal{S}}_2 = -\partial_\alpha\left[C_{\alpha 3}C_{33}^{-1}\vec{\sigma}_z - \vec{\sigma}_\alpha\right] - \vec{F}. \tag{IV-54j}$$

The general solution of equation (IV-54) reads, in analogy with (III-30),

$$\vec{\mathcal{Q}}(z) = \boldsymbol{U}(z,z_o)\vec{\mathcal{Q}}(z_o) + \int_{z_o}^{z} \boldsymbol{U}(z,z')\vec{\mathcal{S}}(z')dz', \tag{IV-55a}$$

where the two-way wave field extrapolation operator $\boldsymbol{U}(z,z_o)$ is formally defined by

$$\boldsymbol{U}(z,z_o) = \sum_{m=o}^{\infty} \frac{(z-z_o)^m}{m!} \boldsymbol{A}_m(z_o), \tag{IV-55b}$$

with $\boldsymbol{A}_m(z_o)$ defined recursively by

$$\boldsymbol{A}_{m+1}(z_o) = \left.\frac{\partial \boldsymbol{A}_m(z)}{\partial z}\right|_{z_o} + \boldsymbol{A}_m(z_o)\boldsymbol{A}_1(z_o) \tag{IV-55c}$$

and

$$\boldsymbol{A}_o(z_o) = \mathbf{I}. \tag{IV-55d}$$

For notational convenience we omitted the variables x,y and ω. Relation (IV-55) is the basis for numerical two-way wave field extrapolation algorithms which are valid for primary and multiply reflected and converted waves in arbitrarily inhomogeneous anisotropic elastic media and which are accurate upto high tilt angles of propagation. A further discussion of the numerical aspects of elastic two-way wave field extrapolation is beyond the scope of this book. The reader is referred to Wapenaar et al. (1987).

IV.3.2 Elastic one-way wave equations in the space-frequency domain

In this section we derive the elastic one-way wave equations in the space-frequency domain. In principle we could follow the same approach as in section III.3.2, i.e., we could use the elastic two-way wave equation (IV-54) as a starting point and perform all our derivations entirely in the space-frequency domain. However, these derivations are rather involved and therefore we follow an alternative approach. We use equations (IV-31), (IV-32) and (IV-35) in the wavenumber-frequency domain as a starting point and transform these equations back to the space-frequency domain. This approach is fully justified as long as the medium parameters do not vary laterally. We also indicate how to modify the results such that lateral variations of the medium parameters are taken into account. This modification involves some approximations. In particular, it is assumed that the lateral derivatives of the medium parameters are negligible in comparison with the lateral derivatives of the elastic wave field.

Consider arbitrary functions $H_1(x,y,z,\omega)$ and $H_2(x,y,z,\omega)$ in the space-frequency domain. The corresponding functions in the wavenumber-frequency domain read $\tilde{H}_1(k_x,k_y,z,\omega)$ and $\tilde{H}_2(k_x k_y,z,\omega)$. Equations (III-4a) and (III-4b), which describe the forward and inverse spatial Fourier transformations, imply that the product

$$\tilde{H}_3(k_x,k_y,z,\omega) = \tilde{H}_1(k_x,k_y,z,\omega)\tilde{H}_2(k_x,k_y,z,\omega) \tag{IV-56a}$$

in the wavenumber-frequency domain corresponds to the convolution integral

$$H_3(x,y,z,\omega) = \int\int_{-\infty}^{\infty} H_1(x-x',y-y',z,\omega)H_2(x',y',z,\omega)dx'dy' \tag{IV-56b}$$

in the space-frequency domain. In the following we make extensively use of this important property.

Consider equations (IV-31b) and (IV-31c),

$$\tilde{\vec{V}} = \tilde{L}_1^+\tilde{\vec{D}}^+ + \tilde{L}_1^-\tilde{\vec{D}}^- \tag{IV-57a}$$

and

$$\tilde{\vec{\tau}}_z = \tilde{L}_2^+\tilde{\vec{D}}^+ + \tilde{L}_2^-\tilde{\vec{D}}^-, \tag{IV-57b}$$

with $\tilde{L}_1^{\pm}$ and $\tilde{L}_2^{\pm}$ defined by (IV-25a) and (IV-25b), respectively. These equations describe, for the isotropic situation, *composition* of the elastic two-way wave field $(\tilde{\vec{V}}, \tilde{\vec{\tau}}_z)$ from its downgoing and upgoing constituents $(\tilde{\vec{D}}^+,\tilde{\vec{D}}^-)$. For the moment we only consider equation (IV-57a), which can be rewritten as

$$\tilde{V}_i(k_x,k_y,z,\omega) = \tilde{L}^+_{1,ij}(k_x,k_y,z,\omega)\tilde{D}^+_j(k_x,k_y,z,\omega) + \tilde{L}^-_{1,ij}(k_x,k_y,z,\omega)\tilde{D}^-_j(k_x,k_y,z,\omega). \tag{IV-58}$$

Here $\tilde{V}_i$ for i=1, 2, 3 represents the particle velocity components $\tilde{V}_x$, $\tilde{V}_y$ and $\tilde{V}_z$, respectively. $\tilde{D}^\pm_j$ for j=1, 2, 3 represents the P- and S-waves potentials $\tilde{\Phi}^\pm$, $\tilde{\Psi}^\pm_x$ and $\tilde{\Psi}^\pm_y$, respectively, see also equation (IV-30). Finally, $\tilde{L}^\pm_{1,ij}$ for i=1, 2, 3 and j=1, 2, 3 represents the elements of matrix $\tilde{\mathbf{L}}^\pm_1$, as defined by (IV-25a).

According to property (IV-56), we may replace the multiplications in (IV-58) by convolution integrals in the space-frequency domain, according to

$$V_i(x,y,z,\omega) = \iint_{-\infty}^{\infty} L^+_{1,ij}(x-x',y-y',z,\omega)D^+_j(x,y,z,\omega)dx'dy' + \iint_{-\infty}^{\infty} L^-_{1,ij}(x-x',y-y',z,\omega)D^-_j(x,y,z,\omega)dx'dy', \tag{IV-59}$$

where $L^\pm_{1,ij}(x,y,z,\omega)$ is obtained by applying a spatially band-limited version of the inverse Fourier transform (III-4b) to the matrix element $\tilde{L}^\pm_{1,ij}(k_x,k_y,z,\omega)$. Equation (IV-59) is exact when the wave field $D^\pm_j(x,y,z,\omega)$ is spatially bandlimited. Rewrite

$$L^\pm_{1,ij}(x-x',y-y',z,\omega) \tag{IV-60a}$$

in the more general notation

$$L^\pm_{1,ij}(x,y,z;x',y',z'=z;\omega). \tag{IV-60b}$$

Now, with the matrix notation discussed in Appendix A, we may rewrite (IV-59) as

$$\vec{V}_i(z) = \mathbf{L}^+_{1,ij}(z)\vec{D}^+_j(z) + \mathbf{L}^-_{1,ij}(z)\vec{D}^-_j(z). \tag{IV-61a}$$

Here the vector $\vec{V}_i(z)$ for i=1, 2, 3 contains the discretized wave field $V_i(x,y,z,\omega)$ at depth level z. Similarly, vector $\vec{D}^\pm_j(z)$ for j=1, 2, 3 contains the discretized wave field $D_j^{\pm}(x,y,z,\omega)$ at depth level z. Finally, matrix $\mathbf{L}_{1,ij}^{\pm}(z)$ for i=1, 2, 3 and j=1, 2, 3 contains the discretized operator $L_{1,ij}^{\pm}(x,y,z;x',y',z'=z;\omega)$ at depth level z'=z.

In a similar way we may derive from (IV-57b),

$$\vec{\tau}_{iz}(z) = \mathbf{L}_{2,ij}^{+}(z)\vec{D}_j^{+}(z) + \mathbf{L}_{2,ij}^{-}(z)\vec{D}_j^{-}(z). \tag{IV-61b}$$

Here the vector $\vec{\tau}_{iz}(z)$ for i=1, 2, 3 contains the discretized wave field $\tau_{iz}(x,y,z,\omega)$ at depth level z and matrix $\mathbf{L}_{2,ij}^{\pm}(z)$ for i=1, 2, 3 and j=1, 2, 3 contains the discretized operator $L_{2,ij}^{\pm}(x,y,z;x',y',z'=z;\omega)$ at depth level z'=z.

Equations (IV-61a) and (IV-61b) for i=1, 2, 3 can be combined according to[1)]

$$\begin{bmatrix} \underset{\sim}{\vec{V}}(z) \\ \underset{\sim}{\vec{\tau}}_z(z) \end{bmatrix} = \begin{bmatrix} \underset{\sim}{\mathbf{L}}_1^{+}(z) & \underset{\sim}{\mathbf{L}}_1^{-}(z) \\ \underset{\sim}{\mathbf{L}}_2^{+}(z) & \underset{\sim}{\mathbf{L}}_2^{-}(z) \end{bmatrix} \begin{bmatrix} \underset{\sim}{\vec{D}}^{+}(z) \\ \underset{\sim}{\vec{D}}^{-}(z) \end{bmatrix}, \tag{IV-62a}$$

where the *three-component data vectors* $\underset{\sim}{\vec{V}}(z)$, $\underset{\sim}{\vec{\tau}}_z(z)$ and $\underset{\sim}{\vec{D}}^{\pm}(z)$ are defined according to

$$\underset{\sim}{\vec{V}}(z) = \begin{bmatrix} \vec{V}_x(z) \\ \vec{V}_y(z) \\ \vec{V}_z(z) \end{bmatrix}, \tag{IV-62b}$$

$$\underset{\sim}{\vec{\tau}}_z(z) = \begin{bmatrix} \vec{\tau}_{xz}(z) \\ \vec{\tau}_{yz}(z) \\ \vec{\tau}_{zz}(z) \end{bmatrix}, \tag{IV-62c}$$

---

1) The tilde (~) underneath the data vectors and the operator matrices denotes that the "elements" are again data vectors and operator matrices

$$\vec{\underset{\sim}{D}}^{\pm}(z) = \begin{bmatrix} \vec{D}_1^{\pm}(z) \\ \vec{D}_2^{\pm}(z) \\ \vec{D}_3^{\pm}(z) \end{bmatrix} = \begin{bmatrix} \vec{\Phi}^{\pm}(z) \\ \vec{\Psi}_x^{\pm}(z) \\ \vec{\Psi}_y^{\pm}(z) \end{bmatrix} \tag{IV-62d}$$

and where matrix $\underset{\sim}{L}_{\alpha}^{\pm}(z)$ for α=1, 2 is defined according to

$$\underset{\sim}{L}_{\alpha}^{\pm}(z) = \begin{bmatrix} L_{\alpha,11}^{\pm}(z) & L_{\alpha,12}^{\pm}(z) & L_{\alpha,13}^{\pm}(z) \\ L_{\alpha,21}^{\pm}(z) & L_{\alpha,22}^{\pm}(z) & L_{\alpha,23}^{\pm}(z) \\ L_{\alpha,31}^{\pm}(z) & L_{\alpha,32}^{\pm}(z) & L_{\alpha,33}^{\pm}(z) \end{bmatrix} . \tag{IV-62e}$$

Consider equations (IV-32b) and (IV-32c),

$$\tilde{\vec{D}}^{+} = \tilde{N}_1^{+} \tilde{\vec{V}} + \tilde{N}_2^{+} \tilde{\vec{\tau}}_z \tag{IV-63a}$$

and

$$\tilde{\vec{D}}^{-} = \tilde{N}_1^{-} \tilde{\vec{V}} + \tilde{N}_2^{-} \tilde{\vec{\tau}}_z \ , \tag{IV-63b}$$

with $\tilde{N}_1^{\pm}$ and $\tilde{N}_2^{\pm}$ defined by (IV-26a) and (IV-26b), respectively. These equations describe, for the isotropic situation, *decomposition* of the elastic two-way wave field into downgoing and upgoing P- and S-waves. In a similar way as above, we can derive the space-frequency domain representation of this decomposition algorithm, yielding

$$\begin{bmatrix} \vec{\underset{\sim}{D}}^{+}(z) \\ \vec{\underset{\sim}{D}}^{-}(z) \end{bmatrix} = \begin{bmatrix} \underset{\sim}{N}_1^{+}(z) & \underset{\sim}{N}_2^{+}(z) \\ \underset{\sim}{N}_1^{-}(z) & \underset{\sim}{N}_2^{-}(z) \end{bmatrix} \begin{bmatrix} \vec{\underset{\sim}{V}}(z) \\ \vec{\underset{\sim}{\tau}}_z(z) \end{bmatrix} , \tag{IV-64a}$$

where matrix $\underset{\sim}{N}_{\alpha}^{\pm}(z)$ for α=1, 2 is defined according to

$$\underset{\sim}{\mathbf{N}}_{\alpha}^{\pm}(z) = \begin{bmatrix} \mathbf{N}_{\alpha,11}^{\pm}(z) & \mathbf{N}_{\alpha,12}^{\pm}(z) & \mathbf{N}_{\alpha,13}^{\pm}(z) \\ \mathbf{N}_{\alpha,21}^{\pm}(z) & \mathbf{N}_{\alpha,22}^{\pm}(z) & \mathbf{N}_{\alpha,23}^{\pm}(z) \\ \mathbf{N}_{\alpha,31}^{\pm}(z) & \mathbf{N}_{\alpha,32}^{\pm}(z) & \mathbf{N}_{\alpha,33}^{\pm}(z) \end{bmatrix}, \tag{IV-64b}$$

where sub-matrix $\mathbf{N}_{\alpha,ij}^{\pm}$ for α=1, 2 and i=1, 2, 3 and j=1, 2, 3 contains the discretized operator $N_{\alpha,ij}^{\pm}(x,y,z,x',y',z'=z,\omega)$ at depth level z'=z.

Note that the decomposition operators are related to the composition operators, according to

$$\underset{\sim}{\mathbf{N}}_{1}^{\pm} = \left(\underset{\sim}{\mathbf{L}}_{1}^{\pm} - \underset{\sim}{\mathbf{L}}_{1}^{\mp}(\underset{\sim}{\mathbf{L}}_{2}^{\mp})^{-1}\underset{\sim}{\mathbf{L}}_{2}^{\pm}\right)^{-1} \tag{IV-64c}$$

and

$$\underset{\sim}{\mathbf{N}}_{2}^{\pm} = \left(\underset{\sim}{\mathbf{L}}_{2}^{\pm} - \underset{\sim}{\mathbf{L}}_{2}^{\mp}(\underset{\sim}{\mathbf{L}}_{1}^{\mp})^{-1}\underset{\sim}{\mathbf{L}}_{1}^{\pm}\right)^{-1}. \tag{IV-64d}$$

Consider the elastic one-way wave equations (IV-35c) and (IV-35d) for the source-free homogeneous isotropic situation. In a similar way as above, we can derive the space-frequency domain representation of these one-way wave equations, yielding

$$\frac{\partial}{\partial z}\begin{bmatrix} \vec{\Phi}^{+}(z) \\ \vec{\Psi}_{x}^{+}(z) \\ \vec{\Psi}_{y}^{+}(z) \end{bmatrix} = \begin{bmatrix} -jH_{1,p} & 0 & 0 \\ 0 & -jH_{1,s} & 0 \\ 0 & 0 & -jH_{1,s} \end{bmatrix}\begin{bmatrix} \vec{\Phi}^{+}(z) \\ \vec{\Psi}_{x}^{+}(z) \\ \vec{\Psi}_{y}^{+}(z) \end{bmatrix} \tag{IV-65}$$

and

$$\frac{\partial}{\partial z}\begin{bmatrix}\vec{\Phi}^-(z)\\ \vec{\Psi}_x^-(z)\\ \vec{\Psi}_y^-(z)\end{bmatrix}=\begin{bmatrix}+j\mathbf{H}_{1,p} & \mathbf{0} & \mathbf{0}\\ \mathbf{0} & +j\mathbf{H}_{1,s} & \mathbf{0}\\ \mathbf{0} & \mathbf{0} & +j\mathbf{H}_{1,s}\end{bmatrix}\begin{bmatrix}\vec{\Phi}^-(z)\\ \vec{\Psi}_x^-(z)\\ \vec{\Psi}_y^-(z)\end{bmatrix}, \tag{IV-66}$$

respectively.

In analogy with matrices $\mathbf{L}^{\pm}_{\alpha,ij}$ and $\mathbf{N}^{\pm}_{\alpha,ij}$, matrices $\mathbf{H}_{1,p}$ and $\mathbf{H}_{1,s}$ are obtained as follows:

- . Apply a band-limited version of the inverse spatial Fourier transformation (III-4b) to $k_{z,p}$ and $k_{z,s}$, yielding $H_{1,p}(x,y,\omega)$ and $H_{1,s}(x,y,\omega)$, respectively.
- . Rewrite $H_{1,p}(x\text{-}x',y\text{-}y',\omega)$ and $H_{1,s}(x\text{-}x',y\text{-}y',\omega)$ in the more general notation $H_{1,p}(x,y,z;x',y',z'=z;\omega)$ and $H_{1,s}(x,y,z;x',y',z'=z;\omega)$, respectively.
- . Store the discretized versions of these operators to the matrices $\mathbf{H}_{1,p}$ and $\mathbf{H}_{1,s}$, respectively, as explained in Appendix A.

The solutions of equations (IV-65) and (IV-66) read

$$\begin{bmatrix}\vec{\Phi}^+(z)\\ \vec{\Psi}_x^+(z)\\ \vec{\Psi}_y^+(z)\end{bmatrix}=\begin{bmatrix}\mathbf{W}^+_{\phi,\phi}(z,z_0) & \mathbf{0} & \mathbf{0}\\ \mathbf{0} & \mathbf{W}^+_{\psi_x,\psi_x}(z,z_0) & \mathbf{0}\\ \mathbf{0} & \mathbf{0} & \mathbf{W}^+_{\psi_y,\psi_y}(z,z_0)\end{bmatrix}\begin{bmatrix}\vec{\Phi}^+(z_0)\\ \vec{\Psi}_x^+(z_0)\\ \vec{\Psi}_y^+(z_0)\end{bmatrix} \tag{IV-67a}$$

and

$$\begin{bmatrix}\vec{\Phi}^-(z)\\ \vec{\Psi}_x^-(z)\\ \vec{\Psi}_y^-(z)\end{bmatrix}=\begin{bmatrix}\mathbf{W}^-_{\phi,\phi}(z,z_0) & \mathbf{0} & \mathbf{0}\\ \mathbf{0} & \mathbf{W}^-_{\psi_x,\psi_x}(z,z_0) & \mathbf{0}\\ \mathbf{0} & \mathbf{0} & \mathbf{W}^-_{\psi_y,\psi_y}(z,z_0)\end{bmatrix}\begin{bmatrix}\vec{\Phi}^-(z_0)\\ \vec{\Psi}_x^-(z_0)\\ \vec{\Psi}_y^-(z_0)\end{bmatrix} \tag{IV-67b}$$

where, in analogy with (III-81b),

$$\mathbf{W}^{\pm}_{\phi,\phi}(z,z_0) = \sum_{m=0}^{\infty} \frac{(z-z_0)^m}{m!} \left(\mp j\right)^m \mathbf{H}^m_{1,p} \tag{IV-67c}$$

and

$$\mathbf{W}^{\pm}_{\psi_x,\psi_x}(z,z_0) = \mathbf{W}^{\pm}_{\psi_y,\psi_y}(z,z_0) = \sum_{m=0}^{\infty} \frac{(z-z_0)^m}{m!} \left(\mp j\right)^m \mathbf{H}^m_{1,s} \; . \tag{IV-67d}$$

The composition algorithm (IV-62), the decomposition algorithm (IV-64), the one-way wave equations (IV-65) and (IV-66) and the one-way wave field extrapolation algorithms (IV-67) have all been derived for the situation of laterally invariant, isotropic media. Lateral variations of the medium parameters can be accounted for by designing the operators $L^{\pm}_{\alpha,ij}(x,y,z;x',y',z'=z;\omega)$, $N^{\pm}_{\alpha,ij}(x,y,z;x',y',z'=z;\omega)$, $H_{1,p}(x,y,z;x',y',z'=z;\omega)$ and $H_{1,s}(x,y,z;x',y',z'=z;\omega)$ in accordance with the *local* medium parameters at (x,y,z) and by storing these operators to the matrices $\mathbf{L}^{\pm}_{\alpha,ij}$, $\mathbf{N}^{\pm}_{\alpha,ij}$, $\mathbf{H}_{1,p}$ and $\mathbf{H}_{1,s}$, respectively, as explained in Appendix A. With this modification, the composition algorithm (IV-62), the decomposition algorithm (IV-64), the one-way wave equations (IV-65) and (IV-66) and the one-way wave field extrapolation algorithms (IV-67) may be applied in smoothly laterally varying, isotropic media.

It is interesting to note that we would have obtained exactly the same results if we would have used the elastic two-way wave equation (IV-54) in the space-frequency domain as our starting point and if we would have neglected the lateral derivatives of the medium parameters in comparison with the lateral derivatives of the elastic wave field.

IV.3.3 Elastic one-way wave fields at interfaces in the space-frequency domain

Solution (IV-67) breaks down whenever the medium contains "interfaces" in the interval (z,z_0).

We consider a horizontal interface at $z=z_1$ between two homogeneous isotropic elastic half-spaces. We use the sub-scripts u and ℓ to distinguish between the upper and lower half-space, respectively. First consider the situation depicted in Figure IV-3a. Downgoing P- and S-waves are incident

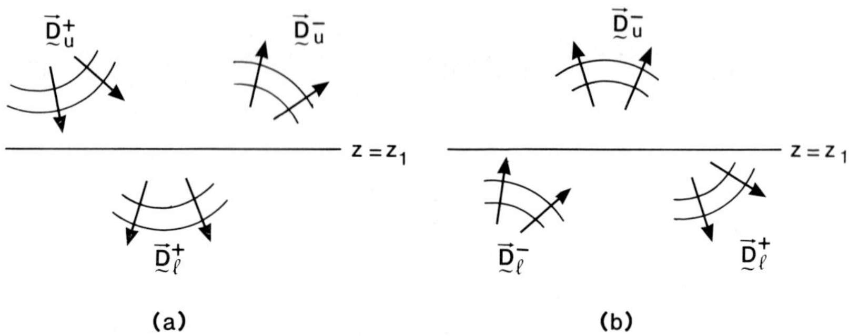

Figure IV-3: Reflection and transmission at an interface between two homogeneous isotropic elastic half-spaces.
a. Situation for incident downgoing P- and S-waves $\vec{D}_u^+=(\vec{\Phi}_u^+, \vec{\Psi}_{x,u}^+, \vec{\Psi}_{y,u}^+)^T$.
b. Situation for incident upgoing P- and S-waves $\vec{D}_\ell^-=(\vec{\Phi}_\ell^-, \vec{\Psi}_{x,\ell}^-, \vec{\Psi}_{y,\ell}^-)^T$.

to the interface from above, upgoing P- and S-waves are reflected into the upper half-space and downgoing P- and S-waves are transmitted into the lower half-space. In the following, the discretized versions of these wave fields are represented by vectors $\vec{D}_u^+(z)$, $\vec{D}_u^-(z)$ and $\vec{D}_\ell^+(z)$, respectively, see also equation (IV-62d). In analogy with (IV-42a) and (IV-42b), we define reflection and transmission matrices $\mathbf{R}^+$ and $\mathbf{T}^+$ for the interface at z_1, according to

$$\vec{D}_u^-(z_1) = \mathbf{R}^+(z_1)\vec{D}_u^+(z_1) \tag{IV-68a}$$

and

$$\vec{D}_\ell^+(z_1) = \mathbf{T}^+(z_1)\vec{D}_u^+(z_1), \tag{IV-68b}$$

respectively. In a similar way as described in section IV.2.3, we may derive

$$\begin{bmatrix} \left(\underset{\sim}{T}^{+}(z_1)\right)^{-1} \\ \underset{\sim}{R}^{+}(z_1)\left(\underset{\sim}{T}^{+}(z_1)\right)^{-1} \end{bmatrix} = \underset{\sim}{L}_u^{-1}\,\underset{\sim}{L}_\ell \begin{bmatrix} I \\ 0 \end{bmatrix}, \tag{IV-69}$$

or

$$\underset{\sim}{R}^{+}(z_1) = \left(\underset{\sim}{N}^{-}_{\alpha,u}\underset{\sim}{L}^{+}_{\alpha,\ell}\right)\left(\underset{\sim}{N}^{+}_{\alpha,u}\underset{\sim}{L}^{+}_{\alpha,\ell}\right)^{-1} \tag{IV-70a}$$

and

$$\underset{\sim}{T}^{+}(z_1) = \left(\underset{\sim}{N}^{+}_{\alpha,u}\underset{\sim}{L}^{+}_{\alpha,\ell}\right)^{-1}, \tag{IV-70b}$$

with $\underset{\sim}{L}^{+}_{\alpha,\ell}$ and $\underset{\sim}{N}^{\pm}_{\alpha,u}$ for α=1, 2 defined by (IV-62e) and (IV-64b), respectively.

For the situation depicted in Figure IV-3b we define reflection and transmission matrices $\underset{\sim}{R}^{-}$ and $\underset{\sim}{T}^{-}$ for the interface at z_1, according to

$$\vec{\underset{\sim}{D}}^{+}_{\ell}(z_1) = \underset{\sim}{R}^{-}(z_1)\vec{\underset{\sim}{D}}^{-}_{\ell}(z_1) \tag{IV-71a}$$

and

$$\vec{\underset{\sim}{D}}^{-}_{u}(z_1) = \underset{\sim}{T}^{-}(z_1)\vec{\underset{\sim}{D}}^{-}_{\ell}(z_1), \tag{IV-71b}$$

respectively. In a similar way as above we obtain

$$\underset{\sim}{R}^{-}(z_1) = \left(\underset{\sim}{N}^{+}_{\alpha,\ell}\underset{\sim}{L}^{-}_{\alpha,u}\right)\left(\underset{\sim}{N}^{-}_{\alpha,\ell}\underset{\sim}{L}^{-}_{\alpha,u}\right)^{-1} \tag{IV-72a}$$

and

$$\underset{\sim}{T}^{-}(z_1) = \left(\underset{\sim}{N}^{-}_{\alpha,\ell}\underset{\sim}{L}^{-}_{\alpha,u}\right)^{-1}, \tag{IV-72b}$$

with $\underset{\sim}{L}^{-}_{\alpha,u}$ and $\underset{\sim}{N}^{\pm}_{\alpha,\ell}$ for α=1, 2 defined by (IV-62e) and (IV-64b), respectively.

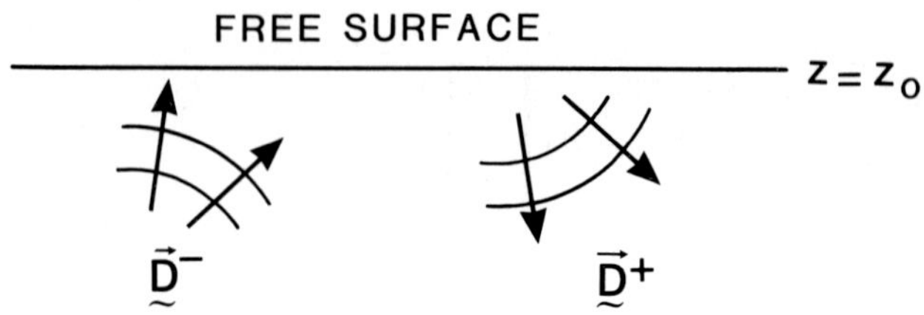

Figure IV-4: Reflection at the free surface of a homogeneous isotropic elastic half-space.

Finally, we consider reflection at the free surface of a homogeneous isotropic elastic half-space. For the situation depicted in Figure IV-4, we define the free surface reflection operator $\underset{\sim}{\mathbf{R}}^{-}_{fr}(z_0)$, according to

$$\vec{\underset{\sim}{D}}^{+}(z_0) = \underset{\sim}{\mathbf{R}}^{-}_{fr}(z_0)\vec{\underset{\sim}{D}}^{-}(z_0). \tag{IV-73a}$$

In analogy with (IV-49b) we find

$$\underset{\sim}{\mathbf{R}}^{-}_{fr}(z_0) = -(\underset{\sim}{\mathbf{L}}^{+}_{2})^{-1}\underset{\sim}{\mathbf{L}}^{-}_{2}, \tag{IV-73b}$$

with $\underset{\sim}{\mathbf{L}}^{\pm}_{2}$ defined by (IV-62e).

The expressions for reflection and transmission, derived in this section, are strictly valid only for the situation of a horizontal interface at $z=z_1$ between two homogeous isotropic elastic half-spaces (Figure IV-3) or for a free surface at $z=z_0$ (Figure IV-4). Because the expressions are formulated in the space-frequency domain they are in principle suited to handle more complex configurations. Of course the accuracy decreases with increasing complexity of the configuration.

IV.4 REFERENCES

Aki, K., and Richards, P.G., 1980, Quantitative seismology: W.H. Freeman and Co.

Daley, P.F., and Hron, F., 1977, Reflection and transmission coefficients for transversely isotropic media: Bull., Seis. Soc. Am., *67*, 661 - 675.

Gilbert, F., and Backus, G.E., 1966, Propagator matrices in elastic wave and vibration problems: Geophysics, *31*, 326 - 332.

Ursin, B., 1983, Review of elastic and electromagnetic wave propagation in horizontally layered media: Geophysics, *48*, 1063 - 1081.

Wapenaar, C.P.A., Kinneging, N.A., and Berkhout, A.J., 1987, Principle of prestack migration based on the full elastic two-way wave equation: Geophysics, *52*, 151 - 173.

Woodhouse, J.H., 1974, Surface waves in a laterally varying layered structure: Geophys. J.R. astr. Soc., *37*, 461 - 490.

V

ACOUSTIC FORWARD WAVE FIELD EXTRAPOLATION

V.1 INTRODUCTION

In this chapter we start with reviewing *Rayleigh's reciprocity theorem.* This theorem gives the mathematical relationship between two independent acoustic wave fields. A special form of the reciprocity theorem is obtained if one of the acoustic wave fields represents the "impulse response" of a reference medium (the *Green's function*) whereas the other wave field represents the physical wave field in the true medium. This special form is commonly referred to as a *representation theorem*: it expresses the acoustic pressure of the physical wave field at any point in the true medium in terms of a closed surface- and a volume-integral over the same physical wave field. It is also known as the *Kirchhoff-Helmholtz integral* formula. We use this formula as the starting point for deriving *Rayleigh integrals* that express the acoustic pressure at any point in the medium in terms of the acoustic pressure at a plane surface. We discuss both *two-way* and *one-way* versions of the Rayleigh integral. The one-way Rayleigh integral is the basis for deriving wave field extrapolation operators. We derive matrix operators for numerical *forward* extrapolation of downgoing and upgoing waves through arbitrarily inhomogeneous acoustic media. These matrices play an important role in chapter XI, where we discuss an acoustic processing scheme for single-component seismic data.

V.2 ACOUSTIC RECIPROCITY THEOREMS

Consider a volume V enclosed by a surface S with outward pointing normal vector $\vec{n}$, see Figure V-1.

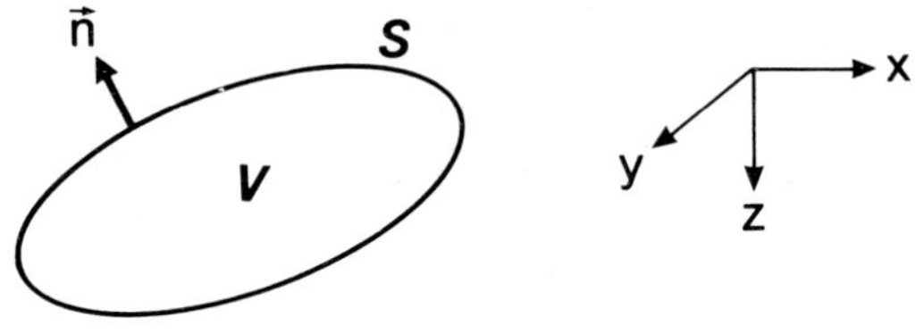

Figure V-1: Volume V enclosed by surface S.

In this volume we define two non-identical acoustic wave fields ("state A" and "state B") and we derive reciprocity relations for these wave fields.

State A:

Define in V

$K^A(\vec{r})$	: adiabatic compression modulus,
$\rho^A(\vec{r})$	: volume density of mass,
$p^A(\vec{r},t)$	: acoustic pressure,
$i_v^A(\vec{r},t)$	: volume density of volume injection,
$\vec{f}^A(\vec{r},t)$	: volume density of external force.

According to equation (I-11), the acoustic pressure in state A satisfies in V the following two-way wave equation

$$\nabla.\left(\frac{1}{\rho^A}\nabla p^A\right) - \frac{1}{K^A}\frac{\partial^2 p^A}{\partial t^2} = -s^A, \tag{V-1a}$$

where the source distribution $s^A(\vec{r},t)$ is given by

$$s^A = \frac{\partial^2 i_v^A}{\partial t^2} - \nabla.\left(\frac{1}{\rho^A}\vec{f}^A\right). \tag{V-1b}$$

State B:

Define in V

$K^B(\vec{r})$	: adiabatic compression modulus,
$\rho^B(\vec{r})$	: volume density of mass,
$p^B(\vec{r},t)$	: acoustic pressure,
$i_v^B(\vec{r},t)$	: volume density of volume injection,
$\vec{f}^B(\vec{r},t)$	: volume density of external force.

According to equation (I-11), the acoustic pressure in state B satisfies in V the following two-way wave equation

$$\nabla \cdot \left(\frac{1}{\rho^B} \nabla p^B\right) - \frac{1}{K^B} \frac{\partial^2 p^B}{\partial t^2} = -s^B, \tag{V-2a}$$

where the source distribution $s^B(\vec{r},t)$ is given by

$$s^B = \frac{\partial^2 i_v^B}{\partial t^2} - \nabla \cdot \left(\frac{1}{\rho^B} \vec{f}^B\right). \tag{V-2b}$$

For the moment we do not specify the initial conditions nor the boundary conditions for the wave fields $p^A(\vec{r},t)$ and $p^B(\vec{r},t)$. We apply the temporal Fourier transformation (III-1a) to equations (V-1) and (V-2). The resulting two-way wave equations in the space-frequency domain read

State A:

$$\nabla \cdot \left(\frac{1}{\rho^A} \nabla P^A\right) + \frac{\omega^2}{K^A} P^A = -S^A, \tag{V-3a}$$

where $P^A(\vec{r},\omega)$ and $S^A(\vec{r},\omega)$ are the space-frequency domain representations of the acoustic pressure and the source distribution, respectively, in state A.

State B:

$$\nabla \cdot \left(\frac{1}{\rho^B} \nabla P^B\right) + \frac{\omega^2}{K^B} P^B = -S^B, \tag{V-3b}$$

where $P^B(\vec{r},\omega)$ and $S^B(\vec{r},\omega)$ are the space-frequency domain representations of the acoustic pressure and the source distribution, respectively, in state B.

We apply the theorem of Gauss to a vector function $\vec{Q}(\vec{r},\omega)$, which we define as

$$\vec{Q} = P^A\left(\frac{1}{\rho^B}\nabla P^B\right) - P^B\left(\frac{1}{\rho^A}\nabla P^A\right). \tag{V-4}$$

The divergence of this vector function reads

$$\nabla.\vec{Q} = \left(\nabla P^A\right).\left(\frac{1}{\rho^B}\nabla P^B\right)+P^A\nabla.\left(\frac{1}{\rho^B}\nabla P^B\right)$$

$$-\left(\nabla P^B\right).\left(\frac{1}{\rho^A}\nabla P^A\right)-P^B\nabla.\left(\frac{1}{\rho^A}\nabla P^A\right), \tag{V-5a}$$

or, upon substitution of (V-3a) and (V-3b),

$$\nabla.\vec{Q} = P^B S^A - P^A S^B + \omega^2\left(\frac{1}{K^A} - \frac{1}{K^B}\right)P^A P^B$$

$$+\left(\frac{1}{\rho^B} - \frac{1}{\rho^A}\right)\left(\nabla P^A\right).\left(\nabla P^B\right). \tag{V-5b}$$

Hence, applying the theorem of Gauss,

$$\oint_S \vec{Q}.\vec{n}\,dS = \int_V \nabla.\vec{Q}\,dV \tag{V-6}$$

yields

$$\oint_S \left[P^A\left(\frac{1}{\rho^B}\nabla P^B\right) - P^B\left(\frac{1}{\rho^A}\nabla P^A\right)\right].\vec{n}\,dS =$$

$$\int_V \left[P^B S^A - P^A S^B + \omega^2\left(\frac{1}{K^A} - \frac{1}{K^B}\right)P^A P^B\right.$$

$$\left. + \left(\frac{1}{\rho^B} - \frac{1}{\rho^A}\right)\left(\nabla P^A\right).\left(\nabla P^B\right)\right]dV.^{1)} \tag{V-7}$$

1) Bear in mind that S^A and S^B denote source distributions, whereas S denotes a closed surface.

This equation is generally known as *Rayleigh's reciprocity theorem* (Rayleigh, 1965). It is the basis for the acoustic representation theorem, which is derived in section V.3. Note that in the theorem of Gauss (V-6) it is assumed that $\vec{Q}$ is differentiable and that the partial derivatives are continuous. Suppose that the acoustic medium in state A and state B consists of piecewise continuous regious V_i, separated by "interfaces", where the medium parameters are discontinuous, see Figure V-2.

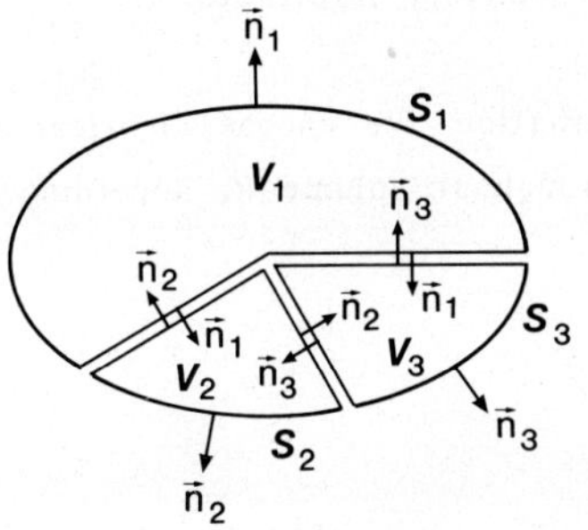

Figure V-2: Volume V, consisting of piecewise continuous regions V_i.

Then the reciprocity theorem may be applied to any of the regions V_i, hence

$$\oint_{S_i} \vec{Q}_i . \vec{n}_i dS_i = \int_{V_i} \nabla . \vec{Q}_i dV_i{}^{1)} \tag{V-8a}$$

and, consequently,

$$\sum_i \oint_{S_i} \vec{Q}_i . \vec{n}_i dS_i = \sum_i \int_{V_i} \nabla . \vec{Q}_i dV_i, \tag{V-8b}$$

with $\vec{Q}_i$ and $\nabla . \vec{Q}_i$ defined by (V-4) and (V-5b), respectively. Any of the interfaces contributes to two surface integrals. Consider the interface between surfaces S_i and S_{i+1}. The boundary conditions on this specific interface imply

---

1) Here the summation convention does not apply to the region index i.

$$\vec{Q}_i.\vec{n}_i = -\vec{Q}_{i+1}.\vec{n}_{i+1}. \tag{V-8c}$$

Hence, the surface integrals cancel at any interface, so the left-hand side in equation (V-8b) may be replaced by one surface integral over the outer surface S. The right-hand side in equation (V-8b) may be replaced by one volume integral over the total volume V. So we obtain again equation (V-7), which has now been shown to be valid for any inhomogeneous acoustic medium, containing arbitrary interfaces.

Let us consider a special situation. We choose identical medium parameters for state A and state B throughout volume V, according to

$$K^A(\vec{r}) = K^B(\vec{r}) \triangleq K(\vec{r}) \tag{V-9a}$$

and

$$\rho^A(\vec{r}) = \rho^B(\vec{r}) \triangleq \rho(\vec{r}). \tag{V-9b}$$

Furthermore, we choose monopole sources in V at $\vec{r}_A$ and $\vec{r}_B$, respectively, according to

$$S^A(\vec{r},\omega) = \delta(\vec{r}-\vec{r}_A)S_0(\omega) \tag{V-10a}$$

and

$$S^B(\vec{r},\omega) = \delta(\vec{r}-\vec{r}_B)S_0(\omega), \tag{V-10b}$$

where $S_0(\omega)$ is the source signature. Thus the reciprocity theorem (V-7) simplifies to

$$\oint_S \left[P^A \left(\frac{1}{\rho}\nabla P^B\right) - P^B \left(\frac{1}{\rho}\nabla P^A\right)\right].\vec{n}\,dS = \left[P^B(\vec{r}_A,\omega) - P^A(\vec{r}_B,\omega)\right] S_0(\omega). \tag{V-11}$$

This reciprocity theorem can be further simplified in the following three situations:

1. *S is a rigid boundary in state A and state B*

In this case the normal component of the particle velocity on S vanishes in both states, or, using the equation of motion (III-11b),

$$\nabla P^A.\vec{n} = 0 \quad \text{on } S \tag{V-12a}$$

and

$$\nabla P^B.\vec{n} = 0 \quad \text{on } S. \tag{V-12b}$$

Hence, the surface integral in the left-hand side of (V-11) vanishes.

2. *S is a free boundary in state A and state B*

In this case the acoustic pressure on S vanishes in both states,

$$P^A = 0 \quad \text{on } S \tag{V-13a}$$

and

$$P^B = 0 \quad \text{on } S. \tag{V-13b}$$

Hence, again the surface integral in the left-hand side of (V-11) vanishes.

3. *V is unbounded in state A and state B*

Assume that the medium is homogeneous outside a sphere with a finite radius and let S_∞ be a sphere with infinite radius. Then at any point on S_∞ the wave fronts are locally plane and propagate parallel to $\vec{n}$. Hence

$$\nabla P^A.\vec{n} = \mp jkP^A \quad \text{on } S_\infty \tag{V-14a}$$

and

$$\nabla P^B.\vec{n} = \mp jkP^B \quad \text{on } S_\infty, \tag{V-14b}$$

where

$$k = \frac{\omega}{c} = \frac{\omega}{\sqrt{K/\rho}}. \qquad \text{(V-14c)}$$

In (V-14a) and (V-14b) the -sign corresponds to waves propagating *outward* through S_∞ whereas the +sign corresponds to waves propagating *inward* through S_∞. When state A and state B represent both a physical situation, then in (V-14a) as well as in (V-14b) the -sign must be chosen and the surface integral in (V-11) vanishes[1)]. This is known as *Sommerfeld's radiation condition.* For a more detailed discussion the reader is referred to Bleistein (1984).

In the three situations described above the reciprocity theorem (V-11) simplifies to

$$P^B(\vec{r}_A,\omega) = P^A(\vec{r}_B,\omega). \qquad \text{(V-15)}$$

This is probably the best known formulation of the reciprocity principle. It states that the acoustic pressure P^B at $\vec{r}_A$ related to a monopole source with signature $S_0(\omega)$ at $\vec{r}_B$ is identical to the acoustic pressure P^A at $\vec{r}_B$ related to a monopole source with the same signature $S_0(\omega)$ at $\vec{r}_A$. Note that this principle holds for arbitrarily inhomogeneous acoustic media.

V. 3 ACOUSTIC REPRESENTATION THEOREMS

V.3.1. Acoustic Green's functions

An acoustic Green's function defines the impulse response of a fluid medium. For an impulse at $\vec{r}_A$, the Green's function satisfies the following two-way wave equation in the space-time domain

$$\nabla.\left(\frac{1}{\rho(\vec{r})}\nabla g(\vec{r},\vec{r}_A,t)\right) - \frac{1}{K(\vec{r})}\frac{\partial^2 g(\vec{r},\vec{r}_A,t)}{\partial t^2} = -\delta(\vec{r}-\vec{r}_A)\delta(t), \qquad \text{(V-16a)}$$

1) In chapter VII, where we discuss inverse wave field extrapolation, one of the states represents the physical situation whereas the other state represents the non-physical situation of back-propagation. As a consequence, the surface integral does not vanish.

(Morse and Feshbach, 1953) with initial conditions

$$g(\vec{r},\vec{r}_A,t) = 0 \qquad \text{for } t < 0 \tag{V-16b}$$

and

$$\frac{\partial g(\vec{r},\vec{r}_A,t)}{\partial t} = 0 \qquad \text{for } t < 0. \tag{V-16c}$$

For the moment we do not specify the boundary conditions for $g(\vec{r},\vec{r}_A,t)$.

In equation (V-16), $g(\vec{r},\vec{r}_A,t)$ denotes the impulse response at observation point $\vec{r}$ as a function of time t, related to an impulse at source point $\vec{r}_A$ at t=0. In this notation, the reciprocity principle (V-15) can be reformulated as

$$g(\vec{r}_A,\vec{r}_B,t) = g(\vec{r}_B,\vec{r}_A,t). \tag{V-17}$$

The initial conditions (V-16b) and (V-16c) ensure that $g(\vec{r},\vec{r}_A,t)$ is a *causal* wave field which propagates away from the source at $\vec{r}_A$. Therefore we will also refer to $g(\vec{r},\vec{r}_A,t)$ as the *forward propagating* acoustic Green's wave field. Opposed to this, we also define an anti-causal or *backward propagating* acoustic Green's wave field $\hat{g}(\vec{r},\vec{r}_A,t)$ which satisfies the same two-way wave equation

$$\nabla.\left(\frac{1}{\rho(\vec{r})}\nabla\hat{g}(\vec{r},\vec{r}_A,t)\right) - \frac{1}{K(\vec{r})}\frac{\partial^2\hat{g}(\vec{r},\vec{r}_A,t)}{\partial t^2} = -\delta(\vec{r}-\vec{r}_A)\delta(t), \tag{V-18a}$$

with final conditions

$$\hat{g}(\vec{r},\vec{r}_A,t) = 0 \qquad \text{for } t > 0 \tag{V-18b}$$

and

$$\frac{\partial \hat{g}(\vec{r},\vec{r}_A,t)}{\partial t} = 0 \quad \text{for } t > 0. \tag{V-18c}$$

In the following we assume that $g(\vec{r},\vec{r}_A,-t)$ and $\hat{g}(\vec{r},\vec{r}_A,t)$ satisfy the same boundary conditions. Then the backward propagating Green's wave field is simply related to the forward propagating Green's wave field, according to

$$\hat{g}(\vec{r},\vec{r}_A,t) = g(\vec{r},\vec{r}_A,-t), \tag{V-19}$$

for all $\vec{r}$, $\vec{r}_A$ and t.

We define the forward and backward propagating Green's wave fields in the space-frequency domain according to

$$G(\vec{r},\vec{r}_A,\omega) \triangleq \int_{-\infty}^{\infty} g(\vec{r},\vec{r}_A,t)e^{-j\omega t}dt \tag{V-20a}$$

and

$$\hat{G}(\vec{r},\vec{r}_A,\omega) \triangleq \int_{-\infty}^{\infty} \hat{g}(\vec{r},\vec{r}_A,t)e^{-j\omega t}dt, \tag{V-20b}$$

respectively. Note that, in analogy with (V-17), the reciprocity principle reads

$$G(\vec{r}_A,\vec{r}_B,\omega) = G(\vec{r}_B,\vec{r}_A,\omega). \tag{V-21}$$

Furthermore, in analogy with (V-19),

$$\hat{G}(\vec{r},\vec{r}_A,\omega) = G^*(\vec{r},\vec{r}_A,\omega). \tag{V-22}$$

We define the forward and backward propagating Green's wave fields in the wavenumber-frequency domain according to

$$\tilde{G}(k_x,k_y,z;x_A,y_A,z_A;\omega) \triangleq$$

$$\iint_{-\infty}^{\infty} G(x,y,z;x_A,y_A,z_A;\omega)e^{j(k_x x+k_y y)}dxdy \tag{V-23a}$$

and

$$\tilde{\hat{G}}(k_x,k_y,z;x_A,y_A,z_A;\omega) \triangleq$$

$$\iint_{-\infty}^{\infty} \hat{G}(x,y,z;x_A,y_A,z_A;\omega)e^{j(k_x x+k_y y)}dxdy, \tag{V-23b}$$

respectively. Note that, in analogy with (V-22),

$$\tilde{\hat{G}}(k_x,k_y,z;x_A,y_A,z_A;\omega) = \tilde{G}^*(-k_x,-k_y,z;x_A,y_A,z_A;\omega). \tag{V-24}$$

As an example, we consider the *free space* Green's wave fields in an unbounded homogeneous fluid. In analogy with (I-17), we may write for the forward and backward propagating Green's wave fields in the space-time domain

$$g(\vec{r},\vec{r}_A,t) = \frac{\rho}{4\pi}\frac{\delta(t-\Delta r/c)}{\Delta r} \tag{V-25a}$$

and

$$\hat{g}(\vec{r},\vec{r}_A,t) = \frac{\rho}{4\pi}\frac{\delta(t+\Delta r/c)}{\Delta r}, \tag{V-25b}$$

respectively, with the propagation velocity c being given by

$$c = \sqrt{K/\rho} \tag{V-25c}$$

and Δr being the distance between the source point $\vec{r}_A$ and the observation point $\vec{r}$, according to

$$\Delta r = | \vec{r} - \vec{r}_A | . \tag{V-25d}$$

Applying (V-20a) and (V-20b) to the Green's wave fields (V-25a) and (V-25b), respectively, yields

$$G(\vec{r},\vec{r}_A,\omega) = \frac{\rho}{4\pi} \frac{e^{-jk\Delta r}}{\Delta r} \tag{V-26a}$$

and

$$\hat{G}(\vec{r},\vec{r}_A,\omega) = \frac{\rho}{4\pi} \frac{e^{+jk\Delta r}}{\Delta r} , \tag{V-26b}$$

with

$$k = \omega/c. \tag{V-26c}$$

Applying (V-23a) and (V-23b) to the Green's wave fields (V-26a) and (V-26b), respectively, yields

$$\tilde{G}(k_x,k_y,z;x_A,y_A,z_A;\omega) = \rho e^{j(k_x x_A + k_y y_A)} \frac{e^{-jk_z|z-z_A|}}{2jk_z} \tag{V-27a}$$

and

$$\tilde{\hat{G}}(k_x,k_y,z;x_A,y_A,z_A;\omega) = \rho e^{j(k_x x_A + k_y y_A)} \frac{e^{+jk_z^*|z-z_A|}}{-2jk_z^*} , \tag{V-27b}$$

where

$$k_z \hat{=} +\sqrt{k^2 - k_x^2 - k_y^2} \qquad \text{for } k_x^2+k_y^2 \leq k^2 \tag{V-27c}$$

and

$$k_z \hat{=} -j\sqrt{k_x^2 + k_y^2 - k^2} \qquad \text{for } k_x^2+k_y^2 > k^2, \tag{V-27d}$$

see Berkhout (1985, Appendix F).

V.3.2 Acoustic Kirchhoff-Helmholtz integral

We return to the inhomogeneous situation. Our starting point is Rayleigh's reciprocity theorem (V-7)

$$\oint_S \left[P^A\left(\frac{1}{\rho^B}\nabla P^B\right) - P^B\left(\frac{1}{\rho^A}\nabla P^A\right)\right].\vec{n}\,dS =$$

$$\int_V \Big[P^B S^A - P^A S^B + \omega^2\left(\frac{1}{K^A} - \frac{1}{K^B}\right)P^A P^B$$

$$+ \left(\frac{1}{\rho^B} - \frac{1}{\rho^A}\right)(\nabla P^A).(\nabla P^B)\Big]\,dV. \qquad \text{(V-28)}$$

In the following, state A will represent a forward or backward propagating Green's wave field in a *reference medium* whereas state B will represent the physical wave field in the true medium. Hence, we make the following substitutions:

State A:

$$K^A(\vec{r}) \rightarrow \bar{K}(\vec{r}),$$

$$\rho^A(\vec{r}) \rightarrow \bar{\rho}(\vec{r}), \qquad \text{(V-29a)}$$

$$S^A(\vec{r},\omega) \rightarrow \delta(\vec{r}-\vec{r}_A), \text{ with } \vec{r}_A \text{ in } V,$$

and

$$P^A(\vec{r},\omega) \rightarrow G(\vec{r},\vec{r}_A,\omega) \text{ or } G^*(\vec{r},\vec{r}_A,\omega).$$

State B:

$$K^B(\vec{r}) \rightarrow K(\vec{r}) = \bar{K}(\vec{r}) + \Delta K(\vec{r}),$$

$$\rho^B(\vec{r}) \rightarrow \rho(\vec{r}) = \bar{\rho}(\vec{r}) + \Delta\rho(\vec{r}), \qquad \text{(V-29b)}$$

$$S^B(\vec{r},\omega) \rightarrow S(\vec{r},\omega)$$

and

$$P^B(\vec{r},\omega) \rightarrow P(\vec{r},\omega).$$

With these substitutions, we obtain the following two *representation theorems*

$$P(\vec{r}_A,\omega) = \oint_S \left[G\left(\frac{1}{\rho}\nabla P\right) - P\left(\frac{1}{\bar{\rho}}\nabla G\right)\right].\vec{n}\,dS + \int_V \left[GS - \omega^2\frac{\Delta K}{\bar{K}K}GP + \frac{\Delta\rho}{\bar{\rho}\rho}(\nabla G).(\nabla P)\right]dV, \tag{V-30a}$$

or, equivalently,

$$P(\vec{r}_A,\omega) = \oint_S \left[G^*\left(\frac{1}{\rho}\nabla P\right) - P\left(\frac{1}{\rho}\nabla G^*\right)\right].\vec{n}\,dS + \int_V \left[G^*S - \omega^2\frac{\Delta K}{\bar{K}K}G^*P + \frac{\Delta\rho}{\bar{\rho}\rho}(\nabla G^*).(\nabla P)\right]dV. \tag{V-30b}$$

These expressions are also known as the acoustic *Kirchhoff-Helmholtz integral* formulas. They are the basis for multi-dimensional acoustic forward and inverse scattering techniques (see, for instance, Clayton and Stolt, 1981).

Here we follow a different approach. By choosing the reference medium $(\bar{K},\bar{\rho})$ equal or close to the actual medium (K,ρ) in V and on S it is justified to ignore the deviation parameters $(\Delta K, \Delta\rho)$ in (V-30a) and (V-30b). When we also assume that volume V is source-free then expressions (V-30a) and (V-30b) simplify to

$$P(\vec{r}_A,\omega) \approx \oint_S \frac{1}{\bar{\rho}}\left[G\nabla P - P\nabla G\right].\vec{n}\,dS, \tag{V-31a}$$

or, equivalently,

$$P(\vec{r}_A,\omega) \approx \oint_S \frac{1}{\bar{\rho}}\left[G^*\nabla P - P\nabla G^*\right].\vec{n}\,dS. \tag{V-31b}$$

These expressions describe the acoustic wave field at $\vec{r}_A$ in V in terms of the wave field and its gradient on S, enclosing V (see Figures V-3a and V-3b). The *scattering* effects related to the deviation parameters $\Delta K(\vec{r})$ and $\Delta\rho(\vec{r})$ are zero or neglected. However, the *propagation* effects related to the reference parameters $\bar{K}(\vec{r})$ and $\bar{\rho}(\vec{r})$ are properly included and therefore expressions (V-31a) and (V-31b) are the basis for multi-dimensional forward and inverse wave field extrapolation techniques.

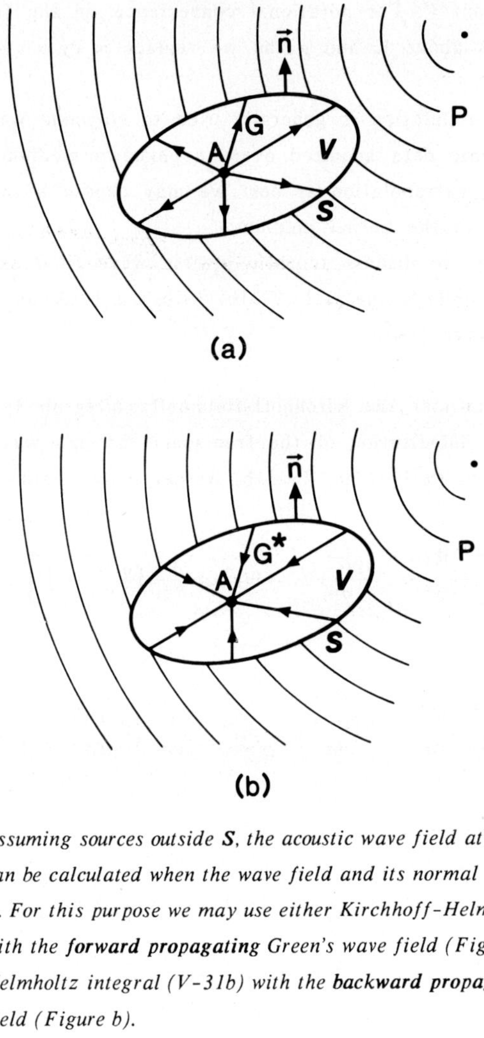

*Figure V-3: Assuming sources outside **S**, the acoustic wave field at any point A inside **S** can be calculated when the wave field and its normal derivative are known on **S**. For this purpose we may use either Kirchhoff-Helmholtz integral (V-31a) with the **forward propagating** Green's wave field (Figure a), or Kirchhoff-Helmholtz integral (V-31b) with the **backward propagating** Green's wave field (Figure b).*

Forward wave field extrapolation is often used to *simulate* wave propagation in a known medium. Therefore, for forward extrapolation we may often choose the reference medium equal to the actual medium. In the following sections of this chapter we discuss acoustic forward wave field extrapolation, based on Kirchhoff-Helmholtz integral (V-31a) with the forward propagating Green's wave fields, which is exact when $\bar{K}(\vec{r})=K(\vec{r})$ and $\bar{\rho}(\vec{r})=\rho(\vec{r})$ throughout V. For notational convenience, in the following we will omit the bars above K and ρ and we replace $\approx$ by =.

Inverse wave field extrapolation is generally used to *eliminate* wave propagation from seismic data acquired over an unknown medium. Therefore, for inverse extrapolation at best we may choose a reference medium that is close to the actual medium.[1)]
In chapters VII and IX we discuss acoustic inverse wave field extrapolation based on Kirchhoff-Helmholtz integral (V-31b) with the backward propagating Green's wave fields.

As an example, we consider the Kirchhoff-Helmholtz integrals for a homogeneous medium. Substitution of the free space Green's wave fields (V-26a) and (V-26b) into (V-31a) and (V-31b), respectively, yields

$$P(\vec{r}_A,\omega) = \frac{1}{4\pi} \oint_S \left[\frac{e^{-jk\Delta r}}{\Delta r} \frac{\partial P(\vec{r},\omega)}{\partial n} - P(\vec{r},\omega)\frac{\partial}{\partial n}\left(\frac{e^{-jk\Delta r}}{\Delta r}\right)\right] dS, \tag{V-32a}$$

or, equivalently,

$$P(\vec{r}_A,\omega) = \frac{1}{4\pi} \oint_S \left[\frac{e^{+jk\Delta r}}{\Delta r} \frac{\partial P(\vec{r},\omega)}{\partial n} - P(\vec{r},\omega)\frac{\partial}{\partial n}\left(\frac{e^{+jk\Delta r}}{\Delta r}\right)\right] dS, \tag{V-32b}$$

where $\partial/\partial n$ stands for $\vec{n}.\nabla$.

V.4 ACOUSTIC TWO-WAY AND ONE-WAY RAYLEIGH INTEGRALS

In this section we review the derivations of the two-way and one-way Rayleigh integrals (Berkhout and Wapenaar, 1989).

---

1) Choosing $\bar{K}(\vec{r})$ and $\bar{\rho}(\vec{r})$ close to $K(\vec{r})$ and $\rho(\vec{r})$ means that the reference medium must be designed in a geologically oriented way. Berkhout (1986) refers to such a reference medium as the *macro subsurface model.*

V.4.1 Boundary conditions for the acoustic Green's functions

We return to the inhomogeneous situation. Consider the Kirchhoff-Helmholtz integral (V-31a) with the forward propagating Green's functions,

$$P(\vec{r}_A,\omega) = \oint_S \frac{1}{\rho(\vec{r})}\left[G(\vec{r},\vec{r}_A,\omega)\frac{\partial P(\vec{r},\omega)}{\partial n} - \frac{\partial G(\vec{r},\vec{r}_A,\omega)}{\partial n}P(\vec{r},\omega)\right]dS. \quad \text{(V-33)}$$

As we showed in the previous section, this expression is exact when the medium parameters for the Green's wave field are identical to the actual medium parameters throughout volume V. Outside V we may choose any convenient medium for the Green's wave field and on (parts of) S we may impose any convenient boundary condition for the Green's wave field. If we choose

$$\frac{\partial G(\vec{r},\vec{r}_A,\omega)}{\partial n} = 0 \quad \text{on } S, \quad \text{(V-34a)}$$

(Neumann boundary condition) then Kirchhoff-Helmholtz integral (V-33) simplifies to

$$P(\vec{r}_A,\omega) = \oint_S \left[\frac{1}{\rho(\vec{r})} G_I(\vec{r},\vec{r}_A,\omega)\frac{\partial P(\vec{r},\omega)}{\partial n}\right]dS. \quad \text{(V-34b)}$$

However, a practical disadvantage is that the Green's wave field G_I may become very complicated because boundary condition (V-34a) means that S is a perfectly reflecting *rigid* surface (with reflection coefficient R=+1). Similarly, if we choose

$$G(\vec{r},\vec{r}_A,\omega) = 0 \quad \text{on } S, \quad \text{(V-35a)}$$

(Dirichlet boundary condition) then Kirchhoff-Helmholtz integral (V-33) simplifies to

$$P(\vec{r}_A,\omega) = -\oint_S \left[\frac{1}{\rho(\vec{r})}\frac{\partial G_{II}(\vec{r},\vec{r}_A,\omega)}{\partial n}P(\vec{r},\omega)\right]dS. \quad \text{(V-35b)}$$

In a manner similar to G_I, $\partial G_{II}/\partial n$ may become very complicated because boundary condition (V-35a) means that S is a perfectly reflecting *free* surface (with R=-1). It is important to bear in mind that the Kirchhoff-Helmholtz integral (V-33) as well as its simplified versions (V-34b) and (V-35b) are in principle *two-way* expressions. In the next section we consider (V-34b) and (V-35b) for a special configuration and we illustrate the two-way properties with an example. Next we divert from what is generally done in the literature and, instead of fully reflecting boundary conditions, we choose fully absorbing boundary conditions for the Green's wave fields on S. Following this alternative route, we show how to derive *one-way* versions of (V-34b) and (V-35b).

V.4.2 Acoustic two-way Rayleigh integrals

Consider the half-space geometry of Figure V-4. Closed surface S consists of a horizontal flat surface S_0 at $z=z_0$ and a hemi-sphere S_1 in the lower half-space $z \geq z_0$, with midpoint A and radius r_1. Assuming that the sources of the acoustic wave field P are situated in the upper half-space $z<z_0$, then the contribution of the Kirchhoff-Helmholtz integral over S_1 to the acoustic pressure at A vanishes if r_1 goes to infinity (Sommerfeld radiation conditions, see also section V-2). Hence, for this situation equation (V-33) may be replaced by

$$P(\vec{r}_A,\omega) = \int\int_{-\infty}^{\infty} \frac{1}{\rho(\vec{r})} \left[\frac{\partial G(\vec{r},\vec{r}_A,\omega)}{\partial z} P(\vec{r},\omega) - G(\vec{r},\vec{r}_A,\omega) \frac{\partial P(\vec{r},\omega)}{\partial z} \right]_{z_0} dxdy. \tag{V-36}$$

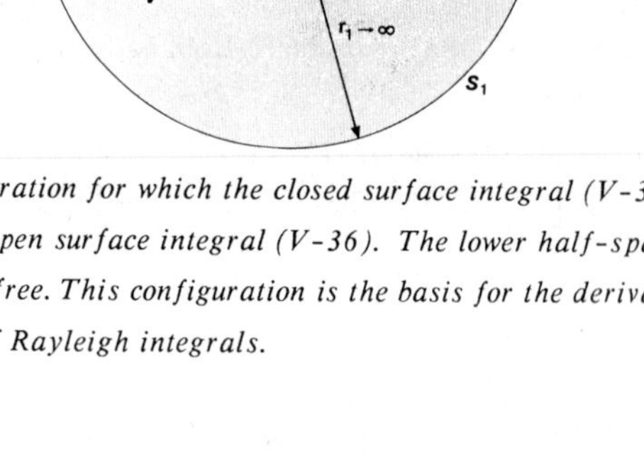

Figure V-4: Configuration for which the closed surface integral (V-33) may be replaced by the open surface integral (V-36). The lower half-space is assumed to be source-free. This configuration is the basis for the derivation of several types of Rayleigh integrals.

We consider again the Neumann and Dirichlet boundary conditions for G,

$$\frac{\partial G(\vec{r},\vec{r}_A,\omega)}{\partial z} = 0 \quad \text{at } z=z_o \tag{V-37a}$$

or

$$G(\vec{r},\vec{r}_A,\omega) = 0 \quad \text{at } z=z_o, \tag{V-37b}$$

respectively. Conditions (V-37) are fulfilled if we assume for G either a rigid surface or a free surface at z_o. In both cases the surface acts as a perfect reflector, so we may alternatively interpret the Green's function as if it was caused by *two* monopoles situated symmetrically with respect to $z=z_o$ in a reference medium which is also symmetric with respect to $z=z_o$ (classical representation). If these monopoles have the same polarity (Figure V-5a), then Neumann's condition (V-37a) is satisfied and Kirchhoff-Helmholtz integral (V-36) may be replaced by the following integral

$$P(\vec{r}_A,\omega) = -\iint_{-\infty}^{\infty}\left[\frac{1}{\rho(\vec{r})}\, G_I(\vec{r},\vec{r}_A,\omega)\,\frac{\partial P(\vec{r},\omega)}{\partial z}\right]_{z_o} dxdy. \tag{V-38a}$$

For the special situation of a homogeneous lower half-space, $G_I(\vec{r},\vec{r}_A,\omega)$ at z_o equals twice the free space solution (V-26a), hence, for this situation (V-38a) reads

$$P(\vec{r}_A,\omega) = -\frac{1}{2\pi}\iint_{-\infty}^{\infty}\left[\frac{e^{-jk\Delta r}}{\Delta r}\,\frac{\partial P(\vec{r},\omega)}{\partial z}\right]_{z_o} dxdy. \tag{V-38b}$$

If the monopoles for the Green's function have opposite polarity (Figure V-5b), then Dirichlet's condition (V-37b) is satisfied and Kirchhoff-Helmholtz integral (V-36) may be replaced by the following integral

$$P(\vec{r}_A,\omega) = \iint_{-\infty}^{\infty}\left[\frac{1}{\rho(\vec{r})}\,\frac{\partial G_{II}(\vec{r},\vec{r}_A,\omega)}{\partial z}\, P(\vec{r},\omega)\right]_{z_o} dxdy. \tag{V-39a}$$

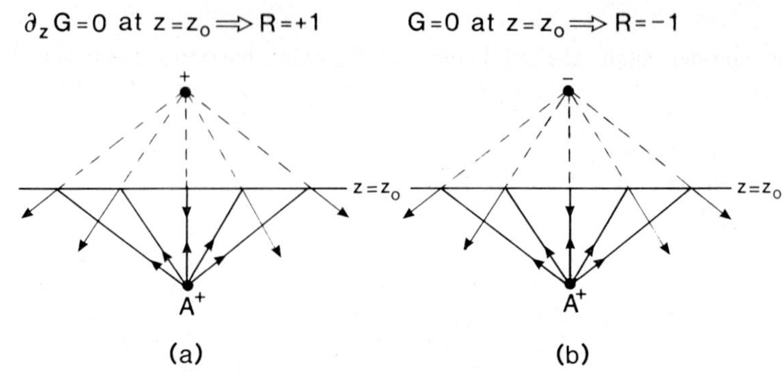

Figure V-5: a. Neumann boundary condition:
surface $z=z_0$ is a perfect reflector with R=+1.
b. Dirichlet boundary condition:
surface $z=z_0$ is a perfect reflector with R=-1.

For the special situation of a homogeneous lower half-space, $\partial G_{II}(\vec{r},\vec{r}_A,\omega)/\partial z$ at z_0 equals twice the free space solution, hence, for this situation (V-39a) reads

$$P(\vec{r}_A,\omega) = \frac{1}{2\pi}\int\int_{-\infty}^{\infty} \left[\frac{\partial}{\partial z}\left(\frac{e^{-jk\Delta r}}{\Delta r}\right) P(\vec{r},\omega)\right]_{z_0} dxdy. \tag{V-39b}$$

Lord Rayleigh derived expressions (V-38b) and (V-39b) to describe the radiated field of a planar vibration source in an infinite baffle (Rayleigh, 1965). Therefore, following Berkhout and Van Wulfften Palthe (1979), we refer to equations (V-38) and (V-39) as the acoustic *Rayleigh I* and *Rayleigh II* integrals, respectively.

For practical use expressions (V-38a) and (V-39a) are rather inconvenient, because in the general inhomogeneous case the *two-way* Green's functions G_I or $\partial G_{II}/\partial z$ may contain strong surface-related multiple reflections. This means that these Green's functions must be designed very accurately, which is best illustrated with a simple example. Referring to Figure V-6a, we consider a 2-D medium which consists of two homogeneous half-spaces $z \geq z_1$ (with propagation velocity c_1) and $z<z_1$ (with propagation velocity c_0). The surface S_0 lies in the homogeneous upper half-space at $z=z_0$,

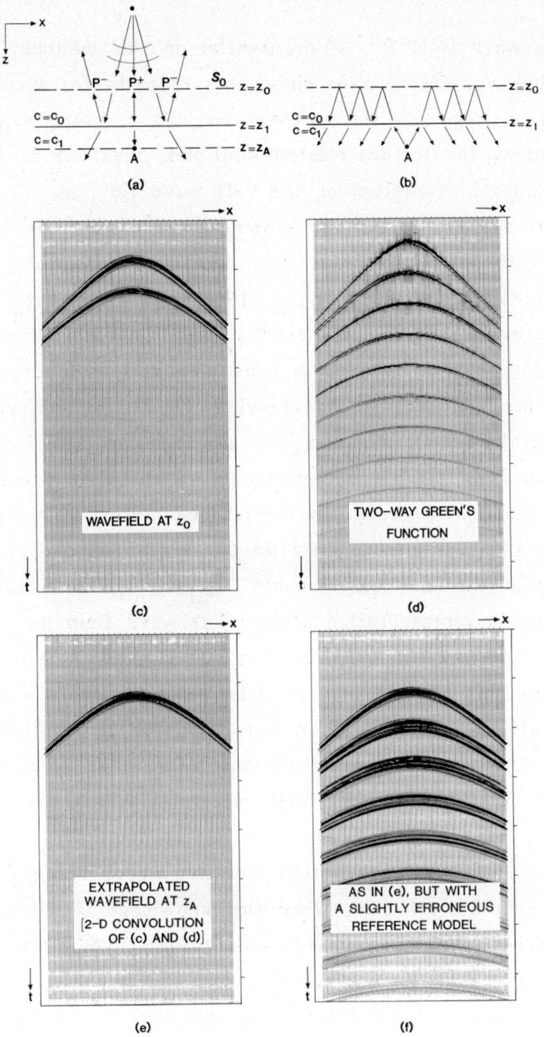

*Figure V-6: Application of the **two-way** Rayleigh II integral*

a. Raypaths in the actual medium for the wave field $P(\vec{r},\omega)$.

b. Raypaths in the reference medium for the two-way Green's function.

c. Wave field at $z=z_0$.

d. Two-way Green's function (band-limited) at $z=z_0$.

e. Exact extrapolation result at $z=z_A$.

f. Extrapolation result at $z=z_A$, when using a reference medium that is slightly in error.

with $z_0<z_1$. The wave field $P(\vec{r},\omega)$ propagating in this medium, is radiated by a source above z_0. Figure V-6a shows the raypaths for $P(\vec{r},\omega)$. In Figure V-6b the raypaths for the two-way Green's function $\partial G_{II}(\vec{r},\vec{r}_A,\omega)/\partial z$ are shown, including the surface-related multiples. Figure V-6c is a space-time domain representation of the 2-D wave field $p(\vec{r},t)$, registered at z_0. Note that the direct and the scattered wave field can be clearly distinguished. Figure V-6d is a space-time domain representation of the 2-D two-way Green's function $\partial g_{II}(\vec{r},\vec{r}_A,t)/\partial z$, registered at z_0. Note that the multiples between the reflector at $z=z_1$ and the free surface are clearly visible. Figure V-6e is a space-time domain representation of the 2-D wave field $p(\vec{r}_A,t)$, obtained by applying the 2-D version of two-way Rayleigh II integral (V-39a) for all x_A at $z=z_A$ and by applying an inverse Fourier transform from the frequency domain to the time domain. Note that this exact result represents the transmitted downgoing wave at $z=z_A$ (see also Figure V-6a). In practice we do not always have available an *exact* description of the 'medium response' $\partial g_{II}(\vec{r},\vec{r}_A,t)/\partial z$. Figure V-6f is a space-time domain representation of the 2-D wave field $p(\vec{r}_A,t)$ which was obtained by applying the two-way Rayleigh II integral with a Green's function that was only slightly in error (the reflector depth $z=z_1$ in Figure V-6b was one half of a wavelength in error for the central frequency). Note that this result contains many spurious reflections. From this example we may conclude that two-way Rayleigh integrals (V-38a) and (V-39a) in general require a very accurate generation of the multiple reflections. This can only be accomplished if the actual medium, where the extrapolation occurs, is very accurately known. When this knowledge is not available, the two-way Rayleigh integrals (V-38a) and (V-39a) have practical use only in the situation where the subsurface does not contain reflectors but smooth transition zones only. In that case, the two-way Green's functions do not include significant multiple reflections.

V.4.3 Acoustic one-way Rayleigh integrals

An important property of the Kirchhoff-Helmholtz integral is that the choice of the medium for the Green's function is not unique:

Inside volume V the medium for the Green's function $G(\vec{r},\vec{r}_A,\omega)$ is equal or close to the medium for the acoustic wave field $P(\vec{r},\omega)$.

Outside volume V the medium for the Green's function $G(\vec{r},\vec{r}_A,\omega)$ may be chosen in any convenient way.

We made use of this property already in the previous section, where we chose a reference medium for G that is different from the actual medium for P (compare Figure V-6b with V-6a). By choosing a reference medium with a fully reflecting boundary, we got a Green's function containing many significant multiple reflections. Let us therefore choose instead a reference medium for G which is fully *non-reflecting* outside V. For the half-space geometry of Figure V-4 this means that we now choose for G a non-reflecting upper half-space $z<z_0$. Hence

$$K(x,y,z<z_0) = K(x,y,z_0) \qquad \text{for all } z<z_0 \tag{V-40a}$$

and

$$\rho(x,y,z<z_0) = \rho(x,y,z_0) \qquad \text{for all } z<z_0, \tag{V-40b}$$

(see also section III.3.2, where it was shown that downgoing and upgoing waves fully decouple in a medium described by (V-40)). With this choice no downgoing waves return from the upper half-space, so G is purely *upgoing* at $z=z_0$:

$$G(\vec{r},\vec{r}_A,\omega) = G^-(\vec{r},\vec{r}_A,\omega) \qquad \text{at } z=z_0. \tag{V-41a}$$

In terms of boundary conditions we may say that surface $z=z_0$ is an *absorbing* boundary for G.

The sources for the acoustic wave field $P(\vec{r},\omega)$ are situated in the upper half-space, so at z_0 this wave field consists of the downgoing incident wave field (including higher order terms) $P^+(\vec{r},\omega)$ and the upgoing scattered wave field (including higher order terms) $P^-(\vec{r},\omega)$, according to

$$P(\vec{r},\omega) = P^+(\vec{r},\omega) + P^-(\vec{r},\omega) \qquad \text{at } z=z_0. \tag{V-41b}$$

Substitution of (V-41a) and (V-41b) into Kirchhoff-Helmholtz integral (V-36) yields

$$P(\vec{r}_A,\omega) = \int\int_{-\infty}^{\infty} \frac{1}{\rho}\left[\frac{\partial G^-}{\partial z}(P^+ + P^-) - G^-\left(\frac{\partial P^+}{\partial z} + \frac{\partial P^-}{\partial z}\right)\right]_{z_0} dxdy, \tag{V-42a}$$

or

$$P(\vec{r}_A,\omega) = \int\int_{-\infty}^{\infty} \frac{1}{\rho}\left[\frac{\partial G^-}{\partial z}P^+ - G^-\frac{\partial P^+}{\partial z}\right]_{z_0} dxdy + \int\int_{-\infty}^{\infty} \frac{1}{\rho}\left[\frac{\partial G^-}{\partial z}P^- - G^-\frac{\partial P^-}{\partial z}\right]_{z_0} dxdy. \tag{V-42b}$$

In Appendix B, section B.2 we show that the only contribution to $P(\vec{r}_A,\omega)$ comes from the *first* integral in the right-hand side of equation (V-42b). In other words, only the integral with the wave fields P^+ and G^- propagating in *opposite* directions through z_0 contribute to the wave field $P(\vec{r}_A,\omega)$, see also Figure V-7.

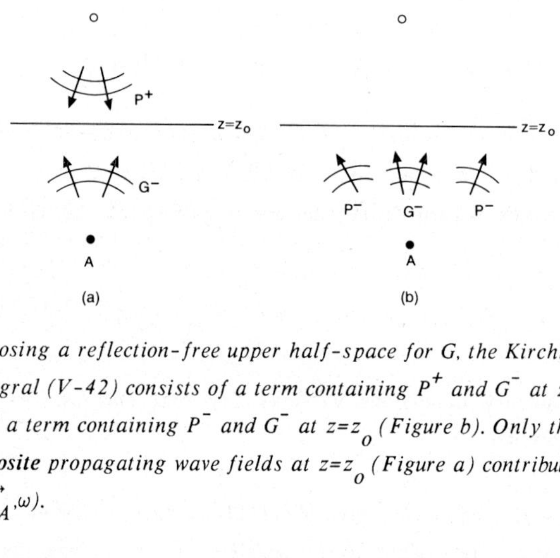

Figure V-7: Choosing a reflection-free upper half-space for G, the Kirchhoff-Helmholtz integral (V-42) consists of a term containing P^+ and G^- at $z=z_0$ (Figure a) and a term containing P^- and G^- at $z=z_0$ (Figure b). Only the term with the ***opposite*** *propagating wave fields at $z=z_0$ (Figure a) contributes to the result $P(\vec{r}_A,\omega)$.*

Hence

$$P(\vec{r}_A,\omega) = \int\int_{-\infty}^{\infty} \frac{1}{\rho}\left[\frac{\partial G^-}{\partial z}P^+ - G^-\frac{\partial P^+}{\partial z}\right]_{z_0} dxdy, \tag{V-43a}$$

or

$$P(\vec{r}_A,\omega) = \iint_{-\infty}^{\infty} \left[\frac{1}{\rho}\frac{\partial G^-}{\partial z}P^+\right]_{z_0} dxdy + \iint_{-\infty}^{\infty} \left[\frac{-1}{\rho}G^-\frac{\partial P^+}{\partial z}\right]_{z_0} dxdy. \tag{V-43b}$$

From Appendix B it also follows that the two integrals in the right-hand side of equation (V-43b) are identical. Hence, equation (V-43b) may finally be rewritten as

$$P(\vec{r}_A,\omega) = -2\iint_{-\infty}^{\infty} \left[\frac{1}{\rho(\vec{r})} G^-(\vec{r},\vec{r}_A,\omega) \frac{\partial P^+(\vec{r},\omega)}{\partial z}\right]_{z_0} dxdy, \tag{V-44a}$$

or, equivalently,

$$P(\vec{r}_A,\omega) = 2\iint_{-\infty}^{\infty} \left[\frac{1}{\rho(\vec{r})} \frac{\partial G^-(\vec{r},\vec{r}_A,\omega)}{\partial z} P^+(\vec{r},\omega)\right]_{z_0} dxdy. \tag{V-44b}$$

In analogy with equations (V-38a) and (V-39a), we call equations (V-44a) and (V-44b) the *one-way versions* of the Rayleigh I and Rayleigh II integrals, respectively. Equations (V-44a) and (V-44b) are valid for an arbitrarily inhomogeneous acoustic medium. The only assumption is that at $z=z_0$ the downgoing and upgoing waves are decoupled. This assumption is validated when the vertical derivatives of the medium parameters vanish at $z=z_0$, which is expressed by equation (B-4). For the special situation of a homogeneous medium, we may substitute for G^- the free space solution (V-26a), yielding

$$P(\vec{r}_A,\omega) = -\frac{1}{2\pi}\iint_{-\infty}^{\infty} \left[\frac{e^{-jk\Delta r}}{\Delta r} \frac{\partial P^+(\vec{r},\omega)}{\partial z}\right]_{z_0} dxdy, \tag{V-45a}$$

or, equivalently,

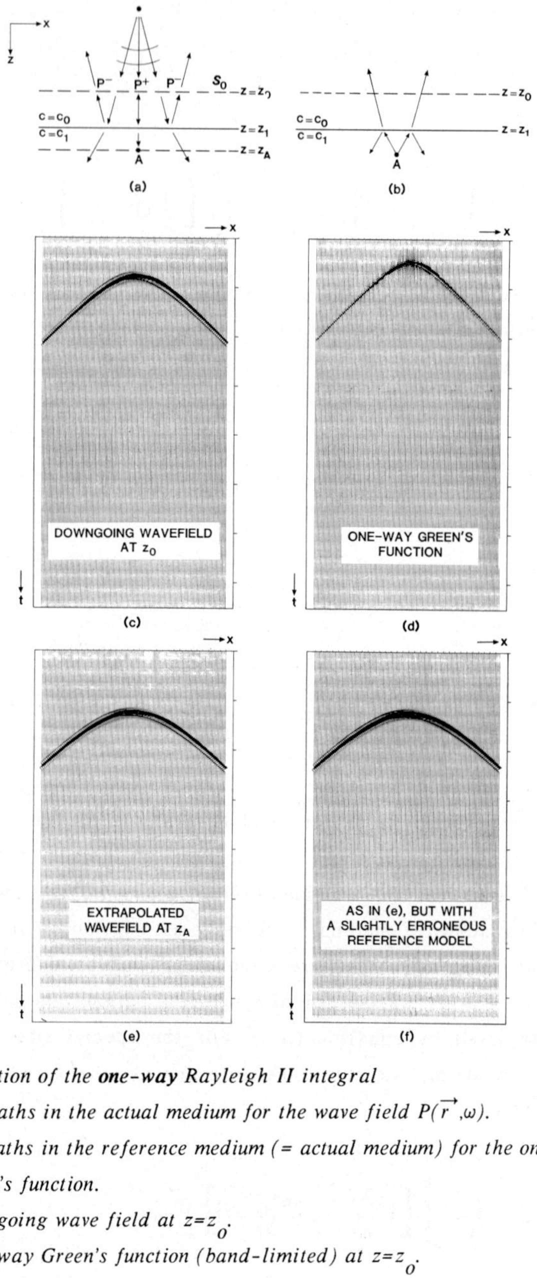

Figure V-8: Application of the ***one-way*** *Rayleigh II integral*

a. Raypaths in the actual medium for the wave field $P(\vec{r},\omega)$.

b. Raypaths in the reference medium (= actual medium) for the one-way Green's function.

c. Downgoing wave field at $z=z_o$.

d. One-way Green's function (band-limited) at $z=z_o$.

e. Exact extrapolation result at $z=z_A$.

f. Extrapolation result at $z=z_A$*, when using a reference medium that is slightly in error.*

$$P(\vec{r}_A,\omega) = \frac{1}{2\pi} \iint_{-\infty}^{\infty} \left[\frac{\partial}{\partial z} \left(\frac{e^{-jk\Delta r}}{\Delta r} \right) P^+(\vec{r},\omega) \right]_{z_0} dxdy. \qquad \text{(V-45b)}$$

Hence, for this situation the two-way Rayleigh integrals (V-38b) and (V-39b) are identical to the one-way Rayleigh integrals (V-45a) and (V-45b), respectively (bear in mind that for this situation $P=P^+$ for $z \geq z_0$).

For arbitrary inhomogeneous media, however, one-way Rayleigh integrals (V-44a) and (V-44b) are very different from two-way Rayleigh integrals (V-38a) and (V-39a). Because the one-way Green's functions do not contain *surface-related multiples*, one-way Rayleigh integrals (V-44a) and (V-44b) are rather insensitive to small errors in the reference model. This is illustrated in Figure V-8, where the experiment of Figure V-6 was repeated for the downgoing wave field P^+, using the 2-D version of the one-way Rayleigh II integral for downgoing wave fields, as given by (V-44b). Note that a small error in the reference medium has only a minor effect on the extrapolation result (Figure V-8f). This is typical for one-way techniques.

V.5 ACOUSTIC FORWARD WAVE FIELD EXTRAPOLATION OPERATORS

V.5.1 Integral formulation of acoustic forward wave field extrapolation

We consider again the inhomogeneous situation. Our starting point for deriving acoustic forward wave field extrapolation operators is the one-way version of the acoustic Rayleigh II integral (V-44b),

$$P(\vec{r}_A,\omega) = 2 \iint_{-\infty}^{\infty} \left[\frac{1}{\rho(\vec{r})} \frac{\partial G^-(\vec{r},\vec{r}_A,\omega)}{\partial z} P^+(\vec{r},\omega) \right]_{z_0} dxdy. \qquad \text{(V-46)}$$

Here $P^+(\vec{r},\omega)$ for $z=z_0$ represents the downgoing part of the total acoustic pressure at z_0, related to sources in the upper half-space, see Figure V-9a. $G^-(\vec{r},\vec{r}_A,\omega)$ for $z=z_0$ represents the upgoing Green's wave field at z_0, related to a monopole source at $\vec{r}_A$ in the lower half-space, see Figure V-9b. $P(\vec{r}_A,\omega)$ represents the total acoustic pressure at $\vec{r}_A$ in the lower half-space. Hence, equation (V-46) describes wave field extrapolation of the *downgoing* (i.e., *one-way*) wave field $P^+(\vec{r},\omega)$ at z_0, yielding the *total* (i.e., *two-way*) wave field at $\vec{r}_A$.

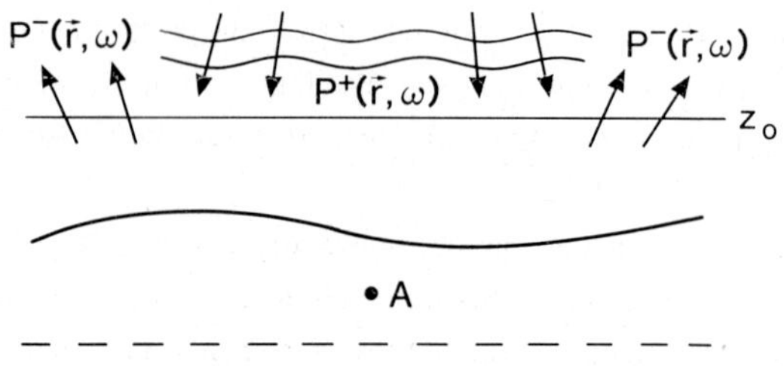

(a)

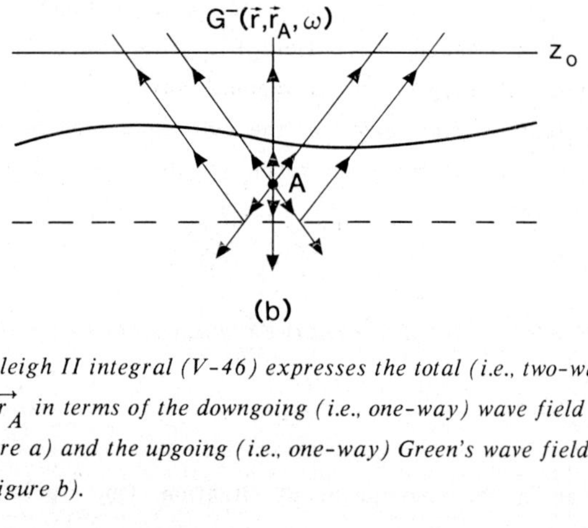

(b)

Figure V-9: The Rayleigh II integral (V-46) expresses the total (i.e., two-way) wave field at $\vec{r}_A$ in terms of the downgoing (i.e., one-way) wave field $P^+(\vec{r},\omega)$ at z_0 (Figure a) and the upgoing (i.e., one-way) Green's wave field $G^-(\vec{r},\vec{r}_A,\omega)$ at z_0 (Figure b).

For the Green's wave field the upper half-space was chosen reflection-free, so

$$G^-(\vec{r},\vec{r}_A,\omega) = G(\vec{r},\vec{r}_A,\omega) \qquad \text{at } z=z_0. \qquad \text{(V-47a)}$$

Based on this property and the reciprocity relation (V-21),

$$G(\vec{r}_A,\vec{r}_B,\omega) = G(\vec{r}_B,\vec{r}_A,\omega), \qquad \text{(V-47b)}$$

we may reformulate equation (V-46) as

$$P(\vec{r}_A,\omega) = 2\iint_{-\infty}^{\infty} \left[\frac{\partial G(\vec{r}_A,\vec{r},\omega)}{\partial z} \frac{1}{\rho(\vec{r})} P^+(\vec{r},\omega)\right]_{z_0} dxdy. \quad \text{(V-48)}$$

Here $G(\vec{r}_A,\vec{r},\omega)$ for $z=z_0$ represents the Green's wave field at $\vec{r}_A$ in the lower half-space related to a *monopole* source at $\vec{r}$ at the surface z_0. Accordingly, $\partial G(\vec{r}_A,\vec{r},\omega)/\partial z$ for $z=z_0$ represents the Green's wave field at $\vec{r}_A$ in the lower half-space, related to a *dipole* source at $\vec{r}$ at the surface z_0 (see also section I.3.1). Hence, equation (V-48) actually states that the total wave field at $\vec{r}_A$ in the lower half-space is obtained as an integral over dipole source responses, the dipole sources being distributed over the surface z_0, and the strength of the dipole sources being given by the (scaled) downgoing wave field at z_0 (Huygen's principle, see Figure V-10).

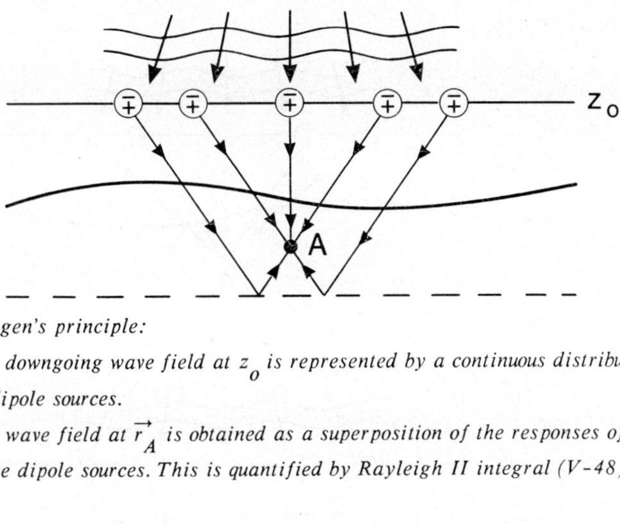

Figure V-10: Huygen's principle:

The downgoing wave field at z_0 is represented by a continuous distribution of dipole sources.

The wave field at $\vec{r}_A$ is obtained as a superposition of the responses of these dipole sources. This is quantified by Rayleigh II integral (V-48).

We split the total wave field at $\vec{r}_A$ into a downgoing and an upgoing part, according to

$$P(\vec{r}_A,\omega) = P^+(\vec{r}_A,\omega) + P^-(\vec{r}_A,\omega). \quad \text{(V-49a)}$$

Similarly, we split the total Green's wave field at $\vec{r}_A$ into a downgoing and an upgoing part, according to

$$G(\vec{r}_A,\vec{r},\omega) = G^+(\vec{r}_A,\vec{r},\omega) + G^-(\vec{r}_A,\vec{r},\omega). \quad \text{(V-49b)}$$

It can now be understood that the downgoing wave field $P^+(\vec{r}_A,\omega)$ is given by

$$P^+(\vec{r}_A,\omega) = 2\int\int_{-\infty}^{\infty}\left[\frac{\partial G^+(\vec{r}_A,\vec{r},\omega)}{\partial z}\frac{1}{\rho(\vec{r})}P^+(\vec{r},\omega)\right]_{z_0} dxdy, \tag{V-50a}$$

see Figure V-11a, whereas the upgoing wave field $P^-(\vec{r},\omega)$ is given by

$$P^-(\vec{r}_A,\omega) = 2\int\int_{-\infty}^{\infty}\left[\frac{\partial G^-(\vec{r}_A,\vec{r},\omega)}{\partial z}\frac{1}{\rho(\vec{r})}P^+(\vec{r},\omega)\right]_{z_0} dxdy, \tag{V-50b}$$

see Figure V-11b.

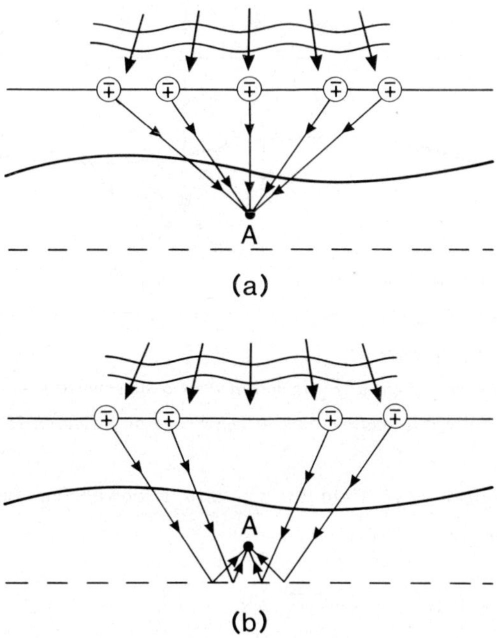

Figure V-11: Huygen's principle (revisited)

a. The downgoing wave field at $\vec{r}_A$ is obtained as a superposition of downgoing dipole source responses at $\vec{r}_A$ (Rayleigh II integral (V-50a)).

b. The upgoing wave field at $\vec{r}_A$ is obtained as a superposition of upgoing dipole source responses at $\vec{r}_A$ (Rayleigh II integral (V-50b)).

From (V-50a) and Figure V-11a we may conclude that $P^+(\vec{r}_A, \omega)$ is mainly determined by the *propagation* properties of the medium between z_o and z_A. From (V-50b) and Figure V-11b we may conclude that $P^-(\vec{r}_A, \omega)$ is mainly determined by the *reflection* properties of the medium below z_A.

In the following we only consider equation (V-50a), which we rewrite in a slightly more general notation, according to

$$P^+(x,y,z_1;\omega) = 2\int\int_{-\infty}^{\infty} \left[\frac{\partial G^+(x,y,z_1;x',y',z'=z_o;\omega)}{\partial z'} \frac{1}{\rho(x',y',z_o)} P^+(x',y',z_o;\omega) \right] dx'dy'. \tag{V-51a}$$

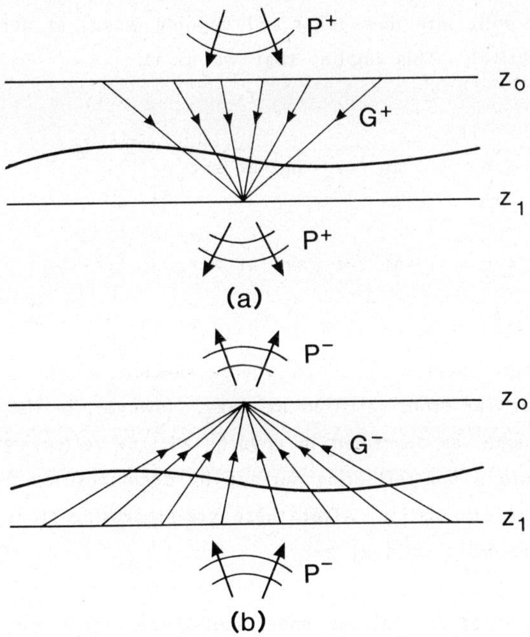

Figure V-12: Acoustic forward wave field extrapolation

a. Downward extrapolation of downgoing waves (one way Rayleigh II integral (V-51a)).

b. Upward extrapolation of upgoing waves (one-way Rayleigh II integral (V-51b)).

This one-way Rayleigh II integral is the basis for acoustic forward extrapolation of downgoing waves from depth level z_0 to depth level z_1 (with $z_1>z_0$, Figure V-12a). A similar expression can be derived for acoustic forward extrapolation of upgoing waves from depth level z_1 to depth level z_0 (Figure V-12b), yielding

$$P^-(x,y,z_0;\omega) = -2\int\int_{-\infty}^{\infty}\left[\frac{\partial G^-(x,y,z_0;x',y',z'=z_1;\omega)}{\partial z'}\frac{1}{\rho(x',y',z_1)}P^-(x',y',z_1;\omega)\right]dx'dy'. \tag{V-51b}$$

Note that both Rayleigh II integrals (V-51a) and (V-51b) are valid for arbitrarily inhomogeneous acoustic media. The assumption is that the wave fields may be split into downgoing and upgoing waves at depth levels z_0 and z_1, respectively. This implies that we assume

$$\frac{\partial K(x,y,z)}{\partial z} = 0 \qquad \text{at } z=z_0 \text{ and at } z=z_1$$

and

$$\frac{\partial \rho(x,y,z)}{\partial z} = 0 \qquad \text{at } z=z_0 \text{ and at } z=z_1.$$

In practice depth level z_0 is often a free surface, so the above assumptions are far from satisfied at $z=z_0$. However, in the seismic processing scheme, as described in chapter XI, the reflecting surface z_0 is transformed into a non-reflecting surface by decomposition and surface-related multiple elimination. After these pre-processing steps the above assumptions are fully valid at $z=z_0$.

In the derivation of (V-51a) the upper half-space ($z \le z_0$) was chosen reflection free for the Green's wave field. Accordingly, in the derivation of (V-51b) the lower half-space ($z \ge z_1$) was chosen reflection free for the Green's wave field. In the following we will choose for both Green's wave fields one and the same reference medium which is reflection free in the upper half-space ($z \le z_0$) as well as in the lower half-space ($z \ge z_1$). With this choice, the reciprocity principle for the Green's wave fields in

(V-51a) and (V-51b) reads

$$G^-(x_0,y_0,z_0;x_1,y_1,z_1;\omega) = G^+(x_1,y_1,z_1;x_0,y_0,z_0;\omega). \tag{V-52}$$

For the special situation that the medium between z_0 and z_1 is laterally invariant, the Green's wave fields are a function of the distances x-x' and y-y' only, hence, for this situation (V-51a) and (V-51b) may be written as spatial convolution integrals (Berkhout, 1985), according to

$$P^+(x,y,z_1;\omega) = \frac{2}{\rho(z_0)} \iint_{-\infty}^{\infty} \left[\frac{\partial G^+(x-x',y-y',z_1;0,0,z'=z_0;\omega)}{\partial z'} P^+(x',y',z_0;\omega) \right] dx'dy' \tag{V-53a}$$

and

$$P^-(x,y,z_0;\omega) = \frac{-2}{\rho(z_1)} \iint_{-\infty}^{\infty} \left[\frac{\partial G^-(x-x',y-y',z_0;0,0,z'=z_1;\omega)}{\partial z'} P^-(x',y',z_1;\omega) \right] dx'dy', \tag{V-53b}$$

respectively, or, symbolically,

$$P^+(x,y,z_1;\omega) = W^+(x,y;z_1,z_0;\omega) * P^+(x,y,z_0;\omega) \tag{V-54a}$$

and

$$P^-(x,y,z_0;\omega) = W^-(x,y;z_0,z_1;\omega) * P^-(x,y,z_1;\omega), \tag{V-54b}$$

respectively, where the one-way wave field extrapolation operators W^+ and W^- are given by

$$W^+(x,y;z_1,z_0;\omega) \triangleq \frac{2}{\rho(z_0)} \frac{\partial G^+(x,y,z_1;0,0,z=z_0;\omega)}{\partial z} \tag{V-55a}$$

and

$$W^-(x,y;z_o,z_1;\omega) \triangleq \frac{-2}{\rho(z_1)} \frac{\partial G^-(x,y,z_o;o,o,z=z_1;\omega)}{\partial z} . \tag{V-55b}$$

When the medium parameters between z_o and z_1 are vertically invariant as well, then we may substitute the free space solution (V-26a), yielding

$$W^+(x,y;z_1,z_o;\omega) = \frac{1}{2\pi} \frac{\partial}{\partial z_o} \left(\frac{e^{-jk\Delta r}}{\Delta r}\right) \tag{V-56a}$$

and

$$W^-(x,y;z_o,z_1;\omega) = \frac{-1}{2\pi} \frac{\partial}{\partial z_1} \left(\frac{e^{-jk\Delta r}}{\Delta r}\right), \tag{V-56b}$$

with

$$\Delta r = \sqrt{x^2 + y^2 + (z_1 - z_o)^2} \tag{V-56c}$$

and

$$k = \omega/c. \tag{V-56d}$$

Note that for this situation of a homogeneous medium the one-way operators for downward and upward extrapolation are identical:

$$W^+(x,y;z_1,z_o;\omega) = W^-(x,y;z_o,z_1;\omega). \tag{V-56e}$$

According to (IV-56), a convolution integral in the space-frequency domain corresponds to a multiplication in the wavenumber-frequency domain. Hence, transforming expressions (V-54a) and (V-54b) for laterally invariant media to the wavenumber-frequency domain yields

$$\tilde{P}^+(k_x,k_y,z_1;\omega) = \tilde{W}^+(k_x,k_y;z_1,z_o;\omega)\tilde{P}^+(k_x,k_y,z_o;\omega) \tag{V-57a}$$

and

$$\tilde{P}^-(k_x,k_y,z_o;\omega) = \tilde{W}^-(k_x,k_y;z_o,z_1;\omega)\tilde{P}^-(k_x,k_y,z_1;\omega), \tag{V-57b}$$

respectively, where

$$\tilde{W}^+(k_x,k_y;z_1,z_o;\omega) = \frac{2}{\rho(z_o)} \frac{\partial\tilde{G}^+(k_x,k_y,z_1;o,o,z=z_o;\omega)}{\partial z} \tag{V-58a}$$

and

$$\tilde{W}^-(k_x,k_y;z_o,z_1;\omega) = \frac{-2}{\rho(z_1)} \frac{\partial\tilde{G}^-(k_x,k_y,z_o;o,o,z=z_1;\omega)}{\partial z} \quad . \tag{V-58b}$$

Note that this is an alternative formulation for the WKB-operators (III-44c) and (III-44d), respectively. Again, when the medium parameters between z_o and z_1 are vertically invariant as well, we may substitute the free space solution (V-27a), yielding

$$\tilde{W}^+(k_x,k_y;z_1,z_o;\omega) = 2\frac{\partial}{\partial z_o}\left[\frac{e^{-jk_z\Delta z}}{2jk_z}\right] = e^{-jk_z\Delta z} \tag{V-59a}$$

and

$$\tilde{W}^-(k_x,k_y;z_o,z_1;\omega) = -2\frac{\partial}{\partial z_1}\left[\frac{e^{-jk_z\Delta z}}{2jk_z}\right] = e^{-jk_z\Delta z} , \tag{V-59b}$$

with

$$\Delta z = |z_1-z_o| = z_1-z_o \tag{V-59c}$$

and k_z defined by (V-27c) and (V-27d).

Note that these operators are identical to the phase-shift operators (III-39b) and (III-39c), respectively.

V.5.2 Matrix formulation of acoustic forward wave field extrapolation

Consider the acoustic one-way Rayleigh II integrals (V-51a) and (V-51b),

$$P^+(x,y,z_1;\omega) =$$

$$2\int\int_{-\infty}^{\infty}\left[\frac{\partial G^+(x,y,z_1;x',y',z'=z_0;\omega)}{\partial z'}\frac{1}{\rho(x',y',z_0)}P^+(x',y',z_0;\omega)\right]dx'dy' \qquad \text{(V-60a)}$$

and

$$P^-(x,y,z_0;\omega) =$$

$$-2\int\int_{-\infty}^{\infty}\left[\frac{\partial G^-(x,y,z_0;x',y',z'=z_1;\omega)}{\partial z'}\frac{1}{\rho(x',y',z_1)}P^-(x',y',z_1;\omega)\right]dx'dy', \qquad \text{(V-60b)}$$

which describe forward extrapolation of downgoing and upgoing acoustic waves, respectively, through arbitrarily inhomogeneous fluid media (Figure V-12).

In practical situations the wave fields are discretized along the x- and y-axes. It is shown in Appendix A that integral expressions of the form (V-60a) and (V-60b) can be replaced by matrix products in the case of discretized wave fields.

We may replace equation (V-60a) by

$$\vec{P}^+(z_1) = \mathbf{W}^+(z_1,z_0)\vec{P}^+(z_0), \qquad \text{(V-61a)}$$

where the one-way wave field extrapolation matrix $\mathbf{W}^+$ is defined as

$$\mathbf{W}^+(z_1,z_0) = 2\,\frac{\partial \mathbf{G}^+(z_1,z=z_0)}{\partial z}\,\mathbf{M}^{-1}(z_0). \qquad \text{(V-61b)}$$

Similarly, we may replace (V-60b) by

$$\vec{P}^-(z_0) = \mathbf{W}^-(z_0,z_1)\vec{P}^-(z_1), \tag{V-62a}$$

where the one-way wave field extrapolation matrix $\mathbf{W}^-$ is defined as

$$\mathbf{W}^-(z_0,z_1) = -2\,\frac{\partial G^-(z_0,z=z_1)}{\partial z}\,\mathbf{M}^{-1}(z_1). \tag{V-62b}$$

Here vectors $\vec{P}^{\pm}(z_0)$ and $\vec{P}^{\pm}(z_1)$ contain the discretized versions of the monochromatic wave fields $P^{\pm}(x,y,z_0;\omega)$ and $P^{\pm}(x,y,z_1;\omega)$, respectively (see Appendix A, section A.2). Matrices $\mathbf{M}(z_0)$ and $\mathbf{M}(z_1)$ are diagonal matrices, the diagonal elements representing the discretized versions of $\rho(x,y,z_0)$ and $\rho(x,y,z_1)$, respectively (see Appendix A, section A.3). Matrix $G^+(z_1,z_0)$ is an operator matrix, each column containing a discretized monochromatic "spatial impulse response" $G^+(x,y,z_1;x',y',z_0;\omega)$ as a function of (x,y) at z_1 for an "impulse" (monopole source) at (x',y',z_0). Similarly, matrix $G^-(z_0,z_1)$ is an operator matrix, each column containing a discretized monochromatic "spatial impulse response" $G^-(x,y,z_0;x',y',z_1;\omega)$ as a function of (x,y) at z_0 for an "impulse" at (x',y',z_1) (see Appendix A, section A.3). In this matrix notation, the reciprocity principle reads

$$G^-(z_0,z_1) = [G^+(z_1,z_0)]^T. \tag{V-63}$$

Note that equations (V-61b) and (V-62b), which define the one-way wave field extrapolation matrices $\mathbf{W}^+(z_1,z_0)$ and $\mathbf{W}^-(z_0,z_1)$, respectively, are an alternative formulation of the extrapolation matrices defined by the Taylor series in equation (III-81b). The Taylor series approach is particularly suited for small extrapolation steps whereas the "discretized Rayleigh operator" approach is particularly suited for large extrapolation steps.

For an arbitrarily inhomogeneous acoustic medium between z_0 and z_1, no analytical expressions are available for the Green's matrices $\mathbf{G}^+(z_1,z_0)$ and $\mathbf{G}^-(z_0,z_1)$. In practice they may be obtained as follows:

1. Define a reference medium which accurately describes the geology between depth levels z_o and z_1 and which is non-reflecting outside this depth interval.

2. Solve numerically the two-way wave equation (V-16a),

$$\nabla.\left(\frac{1}{\rho(\vec{r})}\nabla g(\vec{r},\vec{r}',t)\right)-\frac{1}{K(\vec{r})}\frac{\partial^2 g(\vec{r},\vec{r}',t)}{\partial t^2}=-\delta(\vec{r}-\vec{r}')\delta(t), \tag{V-64a}$$

imposing initial conditions

$$g(\vec{r},\vec{r}',t)=o \quad \text{for } t<o \tag{V-64b}$$

and

$$\frac{\partial g(\vec{r},\vec{r}',t)}{\partial t}=o \quad \text{for } t<o. \tag{V-64c}$$

By choosing the source point $\vec{r}'$ at z_1 one obtains the "spatial impulse response" $g(x,y,z;x',y',z'=z_1;t)$. For $z=z_o$ this Green's wave field is purely upgoing, which we denote by $g^-(x,y,z=z_o;x',y',z'=z_1;t)$.

3. Apply a temporal Fourier transform to this Green's wave field, according to

$$G^-(x,y,z_o;x',y',z_1;\omega)=\int_o^\infty g^-(x,y,z_o;x',y',z_1;t)e^{-j\omega t}dt. \tag{V-65}$$

For a fixed frequency ω, the "spatial impulse response" $G^-(x,y,z_o;x',y',z_1;\omega)$ represents one column of the Green's matrix $\mathbf{G}^-(z_o,z_1)$.

4. Repeat steps 2 and 3 for a range of source points $\vec{r}'$ at z_1, yielding the different columns of the Green's matrix $\mathbf{G}^-(z_o,z_1)$.

5. Determine the Green's matrix $\mathbf{G}^+(z_1,z_o)$ by interchanging the rows and columns of matrix $\mathbf{G}^-(z_o,z_1)$.

In principle any accurate forward modeling scheme can be used to generate the Green's wave fields in step 2.

Here we discuss an example with *Gaussian beam* modeling (Çerveny et al., 1982). For the 2-D configuration of Figure V-13a, Gaussian beam modeling involves shooting a fan of rays from (x',z_1) and solving a high-frequency approximation of the 2-D version of wave equation (V-64) around each ray. The resulting beams exhibit a Gaussian amplitude distribution around the rays. The response at any point (x,z_0) is now found by superposing the contributions of the individual beams at that point. Figure V-13b shows a band-limited version of a 2-D Green's wave field $g^-(x,z_0;x',z_1;t)$ for fixed x'.

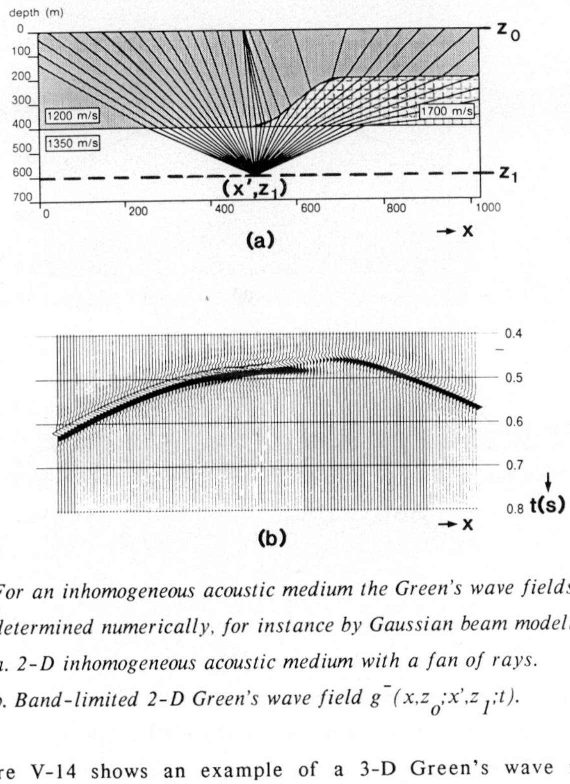

Figure V-13: For an inhomogeneous acoustic medium the Green's wave fields must be determined numerically, for instance by Gaussian beam modeling:
a. 2-D inhomogeneous acoustic medium with a fan of rays.
b. Band-limited 2-D Green's wave field $g^-(x,z_0;x',z_1;t)$.

Finally, Figure V-14 shows an example of a 3-D Green's wave field obtained by raytracing through a 3-D inhomogeneous acoustic subsurface model. In chapter VII, section VII.3.6 we use the Green's wave fields of Figures V-13 and V-14 in examples of acoustic 2-D and 3-D inverse wave field extrapolation.

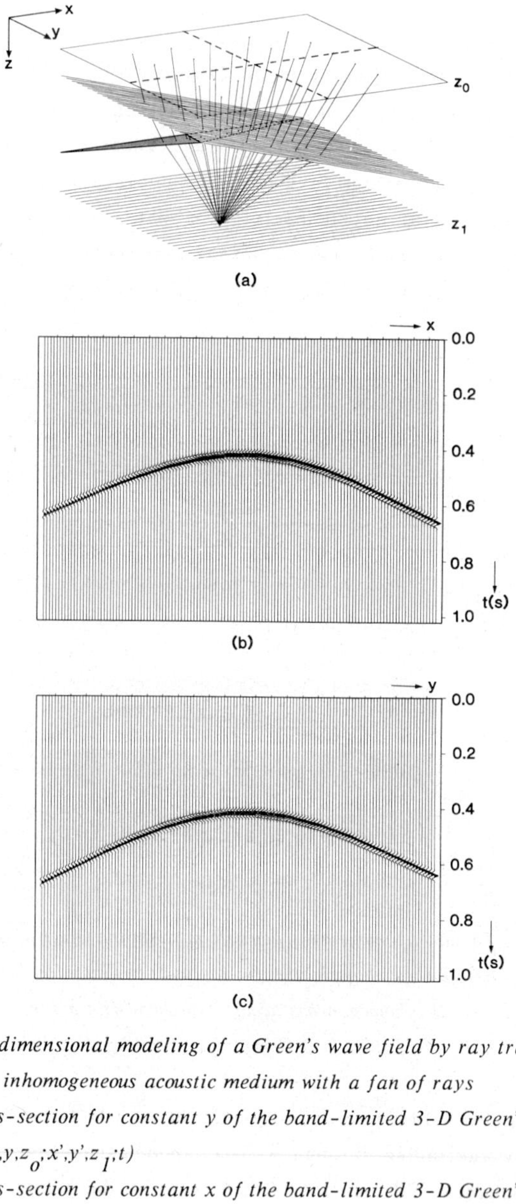

Figure V-14: Three-dimensional modeling of a Green's wave field by ray tracing.

a. 3-D inhomogeneous acoustic medium with a fan of rays

b. Cross-section for constant y of the band-limited 3-D Green's wave field $g^-(x,y,z_o;x',y',z_1;t)$

c. Cross-section for constant x of the band-limited 3-D Green's wave field $g^-(x,y,z_o;x',y',z_1;t)$

V.6 REFERENCES

Berkhout, A.J., 1986, Seismic inversion in terms of pre-stack migration and multiple elimination: Proceedings of the IEEE, *74*, 415 - 427.

Berkhout, A.J., and Van Wulfften Palthe, D.W., 1979, Migration in terms of spatial deconvolution: Geophysical Prospecting, *27*, 261 - 291.

Berkhout, A.J., 1985, Seismic migration: Imaging of acoustic energy by wave field extrapolation. A. Theoretical aspects: Elsevier Science Publ. Co., Inc.

Berkhout, A.J., and Wapenaar, C.P.A., 1989, One-way versions of the Kirchhoff integral: Geophysics, *54*, 460 - 467.

Bleistein, N., 1984, Mathematical methods for wave phenomena: Academic Press, Inc.

Çerveny, V., Popov, M.M., and Psencik, I., 1982, Computation of wave fields in inhomogeneous media - Gaussian beam approach: Geophys. J.R. astr. Soc., *70*, 109 - 128.

Clayton, R.W., and Stolt, R.H., 1981, A Born-WKBJ inversion method for acoustic reflection data: Geophysics, *46*, 1559 - 1567.

Morse, P.M., and Feshbach, H., 1953, Methods of theoretical physics: McGraw-Hill Book Company, Inc.

Rayleigh, J.W.S., 1965, The theory of sound, volume II (first edition printed in 1878): Dover Publications, Inc.

VI

ELASTIC FORWARD WAVE FIELD EXTRAPOLATION

VI.1 INTRODUCTION

In this chapter we start with reviewing *Betti's reciprocity theorem.* This theorem gives the mathematical relationship between two independent elastic wave fields. A special form of the reciprocity theorem is obtained if one of the elastic wave fields represents the "impulse response" of a reference medium (the *Green's function*) whereas the other wave field represents the physical wave field in the true medium. This special form is commonly referred to as a *representation theorem*: it expresses the particle velocity of the physical wave field at any point in the true medium in terms of a closed surface- and a volume-integral over the same physical wave field. It is also known as the *Kirchhoff-Helmholtz integral* formula. We introduce modified Kirchhoff-Helmholtz integrals which express the *P-wave* or *S-wave* potential at any point in the medium in terms of a closed surface integral. We use these integrals as the starting point for deriving *Rayleigh-integrals* that express the P-wave or S-wave potential at any point in the medium in terms of the elastic wave field at a plane surface. We discuss both *two-way* and *one-way* versions of the Rayleigh integral. The one-way Rayleigh integral is the basis for deriving wave field extrapolation operators. We derive matrix operators for numerical *forward* extrapolation of downgoing and upgoing P- and S-waves through arbitrarily inhomogeneous anisotropic elastic media. These matrices play an important role in chapter XII, where we discuss an elastic processing scheme for multi-component seismic data.

VI.2 ELASTIC RECIPROCITY THEOREMS

Consider a volume V enclosed by a surface S with outward pointing normal vector $\vec{n}$, see Figure VI-1.

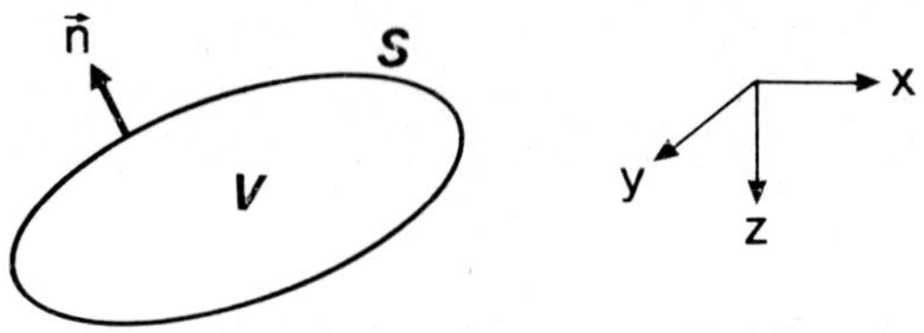

Figure VI-1: Volume V, enclosed by surface S.

In this volume we define two non-identical elastic wave fields ("state A" and "state B") and we derive reciprocity relations for these wave fields.

State A:

Define in V

$\mathbf{c}^A(\vec{r})$: stiffness tensor (components: $c^A_{ijk\ell}(\vec{r})$),
$\rho^A(\vec{r})$: volume density of mass,
$\vec{v}^A(\vec{r},t)$: particle velocity vector (components: $v^A_i(\vec{r},t)$),
$\boldsymbol{\tau}^A(\vec{r},t)$: stress tensor (components: $\tau^A_{ij}(\vec{r},t)$),
$\vec{f}^A(\vec{r},t)$: volume density of external force (components: $f^A_i(\vec{r},t)$),
$\boldsymbol{\sigma}^A(\vec{r},t)$: source stress tensor (components: $\sigma^A_{ij}(\vec{r},t)$).

According to equations (II-14) and (II-21), the elastic wave field in state A satisfies in V the following two equations

$$\partial_j \tau^A_{ij} - \rho^A \partial_t v^A_i = -f^A_i \tag{VI-1a}$$

and

$$\partial_t \tau^A_{ij} - c^A_{ijk\ell} \partial_\ell v^A_k = -\partial_t \sigma^A_{ij} \, . \tag{VI-1b}$$

State B:

Define in V

$\mathbf{c}^B(\vec{r})$: stiffness tensor (components: $c^B_{ijk\ell}(\vec{r})$),
$\rho^B(\vec{r})$: volume density of mass,
$\vec{v}^B(\vec{r},t)$: particle velocity vector (components: $v^B_i(\vec{r},t)$),
$\boldsymbol{\tau}^B(\vec{r},t)$: stress tensor (components: $\tau^B_{ij}(\vec{r},t)$),
$\vec{f}^B(\vec{r},t)$: volume density of external force (components: $f^B_i(\vec{r},t)$),
$\boldsymbol{\sigma}^B(\vec{r},t)$: source stress tensor (components: $\sigma^B_{ij}(\vec{r},t)$).

According to equations (II-14) and (II-21), the elastic wave field in state B satisfies in V the following two equations

$$\partial_j \tau_{ij}^B - \rho^B \partial_t v_i^B = -f_i^B \tag{VI-2a}$$

and

$$\partial_t \tau_{ij}^B - c_{ijk\ell}^B \partial_\ell v_k^B = -\partial_t \sigma_{ij}^B \,. \tag{VI-2b}$$

For the moment we do not specify the initial conditions nor the boundary conditions. We apply the temporal Fourier transformation (III-1a) to equations (VI-1) and (VI-2). The resulting equations in the space-frequency domain read

State A:

$$\partial_j \tau_{ij}^A - j\omega\rho^A V_i^A = -F_i^A \quad {}^{1)} \tag{VI-3a}$$

and

$$j\omega\tau_{ij}^A - c_{ijk\ell}^A \partial_\ell V_k^A = -j\omega\sigma_{ij}^A, \tag{VI-3b}$$

where $V_i^A(\vec{r},\omega)$, $\tau_{ij}^A(\vec{r},\omega)$, $F_i^A(\vec{r},\omega)$ and $\sigma_{ij}^A(\vec{r},\omega)$ are the space-frequency domain representations of $v_i^A(\vec{r},t)$, $\tau_{ij}^A(\vec{r},t)$, $f_i^A(\vec{r},t)$ and $\sigma_{ij}^A(\vec{r},t)$, respectively.

State B:

$$\partial_j \tau_{ij}^B - j\omega\rho^B V_i^B = -F_i^B \tag{VI-4a}$$

and

$$j\omega\tau_{ij}^B - c_{ijk\ell}^B \partial_\ell V_k^B = -j\omega\sigma_{ij}^B, \tag{VI-4b}$$

---

1) Here and in the following equations the symbol j has two different meanings. When used as a factor, it denotes the imaginary unit $\sqrt{-1}$. Otherwise it is an index which may take the values 1, 2 or 3.

where $V_i^B(\vec{r},\omega)$, $\tau_{ij}^B(\vec{r},\omega)$, $F_i^B(\vec{r},\omega)$ and $\sigma_{ij}^B(\vec{r},\omega)$ are the space-frequency domain representations of $v_i^B(\vec{r},t)$, $\tau_{ij}^B(\vec{r},t)$, $f_i^B(\vec{r},t)$ and $\sigma_{ij}^B(\vec{r},t)$, respectively.

We apply the theorem of Gauss to a vector function $\vec{Q}(\vec{r},\omega)$, which we define as

$$\vec{Q} = \boldsymbol{\tau}^A\vec{V}^B - \boldsymbol{\tau}^B\vec{V}^A, \tag{VI-5a}$$

where the components of $\vec{Q}(\vec{r},\omega)$, $\vec{V}^A(\vec{r},\omega)$, $\vec{V}^B(\vec{r},\omega)$, $\boldsymbol{\tau}^A(\vec{r},\omega)$ and $\boldsymbol{\tau}^B(\vec{r},\omega)$ read $Q_j(\vec{r},\omega)$, $V_i^A(\vec{r},\omega)$, $V_i^B(\vec{r},\omega)$, $\tau_{ji}^A(\vec{r},\omega)$ and $\tau_{ji}^B(\vec{r},\omega)$, respectively, hence

$$Q_j = \tau_{ji}^A V_i^B - \tau_{ji}^B V_i^A . \tag{VI-5b}$$

The divergence of $\vec{Q}(\vec{r},\omega)$ reads

$$\nabla.\vec{Q} = \partial_j Q_j, \tag{VI-6a}$$

or, upon substitution of (VI-5b),

$$\nabla.\vec{Q} = \partial_j(\tau_{ji}^A V_i^B - \tau_{ji}^B V_i^A), \tag{VI-6b}$$

or, by applying the product rule for differentiation and using the symmetry properties $\tau_{ji}^A = \tau_{ij}^A$ and $\tau_{ji}^B = \tau_{ij}^B$,

$$\nabla.\vec{Q} = \partial_j\tau_{ij}^A V_i^B - \partial_j\tau_{ij}^B V_i^A + \tau_{ij}^A\partial_j V_i^B - \tau_{ij}^B\partial_j V_i^A, \tag{VI-6c}$$

or, upon substitution of equations (VI-3) and (VI-4),

$$\nabla.\vec{Q} = \left(j\omega\rho^A V_i^A - F_i^A\right)V_i^B - \left(j\omega\rho^B V_i^B - F_i^B\right)V_i^A$$
$$+ \frac{1}{j\omega}c_{ijk\ell}^A\partial_\ell V_k^A\,\partial_j V_i^B - \frac{1}{j\omega}c_{ijk\ell}^B\partial_\ell V_k^B\,\partial_j V_i^A - \sigma_{ij}^A\partial_j V_i^B + \sigma_{ij}^B\partial_j V_i^A . \tag{VI-6d}$$

Hence, applying the theorem of Gauss

$$\oint_S \vec{Q}.\vec{n}\,dS = \int_V \nabla.\vec{Q}\,dV \tag{VI-7}$$

yields

$$\oint_S [\tau^A \vec{V}^B - \tau^B \vec{V}^A].\vec{n}\,dS = \int_V [F_i^B V_i^A - F_i^A V_i^B + \sigma_{ij}^B \partial_j V_i^A - \sigma_{ij}^A \partial_j V_i^B$$

$$+ \frac{1}{j\omega}\left(c_{ijk\ell}^A - c_{ijk\ell}^B\right)\partial_\ell V_k^B \partial_j V_i^A + j\omega\left(\rho^A - \rho^B\right) V_i^A V_i^B\,]\,dV,^{1)} \tag{VI-8}$$

where we made use of the symmetry property $c_{ijk\ell}^A = c_{k\ell ij}^A$. Equation (VI-8) is generally known as *Betti's reciprocity theorem* (Aki and Richards, 1980). It is the basis for the elastic representation theorem, which is derived in section VI.3. Similar arguments as given in section V.2 lead to the conclusion that theorem (VI-8) is valid for any inhomogeneous anisotropic elastic medium, containing arbitrary interfaces.

Let us consider a special situation. We choose identical medium parameters for state A and state B throughout volume V, according to

$$c_{ijk\ell}^A(\vec{r}) = c_{ijk\ell}^B(\vec{r}) \triangleq c_{ijk\ell}(\vec{r}) \tag{VI-9a}$$

and

$$\rho^A(\vec{r}) = \rho^B(\vec{r}) \triangleq \rho(\vec{r}). \tag{VI-9b}$$

Furthermore, we choose in V uni-directional point forces at $\vec{r}_A$ and $\vec{r}_B$, respectively, according to

$$F_i^A(\vec{r},\omega) \triangleq \delta(\vec{r}-\vec{r}_A)\delta_{im} M^A(\omega) \tag{VI-10a}$$

and

---

1) V denotes the volume, depicted in Figure VI-1; V denotes the particle velocity.

$$F_i^B(\vec{r},\omega) \triangleq \delta(\vec{r}-\vec{r}_B)\delta_{in}N^B(\omega), \tag{VI-10b}$$

where $M^A(\omega)$ and $N^B(\omega)$ are the source signatures of the forces in the m- and n-direction at $\vec{r}_A$ and $\vec{r}_B$, respectively.

Finally, we choose in V

$$\sigma_{ij}^A(\vec{r},\omega) = \sigma_{ij}^B(\vec{r},\omega) \triangleq 0. \tag{VI-10c}$$

Thus the reciprocity theorem (VI-8) simplifies to

$$\oint_S [\tau^A \vec{V}^B - \tau^B \vec{V}^A].\vec{n}\,dS = V_n^A(\vec{r}_B,\omega)N^B(\omega) - V_m^B(\vec{r}_A,\omega)M^A(\omega). \tag{VI-11}$$

Similar arguments as given in section V.2 lead to the conclusion that the surface integral vanishes when S is a rigid or a free boundary in both states or when V is unbounded in both states. Hence, in all these situations the reciprocity theorem (VI-11) simplifies to

$$\frac{V_n^A(\vec{r}_B,\omega)}{M^A(\omega)} = \frac{V_m^B(\vec{r}_A,\omega)}{N^B(\omega)} . \tag{VI-12}$$

For identical source signatures $M^A(\omega)$ and $N^B(\omega)$, equation (VI-12) states that the n-component of the particle velocity vector at $\vec{r}_B$, related to a force in the m-direction at $\vec{r}_A$ is identical to the m-component of the particle velocity vector at $\vec{r}_A$, related to a force in the n-direction at $\vec{r}_B$. Note that this principle holds for arbitrarily inhomogeneous anisotropic elastic media.

VI.3 ELASTIC REPRESENTATION THEOREMS

VI.3.1 Elastic Green's functions

An elastic Green's function defines the impulse response of a solid medium. For an impulse in the m-direction at $\vec{r}_A$, the Green's function satisfies the following two equations in the space-time domain

$$\partial_j \theta_{ij,m} - \rho \partial_t g_{i,m} = -\delta_{im} \delta(\vec{r} - \vec{r}_A) \delta(t) \tag{VI-13a}$$

and

$$\partial_t \theta_{ij,m} - c_{ijk\ell} \partial_\ell g_{k,m} = 0, \tag{VI-13b}$$

with initial conditions

$$g_{i,m}(\vec{r}, \vec{r}_A, t) = 0 \qquad \text{for } t < 0 \tag{VI-13c}$$

and

$$\theta_{ij,m}(\vec{r}, \vec{r}_A, t) = 0 \qquad \text{for } t < 0. \tag{VI-13d}$$

For the moment we do not specify the boundary conditions for $g_{i,m}(\vec{r}, \vec{r}_A, t)$ and $\theta_{ij,m}(\vec{r}, \vec{r}_A, t)$. In equation (VI-13), $g_{i,m}(\vec{r}, \vec{r}_A, t)$ denotes the impulse response in terms of the i-component of the particle velocity at observation point $\vec{r}$ as a function of time t, related to an impulse in the m-direction at $\vec{r}_A$ at t=0; $\theta_{ij,m}(\vec{r}, \vec{r}_A, t)$ denotes the stress associated to this impulse response. In this notation, the reciprocity principle (VI-12) can be reformulated as

$$g_{i,m}(\vec{r}_B, \vec{r}_A, t) = g_{m,i}(\vec{r}_A, \vec{r}_B, t). \tag{VI-14}$$

The initial conditions (VI-13c) and (VI-13d) ensure that $g_{i,m}(\vec{r}, \vec{r}_A, t)$ and $\theta_{ij,m}(\vec{r}, \vec{r}_A, t)$ represent a *causal* wave field which propagates away from the source at $\vec{r}_A$. Therefore we will also refer to $g_{i,m}$ and $\theta_{ij,m}$ as the *forward propagating* elastic Green's wave field. Opposed to this we also define an anti-causal or *backward propagating* elastic Green's wave field which satisfies the same equations:

$$\partial_j \hat{\theta}_{ij,m} - \rho \partial_t \hat{g}_{i,m} = -\delta_{im} \delta(\vec{r} - \vec{r}_A) \delta(t) \qquad \text{(VI-15a)}$$

and

$$\partial_t \hat{\theta}_{ij,m} - c_{ijk\ell} \partial_\ell \hat{g}_{k,m} = o, \qquad \text{(VI-15b)}$$

with final conditions

$$\hat{g}_{i,m}(\vec{r}, \vec{r}_A, t) = o \qquad \text{for } t > o \qquad \text{(VI-15c)}$$

and

$$\hat{\theta}_{ij,m}(\vec{r}, \vec{r}_A, t) = o \qquad \text{for } t > o. \qquad \text{(VI-15d)}$$

In the following we assume that $\hat{g}_{i,m}(\vec{r}, \vec{r}_A, t)$ and $\hat{\theta}_{ij,m}(\vec{r}, \vec{r}_A, t)$ satisfy the same boundary conditions as $-g_{i,m}(\vec{r}, \vec{r}_A, -t)$ and $\theta_{ij,m}(\vec{r}, \vec{r}_A, -t)$. Then the backward propagating Green's wave field is simply related to the forward propagating Green's wave field, according to

$$\hat{g}_{i,m}(\vec{r}, \vec{r}_A, t) = -g_{i,m}(\vec{r}, \vec{r}_A, -t) \qquad \text{(VI-16a)}$$

and

$$\hat{\theta}_{ij,m}(\vec{r}, \vec{r}_A, t) = \theta_{ij,m}(\vec{r}, \vec{r}_A, -t), \qquad \text{(VI-16b)}$$

for all $\vec{r}, \vec{r}_A$ and t.

We define the forward propagating Green's wave field in the space-frequency domain according to

$$G_{i,m}(\vec{r},\vec{r}_A,\omega) = \int_{-\infty}^{\infty} g_{i,m}(\vec{r},\vec{r}_A,t)e^{-j\omega t}dt \tag{VI-17a}$$

and

$$\Theta_{ij,m}(\vec{r},\vec{r}_A,\omega) = \int_{-\infty}^{\infty} \theta_{ij,m}(\vec{r},\vec{r}_A,t)e^{-j\omega t}dt, \tag{VI-17b}$$

respectively. Note that, in analogy with (VI-14), the reciprocity principle reads

$$G_{i,m}(\vec{r}_B,\vec{r}_A,\omega) = G_{m,i}(\vec{r}_A,\vec{r}_B,\omega). \tag{VI-18}$$

In the space-frequency domain, the backward propagating Green's wave field is related to the forward propagating Green's wave field, according to

$$\hat{G}_{i,m}(\vec{r},\vec{r}_A,\omega) = -G^*_{i,m}(\vec{r},\vec{r}_A,\omega) \tag{VI-19a}$$

and

$$\hat{\Theta}_{ij,m}(\vec{r},\vec{r}_A,\omega) = \Theta^*_{ij,m}(\vec{r},\vec{r}_A,\omega). \tag{VI-19b}$$

We define the forward propagating Green's wave field in the wavenumber-frequency domain according to

$$\tilde{G}_{i,m}(k_x,k_y,z;x_A,y_A,z_A;\omega) = \iint_{-\infty}^{\infty} G_{i,m}(x,y,z;x_A,y_A,z_A;\omega)e^{j(k_x x+k_y y)}dxdy \tag{VI-20a}$$

and

$$\tilde{\Theta}_{ij,m}(k_x,k_y,z;x_A,y_A,z_A;\omega) = \iint_{-\infty}^{\infty} \Theta_{ij,m}(x,y,z;x_A,y_A,z_A;\omega)e^{j(k_x x+k_y y)}dxdy, \tag{VI-20b}$$

respectively. In the wavenumber-frequency domain, the backward propagating Green's wave field is related to the forward propagating Green's wave field, according to

$$\tilde{\hat{G}}_{i,m}(k_x,k_y,z;x_A,y_A,z_A;\omega) = -\tilde{G}^*_{i,m}(-k_x,-k_y,z;x_A,y_A,z_A;\omega) \tag{VI-21a}$$

and

$$\tilde{\hat{\Theta}}_{ij,m}(k_x,k_y,z;x_A,y_A,z_A;\omega) = \tilde{\Theta}^*_{ij,m}(-k_x,-k_y,z;x_A,y_A,z_A;\omega). \tag{VI-21b}$$

VI.3.2 Elastic Kirchhoff-Helmholtz integral

Our starting point is Betti's reciprocity theorem (VI-8). In the following, state A will represent a forward or backward propagating Green's wave field in a *reference medium* whereas state B will represent the physical wave field in the true medium. Hence, we make the following substitutions:

State A:

$$\begin{aligned}
c^A_{ijk\ell}(\vec{r}) &\rightarrow \bar{c}_{ijk\ell}(\vec{r}),\\
\rho^A(\vec{r}) &\rightarrow \bar{\rho}(\vec{r}),\\
\vec{V}^A(\vec{r},\omega) &\rightarrow \vec{G}_m(\vec{r},\vec{r}_A,\omega) \text{ (components: } G_{i,m}(\vec{r},\vec{r}_A,\omega)\text{), or}\\
&\qquad -\vec{G}^*_m(\vec{r},\vec{r}_A,\omega) \text{ (components: } -G^*_{i,m}(\vec{r},\vec{r}_A,\omega)\text{)},\\
\boldsymbol{\tau}^A(\vec{r},\omega) &\rightarrow \boldsymbol{\Theta}_m(\vec{r},\vec{r}_A,\omega) \text{ (components: } \Theta_{ij,m}(\vec{r},\vec{r}_A,\omega)\text{), or}\\
&\qquad \boldsymbol{\Theta}^*_m(\vec{r},\vec{r}_A,\omega) \text{ (components: } \Theta^*_{ij,m}(\vec{r},\vec{r}_A,\omega)\text{)},\\
F^A_i(\vec{r},\omega) &\rightarrow \delta_{im}\delta(\vec{r}-\vec{r}_A),\\
\sigma^A_{ij}(\vec{r},\omega) &\rightarrow 0.
\end{aligned} \tag{VI-22a}$$

State B:

$$
\begin{aligned}
&c^B_{ijk\ell}(\vec{r}) \rightarrow c_{ijk\ell}(\vec{r}) = \bar{c}_{ijk\ell}(\vec{r}) + \Delta c_{ijk\ell}(\vec{r}),\\
&\rho^B(\vec{r}) \rightarrow \rho(\vec{r}) = \bar{\rho}(\vec{r}) + \Delta\rho(\vec{r}),\\
&\vec{V}^B(\vec{r},\omega) \rightarrow \vec{V}(\vec{r},\omega) \text{ (components: } V_i(\vec{r},\omega)),\\
&\tau^B(\vec{r},\omega) \rightarrow \tau(\vec{r},\omega) \text{ (components: } \tau_{ij}(\vec{r},\omega)),\\
&F^B_i(\vec{r},\omega) \rightarrow F_i(\vec{r},\omega),\\
&\sigma^B_{ij}(\vec{r},\omega) \rightarrow \sigma_{ij}(\vec{r},\omega).
\end{aligned}
\tag{VI-22b}
$$

With these substitutions, we obtain the following two elastic *representation theorems* (De Hoop, 1958; Burridge and Knopoff, 1964; Aki and Richards, 1980):

$$
V_m(\vec{r}_A,\omega) = -\oint_S [\Theta_m\vec{V} - \tau\vec{G}_m].\vec{n}\,dS
$$

$$
+\int_V [F_iG_{i,m} + \sigma_{ij}\partial_jG_{i,m} - j\omega\Delta\rho G_{i,m}V_i - \frac{1}{j\omega}\Delta c_{ijk\ell}\partial_\ell G_{k,m}\partial_jV_i]\,dV, \tag{VI-23a}
$$

or, equivalently,

$$
V_m(\vec{r}_A,\omega) = -\oint_S [\Theta^*_m\vec{V} + \tau\vec{G}^*_m].\vec{n}\,dS
$$

$$
-\int_V [F_iG^*_{i,m} + \sigma_{ij}\partial_jG^*_{i,m} - j\omega\Delta\rho G^*_{i,m}V_i - \frac{1}{j\omega}\Delta c_{ijk\ell}\partial_\ell G^*_{k,m}\partial_jV_i]\,dV. \tag{VI-23b}
$$

These expressions are also known as the elastic *Kirchhoff-Helmholtz integral* formulas (Pao and Varatharajulu, 1976). They are the basis for multi-dimensional elastic forward and inverse scattering techniques.

Here we follow a different approach. By choosing the reference medium $(\bar{c}_{ijk\ell},\bar{\rho})$ equal or close to the actual medium $(c_{ijk\ell},\rho)$ in V and on S it is justified to ignore the deviation parameters $(\Delta c_{ijk\ell},\Delta\rho)$ in (VI-23a) and (VI-23b). When we also assume that volume V is source-free then expressions (VI-23a) and (VI-23b) simplify to

$$V_m(\vec{r}_A,\omega) \approx -\oint_S [\boldsymbol{\Theta}_m\vec{V} - \boldsymbol{\tau}\vec{G}_m].\vec{n}\,dS \qquad \text{(VI-24a)}$$

or, equivalently,

$$V_m(\vec{r}_A,\omega) \approx -\oint_S [\boldsymbol{\Theta}_m^*\vec{V} + \boldsymbol{\tau}\vec{G}_m^*].\vec{n}\,dS. \qquad \text{(VI-24b)}$$

These expressions describe the elastic wave field at $\vec{r}_A$ in V in terms of the elastic wave field on S, enclosing V. The *scattering* effects related to the deviation parameters $\Delta c_{ijk\ell}(\vec{r})$ and $\Delta\rho(\vec{r})$ are zero or neglected. However, the *propagation* effects related to the reference parameters $\bar{c}_{ijk\ell}(\vec{r})$ and $\bar{\rho}(\vec{r})$ are properly included and therefore expressions (VI-24a) and (VI-24b) are the basis for multi-dimensional elastic forward and inverse wave field extrapolation techniques.

Forward wave field extrapolation is often used to *simulate* wave propagation in a known medium. Therefore, for forward wave field extrapolation we may often choose the reference medium equal to the actual medium. In the following sections of this chapter we discuss elastic forward wave field extrapolation, based on Kirchhoff-Helmholtz integral (VI-24a) with the forward propagating Green's wave fields, which is exact when $\bar{c}_{ijk\ell}(\vec{r}) = c_{ijk\ell}(\vec{r})$ and $\bar{\rho}(\vec{r}) = \rho(\vec{r})$ throughout V. For notational convenience, in the following we will omit the bars above $c_{ijk\ell}$ and ρ and we replace $\approx$ by $=$.

Inverse wave field extrapolation is generally used to *eliminate* wave propagation from seismic data acquired over an unknown medium. Therefore, for inverse extrapolation at best we may choose a reference medium that is close to the actual medium[1)]. In chapters VIII and X we discuss elastic inverse wave field extrapolation based on Kirchhoff-Helmholtz integral (VI-24b), with the backward propagating Green's wave fields.

VI.3.3 Modified elastic Green's functions for P- and S-waves

In this section we discuss modified Green's functions for P- and S-waves (Wapenaar and Haimé, 1989). In the next section these Green's functions will be used in the modified elastic Kirchhoff-Helmholtz integrals for P- and S-waves.

First we review the expressions for P- and S-wave sources. Consider the elastic two-way wave equation (II-27) for the particle velocity components $v_i(\vec{r},t)$ in a homogeneous isotropic solid medium,

$$(\lambda+\mu)\partial_i\partial_j v_j + \mu\partial_j\partial_j v_i - \rho\partial_t^2 v_i = -\partial_t(f_i - \partial_j\sigma_{ij}). \qquad \text{(VI-25)}$$

The source term in the right-hand side may represent a point source of *force* in the m-direction if we choose $\sigma_{ij}=0$ and

$$f_i = \delta_{im}\delta(\vec{r}-\vec{r}_A)\delta(t). \qquad \text{(VI-26a)}$$

On the other hand, it may also represent a point source of *volume injection* at $\vec{r}_A$ for P-waves if we choose $f_i=0$ and

$$\sigma_{ij} = \delta_{ij}\delta(\vec{r}-\vec{r}_A)\delta(t), \qquad \text{(VI-26b)}$$

---

1) Choosing $\bar{c}_{ijk\ell}(\vec{r})$ and $\bar{\rho}(\vec{r})$ close to $c_{ijk\ell}(\vec{r})$ and $\rho(\vec{r})$ means that the reference medium must be designed in a geologically oriented way. Berkhout (1986) refers to such a reference medium as the *macro subsurface model.*

see also equations (II-33a) and (II-38a). Finally, it may represent a point source of *moment* at $\vec{r}_A$ for S_h-waves[1] if we choose f_i=o and

$$\sigma_{ij} = \epsilon_{hij}\delta(\vec{r}-\vec{r}_A)\delta(t), \tag{VI-26c}$$

where

$$\epsilon_{hij} = o \text{ if any of h,i,j are equal,} \tag{VI-27a}$$

otherwise

$$\epsilon_{123} = \epsilon_{312} = \epsilon_{231} = -\epsilon_{213} = -\epsilon_{321} = -\epsilon_{132} = 1, \tag{VI-27b}$$

see also equations (II-33b, c, d) and (II-40a). ϵ_{hij} is commonly known as the *alternating tensor.*

Let us now consider the elastic two-way wave equation for the Green's function $g_{i,m}(\vec{r},\vec{r}_A,t)$ in an arbitrarily inhomogeneous anisotropic solid medium. Eliminating $\theta_{ij,m}$ from equations (VI-13a) and (VI-13b) yields

$$\partial_j(c_{ijk\ell}\partial_\ell g_{k,m}) - \rho\partial_t^2 g_{i,m} = -\partial_t\left[\delta_{im}\delta(\vec{r}-\vec{r}_A)\delta(t)\right], \tag{VI-28a}$$

with initial conditions

$$g_{i,m}(\vec{r},\vec{r}_A,t) = o \qquad \text{for } t < o \tag{VI-28b}$$

and

$$\partial_t g_{i,m}(\vec{r},\vec{r}_A,t) = o \qquad \text{for } t < o. \tag{VI-28c}$$

---

1) S_h-waves for h=1, 2, 3 are polarized in the plane perpendicular to the x-, y-, or z-axis, respectively, see section II.3.2.

The Green's function $g_{i,m}(\vec{r},\vec{r}_A,t)$ for i=1, 2, 3 represents the three components of the Green's velocity vector $\vec{g}_m(\vec{r},\vec{r}_A,t)$, which is excited by an impulsive *force* in the m-direction at $\vec{r}_A$ at t=o (see Figure VI-2a). We will now introduce P- and S-wave sources for the Green's function. We assume that the solid medium is locally homogeneous and isotropic in an infinitesimal region around the Green's source point $\vec{r}_A$. At $\vec{r}_A$ we define the constrained compression modulus, according to

$$K_c(\vec{r}_A) = \lambda(\vec{r}_A) + 2\mu(\vec{r}_A), \tag{VI-29a}$$

where the Lamé coefficients λ and μ are related to the stiffness coefficients, according to

$$c_{ijk\ell}(\vec{r}_A) = \lambda(\vec{r}_A)\delta_{ij}\delta_{k\ell} + \mu(\vec{r}_A)(\delta_{ik}\delta_{j\ell} + \delta_{i\ell}\delta_{jk}). \tag{VI-29b}$$

We define an operator

$$-K_c(\vec{r}_A)\partial_{m_A} , \tag{VI-30}$$

where ∂_{m_A} for m=1, 2, 3 denotes differentiation with respect to the Green's source point coordinates x_A, y_A, z_A, respectively. By applying this operator to both sides of equation (VI-28a) we obtain

$$\partial_j(c_{ijk\ell}\partial_\ell g_{k,\phi}) - \rho\partial_t^2 g_{i,\phi} = \partial_t\partial_m\sigma_{im}, \tag{VI-31a}$$

with

$$g_{i,\phi}(\vec{r},\vec{r}_A,t) \triangleq -K_c(\vec{r}_A)\partial_{m_A} g_{i,m}(\vec{r},\vec{r}_A,t) \tag{VI-31b}$$

and

$$\sigma_{im} \triangleq -K_c(\vec{r}_A)\delta_{im}\delta(\vec{r}-\vec{r}_A)\delta(t), \tag{VI-31c}$$

where we made use of the property

$$\partial_{m_A}\delta(\vec{r}-\vec{r}_A) = -\partial_m\delta(\vec{r}-\vec{r}_A). \tag{VI-31d}$$

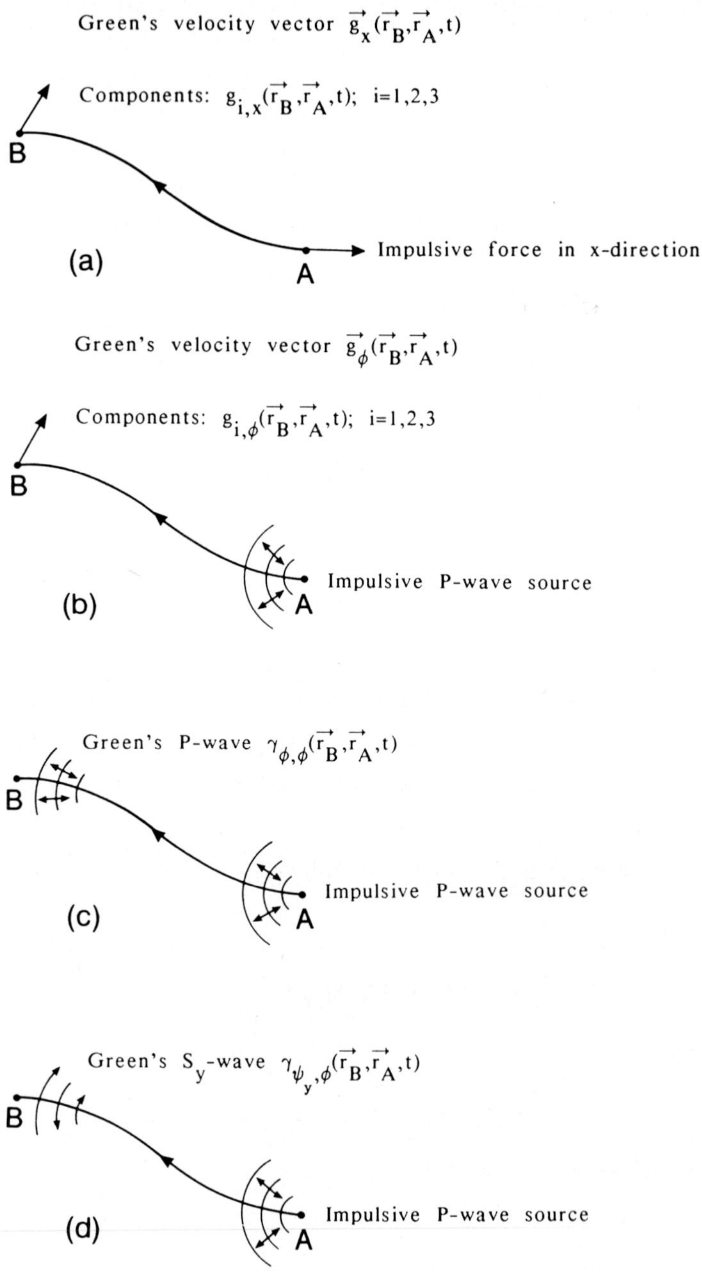

Figure VI-2: Overview of elastic Green's functions.

The source term in the right-hand side of (VI-31a), with σ_{im} defined by (VI-31c), represents a (scaled) point source of volume injection at $\vec{r}_A$ for P-waves. Hence, the Green's function $g_{i,\phi}(\vec{r},\vec{r}_A,t)$ for i = 1, 2, 3 represents the three components of the Green's velocity vector $\vec{g}_\phi(\vec{r},\vec{r}_A,t)$, which is excited by an impulsive P-wave source at $\vec{r}_A$ at t=o. Hence, the sub-script ϕ refers to the P-wave character of the Green's source at $\vec{r}_A$ (see Figure VI-2b).

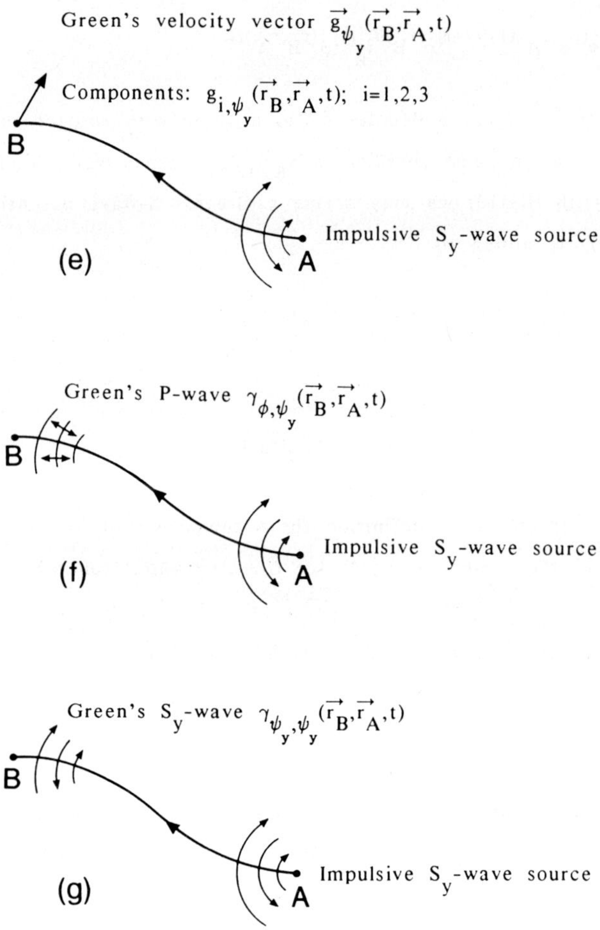

Figure VI-2 (continued)

Of course, at the observation point $\vec{r}$ the Green's function may consist of both P- and S-waves. Let us now assume that the solid medium is also locally homogeneous and isotropic in an infinitesimal region around an observation point $\vec{r} = \vec{r}_B$. Then, in agreement with (II-31c), we may define a Green's P-wave potential $\gamma_{\phi,\phi}(\vec{r}_B,\vec{r}_A,t)$, according to

$$\partial_t \gamma_{\phi,\phi}(\vec{r}_B,\vec{r}_A,t) \hat{=} -K_c(\vec{r}_B)\nabla_B \cdot \vec{g}_\phi (\vec{r}_B,\vec{r}_A,t), \tag{VI-32a}$$

or

$$\partial_t \gamma_{\phi,\phi}(\vec{r}_B,\vec{r}_A,t) \hat{=} -K_c(\vec{r}_B)\partial_{i_B} g_{i,\phi}(\vec{r}_B,\vec{r}_A,t), \tag{VI-32b}$$

where ∂_{i_B} for i = 1, 2, 3 denotes differentiation with respect to the Green's observation point coordinates x_B, y_B, z_B, respectively. Similarly, in agreement with (II-31d) we may define a Green's S-wave potential $\vec{\gamma}_{\psi,\phi}(\vec{r}_B,\vec{r}_A,t)$, according to

$$\partial_t \vec{\gamma}_{\psi,\phi}(\vec{r}_B,\vec{r}_A,t) \hat{=} \mu(\vec{r}_B)\nabla_B \times \vec{g}_\phi(\vec{r}_B,\vec{r}_A,t), \tag{VI-33a}$$

or

$$\partial_t \gamma_{\psi_k,\phi}(\vec{r}_B,\vec{r}_A,t) \hat{=} -\mu(\vec{r}_B)\epsilon_{kij}\partial_{j_B} g_{i,\phi}(\vec{r}_B,\vec{r}_A,t), \tag{VI-33b}$$

where $\gamma_{\psi_k,\phi}$ represents by definition the k-component of vector $\vec{\gamma}_{\psi,\phi}$. In the following, the symbol γ stands for *Green's potential* functions. The first sub-script in $\gamma_{\phi,\phi}(\vec{r}_B,\vec{r}_A,t)$ and $\gamma_{\psi_k,\phi}(\vec{r}_B,\vec{r}_A,t)$ refers to the wave-type at observation point $\vec{r}_B$; the second sub-script refers to the wave-type at source point $\vec{r}_A$ (ϕ refers to P-waves, ψ_k refers to S_k-waves, see Figures VI-2c and VI-2d).

Sofar we only considered modified Green's functions related to a P-wave source at $\vec{r}_A$. Next we follow the same procedure for an S-wave source at $\vec{r}_A$. By applying the operator

$$-\mu(\vec{r}_A)\epsilon_{hmn}\partial_{n_A} \tag{VI-34}$$

to both sides of equation (VI-28a) we obtain

$$\partial_j(c_{ijk\ell}\partial_\ell g_{k,\psi_h}) - \rho\partial_t^2 g_{i,\psi_h} = \partial_t\partial_n\sigma_{in}, \tag{VI-35a}$$

with

$$g_{i,\psi_h}(\vec{r},\vec{r}_A,t) \triangleq -\mu(\vec{r}_A)\epsilon_{hmn}\partial_{n_A} g_{i,m}(\vec{r},\vec{r}_A,t) \tag{VI-35b}$$

and

$$\sigma_{in} \triangleq -\mu(\vec{r}_A)\epsilon_{hin}\delta(\vec{r}-\vec{r}_A)\delta(t). \tag{VI-35c}$$

The source term in the right-hand side of (VI-35a), with σ_{in} defined by (VI-35c) represents a (scaled) point source of moment at $\vec{r}_A$ for S_h-waves. Hence, the Green's function $g_{i,\psi_h}(\vec{r},\vec{r}_A,t)$ for i = 1, 2, 3 represents the three components of the Green's velocity vector $\vec{g}_{\psi_h}(\vec{r},\vec{r}_A,t)$, which is excited by an impulsive S_h-wave source at $\vec{r}_A$ at t=o. Hence, the sub-script ψ_h refers to the S_h-wave character of the Green's source at $\vec{r}_A$ (see Figure VI-2e). At the observation point $\vec{r}=\vec{r}_B$ this Green's function may consist of both P- and S-waves.

In agreement with (II-31c), we may define a Green's P-wave potential $\gamma_{\phi,\psi_h}(\vec{r}_B,\vec{r}_A,t)$, according to

$$\partial_t\gamma_{\phi,\psi_h}(\vec{r}_B,\vec{r}_A,t) \triangleq -K_c(\vec{r}_B)\nabla_B\cdot\vec{g}_{\psi_h}(\vec{r}_B,\vec{r}_A,t), \tag{VI-36a}$$

or

$$\partial_t\gamma_{\phi,\psi_h}(\vec{r}_B,\vec{r}_A,t) \triangleq -K_c(\vec{r}_B)\partial_{i_B} g_{i,\psi_h}(\vec{r}_B,\vec{r}_A,t), \tag{VI-36b}$$

(see Figure VI-2f).

Similarly, in agreement with (II-31d) we may define a Green's S-wave potential $\vec{\gamma}_{\psi,\psi_h}(\vec{r}_B,\vec{r}_A,t)$, according to

$$\partial_t \vec{\gamma}_{\psi,\psi_h}(\vec{r}_B,\vec{r}_A,t) \triangleq \mu(\vec{r}_B)\nabla_B \times \vec{g}_{\psi_h}(\vec{r}_B,\vec{r}_A,t), \tag{VI-37a}$$

or

$$\partial_t \gamma_{\psi_k,\psi_h}(\vec{r}_B,\vec{r}_A,t) \triangleq -\mu(\vec{r}_B)\epsilon_{kij}\partial_{j_B} g_{i,\psi_h}(\vec{r}_B,\vec{r}_A,t), \tag{VI-37b}$$

where γ_{ψ_k,ψ_h} represents by definition the k-component of vector $\vec{\gamma}_{\psi,\psi_h}$, (see Figure VI-2g).

Next we give the expressions for the modified Green's functions for P- and S-waves in the space-frequency domain. Define the Green's wave field $G_{i,m}(\vec{r},\vec{r}_A,\omega)$ by

$$G_{i,m}(\vec{r},\vec{r}_A,\omega) = \int_{-\infty}^{\infty} g_{i,m}(\vec{r},\vec{r}_A,t)e^{-j\omega t}dt, \tag{VI-38}$$

with $g_{i,m}(\vec{r},\vec{r}_A,t)$ being the solution of two-way wave equation (VI-28). In analogy with (VI-31b), (VI-32b) and (VI-33b) we define Green's functions related to a P-wave source at $\vec{r}_A$:

$$G_{i,\phi}(\vec{r},\vec{r}_A,\omega) \triangleq -K_c(\vec{r}_A)\partial_{m_A} G_{i,m}(\vec{r},\vec{r}_A,\omega), \tag{VI-39a}$$

$$\Gamma_{\phi,\phi}(\vec{r}_B,\vec{r}_A,\omega) \triangleq -\frac{1}{j\omega} K_c(\vec{r}_B)\partial_{i_B} G_{i,\phi}(\vec{r}_B,\vec{r}_A,\omega) \tag{VI-39b}$$

and

$$\Gamma_{\psi_k,\phi}(\vec{r}_B,\vec{r}_A,\omega) \triangleq -\frac{1}{j\omega} \mu(\vec{r}_B)\epsilon_{kij}\partial_{j_B} G_{i,\phi}(\vec{r}_B,\vec{r}_A,\omega). \tag{VI-39c}$$

In analogy with (VI-35b), (VI-36b) and (VI-37b) we define Green's functions related to an S_h-wave source at $\vec{r}_A$:

$$G_{i,\psi_h}(\vec{r},\vec{r}_A,\omega) \triangleq -\mu(\vec{r}_A)\epsilon_{hmn}\partial_{n_A}G_{i,m}(\vec{r},\vec{r}_A,\omega), \tag{VI-40a}$$

$$\Gamma_{\phi,\psi_h}(\vec{r}_B,\vec{r}_A,\omega) \triangleq -\frac{1}{j\omega}K_c(\vec{r}_B)\partial_{i_B}G_{i,\psi_h}(\vec{r}_B,\vec{r}_A,\omega) \tag{VI-40b}$$

and

$$\Gamma_{\psi_k,\psi_h}(\vec{r}_B,\vec{r}_A,\omega) \triangleq -\frac{1}{j\omega}\mu(\vec{r}_B)\epsilon_{kij}\partial_{j_B}G_{i,\psi_h}(\vec{r}_B,\vec{r}_A,\omega). \tag{VI-40c}$$

We derive reciprocity relations for the Green's potentials for P- and S-waves. Upon substitution of (VI-39a) into (VI-39b) we obtain

$$\Gamma_{\phi,\phi}(\vec{r}_B,\vec{r}_A,\omega) = \frac{1}{j\omega}K_c(\vec{r}_B)K_c(\vec{r}_A)\partial_{i_B}\partial_{m_A}G_{i,m}(\vec{r}_B,\vec{r}_A,\omega). \tag{VI-41a}$$

In a similar way we may derive

$$\Gamma_{\phi,\phi}(\vec{r}_A,\vec{r}_B,\omega) = \frac{1}{j\omega}K_c(\vec{r}_A)K_c(\vec{r}_B)\partial_{m_A}\partial_{i_B}G_{m,i}(\vec{r}_A,\vec{r}_B,\omega). \tag{VI-41b}$$

Hence, with reciprocity relation (VI-18) we easily find

$$\Gamma_{\phi,\phi}(\vec{r}_B,\vec{r}_A,\omega) = \Gamma_{\phi,\phi}(\vec{r}_A,\vec{r}_B,\omega). \tag{VI-42a}$$

In a similar way we obtain

$$\Gamma_{\psi_k,\psi_h}(\vec{r}_B,\vec{r}_A,\omega) = \Gamma_{\psi_h,\psi_k}(\vec{r}_A,\vec{r}_B,\omega) \tag{VI-42b}$$

and

$$\Gamma_{\psi_k,\phi}(\vec{r}_B,\vec{r}_A,\omega) = \Gamma_{\phi,\psi_k}(\vec{r}_A,\vec{r}_B,\omega). \tag{VI-42c}$$

Reciprocity relation (VI-42a) states that the P-wave potential at $\vec{r}_B$, related to a P-wave source at $\vec{r}_A$, is identical to the P-wave potential at $\vec{r}_A$, related to a P-wave source at $\vec{r}_B$.
Reciprocity relation (VI-42b) states that the k-component of the S-wave potential at $\vec{r}_B$, related to an S_h-wave source at $\vec{r}_A$, is identical to the h-component of the S-wave potential at $\vec{r}_A$, related to an S_k-wave source at $\vec{r}_B$.
Reciprocity relation (VI-42c) states that the k-component of the S-wave potential at $\vec{r}_B$, related to a P-wave source at $\vec{r}_A$, is identical to the P-wave potential at $\vec{r}_A$, related to an S_k-wave source at $\vec{r}_B$.

Note that these reciprocity relations hold for arbitrarily inhomogeneous anisotropic elastic media. The only assumption is that the medium is *locally* homogeneous and isotropic at $\vec{r}_A$ and $\vec{r}_B$.

As an example we consider the *free space* Green's P- and S-wave potentials in an unbounded homogeneous isotropic solid. In analogy with (II-38), (II-40) and (V-25), we may write for the Green's P- and S-wave potentials in the space-time domain

$$\gamma_{\phi,\phi}(\vec{r},\vec{r}_A,t) = -\rho\,\partial_t^2\left(\frac{\delta(t-\Delta r/c_p)}{4\pi\Delta r}\right), \tag{VI-43a}$$

$$\gamma_{\psi_k,\psi_h}(\vec{r},\vec{r}_A,t) = -\left(\delta_{kh}\rho\partial_t^2-\mu\partial_k\partial_h\right)\left(\frac{\delta(t-\Delta r/c_s)}{4\pi\Delta r}\right) \tag{VI-43b}$$

and

$$\gamma_{\psi_k,\phi}(\vec{r},\vec{r}_A,t) = \gamma_{\phi,\psi_h}(\vec{r},\vec{r}_A,t) = 0, \tag{VI-43c}$$

with the propagation velocities c_p and c_s being given by

$$c_p = \sqrt{K_c/\rho} = \sqrt{(\lambda+2\mu)/\rho} \tag{VI-43d}$$

and

$$c_s = \sqrt{\mu/\rho} \ , \tag{VI-43e}$$

respectively, and Δr being the distance between the source point $\vec{r}_A$ and the observation point $\vec{r}$, according to

$$\Delta r = |\vec{r} - \vec{r}_A|. \tag{VI-43f}$$

In the space-frequency domain, the corresponding expressions read, in analogy with (V-26)[1],

$$\Gamma_{\phi,\phi}(\vec{r},\vec{r}_A,\omega) = \rho\omega^2 \left(\frac{e^{-jk_p \Delta r}}{4\pi\Delta r} \right), \tag{VI-44a}$$

$$\Gamma_{\psi_k,\psi_h}(\vec{r},\vec{r}_A,\omega) = \left(\delta_{kh}\rho\omega^2 + \mu\partial_k\partial_h\right) \left(\frac{e^{-jk_s \Delta r}}{4\pi\Delta r} \right) \tag{VI-44b}$$

and

$$\Gamma_{\psi_k,\phi}(\vec{r},\vec{r}_A,\omega) = \Gamma_{\phi,\psi_h}(\vec{r},\vec{r}_A,\omega) = 0, \tag{VI-44c}$$

respectively, with

$$k_p = \omega/c_p \tag{VI-44d}$$

and

$$k_s = \omega/c_s. \tag{VI-44e}$$

Finally, in the wavenumber-frequency domain, the corresponding expressions read, in analogy with (V-27),

$$\tilde{\Gamma}_{\phi,\phi}(k_x,k_y,z;x_A,y_A,z_A;\omega) = \rho\omega^2 e^{j(k_x x_A + k_y y_A)} \frac{e^{-jk_{z,p}|z-z_A|}}{2jk_{z,p}} , \tag{VI-45a}$$

---

[1] In the following the symbol k has two different meanings. When used as a factor, it denotes a wavenumber (k_p etc.). Otherwise it is an index which may take the values 1, 2 or 3.

$$\tilde{\Gamma}_{\psi_k,\psi_h}(k_x,k_y,z;x_A,y_A,z_A;\omega) = (\delta_{kh}\rho\omega^2+\mu d_k d_h)e^{j(k_x x_A+k_y y_A)} \frac{e^{-jk_{z,s}|z-z_A|}}{2jk_{z,s}} \tag{VI-45b}$$

and

$$\tilde{\Gamma}_{\psi_k,\phi}(k_x,k_y,z;x_A,y_A,z_A;\omega) = \tilde{\Gamma}_{\phi,\psi_h}(k_x,k_y,z;x_A,y_A,z_A;\omega) = 0, \tag{VI-45c}$$

respectively, where d_h for h=1, 2, 3 stands for $-jk_x$, $-jk_y$ and ∂_z, respectively, and where

$$k_{z,p} = +\sqrt{k_p^2 - k_x^2 - k_y^2} \qquad \text{for } k_x^2+k_y^2 \le k_p^2, \tag{VI-45d}$$

$$k_{z,p} = -j\sqrt{k_x^2 + k_y^2 - k_p^2} \qquad \text{for } k_x^2+k_y^2 > k_p^2, \tag{VI-45e}$$

$$k_{z,s} = +\sqrt{k_s^2 - k_x^2 - k_y^2} \qquad \text{for } k_x^2+k_y^2 \le k_s^2 \tag{VI-45f}$$

and

$$k_{z,s} = -j\sqrt{k_x^2 + k_y^2 - k_s^2} \qquad \text{for } k_x^2+k_y^2 > k_s^2. \tag{VI-45g}$$

VI.3.4 Modified elastic Kirchhoff-Helmholtz integrals for P- and S-waves

We return to the inhomogeneous anisotropic situation. Consider the elastic Kirchhoff-Helmholtz integral (VI-24a),

$$V_m(\vec{r}_A,\omega) = -\oint_S [\boldsymbol{\Theta}_m(\vec{r},\vec{r}_A,\omega)\vec{V}(\vec{r},\omega) - \tau(\vec{r},\omega)\vec{G}_m(\vec{r},\vec{r}_A,\omega)].\vec{n}\,dS. \tag{VI-46}$$

This integral expresses the velocity component V_m at $\vec{r}_A$ in V in terms of the elastic wave field on S, enclosing V. As we showed in section VI.3.2, this expression is exact when the medium parameters for the Green's wave field are identical to the actual medium parameters throughout volume V. We now derive modified Kirchhoff-Helmholtz integrals that express the P-wave potential or the S-wave potential at $\vec{r}_A$ in terms of the elastic wave field on S. Again we assume that the solid medium is locally homogeneous and isotropic in an infinitesimal region around $\vec{r}_A$. Hence, in agreement with (II-31c), we may define a P-wave potential at $\vec{r}_A$, according to

$$j\omega\Phi(\vec{r}_A,\omega) \triangleq -K_c(\vec{r}_A)\nabla_A \cdot \vec{V}(\vec{r}_A,\omega) \tag{VI-47a}$$

or

$$j\omega\Phi(\vec{r}_A,\omega) \triangleq -K_c(\vec{r}_A)\partial_{m_A} V_m(\vec{r}_A,\omega). \tag{VI-47b}$$

By substituting Kirchhoff-Helmholtz integral (VI-46) into (VI-47b) and interchanging the order of integration and differentiation[1), we obtain

$$j\omega\Phi(\vec{r}_A,\omega) = -\oint_S \left[\boldsymbol{\Theta}_\phi(\vec{r},\vec{r}_A,\omega)\vec{V}(\vec{r},\omega) - \boldsymbol{\tau}(\vec{r},\omega)\vec{G}_\phi(\vec{r},\vec{r}_A,\omega)\right].\vec{n}\,dS, \tag{VI-48a}$$

where

$$\vec{G}_\phi(\vec{r},\vec{r}_A,\omega) \triangleq -K_c(\vec{r}_A)\partial_{m_A}\vec{G}_m(\vec{r},\vec{r}_A,\omega) \tag{VI-48b}$$

and

$$\boldsymbol{\Theta}_\phi(\vec{r},\vec{r}_A,\omega) \triangleq -K_c(\vec{r}_A)\partial_{m_A}\boldsymbol{\Theta}_m(\vec{r},\vec{r}_A,\omega). \tag{VI-48c}$$

The modified Kirchhoff-Helmholtz integral (VI-48a) expresses the P-wave potential at $\vec{r}_A$ in V in terms of the elastic wave field on S, enclosing V.

Note that for the components of $\vec{G}_\phi(\vec{r},\vec{r}_A,\omega)$ we may write

$$G_{i,\phi}(\vec{r},\vec{r}_A,\omega) = -K_c(\vec{r}_A)\partial_{m_A} G_{i,m}(\vec{r},\vec{r}_A,\omega), \tag{VI-49a}$$

which is equivalent to equation (VI-39a). Hence, $\vec{G}_\phi(\vec{r},\vec{r}_A,\omega)$, as defined by

[1) This is justified as the integration takes place along $S(\vec{r})$ whereas the differentiation takes place at $\vec{r}_A$.

(VI-48b), represents the Green's velocity field, related to a P-wave source at $\vec{r}_A$. Similarly, $\boldsymbol{\Theta}_\phi(\vec{r},\vec{r}_A,\omega)$, as defined by (VI-48c), represents the Green's stress field, related to the same P-wave source at $\vec{r}_A$. In analogy with (VI-13b), the components of $\boldsymbol{\Theta}_\phi(\vec{r},\vec{r}_A,\omega)$ are related to the components of $\vec{G}_\phi(\vec{r},\vec{r}_A,\omega)$, according to

$$j\omega\Theta_{ij,\phi}(\vec{r},\vec{r}_A,\omega) = c_{ijk\ell}(\vec{r})\partial_\ell G_{k,\phi}(\vec{r},\vec{r}_A,\omega). \qquad \text{(VI-49b)}$$

Next, in agreement with (II-31d), we define a S-wave potential at $\vec{r}_A$, according to

$$j\omega\vec{\Psi}(\vec{r}_A,\omega) \hat{=} \mu(\vec{r}_A)\nabla_A\times\vec{V}(\vec{r}_A,\omega), \qquad \text{(VI-50a)}$$

or

$$j\omega\Psi_h(\vec{r}_A,\omega) \hat{=} -\mu(\vec{r}_A)\epsilon_{hmn}\partial_{n_A}V_m(\vec{r}_A,\omega). \qquad \text{(VI-50b)}$$

By substituting Kirchhoff-Helmholtz integral (VI-46) into (VI-50b) and interchanging the order of integration (along $S(\vec{r})$) and differentiation (at $\vec{r}_A$), we obtain

$$j\omega\Psi_h(\vec{r}_A,\omega) = -\oint_S [\boldsymbol{\Theta}_{\psi_h}(\vec{r},\vec{r}_A,\omega)\vec{V}(\vec{r},\omega) - \tau(\vec{r},\omega)\vec{G}_{\psi_h}(\vec{r},\vec{r}_A,\omega)].\vec{n}\,dS, \qquad \text{(VI-51a)}$$

where

$$\vec{G}_{\psi_h}(\vec{r},\vec{r}_A,\omega) \hat{=} -\mu(\vec{r}_A)\epsilon_{hmn}\partial_{n_A}\vec{G}_m(\vec{r},\vec{r}_A,\omega) \qquad \text{(VI-51b)}$$

and

$$\boldsymbol{\Theta}_{\psi_h}(\vec{r},\vec{r}_A,\omega) \hat{=} -\mu(\vec{r}_A)\epsilon_{hmn}\partial_{n_A}\boldsymbol{\Theta}_m(\vec{r},\vec{r}_A,\omega). \qquad \text{(VI-51c)}$$

The modified Kirchhoff-Helmholtz integral (VI-51a) expresses the h-component of the S-wave potential at $\vec{r}_A$ in V in terms of the elastic wave field on S, enclosing V. Note that for the components of $\vec{G}_{\psi_h}(\vec{r},\vec{r}_A,\omega)$ we may write

$$G_{i,\psi_h}(\vec{r},\vec{r}_A,\omega) = -\mu(\vec{r}_A)\epsilon_{hmn}\partial_{n_A}G_{i,m}(\vec{r},\vec{r}_A,\omega), \tag{VI-52a}$$

which is equivalent to equation (VI-40a). Hence, $\vec{G}_{\psi_h}(\vec{r},\vec{r}_A,\omega)$, as defined by (VI-51b), represents the Green's velocity field, related to an S_h-wave source at $\vec{r}_A$. Similarly, $\boldsymbol{\Theta}_{\psi_h}(\vec{r},\vec{r}_A,\omega)$, as defined by (VI-51c), represents the Green's stress field, related to the same S_h-wave source at $\vec{r}_A$. In analogy with (VI-13b), the components of $\boldsymbol{\Theta}_{\psi_h}(\vec{r},\vec{r}_A,\omega)$ are related to the components of $\vec{G}_{\psi_h}(\vec{r},\vec{r}_A,\omega)$, according to

$$j\omega\Theta_{ij,\psi_h}(\vec{r},\vec{r}_A,\omega) = c_{ijk\ell}(\vec{r})\partial_\ell G_{k,\psi_h}(\vec{r},\vec{r}_A,\omega). \tag{VI-52b}$$

Kirchhoff-Helmholtz integrals (VI-46), (VI-48a) and (VI-51a) can be summarized by

$$\Omega(\vec{r}_A,\omega) = -\oint_S [\boldsymbol{\Theta}_\Omega(\vec{r},\vec{r}_A,\omega)\vec{V}(\vec{r},\omega) - \boldsymbol{\tau}(\vec{r},\omega)\vec{G}_\Omega(\vec{r},\vec{r}_A,\omega)].\vec{n}\,dS, \tag{VI-53}$$

see Figure VI-3.

1. If we choose for the Green's wave field an impulsive force in the m-direction at $\vec{r}_A$ (i.e., $\vec{G}_\Omega=\vec{G}_m$; $\boldsymbol{\Theta}_\Omega=\boldsymbol{\Theta}_m$), then $\Omega(\vec{r}_A,\omega)$ represents the m-component of the velocity $\vec{V}$ at $\vec{r}_A$.

2. If we choose for the Green's wave field an impulsive P-wave source at $\vec{r}_A$ (i.e., $\vec{G}_\Omega=\vec{G}_\phi$; $\boldsymbol{\Theta}_\Omega=\boldsymbol{\Theta}_\phi$), then $\Omega(\vec{r}_A,\omega)$ represents the scaled P-wave potential $j\omega\Phi$ at $\vec{r}_A$.

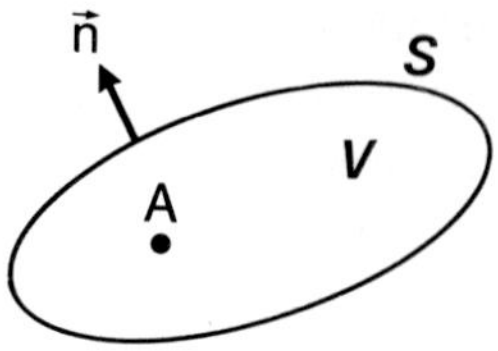

$$\Omega(\vec{r}_A,\omega) = -\oint_S [\boldsymbol{\Theta}_\Omega \vec{V} - \boldsymbol{\tau}\vec{G}_\Omega].\vec{n}\,dS$$

Figure VI-3: Elastic Kirchhoff-Helmholtz integral. The Green's wave field $(\vec{G}_\Omega, \boldsymbol{\Theta}_\Omega)$ *may be excited either by an impulsive force, an impulsive P-wave source or an impulsive* S_h*-wave source at* $\vec{r}_A$*. Accordingly,* $\Omega(\vec{r}_A,\omega)$ *may represent either the velocity, the P-wave potential or the* S_h*-wave potential at* $\vec{r}_A$*.*

3. If we choose for the Green's wave field an impulsive S_h-wave source at $\vec{r}_A$ (i.e., $\vec{G}_\Omega = \vec{G}_{\psi_h}$; $\boldsymbol{\Theta}_\Omega = \boldsymbol{\Theta}_{\psi_h}$), then $\Omega(\vec{r}_A,\omega)$ represents the h-component of the scaled S-wave potential $j\omega\vec{\Psi}$ at $\vec{r}_A$.

VI.4 ELASTIC TWO-WAY AND ONE-WAY RAYLEIGH INTEGRALS

VI.4.1 Boundary conditions for the elastic Green's functions

Consider the Kirchhoff-Helmholtz integral (VI-53). Using the symmetry property of the stress tensors $\boldsymbol{\tau}$ and $\boldsymbol{\Theta}_\Omega$ we may rewrite (VI-53) as

$$\Omega(\vec{r}_A,\omega) = -\oint_S \Big[\big(\boldsymbol{\Theta}_\Omega(\vec{r},\vec{r}_A,\omega)\vec{n}\big).\vec{V}(\vec{r},\omega) - \vec{G}_\Omega(\vec{r},\vec{r}_A,\omega).\big(\boldsymbol{\tau}(\vec{r},\omega)\vec{n}\big)\Big]\,dS, \qquad \text{(VI-54a)}$$

or, using Cauchy's formula (II-11b) for the *tractions* $\vec{\tau}_n$ and $\vec{\Theta}_{n,\Omega}$,

$$\Omega(\vec{r}_A,\omega) = -\oint_S \big[\vec{\Theta}_{n,\Omega}(\vec{r},\vec{r}_A,\omega).\vec{V}(\vec{r},\omega) - \vec{G}_\Omega(\vec{r},\vec{r}_A,\omega).\vec{\tau}_n(\vec{r},\omega)\big]\,dS. \qquad \text{(VI-54b)}$$

These expressions are exact when the medium parameters for the Green's wave field are identical to the actual medium parameters throughout volume

V. Outside *V* we may choose any convenient medium for the Green's wave field and on (parts of) *S* we may impose any convenient boundary condition for the Green's wave field. If we choose

$$\vec{G}_{\Omega}(\vec{r},\vec{r}_{A},\omega) = \vec{o} \text{ on } S, \tag{VI-55a}$$

then Kirchhoff-Helmholtz integral (VI-54b) simplifies to

$$\Omega(\vec{r}_{A},\omega) = -\oint_{S} [\vec{\Theta}_{n,\Omega}(\vec{r},\vec{r}_{A},\omega).\vec{V}(\vec{r},\omega)] \, dS. \tag{VI-55b}$$

However, a practical disadvantage is that the Green's wave field $\vec{\Theta}_{n,\Omega}$ may become very complicated because boundary condition (VI-55a) means that *S* is a *rigid* surface, which is perfectly reflecting. Similarly, if we choose

$$\vec{\Theta}_{n,\Omega}(\vec{r},\vec{r}_{A},\omega) = \vec{o} \quad \text{on } S, \tag{VI-56a}$$

then Kirchhoff-Helmholtz integral (VI-54b) simplifies to

$$\Omega(\vec{r}_{A},\omega) = \oint_{S} [\vec{G}_{\Omega}(\vec{r},\vec{r}_{A},\omega).\vec{\tau}_{n}(\vec{r},\omega)] \, dS. \tag{VI-56b}$$

Again, the Green's wave field $\vec{G}_{\Omega}$ may become very complicated because boundary condition (VI-56a) means that *S* is a *free* surface, which is perfectly reflecting. The Kirchhoff-Helmholtz integral (VI-54b) as well as its simplified versions (VI-55b) and (VI-56b) are in principle *two-way* expressions. In the next section we consider (VI-55b) and (VI-56b) for a special configuration. Next we divert from what is generally done in the literature and, instead of fully reflecting boundary conditions, we choose fully absorbing boundary conditions for the Green's wave fields on S. Following this alternative route, we show how to derive elegant *one-way* expressions.

VI.4.2 Elastic two-way Rayleigh integrals

Consider the half-space geometry of Figure VI-4. Closed surface S consists of a horizontal flat surface S_0 at $z=z_0$ and a hemi-sphere S_1 in the lower half-space $z \geq z_0$, with midpoint A and radius r_1. Assuming that the sources of the elastic wave field are situated in the upper half-space $z<z_0$, then the contribution of the Kirchhoff-Helmholtz integral over S_1 to the elastic wave field at A vanishes if r_1 goes to infinity (Sommerfeld radiation conditions, see also section V-2 and Pao and Varatharajulu, 1976). Hence, for this situation equation (VI-54b) may be replaced by

$$\Omega(\vec{r}_A,\omega) = \int\int_{-\infty}^{\infty} [\vec{\Theta}_{z,\Omega}(\vec{r},\vec{r}_A,\omega).\vec{V}(\vec{r},\omega) - \vec{G}_{\Omega}(\vec{r},\vec{r}_A,\omega).\vec{\tau}_z(\vec{r},\omega)]_{z_0} dxdy, \qquad \text{(VI-57)}$$

where we made use of the fact that $\vec{n}$ on S_0 is a unit vector in the *negative* z-direction. Again we consider rigid and free boundary conditions for the Green's wave field.

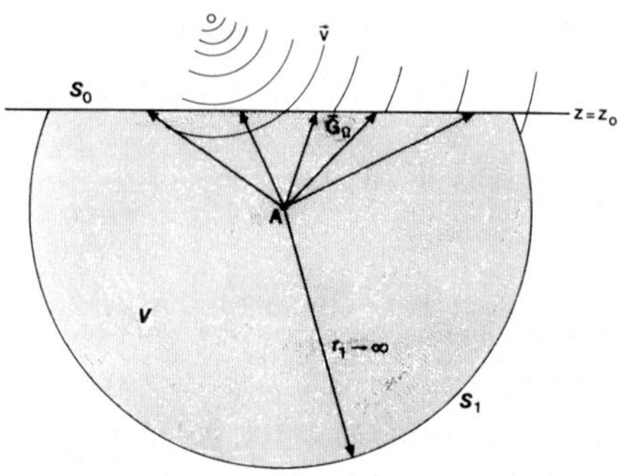

Figure VI-4: Configuration for which the closed surface integral (VI-54b) may be replaced by the open surface integral(VI-57). The lower half-space is assumed to be source-free. This configuration is the basis for the derivation of several types of Rayleigh integrals.

If we choose

$$\vec{G}_{\Omega}(\vec{r},\vec{r}_A,\omega) = \vec{o} \quad \text{at } z=z_0 \tag{VI-58a}$$

then Kirchhoff-Helmholtz integral (VI-57) simplifies to

$$\Omega(\vec{r}_A,\omega) = \int\int_{-\infty}^{\infty} [\vec{\Theta}_{z,\Omega}(\vec{r},\vec{r}_A,\omega).\vec{V}(\vec{r},\omega)]_{z_0} dxdy. \tag{VI-58b}$$

Similarly, if we choose

$$\vec{\Theta}_{z,\Omega}(\vec{r},\vec{r}_A,\omega) = \vec{o} \quad \text{at } z=z_0, \tag{VI-59a}$$

then Kirchhoff-Helmholtz integral (VI-57) simplifies to

$$\Omega(\vec{r}_A,\omega) = -\int\int_{-\infty}^{\infty} [\vec{G}_{\Omega}(\vec{r},\vec{r}_A,\omega).\vec{\tau}_z(\vec{r},\omega)]_{z_0} dxdy. \tag{VI-59b}$$

In analogy with section V.4.2, we refer to expressions (VI-58b) and (VI-59b) as the elastic two-way Rayleigh I and Rayleigh II integrals, respectively. For a further discussion on the two-way aspects of the Rayleigh integrals we refer to section V.4.2.

VI.4.3 Elastic one-way Rayleigh integrals

We consider again the half-space geometry of Figur VI-4. Unlike in the previous section, we now choose for the Green's wave field a fully non-reflecting upper half-space $z<z_0$. Hence,

$$c_{ijk\ell}(x,y,z<z_0) = c_{ijk\ell}(x,y,z_0) \quad \text{for all } z<z_0 \tag{VI-60a}$$

and

$$\rho(x,y,z<z_0) = \rho(x,y,z_0) \quad \text{for all } z<z_0. \tag{VI-60b}$$

With this choice no downgoing Green's waves return from the upper half-space, so the Green's wave field is purely *upgoing* at $z=z_0$:

$$\vec{G}_\Omega(\vec{r},\vec{r}_A,\omega) = \vec{G}_\Omega^-(\vec{r},\vec{r}_A,\omega) \qquad \text{at } z=z_0 \tag{VI-61a}$$

and

$$\vec{\Theta}_{z,\Omega}(\vec{r},\vec{r}_A,\omega) = \vec{\Theta}_{z,\Omega}^-(\vec{r},\vec{r}_A,\omega) \qquad \text{at } z=z_0. \tag{VI-61b}$$

In terms of boundary conditions we may say that surface $z=z_0$ is an *absorbing* boundary for the Green's wave field.

The sources for the elastic wave field are situated in the upper half-space, so at z_0 this wave field consists of the downgoing incident wave field (including higher order terms) and the upgoing scattered wave field (including higher order terms), according to

$$\vec{V}(\vec{r},\omega) = \vec{V}^+(\vec{r},\omega) + \vec{V}^-(\vec{r},\omega) \qquad \text{at } z=z_0 \tag{VI-62a}$$

and

$$\vec{\tau}_z(\vec{r},\omega) = \vec{\tau}_z^+(\vec{r},\omega) + \vec{\tau}_z^-(\vec{r},\omega) \qquad \text{at } z=z_0. \tag{VI-62b}$$

Substitution of (VI-61) and (VI-62) into Kirchhoff-Helmholtz integral (VI-57) yields

$$\Omega(\vec{r}_A,\omega) = \iint_{-\infty}^{\infty} [\vec{\Theta}_{z,\Omega}^- \cdot (\vec{V}^+ + \vec{V}^-) - \vec{G}_\Omega^- \cdot (\vec{\tau}_z^+ + \vec{\tau}_z^-)]_{z_0}\, dxdy. \tag{VI-63}$$

Our aim is to simplify this expression by using the *one-way wave equations* for the downgoing and upgoing wave fields at z_0. For simplicity we assume that the solid medium is locally homogeneous and isotropic (described by λ, μ and ρ) in an infinitesimal region around z_0. Hence, in agreement with (II-31a) and (II-31b) we may define P- and S-wave potentials for the downgoing and upgoing elastic wave fields at z_0, according to

$$[\vec{V}^{\pm}(\vec{r},\omega)]_{z_0} \triangleq \frac{-1}{j\omega\rho}\,[\nabla\Phi^{\pm}(\vec{r},\omega) + \nabla\times\vec{\Psi}^{\pm}(\vec{r},\omega)]_{z_0}, \tag{VI-64a}$$

with (for the source-free situation at z_0)

$$[\nabla.\vec{\Psi}^{\pm}(\vec{r},\omega)]_{z_0} \triangleq 0, \tag{VI-64b}$$

see Figure VI-5a. Similarly, we may define P- and S-wave potentials for the upgoing Green's wave field at z_0, according to

$$[\vec{G}^{-}_{\Omega}(\vec{r},\vec{r}_A,\omega)]_{z_0} \triangleq \frac{-1}{j\omega\rho}[\nabla\Gamma^{-}_{\phi,\Omega}(\vec{r},\vec{r}_A,\omega) + \nabla\times\vec{\Gamma}^{-}_{\psi,\Omega}(\vec{r},\vec{r}_A,\omega)]_{z_0}, \tag{VI-65a}$$

with (for the source-free situation at z_0)

$$[\nabla.\vec{\Gamma}^{-}_{\psi,\Omega}(\vec{r},\vec{r}_A,\omega)]_{z_0} \triangleq 0, \tag{VI-65b}$$

see Figure VI-5b. In Appendix C, section C.2, we show that by applying the one-way wave equations for the P- and S-wave potentials at z_0, the Kirchhoff-Helmholtz integral (VI-63) can be transformed to

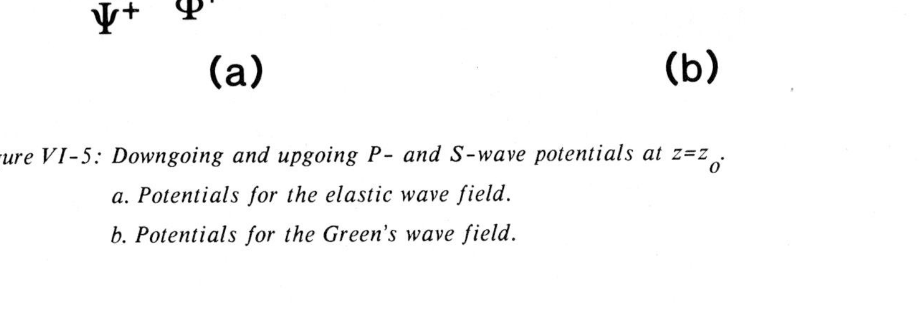

Figure VI-5: Downgoing and upgoing P- and S-wave potentials at $z=z_0$.
a. Potentials for the elastic wave field.
b. Potentials for the Green's wave field.

$$\Omega(\vec{r}_A,\omega) = \frac{2}{j\omega} \iint_{-\infty}^{\infty} \frac{1}{\rho(z_0)} \left[\Gamma^-_{\phi,\Omega} \frac{\partial \Phi^+}{\partial z} + \vec{\Gamma}^-_{\psi,\Omega} \cdot \frac{\partial \vec{\Psi}^+}{\partial z} \right]_{z_0} dxdy, \tag{VI-66a}$$

or, alternatively, to

$$\Omega(\vec{r}_A,\omega) = \frac{-2}{j\omega} \iint_{-\infty}^{\infty} \frac{1}{\rho(z_0)} \left[\frac{\partial \Gamma^-_{\phi,\Omega}}{\partial z} \Phi^+ + \frac{\partial \vec{\Gamma}^-_{\psi,\Omega}}{\partial z} \cdot \vec{\Psi}^+ \right]_{z_0} dxdy. \tag{VI-66b}$$

In analogy with section V.4.3, we refer to expressions (VI-66a) and (VI-66b) as the elastic one-way Rayleigh I and Rayleigh II integrals, respectively. For a discussion on the one-way aspects of the Rayleigh integrals we refer to section V.4.3. Note that in both expressions (VI-66a) and (VI-66b) only the downgoing P-wave Φ^+ interacts with the upgoing Green's P-wave $\Gamma^-_{\phi,\Omega}$ and the downgoing S-wave $\vec{\Psi}^+$ interacts with the upgoing Green's S-wave $\vec{\Gamma}^-_{\psi,\Omega}$. No interaction occurs between the P-wave and the Green's S-wave nor between the S-wave and the Green's P-wave. Finally, note that for the situation of Figure VI-4 equations (VI-66a) and (VI-66b) are valid for an arbitrarily inhomogeneous anisotropic solid medium; the only assumption is that the solid medium is *locally* homogeneous and isotropic in infinitesimal regions around z_0 and $\vec{r}_A$.

Depending on the choice of the source for the Green's wave fields, Ω in equations (VI-66a) and (VI-66b) may represent either V_m for m = 1, 2, 3 or $j\omega\Phi$, or $j\omega\Psi_h$ for h = 1, 2, 3, see also section VI.3.4. We consider the latter two cases. If the Green's wave fields have an impulsive P-wave source at $\vec{r}_A$, then Ω represents the scaled P-wave potential $j\omega\Phi$ at $\vec{r}_A$, hence

$$\Phi(\vec{r}_A,\omega) = \frac{-2}{\omega^2} \iint_{-\infty}^{\infty} \frac{1}{\rho(z_0)} \left[\Gamma^-_{\phi,\phi} \frac{\partial \Phi^+}{\partial z} + \vec{\Gamma}^-_{\psi,\phi} \cdot \frac{\partial \vec{\Psi}^+}{\partial z} \right]_{z_0} dxdy, \tag{VI-67a}$$

or, alternatively,

$$\Phi(\vec{r}_A,\omega) = \frac{2}{\omega^2} \iint_{-\infty}^{\infty} \frac{1}{\rho(z_0)} \left[\frac{\partial \Gamma^-_{\phi,\phi}}{\partial z} \Phi^+ + \frac{\partial \vec{\Gamma}^-_{\psi,\phi}}{\partial z} \cdot \vec{\Psi}^+ \right]_{z_0} dxdy. \tag{VI-67b}$$

On the other hand, if the Green's wave fields have an impulsive S_h-wave source at $\vec{r}_A$, then Ω represents the h-component of the scaled S-wave potential $j\omega\vec{\Psi}$ at $\vec{r}_A$, hence,

$$\Psi_h(\vec{r}_A,\omega) = \frac{-2}{\omega^2} \iint_{-\infty}^{\infty} \frac{1}{\rho(z_0)} \left[\Gamma^-_{\phi,\psi_h} \frac{\partial \Phi^+}{\partial z} + \vec{\Gamma}^-_{\psi,\psi_h} \cdot \frac{\partial \vec{\Psi}^+}{\partial z} \right]_{z_0} dxdy, \tag{VI-68a}$$

or, alternatively,

$$\Psi_h(\vec{r}_A,\omega) = \frac{2}{\omega^2} \iint_{-\infty}^{\infty} \frac{1}{\rho(z_0)} \left[\frac{\partial \Gamma^-_{\phi,\psi_h}}{\partial z} \Phi^+ + \frac{\partial \vec{\Gamma}^-_{\psi,\psi_h}}{\partial z} \cdot \vec{\Psi}^+ \right]_{z_0} dxdy. \tag{VI-68b}$$

For the special situation of a homogeneous isotropic medium, we may substitute the free space solutions (VI-44a) and (VI-44b) for the Green's wave fields $\Gamma^-_{\phi,\phi}$ and $\Gamma^-_{\psi_k,\psi_h}$ (bear in mind that $\Gamma^-_{\psi_k,\psi_h}$ denotes the k-component of $\vec{\Gamma}^-_{\psi,\psi_h}$). Hence, for this situation equations (VI-67a) and (VI-67b) read

$$\Phi(\vec{r}_A,\omega) = \frac{-1}{2\pi} \iint_{-\infty}^{\infty} \left[\frac{e^{-jk_p\Delta r}}{\Delta r} \frac{\partial \Phi^+(\vec{r},\omega)}{\partial z} \right]_{z_0} dxdy, \tag{VI-69a}$$

or

$$\Phi(\vec{r}_A,\omega) = \frac{1}{2\pi} \iint_{-\infty}^{\infty} \left[\frac{\partial}{\partial z} \left(\frac{e^{-jk_p\Delta r}}{\Delta r} \right) \Phi^+(\vec{r},\omega) \right]_{z_0} dxdy, \tag{VI-69b}$$

respectively, with

$$k_p = \omega/c_p \tag{VI-69c}$$

and

$$\Delta r = |\vec{r}-\vec{r}_A| . \tag{VI-69d}$$

Note that the elastic one-way Rayleigh I and Rayleigh II integrals (VI-69a) and (VI-69b) are identical to the acoustic one-way Rayleigh I and Rayleigh II integrals (V-45a) and (V-45b), respectively, with k and P replaced by k_p and Φ, respectively.

For the same situation equations (VI-68a) and (VI-68b) read

$$\Psi_h(\vec{r}_A,\omega) = \frac{-1}{2\pi} \iint_{-\infty}^{\infty} \left[\left((\delta_{kh}+\frac{1}{k_s^2}\,\partial_k\partial_h)\frac{e^{-jk_s\Delta r}}{\Delta r}\right) \frac{\partial\Psi_k^+(\vec{r},\omega)}{\partial z} \right]_{z_0} dxdy, \tag{VI-70a}$$

or

$$\Psi_h(\vec{r}_A,\omega) = \frac{1}{2\pi} \iint_{-\infty}^{\infty} \left[\frac{\partial}{\partial z}\left((\delta_{kh}+\frac{1}{k_s^2}\,\partial_k\partial_h)\frac{e^{-jk_s\Delta r}}{\Delta r}\right) \Psi_k^+(\vec{r},\omega) \right]_{z_0} dxdy, \tag{VI-70b}$$

respectively, with

$$k_s = \omega/c_s . \tag{VI-70c}$$

Note that the elastic one-way Rayleigh I and Rayleigh II integrals (VI-70a) and (VI-70b) are somewhat more complicated than the acoustic one-way Rayleigh I and Rayleigh II integrals (V-45a) and (V-45b). However, in the next section we derive elastic extrapolation operators which, for the homogeneous isotropic situation, are again identical to the acoustic extrapolation operators.

VI.5 ELASTIC FORWARD WAVE FIELD EXTRAPOLATION OPERATORS

VI.5.1 Integral formulation of elastic forward wave field extrapolation

We return to the inhomogeneous anisotropic situation. Our starting point for deriving elastic forward wave field extrapolation operators is the one-way version of the elastic Rayleigh II integral (VI-66b), which may be rewritten as

$$\frac{1}{j\omega}\Omega(\vec{r}_A,\omega) = \frac{2}{\omega^2}\iint_{-\infty}^{\infty}\frac{1}{\rho(z_0)}\left[\frac{\partial\Gamma^-_{\phi,\Omega}(\vec{r},\vec{r}_A,\omega)}{\partial z}\Phi^+(\vec{r},\omega) + \mathscr{L}^-_{\psi_\alpha,\Omega}(\vec{r},\vec{r}_A,\omega)\Psi^+_\alpha(\vec{r},\omega)\right]_{z_0} dxdy, \qquad \text{(VI-71a)}$$

where the modified Green's wave field reads, according to equation (C-21b) in Appendix C,

$$\mathscr{L}^-_{\psi_\alpha,\Omega}(\vec{r},\vec{r}_A,\omega) \triangleq \epsilon_{3\alpha j}\epsilon_{jik}\partial_i\Gamma^-_{\psi_k,\Omega}(\vec{r},\vec{r}_A,\omega). \qquad \text{(VI-71b)}$$

Bear in mind that Greek indices may only take the values 1 or 2. Hence, just as in section IV.2.2, we eliminated the z-component of the S-wave potential. In equation (VI-71), $\Phi^+(\vec{r},\omega)$ and $\Psi^+_\alpha(\vec{r},\omega)$ for $z=z_0$ represent the downgoing parts of the potentials for the elastic P- and S-wave fields at z_0, related to sources in the upper half-space (see Figure V-9a for the acoustic equivalence, where $\Phi^+=P^+$ and $\Psi^+_\alpha=o$). $\Gamma^-_{\phi,\Omega}(\vec{r},\vec{r}_A,\omega)$ and $\Gamma^-_{\psi_k,\Omega}(\vec{r},\vec{r}_A,\omega)$ for $z=z_0$ represent the upgoing parts of the potentials for the Green's P- and S-wave fields at z_0, related to a monopole Ω-wave source (i.e., a P-wave source, or an S_β-wave[1)] source) at $\vec{r}_A$ in the lower half-space (see Figure V-9b for the acoustic equivalence, where $\Gamma^-_{\phi,\Omega}=\omega^2G^-$ and $\Gamma^-_{\psi_k,\Omega}=o$). $\frac{1}{j\omega}\Omega(\vec{r}_A,\omega)$ represents the potential $\Phi(\vec{r}_A,\omega)$ or $\Psi_\beta(\vec{r}_A,\omega)$ for the total elastic P-wave or S_β-wave, respectively, at $\vec{r}_A$ in the lower half-space ($\Phi=P$ and $\Psi_\beta=o$ in the acoustic equivalence). Hence, equation (VI-71) describes wave field extrapolation of the *downgoing* (i.e., *one-way*) potentials $\Phi^+(\vec{r},\omega)$ and $\Psi^+_\alpha(\vec{r},\omega)$ at z_0, yielding the *total* (i.e., *two-way*) potential Φ or Ψ_β at $\vec{r}_A$. For the Green's wave field the upper half-space was chosen reflection-free, so

$$\Gamma^-_{\phi,\Omega}(\vec{r},\vec{r}_A,\omega) = \Gamma_{\phi,\Omega}(\vec{r},\vec{r}_A,\omega) \qquad \text{at } z=z_0, \qquad \text{(VI-72a)}$$

---

1) S_β-waves for $\beta=1$, 2 are polarized in the plane perpendicular to the x- or y-axis, respectively, see section II.3.2.

$$\Gamma^{-}_{\psi_k,\Omega}(\vec{r},\vec{r}_A,\omega) = \Gamma_{\psi_k,\Omega}(\vec{r},\vec{r}_A,\omega) \qquad \text{at } z=z_o, \tag{VI-72b}$$

and, consequently,

$$\mathcal{L}^{-}_{\psi_\alpha,\Omega}(\vec{r},\vec{r}_A,\omega) = \mathcal{L}_{\psi_\alpha,\Omega}(\vec{r},\vec{r}_A,\omega) \qquad \text{at } z=z_o. \tag{VI-72c}$$

Based on this property and the reciprocity relations (VI-42),

$$\Gamma_{\phi,\Omega}(\vec{r}_B,\vec{r}_A,\omega) = \Gamma_{\Omega,\phi}(\vec{r}_A,\vec{r}_B,\omega) \tag{VI-73a}$$

and

$$\Gamma_{\psi_k,\Omega}(\vec{r}_B,\vec{r}_A,\omega) = \Gamma_{\Omega,\psi_k}(\vec{r}_A,\vec{r}_B,\omega), \tag{VI-73b}$$

we may reformulate (VI-71) as

$$\frac{1}{j\omega}\,\Omega(\vec{r}_A,\omega) = \frac{2}{\omega^2}\int\!\!\int_{-\infty}^{\infty}\Bigg[\frac{\partial\Gamma_{\Omega,\phi}(\vec{r}_A,\vec{r},\omega)}{\partial z}\;\frac{1}{\rho(z_o)}\;\Phi^{+}(\vec{r},\omega)$$

$$+\;\underline{\mathcal{L}}_{\Omega,\psi_\alpha}(\vec{r}_A,\vec{r},\omega)\;\frac{1}{\rho(z_o)}\;\Psi^{+}_{\alpha}(\vec{r},\omega)\Bigg]_{z_o} dxdy, \tag{VI-74a}$$

where

$$\underline{\mathcal{L}}_{\Omega,\psi_\alpha}(\vec{r}_A,\vec{r},\omega) \triangleq \epsilon_{3\alpha j}\epsilon_{jik}\partial_i\Gamma_{\Omega,\psi_k}(\vec{r}_A,\vec{r},\omega). \tag{VI-74b}$$

Here $\Gamma_{\Omega,\phi}(\vec{r}_A,\vec{r},\omega)$ and $\Gamma_{\Omega,\psi_k}(\vec{r}_A,\vec{r},\omega)$ for $z=z_o$ represent the potentials for the Green's Ω-wave field (i.e., the P-wave field or the S_β-wave field) at $\vec{r}_A$ in the lower half-space, related to a *monopole* P- or S_k-wave source at $\vec{r}$ at the surface z_o. Accordingly, $\partial_z\Gamma_{\Omega,\phi}(\vec{r}_A,\vec{r},\omega)$ and $\partial_i\Gamma_{\Omega,\psi_k}(\vec{r}_A,\vec{r},\omega)$ for $z=z_o$ represent the potentials for the Green's Ω-wave field at $\vec{r}_A$ in the lower half-space, related to a *dipole* P- or S_k-wave source at $\vec{r}$ at the surface z_o (see also section I.3.1). Hence, equation (VI-74) actually states that the two-way Ω-wave field at $\vec{r}_A$ in the lower half-space is obtained as an integral over dipole source responses, the dipole sources being distributed over the surface z_o, and the strength of the dipole sources being given by the downgoing P- and S_α-wave potentials $\Phi^+(\vec{r},\omega)$ and $\Psi^+_\alpha(\vec{r},\omega)$ at $z=z_0$ (Huygen's principle, see Figure V-10 for the acoustic equivalence).

We split the two-way Ω-wave field at $\vec{r}_A$ into a downgoing and an upgoing part, according to

$$\Omega(\vec{r}_A,\omega) = \Omega^+(\vec{r}_A,\omega) + \Omega^-(\vec{r}_A,\omega). \tag{VI-75a}$$

Similarly, we split the two-way Green's wave fields at $\vec{r}_A$ into a downgoing and an upgoing part, according to

$$\Gamma_{\Omega,\phi}(\vec{r}_A,\vec{r},\omega) = \Gamma^+_{\Omega,\phi}(\vec{r}_A,\vec{r},\omega) + \Gamma^-_{\Omega,\phi}(\vec{r}_A,\vec{r},\omega), \tag{VI-75b}$$

$$\Gamma_{\Omega,\psi_k}(\vec{r}_A,\vec{r},\omega) = \Gamma^+_{\Omega,\psi_k}(\vec{r}_A,\vec{r},\omega) + \Gamma^-_{\Omega,\psi_k}(\vec{r}_A,\vec{r},\omega) \tag{VI-75c}$$

and, consequently,

$$\underline{\mathscr{L}}_{\Omega,\psi_\alpha}(\vec{r}_A,\vec{r},\omega) = \underline{\mathscr{L}}^+_{\Omega,\psi_\alpha}(\vec{r}_A,\vec{r},\omega) + \underline{\mathscr{L}}^-_{\Omega,\psi_\alpha}(\vec{r}_A,\vec{r},\omega). \tag{VI-75d}$$

It can now be understood that the downgoing wave field $\Omega^+(\vec{r}_A,\omega)$ is given by

$$\frac{1}{j\omega}\Omega^+(\vec{r}_A,\omega) = \frac{2}{\omega^2}\iint_{-\infty}^{\infty}\left[\frac{\partial\Gamma^+_{\Omega,\phi}(\vec{r}_A,\vec{r},\omega)}{\partial z}\frac{1}{\rho(z_0)}\Phi^+(\vec{r},\omega)\right.$$

$$\left. + \underline{\mathscr{L}}^+_{\Omega,\psi_\alpha}(\vec{r}_A,\vec{r},\omega)\frac{1}{\rho(z_0)}\Psi^+_\alpha(\vec{r},\omega)\right]_{z_0} dxdy, \tag{VI-76a}$$

(see Figure V-11a for the acoustic equivalence, where $\frac{1}{j\omega}\Omega^+=P^+$), whereas the upgoing wave field $\Omega^-(\vec{r}_A,\omega)$ is given by

$$\frac{1}{j\omega}\,\Omega^-(\vec{r}_A,\omega) = \frac{2}{\omega^2}\iint_{-\infty}^{\infty}\left[\frac{\partial\Gamma^-_{\Omega,\phi}(\vec{r}_A,\vec{r},\omega)}{\partial z}\,\frac{1}{\rho(z_0)}\,\Phi^+(\vec{r},\omega)\right.$$

$$\left. + \mathscr{L}^-_{\Omega,\psi_\alpha}(\vec{r}_A,\vec{r},\omega)\,\frac{1}{\rho(z_0)}\,\Psi^+_\alpha(\vec{r},\omega)\right]_{z_0} dxdy, \quad \text{(VI-76b)}$$

(see Figure V-11b for the acoustic equivalence, where $\frac{1}{j\omega}\Omega^-=P^-$). In the following we only consider equation (VI-76a), which we rewrite in a slightly more general notation, according to

$$\frac{1}{j\omega}\,\Omega^+(x,y,z_1;\omega) = \frac{2}{\omega^2}\iint_{-\infty}^{\infty}\left[\frac{\partial\Gamma^+_{\Omega,\phi}(x,y,z_1;x',y',z'=z_0;\omega)}{\partial z'}\,\frac{1}{\rho(x',y',z_0)}\Phi^+(x',y',z_0;\omega)\right.$$

$$\left. + \mathscr{L}^+_{\Omega,\psi_\alpha}(x,y,z_1;x',y',z_0;\omega)\frac{1}{\rho(x',y',z_0)}\,\Psi^+_\alpha(x',y',z_0;\omega)\right] dx'dy'. \quad \text{(VI-77a)}$$

Note that we replaced the laterally invariant mass density at z_0 by the laterally variant mass density $\rho(x',y',z_0)$. We did this to enlarge the analogy with the acoustic one-way Rayleigh II integral (V-51a). Opposed to (V-51a), the elastic one-way Rayleigh II integral (VI-77a) is not longer exact when the medium parameters at z_0 are laterally variant and/or anisotropic, because the separation into independent pure P- and S-waves at z_0 is not longer possible. Despite of this small shortcoming, we will use expression (VI-77a) as the basis for elastic forward extrapolation of downgoing waves from depth level z_0 to depth level z_1 (with $z_1>z_0$, see Figure V-12a for the acoustic equivalence, where $\frac{1}{j\omega}\Omega^+=P^+$). A similar expression can be derived for elastic forward extrapolation of upgoing waves from depth level z_1 to depth level z_0 (see Figure V-12b for the acoustic equivalence, where $\frac{1}{j\omega}\Omega^-=P^-$), yielding

$$\frac{1}{j\omega}\,\Omega^-(x,y,z_0;\omega) = \frac{-2}{\omega^2}\iint_{-\infty}^{\infty}\left[\frac{\partial\Gamma^-_{\Omega,\phi}(x,y,z_0;x',y',z'=z_1;\omega)}{\partial z'}\,\frac{1}{\rho(x',y',z_1)}\Phi^-(x',y',z_1;\omega)\right.$$

$$\left. + \mathscr{L}^-_{\Omega,\psi_\alpha}(x,y,z_0;x',y',z_1;\omega)\frac{1}{\rho(x',y',z_1)}\,\Psi^-_\alpha(x',y',z_1;\omega)\right] dx'dy'. \quad \text{(VI-77b)}$$

In the derivation of (VI-77a) the upperhalf-space ($z \le z_0$) was chosen reflection free for the Green's wave field. Accordingly, in the derivation of (VI-77b) the lower half-space ($z \ge z_1$) was chosen reflection free for the Green's wave field. In the following we will choose for both Green's wave fields one and the same reference medium which is reflection free in the upper half-space ($z \le z_0$) as well as in the lower half-space ($z \ge z_1$). With this choice, the reciprocity principle for the Green's wave fields reads

$$\Gamma^-_{\phi,\phi}(x_0,y_0,z_0;x_1,y_1,z_1;\omega) = \Gamma^+_{\phi,\phi}(x_1,y_1,z_1;x_0,y_0,z_0;\omega), \quad \text{(VI-78a)}$$

$$\Gamma^-_{\psi_k,\phi}(x_0,y_0,z_0;x_1,y_1,z_1;\omega) = \Gamma^+_{\phi,\psi_k}(x_1,y_1,z_1;x_0,y_0,z_0;\omega), \quad \text{(VI-78b)}$$

$$\Gamma^-_{\phi,\psi_h}(x_0,y_0,z_0;x_1,y_1,z_1;\omega) = \Gamma^+_{\psi_h,\phi}(x_1,y_1,z_1;x_0,y_0,z_0;\omega) \quad \text{(VI-78c)}$$

and

$$\Gamma^-_{\psi_k,\psi_h}(x_0,y_0,z_0;x_1,y_1,z_1;\omega) = \Gamma^+_{\psi_h,\psi_k}(x_1,y_1,z_1;x_0,y_0,z_0;\omega). \quad \text{(VI-78d)}$$

For the special situation that the medium between z_0 and z_1 is laterally invariant, equations (VI-77a) and (VI-77b) represent spatial convolutions, in analogy with (V-54a) and (V-54b) denoted as

$$\frac{1}{j\omega}\Omega^+(x,y,z_1;\omega) = W^+_{\Omega,\phi}(x,y;z_1,z_0;\omega) * \Phi^+(x,y,z_0;\omega)$$

$$+ W^+_{\Omega,\psi_\alpha}(x,y;z_1,z_0;\omega) * \Psi^+_\alpha(x,y,z_0;\omega) \quad \text{(VI-79a)}$$

and

$$\frac{1}{j\omega}\Omega^-(x,y,z_0;\omega) = W^-_{\Omega,\phi}(x,y;z_0,z_1;\omega) * \Phi^-(x,y,z_1;\omega)$$

$$+ W^-_{\Omega,\psi_\alpha}(x,y;z_0,z_1;\omega) * \Psi^-_\alpha(x,y,z_1;\omega), \quad \text{(VI-79b)}$$

respectively, where the one-way wave field extrapolation operators are given by

$$W^{+}_{\Omega,\phi}(x,y;z_1,z_0;\omega) \triangleq \frac{2}{\omega^2\rho(z_0)} \frac{\partial\Gamma^{+}_{\Omega,\phi}(x,y,z_1;o,o,z=z_0;\omega)}{\partial z}, \tag{VI-80a}$$

$$W^{+}_{\Omega,\psi_\alpha}(x,y;z_1,z_0;\omega) \triangleq \frac{2}{\omega^2\rho(z_0)} \mathscr{L}^{+}_{\Omega,\psi_\alpha}(x,y,z_1;o,o,z_0;\omega), \tag{VI-80b}$$

$$W^{-}_{\Omega,\phi}(x,y;z_0,z_1;\omega) \triangleq \frac{-2}{\omega^2\rho(z_1)} \frac{\partial\Gamma^{-}_{\Omega,\phi}(x,y,z_0;o,o,z=z_1;\omega)}{\partial z} \tag{VI-80c}$$

and

$$W^{-}_{\Omega,\psi_\alpha}(x,y;z_0,z_1;\omega) \triangleq \frac{-2}{\omega^2\rho(z_1)} \mathscr{L}^{-}_{\Omega,\psi_\alpha}(x,y,z_0;o,o,z_1;\omega). \tag{VI-80d}$$

When the medium parameters between z_0 and z_1 are homogeneous and isotropic, then we may substitute the free space solutions (VI-44), yielding

$$W^{+}_{\phi,\phi}(x,y;z_1,z_0;\omega) = \frac{1}{2\pi}\frac{\partial}{\partial z_0}\left(\frac{e^{-jk_p\Delta r}}{\Delta r}\right), \tag{VI-81a}$$

$$W^{+}_{\psi_\beta,\psi_\alpha}(x,y;z_1,z_0;\omega) = \delta_{\alpha\beta}\frac{1}{2\pi}\frac{\partial}{\partial z_0}\left(\frac{e^{-jk_s\Delta r}}{\Delta r}\right)^{1)}, \tag{VI-81b}$$

$$W^{+}_{\psi_\beta,\phi}(x,y;z_1,z_0;\omega) = W^{+}_{\phi,\psi_\alpha}(x,y;z_1,z_0;\omega) = o, \tag{VI-81c}$$

$$W^{-}_{\phi,\phi}(x,y;z_0,z_1;\omega) = \frac{-1}{2\pi}\frac{\partial}{\partial z_1}\left(\frac{e^{-jk_p\Delta r}}{\Delta r}\right), \tag{VI-81d}$$

$$W^{-}_{\psi_\beta,\psi_\alpha}(x,y;z_0,z_1;\omega) = -\delta_{\alpha\beta}\frac{1}{2\pi}\frac{\partial}{\partial z_1}\left(\frac{e^{-jk_s\Delta r}}{\Delta r}\right)^{1)} \tag{VI-81e}$$

and

1) Here we made use of the properties $\epsilon_{3\alpha j}\epsilon_{jik}\delta_{k\beta}\partial_i = \delta_{\alpha\beta}\partial_3 \triangleq \delta_{\alpha\beta}\partial_z$ and $\epsilon_{3\alpha j}\epsilon_{jik}\partial_i\partial_k\partial_\beta = o$.

$$W^{-}_{\psi_\beta,\phi}(x,y;z_o,z_1;\omega) = W^{-}_{\phi,\psi_\alpha}(x,y;z_o,z_1;\omega) = o, \tag{VI-81f}$$

with

$$\Delta r = \sqrt{x^2 + y^2 + (z_1-z_o)^2}\ , \tag{VI-81g}$$

$$k_p = \omega/c_p \tag{VI-81h}$$

and

$$k_s = \omega/c_s. \tag{VI-81i}$$

Note that for the situation of a homogeneous isotropic elastic medium these P- and S-wave field extrapolation operators are identical to the acoustic wave field extrapolation operators (V-56a) and (V-56b), with k replaced by k_p or k_s. Also note that for this situation the following relations exist between the one-way operators for downward and upward extrapolation:

$$W^{+}_{\phi,\phi}(x,y;z_1,z_o;\omega) = W^{-}_{\phi,\phi}(x,y;z_o,z_1;\omega) \tag{VI-82a}$$

and

$$W^{+}_{\psi_\beta,\psi_\alpha}(x,y;z_1,z_o;\omega) = W^{-}_{\psi_\beta,\psi_\alpha}(x,y;z_o,z_1;\omega). \tag{VI-82b}$$

According to (IV-56), a convolution in the space-frequency domain corresponds to a multiplication in the wavenumber-frequency domain. Hence, transforming expressions (VI-79a) and (VI-79b) for laterally invariant media to the wavenumber-frequency domain yields

$$\frac{1}{j\omega}\,\tilde{\Omega}^{+}(k_x,k_y,z_1;\omega) = \tilde{W}^{+}_{\Omega,\phi}(k_x,k_y;z_1,z_o;\omega)\,\tilde{\Phi}^{+}(k_x,k_y,z_o;\omega)$$

$$+\tilde{W}^{+}_{\Omega,\psi_\alpha}(k_x,k_y;z_1,z_o;\omega)\,\tilde{\Psi}^{+}_{\alpha}(k_x,k_y,z_o;\omega) \tag{VI-83a}$$

and

$$\frac{1}{j\omega}\tilde{\Omega}^-(k_x,k_y,z_o;\omega) = \tilde{W}^-_{\Omega,\phi}(k_x,k_y;z_o,z_1;\omega)\,\tilde{\Phi}^-(k_x,k_y,z_1;\omega)$$

$$+\ \tilde{W}^-_{\Omega,\psi_\alpha}(k_x,k_y;z_o,z_1;\omega)\,\tilde{\Psi}^-_\alpha(k_x,k_y,z_1;\omega), \tag{VI-83b}$$

respectively, where

$$\tilde{W}^+_{\Omega,\phi}(k_x,k_y;z_1,z_o;\omega) \triangleq \frac{2}{\omega^2\rho(z_o)}\ \frac{\partial\tilde{\Gamma}^+_{\Omega,\phi}(k_x,k_y,z_1;o,o,z=z_o;\omega)}{\partial z}, \tag{VI-84a}$$

$$\tilde{W}^+_{\Omega,\psi_\alpha}(k_x,k_y;z_1,z_o;\omega) \triangleq \frac{2}{\omega^2\rho(z_o)}\ \tilde{\mathscr{L}}^+_{\Omega,\psi_\alpha}(k_x,k_y,z_1;o,o,z_o;\omega), \tag{VI-84b}$$

$$\tilde{W}^-_{\Omega,\phi}(k_x,k_y;z_o,z_1;\omega) \triangleq \frac{-2}{\omega^2\rho(z_1)}\ \frac{\partial\tilde{\Gamma}^-_{\Omega,\phi}(k_x,k_y,z_o;o,o,z=z_1;\omega)}{\partial z} \tag{VI-84c}$$

and

$$\tilde{W}^-_{\Omega,\psi_\alpha}(k_x,k_y;z_o,z_1;\omega) \triangleq \frac{-2}{\omega^2\rho(z_1)}\ \tilde{\mathscr{L}}^-_{\Omega,\psi_\alpha}(k_x,k_y,z_o;o,o,z_1;\omega). \tag{VI-84d}$$

When the medium parameters between z_o and z_1 are homogeneous and isotropic, then we may substitute the free space solutions (VI-45), yielding

$$\tilde{W}^+_{\phi,\phi}(k_x,k_y;z_1,z_o;\omega) = 2\,\frac{\partial}{\partial z_o}\left(\frac{e^{-jk_{z,p}\Delta z}}{2jk_{z,p}}\right) = e^{-jk_{z,p}\Delta z}, \tag{VI-85a}$$

$$\tilde{W}^+_{\psi_\beta,\psi_\alpha}(k_x,k_y;z_1,z_o;\omega) = \delta_{\alpha\beta}\ 2\,\frac{\partial}{\partial z_o}\left(\frac{e^{-jk_{z,s}\Delta z}}{2jk_{z,s}}\right) = \delta_{\alpha\beta}\,e^{-jk_{z,s}\Delta z}, \tag{VI-85b}$$

$$\tilde{W}^+_{\psi_\beta,\phi}(k_x,k_y;z_1,z_o;\omega) = \tilde{W}^+_{\phi,\psi_\alpha}(k_x,k_y;z_1,z_o;\omega) = o, \tag{VI-85c}$$

$$\tilde{W}^-_{\phi,\phi}(k_x,k_y;z_o,z_1;\omega) = -2\frac{\partial}{\partial z_1}\left(\frac{e^{-jk_{z,p}\Delta z}}{2jk_{z,p}}\right) = e^{-jk_{z,p}\Delta z}, \tag{VI-85d}$$

$$\tilde{W}^-_{\psi_\beta,\psi_\alpha}(k_x,k_y;z_o,z_1;\omega) = -\delta_{\alpha\beta}\, 2\frac{\partial}{\partial z_1}\left(\frac{e^{-jk_{z,s}\Delta z}}{2jk_{z,s}}\right) = \delta_{\alpha\beta}\, e^{-jk_{z,s}\Delta z} \tag{VI-85e}$$

and

$$\tilde{W}^-_{\psi_\beta,\phi}(k_x,k_y;z_o,z_1;\omega) = \tilde{W}^-_{\phi,\psi_\alpha}(k_x,k_y;z_o,z_1;\omega) = o, \tag{VI-85f}$$

with

$$\Delta z = |z_1 - z_o| = z_1 - z_o, \tag{VI-85g}$$

and $k_{z,p}$ and $k_{z,s}$ defined as in (VI-45).

Note that these operators are identical to the phase-shift operators defined in (IV-36).

VI.5.2 Matrix formulation of elastic forward wave field extrapolation

Consider the elastic one-way Rayleigh II integrals (VI-77a) and (VI-77b) which describe forward extrapolation of downgoing and upgoing elastic waves, respectively, through arbitrarily inhomogeneous anisotropic solid media. In practical situations the wave fields are discretized along the x- and y-axes. It is shown in Appendix A that integral expressions of the form (VI-77a) and (VI-77b) can be replaced by matrix products in the case of discretized wave fields.

We may replace equation (VI-77a) by

$$\frac{1}{j\omega}\vec{\Omega}^+(z_1) = \mathbf{W}^+_{\Omega,\phi}(z_1,z_o)\vec{\Phi}^+(z_o) + \mathbf{W}^+_{\Omega,\psi_\alpha}(z_1,z_o)\vec{\Psi}^+_\alpha(z_o), \tag{VI-86a}$$

where $\frac{1}{j\omega}\vec{\Omega}^+(z_1)$ may stand for $\vec{\Phi}^+(z_1)$ or $\vec{\Psi}^+_\beta(z_1)$ and where the one-way wave field extrapolation matrices are defined as

$$\mathbf{W}^{+}_{\Omega,\phi}(z_1,z_0) = \frac{2}{\omega^2} \frac{\partial \mathbf{\Gamma}^{+}_{\Omega,\phi}(z_1,z=z_0)}{\partial z} \mathbf{M}^{-1}(z_0) \tag{VI-86b}$$

and

$$\mathbf{W}^{+}_{\Omega,\psi_\alpha}(z_1,z_0) = \frac{2}{\omega^2} \underline{\mathcal{L}}^{+}_{\Omega,\psi_\alpha}(z_1,z_0)\, \mathbf{M}^{-1}(z_0). \tag{VI-86c}$$

In a similar way, we may replace (VI-77b) by

$$\frac{1}{j\omega} \vec{\Omega}^{-}(z_0) = \mathbf{W}^{-}_{\Omega,\phi}(z_0,z_1)\vec{\Phi}^{-}(z_1) + \mathbf{W}^{-}_{\Omega,\psi_\alpha}(z_0,z_1)\vec{\Psi}^{-}_{\alpha}(z_1), \tag{VI-87a}$$

where $\frac{1}{j\omega}\vec{\Omega}^{-}(z_0)$ may stand for $\vec{\Phi}^{-}(z_0)$ or $\vec{\Psi}^{-}_{\beta}(z_0)$ and where the one-way wave field extrapolation matrices are defined as

$$\mathbf{W}^{-}_{\Omega,\phi}(z_0,z_1) = \frac{-2}{\omega^2} \frac{\partial \mathbf{\Gamma}^{-}_{\Omega,\phi}(z_0,z=z_1)}{\partial z} \mathbf{M}^{-1}(z_1) \tag{VI-87b}$$

and

$$\mathbf{W}^{-}_{\Omega,\psi_\alpha}(z_0,z_1) = \frac{-2}{\omega^2} \underline{\mathcal{L}}^{-}_{\Omega,\psi_\alpha}(z_0,z_1)\, \mathbf{M}^{-1}(z_1). \tag{VI-87c}$$

In (VI-86) and (VI-87) vectors $\vec{\Phi}^{\pm}(z_0)$ and $\vec{\Phi}^{\pm}(z_1)$ contain the discretized versions of the monochromatic scalar P-wave fields $\Phi^{\pm}(x,y,z_0;\omega)$ and $\Phi^{\pm}(x,y,z_1;\omega)$, respectively (see Appendix A, section A.2). Similarly, vectors $\vec{\Psi}^{\pm}_{\alpha}(z_0)$ and $\vec{\Psi}^{\pm}_{\alpha}(z_1)$ contain the discretized versions of the monochromatic S_α-wave fields $\Psi^{\pm}_{\alpha}(x,y,z_0;\omega)$ and $\Psi^{\pm}_{\alpha}(x,y,z_1;\omega)$, respectively. Matrices $\mathbf{M}(z_0)$ and $\mathbf{M}(z_1)$ are diagonal matrices, the diagonal elements representing the discretized versions of $\rho(x,y,z_0)$ and $\rho(x,y,z_1)$, respectively (see Appendix A, section A.3). Matrices $\mathbf{\Gamma}^{+}_{\Omega,\phi}(z_1,z_0)$ and $\underline{\mathcal{L}}^{+}_{\Omega,\psi_\alpha}(z_1,z_0)$ are operator matrices, each column containing a discretized monochromatic "spatial impulse response" $\Gamma^{+}_{\Omega,\phi}(x,y,z_1;x',y',z_0;\omega)$ or $\mathcal{L}^{+}_{\Omega,\psi_\alpha}(x,y,z_1;x',y',z_0;\omega)$, respectively, as a function of (x,y) at z_1 for an impulsive source at (x',y',z_0), (see Appendix A, section A.3). Similarly, matrices $\Gamma^{-}_{\Omega,\phi}(z_0,z_1)$ and $\mathcal{L}^{-}_{\Omega,\psi_\alpha}(z_0,z_1)$ are operator matrices, each column containing a discretized monochromatic "spatial impulse response" $\Gamma^{-}_{\Omega,\phi}(x,y,z_0;x',y',z_1;\omega)$ or $\mathcal{L}^{-}_{\Omega,\psi_\alpha}(x,y,z_0;x',y',z_1;\omega)$, respectively, as a function of (x,y) at z_0 for an impulsive source at (x',y',z_1).

Note that equations (VI-86a) and (VI-87a) for elastic downward and upward extrapolation may be elegantly rewritten as

$$\vec{\underset{\sim}{D}}^{+}(z_1) = \underset{\sim}{W}^{+}(z_1,z_0)\vec{\underset{\sim}{D}}^{+}(z_0) \tag{VI-88a}$$

and

$$\vec{\underset{\sim}{D}}^{-}(z_0) = \underset{\sim}{W}^{-}(z_0,z_1)\vec{\underset{\sim}{D}}^{-}(z_1), \tag{VI-88b}$$

respectively, where the *multi-component* data vectors are defined as

$$\vec{\underset{\sim}{D}}^{\pm}(z_0) = \begin{bmatrix} \vec{\Phi}^{\pm}(z_0) \\ \vec{\Psi}^{\pm}_x(z_0) \\ \vec{\Psi}^{\pm}_y(z_0) \end{bmatrix} \quad \text{and} \quad \vec{\underset{\sim}{D}}^{\pm}(z_1) = \begin{bmatrix} \vec{\Phi}^{\pm}(z_1) \\ \vec{\Psi}^{\pm}_x(z_1) \\ \vec{\Psi}^{\pm}_y(z_1) \end{bmatrix} \tag{VI-88c}$$

and where the *multi-component* extrapolation operators are defined as

$$\underset{\sim}{W}^{+}(z_1,z_0) = \begin{bmatrix} W^{+}_{\phi,\phi}(z_1,z_0) & W^{+}_{\phi,\psi_x}(z_1,z_0) & W^{+}_{\phi,\psi_y}(z_1,z_0) \\ W^{+}_{\psi_x,\phi}(z_1,z_0) & W^{+}_{\psi_x,\psi_x}(z_1,z_0) & W^{+}_{\psi_x,\psi_y}(z_1,z_0) \\ W^{+}_{\psi_y,\phi}(z_1,z_0) & W^{+}_{\psi_y,\psi_x}(z_1,z_0) & W^{+}_{\psi_y,\psi_y}(z_1,z_0) \end{bmatrix} \tag{VI-88d}$$

and

$$\underset{\sim}{W}^{-}(z_0,z_1) = \begin{bmatrix} W^{-}_{\phi,\phi}(z_0,z_1) & W^{-}_{\phi,\psi_x}(z_0,z_1) & W^{-}_{\phi,\psi_y}(z_0,z_1) \\ W^{-}_{\psi_x,\phi}(z_0,z_1) & W^{-}_{\psi_x,\psi_x}(z_0,z_1) & W^{-}_{\psi_x,\psi_y}(z_0,z_1) \\ W^{-}_{\psi_y,\phi}(z_0,z_1) & W^{-}_{\psi_y,\psi_x}(z_0,z_1) & W^{-}_{\psi_y,\psi_y}(z_0,z_1) \end{bmatrix} . \qquad \text{(VI-88e)}$$

Note the high degree of similarity with equations (V-61a) and (V-62a) for acoustic downward and upward extrapolation:

$$\vec{P}^{+}(z_1) = W^{+}(z_1,z_0)\vec{P}^{+}(z_0) \qquad \text{(VI-89a)}$$

and

$$\vec{P}^{-}(z_0) = W^{-}(z_0,z_1)\vec{P}^{-}(z_1). \qquad \text{(VI-89b)}$$

For a homogeneous isotropic elastic medium the expressions for the multi-component extrapolation operators simplify to

$$\underset{\sim}{W}^{+}(z_1,z_0) = \begin{bmatrix} W^{+}_{\phi,\phi}(z_1,z_0) & O & O \\ O & W^{+}_{\psi_x,\psi_x}(z_1,z_0) & O \\ O & O & W^{+}_{\psi_y,\psi_y}(z_1,z_0) \end{bmatrix} \qquad \text{(VI-90a)}$$

and

$$\underset{\sim}{W}^{-}(z_0,z_1) = \begin{bmatrix} W^{-}_{\phi,\phi}(z_0,z_1) & O & O \\ O & W^{-}_{\psi_x,\psi_x}(z_0,z_1) & O \\ O & O & W^{-}_{\psi_y,\psi_y}(z_0,z_1) \end{bmatrix} , \qquad \text{(VI-90b)}$$

where the sub-matrices contain the discretized versions of the *free space* Green's wave fields.

For an arbitrarily inhomogeneous anisotropic elastic medium between z_0 and z_1, no analytical expressions are available for the Green's matrices. In practice they are obtained as follows:

1. Define a reference medium which accurately describes the geology between depth levels z_o and z_1 and which in non-reflecting outside this depth interval.

2a. Solve numerically the two-way wave equation (VI-31),

$$\partial_j(c_{ijk\ell}\partial_\ell g_{k,\phi}) - \rho\partial_t^2 g_{i,\phi} = -K_c(\vec{r}\,')\partial_i\delta(\vec{r}-\vec{r}\,')\partial_t\delta(t), \qquad \text{(VI-91a)}$$

imposing initial conditions

$$g_{i,\phi}(\vec{r},\vec{r}\,',t) = o \qquad \text{for } t<o \qquad \text{(VI-91b)}$$

and

$$\partial_t g_{i,\phi}(\vec{r},\vec{r}\,',t) = o \qquad \text{for } t<o. \qquad \text{(VI-91c)}$$

The source-term in the right-hand side of equation (VI-91a) represents an impulsive monopole P-wave source at $\vec{r}\,'$ (see Figure VI-2b). By choosing $\vec{r}\,'$ at z_1 one obtains the "spatial impulse response" $g_{i,\phi}(x,y,z;x',y',z'=z_1;t)$. For $z=z_o$ this Green's wave field is purely upgoing, which we denote by $g^-_{i,\phi}(x,y,z=z_o;x',y',z'=z_1;t)$.

2b. Solve numerically the two-way wave equation (VI-35),

$$\partial_j(c_{ijk\ell}\partial_\ell g_{k,\psi_h}) - \rho\partial_t^2 g_{i,\psi_h} = -\mu(\vec{r}\,')\epsilon_{hin}\partial_n\delta(\vec{r}-\vec{r}\,')\partial_t\delta(t), \qquad \text{(VI-92a)}$$

imposing inititial conditions

$$g_{i,\psi_h}(\vec{r},\vec{r}\,',t) = o \qquad \text{for } t<o \qquad \text{(VI-92b)}$$

and

$$\partial_t g_{i,\psi_h}(\vec{r},\vec{r}\,',t) = o \qquad \text{for } t<o. \qquad \text{(VI-92c)}$$

The source-term in the right-hand side of equation (VI-92a) represents an impulsive monopole S_h-wave source at $\vec{r}\,'$ (see Figure VI-2e). By choosing $\vec{r}\,'$ at z_1 one obtains the "spatial impulse response" $g_{i,\psi_h}(x,y,z;x',y',z'=z_1;t)$. For $z=z_0$ this Green's wave field is purely upgoing, which we denote by $g^-_{i,\psi_h}(x,y,z=z_0;x',y',z'=z_1;t)$.

3. Apply equations (VI-32b), (VI-33b), (VI-36b) and (VI-37b) to compute the Green's potentials at $\vec{r}=(x,y,z=z_0)$, related to the sources at $\vec{r}\,'=(x',y',z'=z_1)$, according to

$$\partial_t \gamma^-_{\phi,\phi}(\vec{r},\vec{r}\,',t) = -K_c(\vec{r})\partial_i g^-_{i,\phi}(\vec{r},\vec{r}\,',t), \tag{VI-93a}$$

(see Figure VI-2c),

$$\partial_t \gamma^-_{\psi_k,\phi}(\vec{r},\vec{r}\,',t) = -\mu(\vec{r})\epsilon_{kij}\partial_j g^-_{i,\phi}(\vec{r},\vec{r}\,',t), \tag{VI-93b}$$

(see Figure VI-2d),

$$\partial_t \gamma^-_{\phi,\psi_h}(\vec{r},\vec{r}\,',t) = -K_c(\vec{r})\partial_i g^-_{i,\psi_h}(\vec{r},\vec{r}\,',t), \tag{VI-93c}$$

(see Figure VI-2f) and

$$\partial_t \gamma^-_{\psi_k,\psi_h}(\vec{r},\vec{r}\,',t) = -\mu(\vec{r})\epsilon_{kij}\partial_j g^-_{i,\psi_h}(\vec{r},\vec{r}\,',t), \tag{VI-93d}$$

(see Figure VI-2g).

4. Apply a temporal Fourier transform to these Green's wave fields, according to

$$\Gamma^-_{\Omega,\phi}(\vec{r},\vec{r}\,',\omega) = \int_0^\infty \gamma^-_{\Omega,\phi}(\vec{r},\vec{r}\,',t)e^{-j\omega t}dt \tag{VI-94a}$$

and

$$\Gamma^-_{\Omega,\psi_h}(\vec{r},\vec{r}\,',\omega) = \int_0^\infty \gamma^-_{\Omega,\psi_h}(\vec{r},\vec{r}\,',t)e^{-j\omega t}dt, \tag{VI-94b}$$

where Ω stands for ϕ or ψ_k, and determine the modified Green's wave field, according to

$$\underline{\mathscr{L}}^-_{\Omega,\psi_\alpha}(\vec{r},\vec{r}\,',\omega) = \epsilon_{3\alpha j}\epsilon_{jih}\partial'_i\Gamma^-_{\Omega,\psi_h}(\vec{r},\vec{r}\,',\omega), \tag{VI-95a}$$

where ∂'_i for i=1, 2, 3 denotes differentiation with respect to x',y',z', respectively. For a fixed frequency ω and for fixed $\vec{r}\,'$ at z_1, the Green's wave fields $\Gamma^-_{\Omega,\phi}(\vec{r},\vec{r}\,',\omega)$ and $\underline{\mathscr{L}}^-_{\Omega,\psi_\alpha}(\vec{r},\vec{r}\,',\omega)$ at z_o each represent one column of the Green's matrices $\Gamma^-_{\Omega,\phi}(z_o,z_1)$ and $\underline{\mathscr{L}}^-_{\Omega,\psi_\alpha}(z_o,z_1)$, respectively.

5. Repeat steps 2, 3 and 4 for a range of source points $\vec{r}\,'$ at z_1, yielding the different columns of the Green's matrices.

6. Apply the one-way reciprocity relations (VI-78), yielding $\Gamma^+_{\Omega,\phi}(\vec{r}\,',\vec{r},\omega)$ and $\Gamma^+_{\Omega,\psi_k}(\vec{r}\,',\vec{r},\omega)$, where Ω stands for ϕ or ψ_h, and determine the modified Green's wave field according to

$$\underline{\mathscr{L}}^+_{\Omega,\psi_\alpha}(\vec{r}\,',\vec{r},\omega) = \epsilon_{3\alpha j}\epsilon_{jik}\partial_i\Gamma^+_{\Omega,\psi_k}(\vec{r}\,',\vec{r},\omega). \tag{VI-95b}$$

For a fixed frequency ω and for fixed $\vec{r}$ at z_0, the Green's wave fields $\Gamma^+_{\Omega,\phi}(\vec{r}\,',\vec{r},\omega)$ and $\underline{\mathscr{L}}^+_{\Omega,\psi_\alpha}(\vec{r}\,',\vec{r},\omega)$ at z_1 each represent one column of the Green's matrices $\Gamma^+_{\Omega,\phi}(z_1,z_o)$ and $\underline{\mathscr{L}}^+_{\Omega,\psi_\alpha}(z_1,z_o)$, respectively.

In principle any accurate forward modeling scheme can be used to generate the Green's wave fields in step 2. Here we discuss an example with *finite difference* modeling (Kelly et al., 1976; Haimé, 1987). Consider the 2-D configuration of Figure VI-6a. A P-wave source is defined at $\vec{r}\,'=(x',z_1)$. Subsequently, the discretized 2-D version of wave equation (VI-91) is solved numerically by a "marching on in time" procedure. The waves arriving at the surface z_o represent the Green's velocity vector $\vec{g}^{\,-}_\phi(x,z_o;x',z_1;t)$.

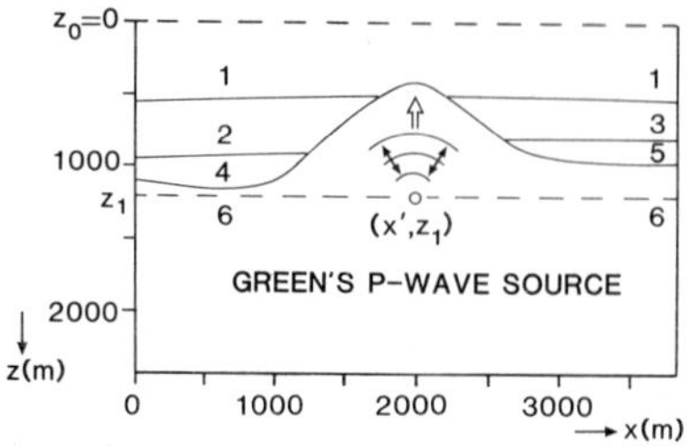

layer nr	velocities	layer nr	velocities
1	Cp=2600 m/s Cs=1500 m/s	4	Cp=2800 m/s Cs=1600 m/s
2	Cp=3200 m/s Cs=1900 m/s	5	Cp=3000 m/s Cs=1800 m/s
3	Cp=3200 m/s Cs=1900 m/s	6	Cp=3600 m/s Cs=2078 m/s

(a)

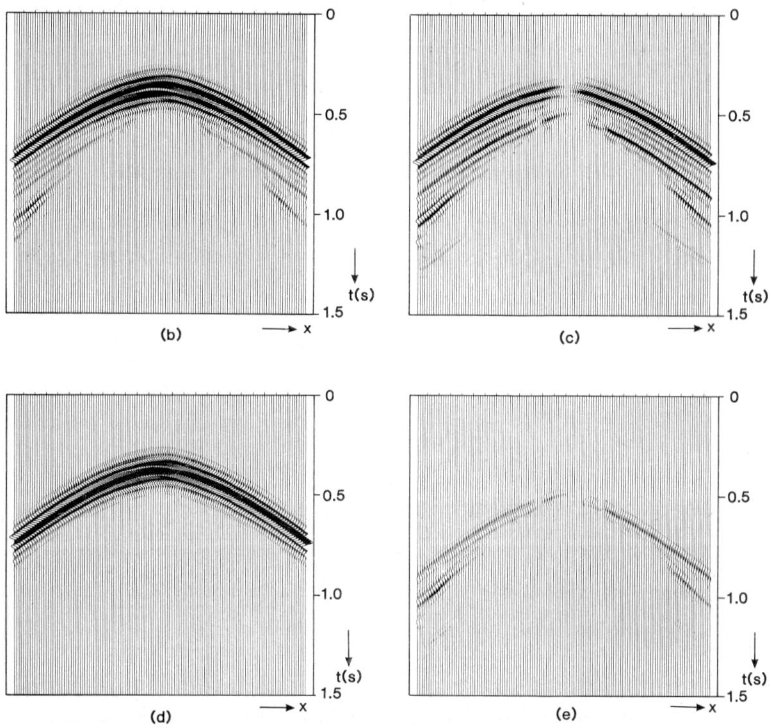

Figure VI-6: For an inhomogeneous elastic medium, the Green's wave fields must be determined numerically, for instance by finite difference modeling:

a. 2-D inhomogeneous medium with a Green's P-wave source at (x',z_1)

b. Vertical component of the band-limited Green's velocity at z_o

c. Horizontal component of the band-limited Green's velocity at z_o

d. Band-limited Green's P-wave potential at z_o

e. Band-limited Green's SV-wave potential at z_o.

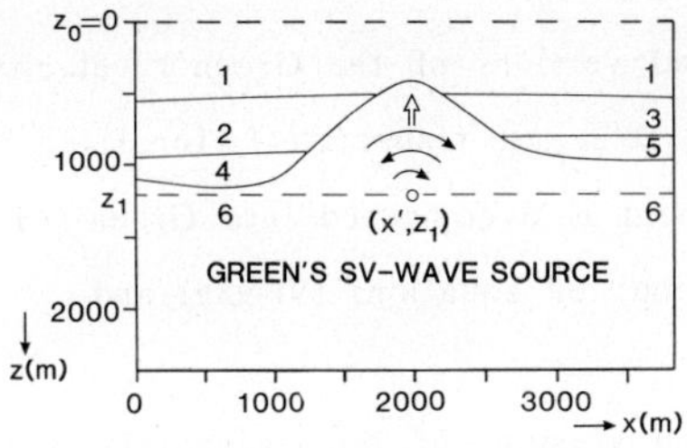

(a)

layer nr	velocities	layer nr	velocities
1	Cp=2600 m/s Cs=1500 m/s	4	Cp=2800 m/s Cs=1600 m/s
2	Cp=3200 m/s Cs=1900 m/s	5	Cp=3000 m/s Cs=1800 m/s
3	Cp=3200 m/s Cs=1900 m/s	6	Cp=3600 m/s Cs=2078 m/s

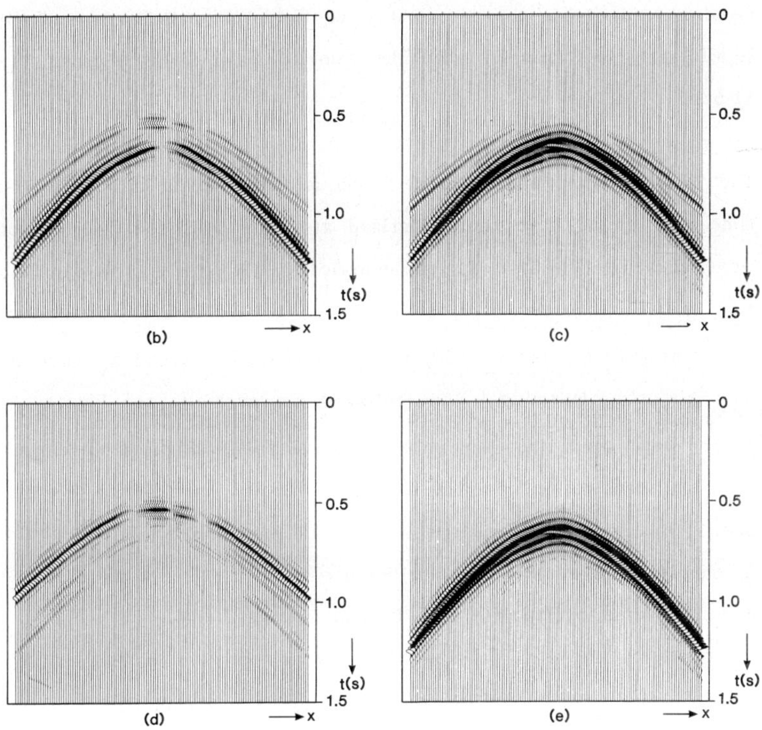

Figure VI-7: a. 2-D inhomogeneous medium of Figure VI-6a, this time with an SV-wave source at (x',z_1).

b. Vertical component of the band-limited Green's velocity at z_0

c. Horizontal component of the band-limited Green's velocity at z_0

d. Band-limited Green's P-wave potential at z_0

e. Band-limited Green's SV-wave potential at z_0.

Figures VI-6b and VI-6c show band-limited versions of the Green's velocity components $\bar{g}_{z,\phi}(x,z_0;x',z_1;t)$ and $\bar{g}_{x,\phi}(x,z_0;x',z_1;t)$, respectively, for fixed x'. Finally, the Green's velocity wave field is decomposed into Green's P- and S_y-wave potentials[1)], using 2-D versions of equations (VI-93a) and (VI-93b), respectively.

Figures VI-6d and VI-6e show band-limited versions of the Green's P- and S_y-wave potentials $\bar{\gamma}_{\phi,\phi}(x,z_0;x',z_1;t)$ and $\bar{\gamma}_{\psi_y,\phi}(x,z_0;x',z_1;t)$, respectively, for fixed x'. Note that the potential wave fields (Figures VI-6d and VI-6e) are significantly less complex than the velocity wave fields (Figures VI-6b and VI-6c).

The 2-D configuration of Figure VI-6a is shown again in Figure VI-7a, this time with an S_y-wave source defined at $\vec{r}\,'=(x',z_1)$. The 2-D version of wave equation (VI-92) is solved numerically, yielding $\bar{g}_{z,\psi_y}(x,z_0;x',z_1;t)$ and $\bar{g}_{x,\psi_y}(x,z_0;x',z_1;t)$, see Figures VI-7b and VI-7c. The 2-D versions of equations (VI-93c) and (VI-93d) yield the Green's P- and S_y-wave potentials $\bar{\gamma}_{\phi,\psi_y}(x,z_0;x',z_1;t)$ and $\bar{\gamma}_{\psi_y,\psi_y}(x,z_0;x',z_1;t)$, respectively, see Figures VI-7d and VI-7e. Note again that the potential wave fields (Figures VI-7d and VI-7e) are significantly less complex than the velocity wave fields (Figures VI-7b and VI-7c).

In chapter VIII, section VIII.4.6, we use these results in an example of elastic inverse extrapolation of P- and S-wave fields.

In this example we used the finite difference approach for illustrative purposes only. For practical applications we recommend to generate the Green's wave fields by elastic Gaussian beam modeling, similarly as discussed in section V.5.2 for the acoustic situation.

---

1) In the 2-D situation S_y-wave potentials are equivalent with SV-wave potentials.

VI.6 REFERENCES

Aki, K., and Richards, P.G., 1980, Quantitative seismology: W.H. Freeman and Co.

Berkhout, A.J., 1986, Seismic inversion in terms of pre-stack migration and multiple elimination: Proceedings of the IEEE, *74*, 415 - 427.

Burridge, R., and Knopoff, L., 1964, Body force equivalents for seismic dislocations: Bull. Seis. Soc. Amer., *54*, 1875 - 1888.

De Hoop, A.T., 1958, Representation theorems for the displacement in an elastic solid and their application to elastodynamic diffraction theory: D.Sc. thesis, Delft University of Technology.

Haimé, G.C., 1987, Full elastic inverse wave field extrapolation through an inhomogeneous medium: M.Sc. thesis, Delft University of Technology.

Kelly, K.R., Ward, R.W., Treitel, S., and Alford, R.M., 1976, Synthetic seismograms: a finite difference approach: Geophysics, *41*, 2 - 27.

Kuo, J.T., and Dai, T.F., 1984, Kirchhoff elastic wave migration for the case of noncoincident source and receiver: Geophysics, *49*, 1223 - 1238.

Pao, Y.H., and Varatharajulu, V., 1976, Huygens' principle, radiation conditions, and integral formulas for the scattering of elastic waves: J. Acoust. Soc. Am., *59*, 1361 - 1371.

Wapenaar, C.P.A., and Haimé, G.C., 1989, Elastic extrapolation of primary seismic P and S waves: Geophysical Prospecting, *37*, in press.

VII

ACOUSTIC INVERSE WAVE FIELD EXTRAPOLATION IN LOW CONTRAST MEDIA

VII.1 INTRODUCTION

In chapter V we discussed acoustic *forward* wave field extrapolation operators $\mathbf{W}^+(z_1,z_0)$ and $\mathbf{W}^-(z_0,z_1)$ that *simulate* the downward and upward propagation effects, respectively (see Figure VII-1). These operators were derived from the acoustic Kirchhoff-Helmholtz integral (V-31a), which contains a *forward* propagating Green's wave field.

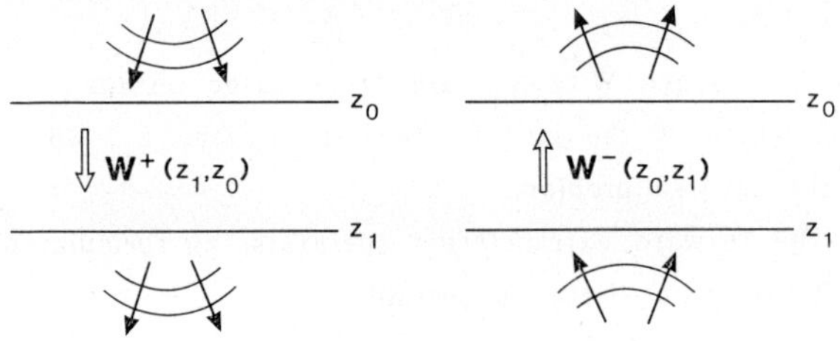

(a) (b)

Figure VII-1: Forward wave field extrapolation operators simulate propagation effects of downgoing waves (Figure a) and upgoing waves (Figure b).

One-way seismic inversion techniques, such as migration (Berkhout and Van Wulfften Palthe, 1979; Berkhout, 1985) or redatuming (Berryhill, 1984; chapters XI and XII of this book) are essentially based on inverse wave field extrapolation. In this chapter we discuss acoustic *inverse* wave field extrapolation operators $\mathbf{F}^+(z_0,z_1)$ and $\mathbf{F}^-(z_1,z_0)$ that *eliminate* the downward and upward propagation effects, respectively (see Figure VII-2).

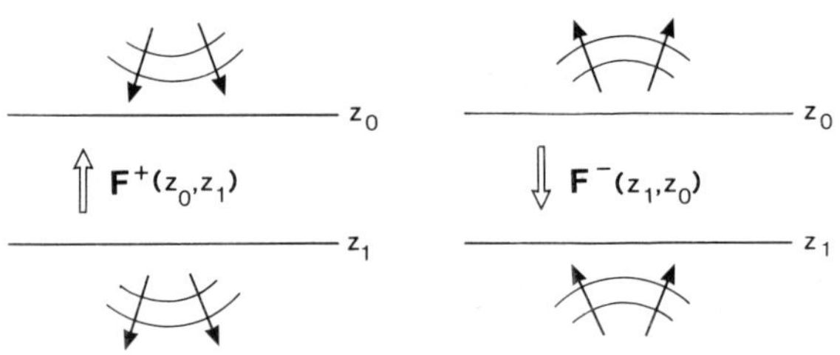

(a) (b)

Figure VII-2: Inverse wave field extrapolation operators eliminate propagation effects of downgoing waves (Figure a) and upgoing waves (Figure b).

Ideally, these inverse wave field extrapolation operators are directly related to the forward wave field extrapolation operators, according to

$$\mathbf{F}^{+}(z_0,z_1) = [\mathbf{W}^{+}(z_1,z_0)]^{-1} \tag{VII-1a}$$

and

$$\mathbf{F}^{-}(z_1,z_0) = [\mathbf{W}^{-}(z_0,z_1)]^{-1}. \tag{VII-1b}$$

There are two important reasons, however, why we do not use these equations in practice:

1. Operators $\mathbf{W}^{+}(z_1,z_0)$ and $\mathbf{W}^{-}(z_0,z_1)$ are often based on the true parameters K and ρ of the medium between z_0 and z_1, which are unknown in the inverse problem.
2. Inversion of the forward extrapolation operators, as formulated by equations (VII-1a) and (VII-1b), is unstable.

Fortunately, since operators $\mathbf{F}^{+}$ and $\mathbf{F}^{-}$ need to eliminate *propagation* effects only, it is justified to replace the true medium (K, ρ) by a geologically oriented reference medium ($\bar{K}$,$\bar{\rho}$), thus ignoring *scattering* effects related to the deviation parameters (ΔK, $\Delta\rho$). Hence, in order to perform inverse wave field extrapolation it is not necessary to have detailed knowledge of the true medium: knowledge of a geologically oriented reference medium (the macro subsurface model, Berkhout, 1986), will suffice. Although this demand is much less severe, it should not be underestimated. In fact, the success of seismic processing depends largely on the accuracy of the available macro subsurface model. Therefore much research is being carried out on improving macro model estimation techniques (Faye and Jeannot, 1986; van der Made et al., 1984; Cox et al., 1988). A discussion on macro model estimation is beyond the scope of this book. We will assume throughout that an accurate description of the macro subsurface model is available. Moreover, to concentrate on the inverse problem as defined by equations (VII-1a) and (VII-1b), in the following derivations we will not distinguish explicitly between the macro subsurface model and the true medium. In this chapter we discuss the acoustic inverse extrapolation problem for increasing complexity of the medium. First we show for a homogeneous medium that stable inverse extrapolation operators may be obtained simply by taking the complex conjugate of the forward operators. It is shown that this elegant solution, which is commonly known as the *matched filter approach*, imposes a restriction to the maximum obtainable spatial resolution. Next we consider the situation in which the medium parameters depend on the depth coordinate only. We show that, with a small modification, the matched filter approach is valid for smoothly varying medium parameters. We also analyse the limitations of the matched filter approach for a medium that contains high contrasts. Finally, we consider the situation in which the medium parameters depend on the lateral coordinates as well as on the depth coordinate. Our starting point is the acoustic Kirchhoff-Helmholtz integral (V-31b) that contains a *backward* propagating Green's wave field. Although the derivation is completely different, we arrive again at inverse extrapolation operators which are the complex conjugate of the forward extrapolation operators and which have restrictions with respect to the maximum obtainable spatial resolution. Again we analyse the limitations for a medium that contains high contrasts. In chapter IX we will modify the acoustic inverse extrapolation operators for the latter situation. The operators derived in chapters VII and IX play an essential role in chapter XI, where we discuss an acoustic processing scheme for single-component seismic data.

VII.2 ACOUSTIC INVERSE WAVE FIELD EXTRAPOLATION IN LATERALLY INVARIANT MEDIA

VII.2.1. Homogeneous media

For a homogeneous acoustic medium, forward extrapolation of downgoing waves (Figure VII-1a) may be described by a *spatial convolution* along the x- and y-axis, according to

$$P^+(x,y,z_1;\omega) = \int\!\!\int_{-\infty}^{\infty} W^+(x-x',y-y';z_1,z_0;\omega)P^+(x',y',z_0;\omega)dx'dy', \qquad \text{(VII-2a)}$$

or, symbolically,

$$P^{+}(x,y,z_1;\omega) = W^{+}(x,y;z_1,z_0;\omega) * P^{+}(x,y,z_0;\omega), \quad \text{(VII-2b)}$$

where

$$W^{+}(x,y;z_1,z_0;\omega) = \frac{1}{2\pi}\frac{\partial}{\partial z_0}\left(\frac{e^{-jk\Delta r}}{\Delta r}\right), \quad \text{(VII-2c)}$$

with

$$\Delta r = \sqrt{x^2 + y^2 + (z_1 - z_0)^2} \quad \text{(VII-2d)}$$

and

$$k = \omega/c, \quad \text{(VII-2e)}$$

see equations (V-53) to (V-56). Similarly, the expression for forward extrapolation of upgoing waves (Figure VII-1b) reads

$$P^{-}(x,y,z_0;\omega) = \iint_{-\infty}^{\infty} W^{-}(x-x',y-y';z_0,z_1;\omega)P^{-}(x',y',z_1;\omega)dx'dy', \quad \text{(VII-3a)}$$

or, symbolically,

$$P^{-}(x,y,z_0;\omega) = W^{-}(x,y;z_0,z_1;\omega) * P^{-}(x,y,z_1;\omega), \quad \text{(VII-3b)}$$

where

$$W^{-}(x,y;z_0,z_1;\omega) = W^{+}(x,y;z_1,z_0;\omega). \quad \text{(VII-3c)}$$

Following Berkhout and Van Wulfften Palthe (1979), we describe inverse extrapolation of downgoing and upgoing waves (Figures VII-2a and VII-2b) as *spatial deconvolution* processes along the x- and y-axis, in the symbolic notation

$$P^{+}(x,y,z_0;\omega) = F^{+}(x,y;z_0,z_1;\omega) * P^{+}(x,y,z_1;\omega) \tag{VII-4a}$$

and

$$P^{-}(x,y,z_1;\omega) = F^{-}(x,y;z_1,z_0;\omega) * P^{-}(x,y,z_0;\omega), \tag{VII-4b}$$

where the deconvolution operators F^{+} and F^{-} are defined implicitly by

$$F^{+}(x,y;z_0,z_1;\omega) * W^{+}(x,y;z_1,z_0;\omega) \hat{=} \delta(x)\delta(y) \tag{VII-5a}$$

and

$$F^{-}(x,y;z_1,z_0;\omega) * W^{-}(x,y;z_0,z_1;\omega) \hat{=} \delta(x)\delta(y) . \tag{VII-5b}$$

Note that, in analogy with (VII-3c),

$$F^{-}(x,y;z_1,z_0;\omega) = F^{+}(x,y;z_0,z_1;\omega). \tag{VII-5c}$$

To find explicit expressions for operators F^{+} and F^{-}, we make use of equation (IV-56), which states that a convolution integral in the space-frequency domain corresponds to a multiplication in the wavenumber-frequency domain. Hence, transforming the expression for *forward* extrapolation of downgoing waves to the wavenumber-frequency domain yields

$$\tilde{P}^{+}(k_x,k_y,z_1;\omega) = \tilde{W}^{+}(k_x,k_y;z_1,z_0;\omega)\tilde{P}^{+}(k_x,k_y,z_0;\omega), \tag{VII-6}$$

where, according to (V-59),

$$\tilde{W}^{+}(k_x,k_y;z_1,z_0;\omega) = \exp(-jk_z\Delta z), \tag{VII-7a}$$

with

$$k_z \triangleq +\sqrt{k^2 - k_x^2 - k_y^2} \qquad \text{for } k_x^2+k_y^2 \leq k^2, \tag{VII-7b}$$

$$k_z \triangleq -j\sqrt{k_x^2 + k_y^2 - k^2} \qquad \text{for } k_x^2+k_y^2 > k^2 \tag{VII-7c}$$

and

$$\Delta z = |z_1 - z_0| = z_1 - z_0. \tag{VII-7d}$$

The amplitude spectrum of operator $\tilde{W}^+$ is shown in Figure VII-3. Note that this spectrum is constant for the *propagating wavenumber area*:

$$|\tilde{W}^+(k_x,k_y;z_1,z_0;\omega)| = 1 \qquad \text{for } k_x^2+k_y^2 \leq k^2, \tag{VII-8a}$$

whereas it is exponentially decaying for the *evanescent wavenumber area*:

$$|\tilde{W}^+(k_x,k_y;z_1,z_0;\omega)| = \exp\left[-\sqrt{k_x^2 + k_y^2 - k^2}\,\Delta z\right] \quad \text{for } k_x^2+k_y^2 > k^2. \tag{VII-8b}$$

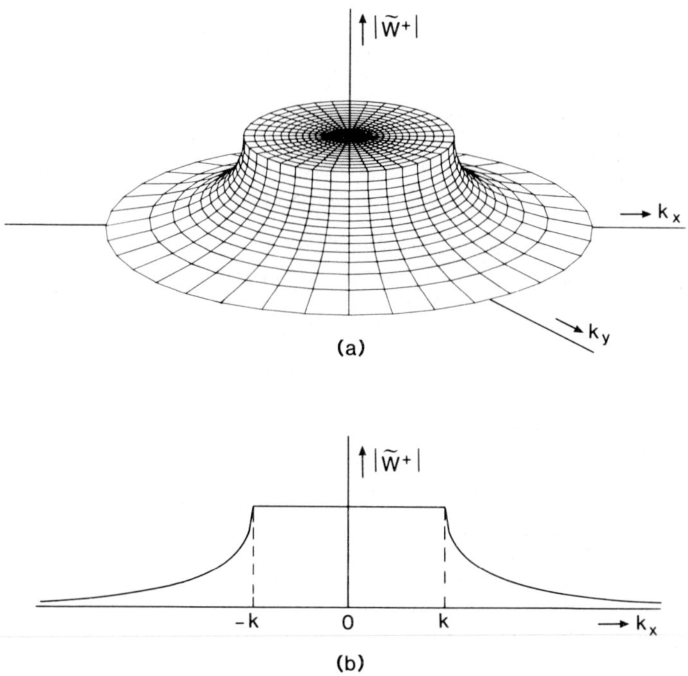

Figure VII-3: a. Amplitude spectrum of the forward wave field extrapolation operator $\tilde{W}^+(k_x,k_y;z_1,z_0;\omega)$
b. Cross-section for k_y=o.

Similarly, transforming the expression for *inverse* extrapolation of downgoing waves to the wavenumber-frequency domain yields

$$\tilde{P}^+(k_x,k_y,z_o;\omega) = \tilde{F}^+(k_x,k_y;z_o,z_1;\omega)\tilde{P}^+(k_x,k_y,z_1;\omega), \tag{VII-9}$$

where the operator $\tilde{F}^+$ is defined by

$$\tilde{F}^+(k_x,k_y;z_o,z_1;\omega)\tilde{W}^+(k_x,k_y;z_1,z_o;\omega) \triangleq 1. \tag{VII-10}$$

Hence, the explicit expression for $\tilde{F}^+$ reads

$$\tilde{F}^+(k_x,k_y;z_o,z_1;\omega) \triangleq 1/\tilde{W}^+(k_x,k_y;z_1,z_o;\omega) = \exp(+jk_z\Delta z), \tag{VII-11}$$

with k_z and Δz as defined in (VII-7). The amplitude spectrum of operator $\tilde{F}^+$ is shown in Figure VII-4. Note that this spectrum is constant for the propagating wavenumber area:

$$|\tilde{F}^+(k_x,k_y;z_o,z_1;\omega)| = 1 \qquad \text{for } k_x^2+k_y^2 \le k^2, \tag{VII-12a}$$

whereas it is exponentially growing for the evanescent wavenumber area:

$$|\tilde{F}^+(k_x,k_y;z_o,z_1;\omega)| = \exp\left[+\sqrt{k_x^2 + k_y^2 - k^2}\;\Delta z\right] \quad \text{for } k_x^2+k_y^2 > k^2. \tag{VII-12b}$$

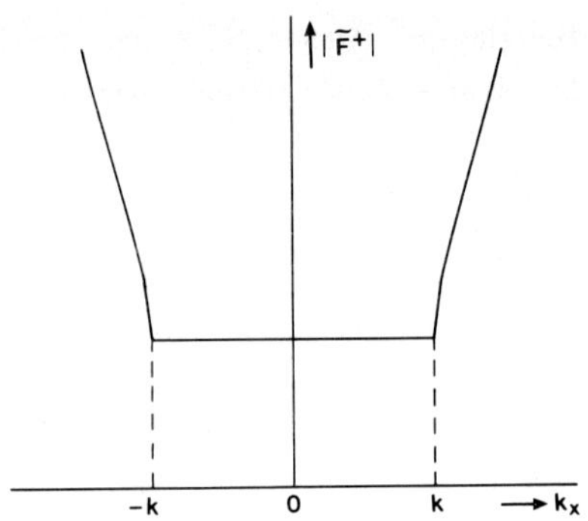

Figure VII-4: Amplitude spectrum (cross-section for k_y=o) of the ***exact*** *inverse wave field extrapolation operator $\tilde{F}^+(k_x,k_y;z_o,z_1;\omega)$. Note that this exact inverse operator is unstable.*

Expression (VII-11) for the inverse extrapolation operator $\tilde{F}^+$ is exact, but it is unpractical for the following two reasons:

1. In practical situations seismic data contain noise. Hence, expression (VII-9) for inverse extrapolation should be modified according to

$$\tilde{P}^+(k_x,k_y,z_o;\omega) = \tilde{F}^+(k_x,k_y;z_o,z_1;\omega)\left[\tilde{P}^+(k_x,k_y,z_1;\omega) + \tilde{N}(k_x,k_y,z_1;\omega)\right], \quad \text{(VII-13)}$$

   where $\tilde{N}$ represents the noise spectrum in the wavenumber-frequency domain. In the evanescent wavenumber area the amplitude of the wave field is generally far below the noise level:

$$|\tilde{P}^+(k_x,k_y,z_1;\omega)| << |\tilde{N}(k_x,k_y,z_1;\omega)| \qquad \text{for } k_x^2+k_y^2>k^2. \quad \text{(VII-14)}$$

   Hence, in the evanescent wavenumber area the exponentially growing inverse operator acts mainly on the noise term. This is unacceptable.

2. The inverse extrapolation operator F^+ in the space-frequency domain is related to the inverse extrapolation operator $\tilde{F}^+$ in the wavenumber-frequency domain via the inverse double spatial Fourier transform:

$$F^+(x,y;z_o,z_1;\omega) = \frac{1}{4\pi^2}\iint_{-\infty}^{\infty} \tilde{F}^+(k_x,k_y;z_o,z_1;\omega)e^{-j(k_x x+k_y y)}dk_x dk_y. \quad \text{(VII-15)}$$

   The integral in the right-hand side of this equation "explodes" for the exponentially growing function $\tilde{F}^+$. Hence, it is not possible to find an exact expression for the inverse extrapolation operator F^+ in the space-frequency domain.

Berkhout (1985, chapter VII) discusses three methods to stabilize the inverse extrapolation operator, namely:

. band-limited inversion,
. least-squares inversion,
. matched filtering.

Here we only discuss the matched filter approach. In this approach the inverse operator is approximated by taking the complex conjugate of the forward operator, hence

$$<\tilde{F}^{+}(k_x,k_y;z_o,z_1;\omega)> \triangleq [\tilde{W}^{+}(k_x,k_y;z_1,z_o;\omega)]^{*}. \qquad \text{(VII-16)}$$

Here the notation $<\tilde{F}^{+}>$ denotes that the matched filter is an approximation of $\tilde{F}^{+}$.

Note that for the propagating wavenumber area, where k_z is real (see equation (VII-7b)), this operator reads

$$<\tilde{F}^{+}(k_x,k_y;z_o,z_1;\omega)> = \exp(+jk_z\Delta z) \qquad \text{for } k_x^2+k_y^2 \leq k^2. \qquad \text{(VII-17)}$$

Comparing this expression with (VII-11) learns that operator $<\tilde{F}^{+}>$ is exact for the propagating wavenumber area. The amplitude spectrum of operator $<\tilde{F}^{+}>$ is shown in Figure VII-5. Note that this spectrum is exponentially decaying for the evanescent wavenumber area. Hence, the matched inverse operator $<\tilde{F}^{+}>$, as defined by (VII-16) is a *stable, spatially band-limited*[1)] approximation of the exact inverse operator $\tilde{F}^{+}$ as defined by (VII-11).

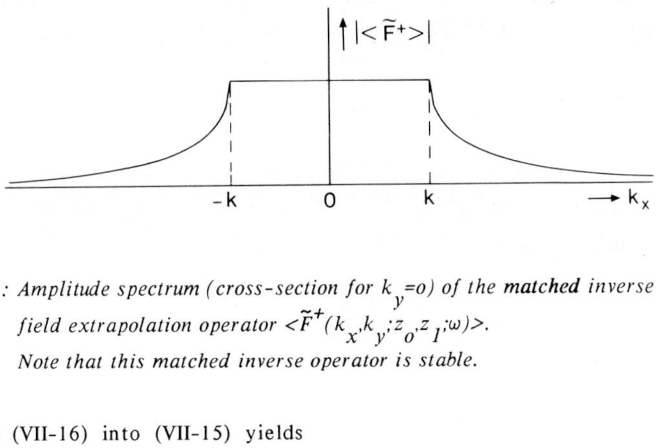

Figure VII-5: Amplitude spectrum (cross-section for k_y=o) of the **matched** *inverse wave field extrapolation operator $<\tilde{F}^{+}(k_x,k_y;z_o,z_1;\omega)>$. Note that this matched inverse operator is stable.*

Substituting (VII-16) into (VII-15) yields

$$<F^{+}(x,y;z_o,z_1;\omega)> = \left[\frac{1}{4\pi^2}\iint_{-\infty}^{\infty} \tilde{W}^{+}(k_x,k_y;z_1,z_o;\omega)e^{+j(k_xx+k_yy)}dk_xdk_y\right]^{*}, \qquad \text{(VII-18a)}$$

---

1) We speak of a stable spatially *band-limited* approximation because the matched inverse operator is exact only for the propagating wavenumber area $k_x^2+k_y^2 \leq k^2$ and attenuates for $k_x^2+k_y^2 > k^2$.

or

$$\langle F^{+}(x,y;z_0,z_1;\omega)\rangle = \left[W^{+}(-x,-y;z_1,z_0;\omega)\right]^{*}, \tag{VII-18b}$$

or, since W^{+} is an even function of x and y (see equations (VII-2c) and (VII-2d)),

$$\langle F^{+}(x,y;z_0,z_1;\omega)\rangle = \left[W^{+}(x,y;z_1,z_0;\omega)\right]^{*}. \tag{VII-19a}$$

Similarly,

$$\langle F^{-}(x,y;z_1,z_0;\omega)\rangle = \left[W^{-}(x,y;z_0,z_1;\omega)\right]^{*}. \tag{VII-19b}$$

Hence, also in the space-frequency domain, stable inverse operators are obtained simply by taking the complex conjugate of the forward operators. This is also true if we use the matrix notation, described in Appendix A. In this notation, equations (VII-2) and (VII-3) for forward extrapolation read

$$\vec{P}^{+}(z_1) = \mathbf{W}^{+}(z_1,z_0)\vec{P}^{+}(z_0) \tag{VII-20a}$$

and

$$\vec{P}^{-}(z_0) = \mathbf{W}^{-}(z_0,z_1)\vec{P}^{-}(z_1), \tag{VII-20b}$$

respectively, see section V.5.2. Similarly, in this notation equations (VII-4a) and (VII-4b) for inverse extrapolation read

$$\vec{P}^{+}(z_0) = \mathbf{F}^{+}(z_0,z_1)\vec{P}^{+}(z_1) \tag{VII-21a}$$

and

$$\vec{P}^{-}(z_1) = \mathbf{F}^{-}(z_1,z_0)\vec{P}^{-}(z_0), \tag{VII-21b}$$

respectively, where ideally operators $\mathbf{F}^+$ and $\mathbf{F}^-$ are the inverse versions of operators $\mathbf{W}^+$ and $\mathbf{W}^-$, see equations (VII-1a) and (VII-1b). In analogy with (VII-19), however, stable spatially band-limited inverse operators are given by

$$<\mathbf{F}^+(z_0,z_1)> = [\mathbf{W}^+(z_1,z_0)]^* \tag{VII-22a}$$

and

$$<\mathbf{F}^-(z_1,z_0)> = [\mathbf{W}^-(z_0,z_1)]^*, \tag{VII-22b}$$

respectively.

Finally, following Berkhout (1984), we show that the spatial band-limitation of the matched inverse operators imposes a restriction to the maximum obtainable spatial resolution.

Equation (VII-5a),

$$F^+(x,y;z_0,z_1;\omega) * W^+(x,y;z_1,z_0;\omega) \triangleq \delta(x)\delta(y), \tag{VII-23a}$$

states that the ideal inverse extrapolation operator F^+ perfectly compensates for the lateral smearing effect caused by the forward extrapolation operator W^+ (the "spatial wavelet"). If we replace F^+ by the matched inverse operator $<F^+>$, then the spatial delta function in the right-hand side of equation (VII-23a) is "smeared" along the x- and y-axis, hence

$$<F^+(x,y;z_0,z_1;\omega)> * W^+(x,y;z_1,z_0;\omega) = d_0(x,y;\omega), \tag{VII-23b}$$

where $d_0(x,y;\omega)$ is a spatially band-limited version of $\delta(x)\delta(y)$. The "width" of this spatially band-limited delta function determines the maximum obtainable spatial resolution. To find an explicit expression for d_0, we transform equation (VII-23b) to the wavenumber-frequency domain, yielding

$$<\tilde{F}^{+}(k_x,k_y;z_o,z_1;\omega)>.\tilde{W}^{+}(k_x,k_y;z_1,z_o;\omega) = \tilde{d}_o(k_x,k_y;\omega), \tag{VII-24a}$$

or, upon substitution of (VII-7) and (VII-16),

$$\tilde{d}_o(k_x,k_y;\omega) = 1 \qquad \text{for } k_x^2+k_y^2 \leq k^2 \tag{VII-24b}$$

and

$$\tilde{d}_o(k_x,k_y;\omega) = \exp\left[-2\sqrt{k_x^2 + k_y^2 - k^2}\ \Delta z\right] \quad \text{for } k_x^2+k_y^2>k^2, \tag{VII-24c}$$

with

$$k = \omega/c \tag{VII-24d}$$

and

$$\Delta z = |z_1-z_o| . \tag{VII-24e}$$

We apply an inverse double spatial Fourier transform to $\tilde{d}_o$ to obtain an expression for d_o in the space-frequency domain. Hence

$$d_o(x,y;\omega) = \frac{1}{4\pi^2}\iint_{-\infty}^{\infty} \tilde{d}_o(k_x,k_y;\omega)e^{-j(k_x x+k_y y)} dk_x dk_y. \tag{VII-25}$$

In the following we assume that the extrapolation stepsize Δz is sufficiently large ($\Delta z>\lambda$, where λ represents the wavelength) so that $\tilde{d}_o(k_x,k_y;\omega)\approx o$ for $k_x^2+k_y^2>k^2$, see Figure VII-6a.

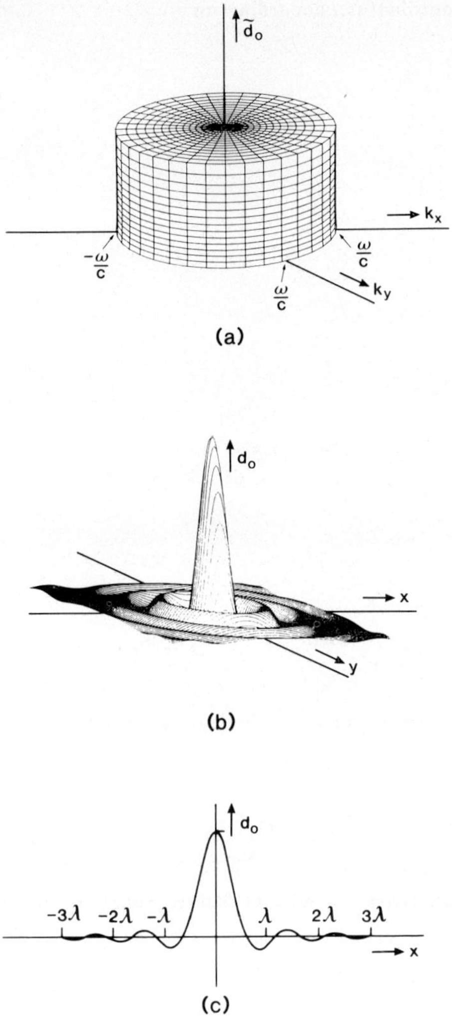

Figure VII-6: Spatial resolution of "matched inverse extrapolation".

a. Spectrum of the spatially band-limited inversion result.

b. Spatially band-limited inversion result.

c. Cross-section for y=o.

We introduce polar coordinates, according to

$$x = r\cos\phi, \tag{VII-26a}$$

$$y = r\sin\phi \tag{VII-26b}$$

and

$$k_x = k_r\cos\theta, \tag{VII-26c}$$

$$k_y = k_r\sin\theta. \tag{VII-26d}$$

Then

$$d_o(r,\phi;\omega) = \frac{1}{4\pi^2}\int_o^k \left[\int_o^{2\pi} e^{-jk_r r\cos(\theta-\phi)} d\theta\right] k_r dk_r, \tag{VII-27a}$$

or

$$d_o(r,\phi;\omega) = \frac{1}{2\pi}\int_o^k \left[\frac{1}{\pi}\int_o^{\pi} \cos(k_r r\cos\theta)d\theta\right] k_r dk_r, \tag{VII-27b}$$

or, using Abramowitz and Stegun (1970, equation 9.1.18),

$$d_o(r,\phi;\omega) = \frac{1}{2\pi}\int_o^k k_r J_o(k_r r)dk_r, \tag{VII-27c}$$

where J_o is the zeroth-order cylindrical Bessel function, or, using Abramowitz and Stegun (1970, equation 9.1.30),

$$d_o(r,\phi;\omega) = k\,\frac{J_1(kr)}{2\pi r}, \tag{VII-27d}$$

where J_1 is the first-order cylindrical Bessel function.
Finally, returning to Cartesian coordinates, we obtain

$$d_o(x,y;\omega) = \frac{\omega}{c}\,\frac{J_1\left(\omega\sqrt{x^2+y^2}/c\right)}{2\pi\sqrt{x^2+y^2}}, \tag{VII-28}$$

see Figure VII-6b. Note that the width of the main lobe of this spatially band-limited inversion result equals approximately $6\lambda/5$ (Figure VII-6c).

VII.2.2 Vertically inhomogeneous media

For a vertically inhomogeneous (i.e., laterally invariant) acoustic medium, forward extrapolation of downgoing and upgoing waves (Figure VII-1) may again be described by spatial convolutions along the x- and y-axis, in the symbolic notation

$$P^+(x,y,z_1;\omega) = W^+(x,y;z_1,z_0;\omega) * P^+(x,y,z_0;\omega) \tag{VII-29a}$$

and

$$P^-(x,y,z_0;\omega) = W^-(x,y;z_0,z_1;\omega) * P^-(x,y,z_1;\omega), \tag{VII-29b}$$

respectively, where

$$W^+(x,y;z_1,z_0;\omega) \triangleq \frac{2}{\rho(z_0)} \frac{\partial G^+(x,y,z_1;0,0,z=z_0;\omega)}{\partial z} \tag{VII-30a}$$

and

$$W^-(x,y;z_0,z_1;\omega) \triangleq \frac{-2}{\rho(z_1)} \frac{\partial G^-(x,y,z_0;0,0,z=z_1;\omega)}{\partial z}, \tag{VII-30b}$$

see equations (V-53) to (V-55). In equation (VII-30), G^+ and G^- represent the one-way Green's wave fields, see section V.5.1. For a laterally invariant medium, they satisfy the following reciprocity relation

$$G^-(x,y,z_0;0,0,z_1;\omega) = G^+(x,y,z_1;0,0,z_0;\omega). \tag{VII-31a}$$

Note, however, that unlike for the homogeneous situation,

$$W^-(x,y;z_0,z_1;\omega) \neq W^+(x,y;z_1,z_0;\omega). \tag{VII-31b}$$

In analogy with (VII-4), inverse extrapolation of downgoing and upgoing waves (Figure VII-2) may be described by spatial deconvolution processes along the x- and y-axis, according to

$$P^+(x,y,z_0;\omega) = F^+(x,y;z_0,z_1;\omega) * P^+(x,y,z_1;\omega) \tag{VII-32a}$$

and

$$P^-(x,y,z_1;\omega) = F^-(x,y;z_1,z_0;\omega) * P^-(x,y,z_0;\omega), \tag{VII-32b}$$

respectively. We follow a similar approach as in section VII.2.1 to find explicit expressions for operators F^+ and F^-. In the wavenumber-frequency domain we write, in analogy with equation (VII-11),

$$\tilde{F}^+(k_x,k_y;z_0,z_1;\omega) \triangleq 1/\tilde{W}^+(k_x,k_y;z_1,z_0;\omega) \tag{VII-33a}$$

and

$$\tilde{F}^-(k_x,k_y;z_1,z_0;\omega) \triangleq 1/\tilde{W}^-(k_x,k_y;z_0,z_1;\omega). \tag{VII-33b}$$

Again these operators are unstable for the evanescent wavenumber area and therefore alternative solutions must be looked for. We assume for the moment that the medium parameters are smoothly varying with depth between z_0 and z_1. Then the forward extrapolation operators $\tilde{W}^+$ and $\tilde{W}^-$ may be approximated by the WKB-solutions (III-44c) and (III-44d), respectively, according to

$$\tilde{W}^+(k_x,k_y;z_1,z_0;\omega) \approx \left[\frac{k_z(z_0)}{\rho(z_0)}\right]^{\frac{1}{2}} \left[\frac{k_z(z_1)}{\rho(z_1)}\right]^{-\frac{1}{2}} \exp \int_{z_0}^{z_1} -jk_z(z)dz \tag{VII-34a}$$

and

$$\tilde{W}^-(k_x,k_y;z_o,z_1;\omega) \approx \left[\frac{k_z(z_1)}{\rho(z_1)}\right]^{\frac{1}{2}} \left[\frac{k_z(z_o)}{\rho(z_o)}\right]^{-\frac{1}{2}} \exp \int_{z_1}^{z_o} +jk_z(z)dz, \qquad \text{(VII-34b)}$$

where

$$k_z(z) \triangleq +\sqrt{k^2(z) - k_x^2 - k_y^2} \qquad \text{for } k_x^2+k_y^2 \leq k^2(z) \qquad \text{(VII-34c)}$$

and

$$k_z(z) \triangleq -j\sqrt{k_x^2 + k_y^2 - k^2(z)} \qquad \text{for } k_x^2+k_y^2 > k^2(z). \qquad \text{(VII-34d)}$$

These expressions break down when $k_z(z) \to o$, i.e., when the waves propagate nearly horizontally. Consider the following *modified*[1)] matched inverse operators

$$<\tilde{F}^+(k_x,k_y;z_o,z_1;\omega)> \triangleq [\tilde{W}^-(k_x,k_y;z_o,z_1;\omega)]^* \qquad \text{(VII-35a)}$$

and

$$<\tilde{F}^-(k_x,k_y;z_1,z_o;\omega)> \triangleq [\tilde{W}^+(k_x,k_y;z_1,z_o;\omega)]^*, \qquad \text{(VII-35b)}$$

respectively. It can be easily seen from equations (VII-34a) and (VII-34b) that operators $<\tilde{F}^+>$ and $<\tilde{F}^->$ are identical to operators $\tilde{F}^+$ and $\tilde{F}^-$, respectively, for propagating waves, i.e., for real $k_z(z)$ on the entire interval (z_o,z_1). Furthermore, for evanescent waves they have the same stable amplitude behaviour as the forward extrapolation operators $\tilde{W}^-$ and $\tilde{W}^+$, respectively. Hence, for smoothly varying vertically inhomogeneous media, the modified matched inverse operators $<\tilde{F}^+>$ and $<\tilde{F}^->$, as defined by (VII-35a) and (VII-35b), respectively, represent stable spatially band-limited approximations of the exact inverse operators $\tilde{F}^+$ and $\tilde{F}^-$, as defined by (VII-33a) and (VII-33b), respectively. Transforming equations

---

1) We speak of *modified* matched inverse operators because, unlike in the homogeneous situation (equation (VII-16)), the inverse operator for *downgoing* waves is approximated by taking the complex conjugate of the forward operator for *upgoing* waves and vice versa. Of course equation (VII-16) is not in contradiction with equation (VII-35a), because in the homogeneous situation the forward operators for downgoing and upgoing waves are identical.

(VII-35a) and (VII-35b) back to the space-frequency domain yields

$$<F^+(x,y;z_0,z_1;\omega)> \;=\; \left[W^-(x,y;z_0,z_1;\omega)\right]^* \tag{VII-36a}$$

and

$$<F^-(x,y;z_1,z_0;\omega)> \;=\; \left[W^+(x,y;z_1,z_0;\omega)\right]^*, \tag{VII-36b}$$

respectively. Similarly, in the matrix notation we obtain

$$<\mathbf{F}^+(z_0,z_1)> \;=\; \left[\mathbf{W}^-(z_0,z_1)\right]^* \tag{VII-37a}$$

and

$$<\mathbf{F}^-(z_1,z_0)> \;=\; \left[\mathbf{W}^+(z_1,z_0)\right]^*. \tag{VII-37b}$$

Let us now investigate the behaviour of the modified matched inverse operators for the situation where the medium contains interfaces. For simplicity we consider two homogeneous half-spaces, separated by an interface at z_1 (Figure VII-7). In the wavenumber-frequency domain, the forward extrapolation operators for downgoing waves (Figure VII-7a) and upgoing waves (Figure VII-7b) read, respectively,

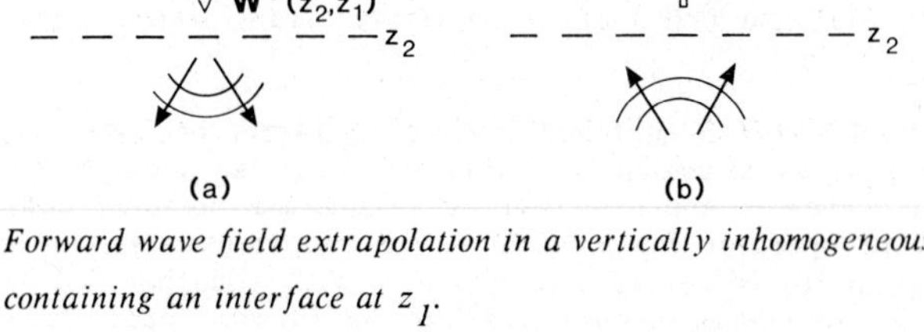

Figure VII-7: Forward wave field extrapolation in a vertically inhomogeneous medium, containing an interface at z_1.

$$\tilde{W}^+(z_2,z_o) = \tilde{W}^+(z_2,z_1)\tilde{T}^+(z_1)\tilde{W}^+(z_1,z_o) \tag{VII-38a}$$

and

$$\tilde{W}^-(z_o,z_2) = \tilde{W}^-(z_o,z_1)\tilde{T}^-(z_1)\tilde{W}^-(z_1,z_2), \tag{VII-38b}$$

where the extrapolation operators for the homogeneous layers are given by

$$\tilde{W}^+(z_1,z_o) = \tilde{W}^-(z_o,z_1) = \exp(-jk_{z,1}|z_1-z_o|) \tag{VII-39a}$$

and

$$\tilde{W}^+(z_2,z_1) = \tilde{W}^-(z_1,z_2) = \exp(-jk_{z,2}|z_2-z_1|) \tag{VII-39b}$$

and where the transmission operators for the interface at z_1 are given by

$$\tilde{T}^+(z_1) = 1 + \tilde{R}^+(z_1) \tag{VII-40a}$$

and

$$\tilde{T}^-(z_1) = 1 - \tilde{R}^+(z_1), \tag{VII-40b}$$

with reflection operator $\tilde{R}^+(z_1)$ given by equation (III-50a). Comparing the modified matched inverse operator

$$<\tilde{F}^+(z_o,z_2)> \hat{=} [\tilde{W}^-(z_o,z_2)]^*, \tag{VII-41a}$$

with the exact inverse operator

$$\tilde{F}^+(z_o,z_2) \hat{=} 1/\tilde{W}^+(z_2,z_o), \tag{VII-41b}$$

yields for propagating waves (i.e., for real $k_{z,1}$ and $k_{z,2}$) the following relation:

$$\langle \tilde{F}^{+}(z_0,z_2) \rangle = \left[\tilde{T}^{-}(z_1)\tilde{T}^{+}(z_1)\right] \tilde{F}^{+}(z_0,z_2), \tag{VII-41c}$$

or

$$\langle \tilde{F}^{+}(z_0,z_2) \rangle = \left[1 - \left(\tilde{R}^{+}(z_1)\right)^2\right] \tilde{F}^{+}(z_0,z_2). \tag{VII-42a}$$

Similarly,

$$\langle \tilde{F}^{-}(z_2,z_0) \rangle = \left[1 - \left(\tilde{R}^{+}(z_1)\right)^2\right] \tilde{F}^{-}(z_2,z_0). \tag{VII-42b}$$

Hence, when the medium contains one or more interfaces, the matched inverse operators deviate from the exact inverse operators, even when we consider propagating waves only. To be more specific, by applying the matched inverse operators, amplitude errors are introduced which are proportional to the *squared reflectivity* of the interfaces (equation (VII-42)). In the following we will refer to these errors as *second order amplitude errors*. When the contrasts at the interfaces are low[1)], these second order amplitude errors are negligible, so the matched inverse operators are applicable for media with low contrasts. On the other hand, when the contrasts at the interfaces are significant, the matched inverse operators are no longer accurate, so an alternative procedure must be followed. In chapter IX we will discuss acoustic inverse wave field extrapolation operators for media with high contrasts.

---

1) As an example, consider a density contrast $\Delta\rho = \rho_2 - \rho_1 =$ (2750-2250) kg/m^3 = 500 kg/m^3. Then, according to equation (III-50a), $\tilde{R}^{+}(z_1) = (2750-2250)/(2750+2250) = 0.1$. Hence, $\left[1-(\tilde{R}^{+}(z_1))^2\right] = 0.99$, meaning that the relative amplitude error of the matched inverse operators is 1% for this situation. On the other hand, for a velocity contrast of the same magnitude, the relative amplitude error ranges from 1% at normal incidence to 100% at the critical angle (55 degrees), meaning that for this situation the matched inverse operators are only valid for moderate propagation angles.

VII.3 ACOUSTIC INVERSE WAVE FIELD EXTRAPOLATION IN ARBITRARILY INHOMOGENEOUS MEDIA

VII.3.1 Introduction

For an arbitrarily inhomogeneous acoustic medium, forward extrapolation of downgoing and upgoing waves (Figure VII-1) may be described by the acoustic one-way Rayleigh II integrals (V-51a) and (V-51b), according to

$$P^+(x,y,z_1;\omega) =$$

$$2\int\int_{-\infty}^{\infty}\left[\frac{\partial G^+(x,y,z_1;x',y',z'=z_0;\omega)}{\partial z'}\ \frac{1}{\rho(x',y',z_0)}\ P^+(x',y',z_0;\omega)\right]dx'dy' \qquad \text{(VII-43a)}$$

and

$$P^-(x,y,z_0;\omega) =$$

$$-2\int\int_{-\infty}^{\infty}\left[\frac{\partial G^-(x,y,z_0;x',y',z'=z_1;\omega)}{\partial z'}\ \frac{1}{\rho(x',y',z_1)}\ P^-(x',y',z_1;\omega)\right]dx'dy', \qquad \text{(VII-43b)}$$

respectively, where G^+ and G^- represent the one-way Green's wave fields, see section V.5.1. For the general inhomogeneous situation, inversion of these integral equations is not straightforward. In the matrix notation of Appendix A, equations (VII-43a) and (VII-43b) may be replaced by

$$\vec{P}^+(z_1) = \mathbf{W}^+(z_1,z_0)\vec{P}^+(z_0) \qquad \text{(VII-44a)}$$

and

$$\vec{P}^-(z_0) = \mathbf{W}^-(z_0,z_1)\vec{P}^-(z_1), \qquad \text{(VII-44b)}$$

respectively, where the forward wave field extrapolation matrices are defined by

$$\mathbf{W}^+(z_1,z_o) = 2\,\frac{\partial \mathbf{G}^+(z_1,z=z_o)}{\partial z}\,\mathbf{M}^{-1}(z_o) \tag{VII-45a}$$

and

$$\mathbf{W}^-(z_o,z_1) = -2\,\frac{\partial \mathbf{G}^-(z_o,z=z_1)}{\partial z}\,\mathbf{M}^{-1}(z_1), \tag{VII-45b}$$

respectively, see section V.5.2. In this matrix notation, the inverse extrapolation problem can be formulated as follows

$$\vec{P}^+(z_o) = \mathbf{F}^+(z_o,z_1)\vec{P}^+(z_1) \tag{VII-46a}$$

and

$$\vec{P}^-(z_1) = \mathbf{F}^-(z_1,z_o)\vec{P}^-(z_o), \tag{VII-46b}$$

where

$$\mathbf{F}^+(z_o,z_1) \triangleq [\mathbf{W}^+(z_1,z_o)]^{-1} \tag{VII-47a}$$

and

$$\mathbf{F}^-(z_1,z_o) \triangleq [\mathbf{W}^-(z_o,z_1)]^{-1}. \tag{VII-47b}$$

However, we have seen already for the homogeneous situation that this inversion is unstable, so for the general inhomogeneous situation the problems may be even larger. Therefore, rather than trying to invert the forward problem, in this section we discuss the inverse extrapolation problem step by step, using the acoustic Kirchhoff-Helmholtz integral as the starting point (Wapenaar et al., 1989). After a number of necessary approximations we finally arrive at stable, spatially band-limited inverse extrapolation operators which read in the matrix notation

$$<\mathbf{F}^+(z_o,z_1)> = [\mathbf{W}^-(z_o,z_1)]^* \tag{VII-48a}$$

and

$$\langle F^-(z_1,z_0)\rangle = [W^+(z_1,z_0)]^*, \tag{VII-48b}$$

a result which was already found in section VII.2.2 for the situation of laterally invariant media.

VII.3.2 Acoustic Kirchhoff-Helmholtz integrals for inverse extrapolation

In section V.3.2 we have derived the following two equivalent versions of the acoustic Kirchhoff-Helmholtz integral,

$$P(\vec{r}_A,\omega) = \oint_S \frac{1}{\rho(\vec{r})} \left[G(\vec{r},\vec{r}_A,\omega)\nabla P(\vec{r},\omega) - \nabla G(\vec{r},\vec{r}_A,\omega)P(\vec{r},\omega)\right].\vec{n}\,dS \tag{VII-49a}$$

and

$$P(\vec{r}_A,\omega) = \oint_S \frac{1}{\rho(\vec{r})} \left[G^*(\vec{r},\vec{r}_A,\omega)\nabla P(\vec{r},\omega) - \nabla G^*(\vec{r},\vec{r}_A,\omega)P(\vec{r},\omega)\right].\vec{n}\,dS. \tag{VII-49b}$$

These expressions describe the acoustic wave field at $\vec{r}_A$ in a source-free volume V in terms of the acoustic wave field and its gradient on surface S, enclosing V (Figure V-3). In (VII-49a), G represents a *forward* propagating Green's wave field (Figure V-3a); in (VII-49b), G^* represents a *backward* propagating Green's wave field (Figure V-3b).

In chapter V we used the Kirchhoff-Helmholtz integral (VII-49a) with the forward propagating Green's wave field as a starting point for deriving forward wave field extrapolation operators. Here we use the Kirchhoff-Helmholtz integral (VII-49b) with the backward propagating Green's wave field as a starting point for deriving inverse wave field extrapolation operators.

For the half-space geometry of Figure V-4, the backward propagating Green's wave field does not obey Sommerfeld's radiation condition on S_1 (see section V.2). Therefore, consider the modified configuration shown in Figure VII-8. We assume that an acoustic wave field, radiated by (secondary) sources in the subsurface, has been measured on "acquisition surface" S_0. Our aim is to find an expression which describes the acoustic wave field at $\vec{r}_A$ in the subsurface in terms of the acoustic wave field and its gradient on S_0. We constructed a surface S enclosing a volume V such that $\vec{r}_A$ lies in V and such that V is source-free. Closed surface S consists of "acquisition surface" S_0, a plane horizontal "reference surface" S_1 at $z=z_1$ (between $\vec{r}_A$ and the sources of the acoustic wave field) and a cylindrical surface S_2 with a vertical axis through $\vec{r}_A$ and radius r. For this configuration we analyse Kirchhoff-Helmholtz integral (VII-49b), with the backward propagating Green's wave field. The contribution of this integral over S_2 vanishes when r goes to infinity (the cylindrical surface is proportional to r, the integrand is proportional to $1/r^2$). So for the geometry of Figure VII-8, Kirchhoff-Helmholtz integral (VII-49b) may be replaced by

$$P(\vec{r}_A,\omega) = P_0(\vec{r}_A,\omega) + \Delta P(\vec{r}_A,\omega), \tag{VII-50a}$$

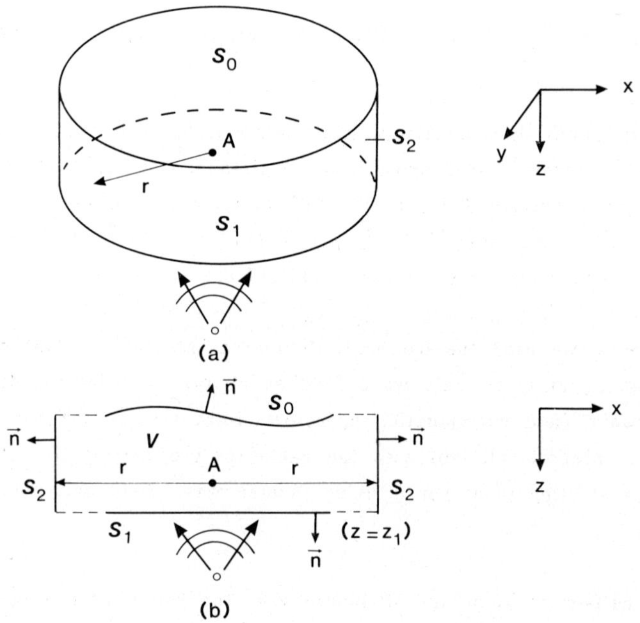

Figure VII-8: Modified geometry for the Kirchhoff-Helmholtz integral (VII-49b) with the backward propagating Green's wave field. Under certain conditions (discussed in the text) the contribution of this integral over S_1 and S_2 can be neglected.

a. Perspective view. b. Cross section for y=o.

where

$$P_o(\vec{r}_A,\omega) \triangleq \int_{S_o} \frac{1}{\rho} \left[G^* \nabla P - \nabla G^* P\right].\vec{n}\, dS_o \qquad \text{(VII-50b)}$$

and

$$\Delta P(\vec{r}_A,\omega) \triangleq \int_{S_1} \frac{1}{\rho} \left[G^* \nabla P - \nabla G^* P\right].\vec{n}\, dS_1. \qquad \text{(VII-50c)}$$

When $\Delta P(\vec{r}_A,\omega)$, as defined in (VII-50c), may be neglected, then equation (VII-50b) describes *inverse* wave field extrapolation (towards the sources) from acquisition surface S_o to subsurface point $\vec{r}_A$. In the next section we investigate under which conditions $\Delta P(\vec{r}_A,\omega)$ may be neglected.

VII.3.3 Error analysis

At $z=z_1$ (the depth of "reference surface" S_1), the acoustic wave field consists of upgoing waves $P^-(\vec{r},\omega)$ (including higher order terms), related to the sources below z_1, and downgoing waves $P^+(\vec{r},\omega)$ (including higher order terms), caused by scattering above z_1, hence

$$P(\vec{r},\omega) = P^+(\vec{r},\omega) + P^-(\vec{r},\omega) \qquad \text{at } z=z_1. \qquad \text{(VII-51a)}$$

For the Green's wave field G we may choose below $z=z_1$ (outside V) any convenient reference medium. We choose $c(x,y,z>z_1) = c(x,y,z_1)$ and $\rho(x,y,z>z_1) = \rho(x,y,z_1)$. With this choice the Green's wave field at z_1 is purely downgoing (the Green's source is situated at $\vec{r}_A$ above z_1; the half-space below z_1 is reflection free), hence

$$G(\vec{r},\vec{r}_A,\omega) = G^+(\vec{r},\vec{r}_A,\omega) \qquad \text{at } z=z_1. \qquad \text{(VII-51b)}$$

Thus we may rewrite equation (VII-50c) as

$$\Delta P(\vec{r}_A,\omega) = \iint_{-\infty}^{\infty} \frac{1}{\rho}\left[(G^+)^* \left(\frac{\partial P^+}{\partial z} + \frac{\partial P^-}{\partial z}\right) - \left(\frac{\partial G^+}{\partial z}\right)^* (P^+ + P^-)\right]_{z_1} dxdy. \qquad \text{(VII-52)}$$

In Appendix B, section B.3, we analyse the interactions between one-way acoustic wave fields (P^+,P^-) and one-way backward propagating Green's wave fields $\left[(G^+)^*, (G^-)^*\right]$.

The starting point of this analysis is equation (B-26), which is identical to equation (VII-52) if we choose $G^-=o$ and if we replace P_o by $-\Delta P$ and z_o by z_1. Hence, if we make the same alterations in the final result (equation (B-34b), generalized for laterally varying medium parameters) we obtain

$$\Delta P(\vec{r}_A,\omega) \approx -2\iint_{-\infty}^{\infty} \left[\frac{1}{\rho(\vec{r})} \left(\frac{\partial G^+(\vec{r},\vec{r}_A,\omega)}{\partial z}\right)^* P^+(\vec{r},\omega)\right]_{z_1} dxdy. \qquad \text{(VII-53)}$$

Apparently, only the wave fields $(G^+)^*$ and P^+ which propagate in *opposite* directions through $z=z_1$ contribute to the error term $\Delta P(\vec{r}_A,\omega)$, see Figure VII-9. The underlying assumption is that $\tilde{G}^+(k_x,k_y,z_1;x_A,y_A,z_A;\omega)$ or $\tilde{P}^-(k_x,k_y,z_1;\omega)$, or both, are negligible in the evanescent wavenumber area

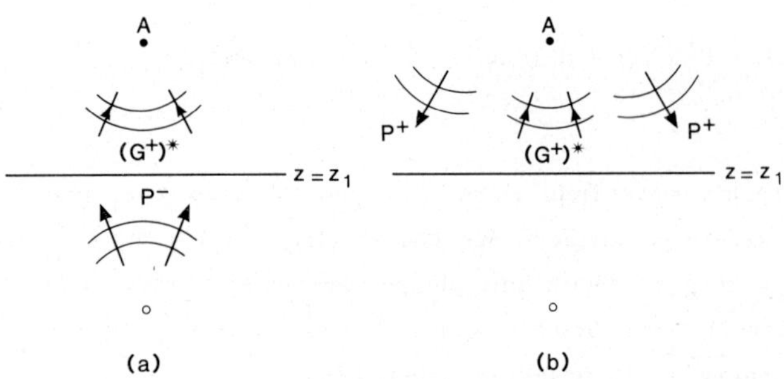

Figure VII-9: Choosing a reflection free lower half-space for G, the Kirchhoff-Helmholtz integral over z_1 (VII-52) consists of terms containing the products of $(G^+)^*$ and P^- at $z=z_1$ (Figure a) and terms containing the products of $(G^+)^*$ and P^+ at $z=z_1$ (Figure b). Only the terms with the ***opposite*** *propagating wave fields at $z=z_1$ (Figure b) contribute to the result $\Delta P(\vec{r}_A,\omega)$ [compare with Figure V-7].*

$k_x^2+k_y^2>k^2(z_1)$. This assumption is satisfied when the source of the acoustic wave field P and the source (at $\vec{r}_A$) of the Green's wave field G are not both in the direct vicinity of the "reference surface" S_1, or in other words, when $\vec{r}_A$ is not too close to the source of the acoustic wave field P. In the following we replace "≈" by "=" whenever the only approximation concerns the negligence of evanescent waves.

Let us first discuss the consequence of this important result for the situation of a homogeneous medium. Then no scattering occurs, hence

$$P^+(\vec{r},\omega) = 0 \quad \text{at } z=z_1 \tag{VII-54a}$$

and, consequently,

$$\Delta P(\vec{r}_A,\omega) = 0. \tag{VII-54b}$$

With this result we may rewrite equation (VII-50) as

$$P(\vec{r}_A,\omega) = P_0(\vec{r}_A,\omega) = \int_{S_0} \frac{1}{\rho} \left[G^*(\vec{r},\vec{r}_A,\omega)\nabla P(\vec{r},\omega) - \nabla G^*(\vec{r},\vec{r}_A,\omega)P(\vec{r},\omega)\right].\vec{n}\,dS_0, \tag{VII-55a}$$

where, according to equations (V-22) and (V-26b),

$$G^*(\vec{r},\vec{r}_A,\omega) = \frac{\rho}{4\pi}\frac{e^{+jk\Delta r}}{\Delta r}, \tag{VII-55b}$$

with

$$k = \omega/c \tag{VII-55c}$$

and

$$\Delta r = |\vec{r}-\vec{r}_A|. \tag{VII-55d}$$

This Kirchhoff-Helmholtz integral describes *inverse* wave field extrapolation from acquisition surface S_0 to subsurface point $\vec{r}_A$. It is interesting to note that, in order to arrive at this result, it was essential to make use of the backward-propagating Green's wave field G^* (if we had used G instead of G^*, then $P_0(\vec{r}_A,\omega)$ would vanish instead of $\Delta P(\vec{r}_A,\omega)$, so we would have obtained a Kirchhoff-Helmholtz integral for *forward* extrapolation from S_1 to $\vec{r}_A$). The only approximation in equation (VII-55) is a spatial bandlimitation (the negligence of evanescent waves at z_1). This imposes a restriction to the maximum obtainable spatial resolution, see also section VII.2.

The validity of Kirchhoff-Helmholtz integral (VII-55) is demonstrated with a simple example (Peels, 1988). We consider 2-D wave propagation in the 2-D configuration, shown in Figure VII-10a. The propagation velocity equals 2000 m/s. The acoustic pressure field of a burried dipole source, measured at the curved surface S_0, is shown in Figure VII-10b as a function of space and time. The normal derivative of the wave field at S_0 is shown in Figure VII-10c. Inverse wave field extrapolation to depth level z_A is carried out by transforming the data from the time domain to the frequency domain and by applying the 2-D version of equation (VII-55) for all points $\vec{r}_A$ at depth level z_A and for all frequencies within the seismic band. The result, transformed back to the time domain, is shown in Figure VII-11a. It represents a hyperbolically shaped dipole response. In Figure VII-11b the maximum amplitude of each trace is shown as a function of the lateral position (dotted line). Note the perfect match with the analytically computed response (solid line). The very small deviations at the edges are due to the limited aperture (ideally S_0 should be of infinite extent). Next we used equation (VII-55) to extrapolate the data inversely from S_0 to the source depth level z'_A, thus violating the condition that $\vec{r}_A$ should not be close to the source.
Figure VII-11c shows the result in the space-time domain. The real part of the data at 35 Hz is shown as a function of the lateral position in Figure VII-11d (the imaginary part is approximately zero). Note that the dipole source (ideally represented by a spatial delta function) is smeared out in space, due to the inevitable negligence of the evanescent wave field. Theoretically the width of the mainlobe should be equal to the wavelength

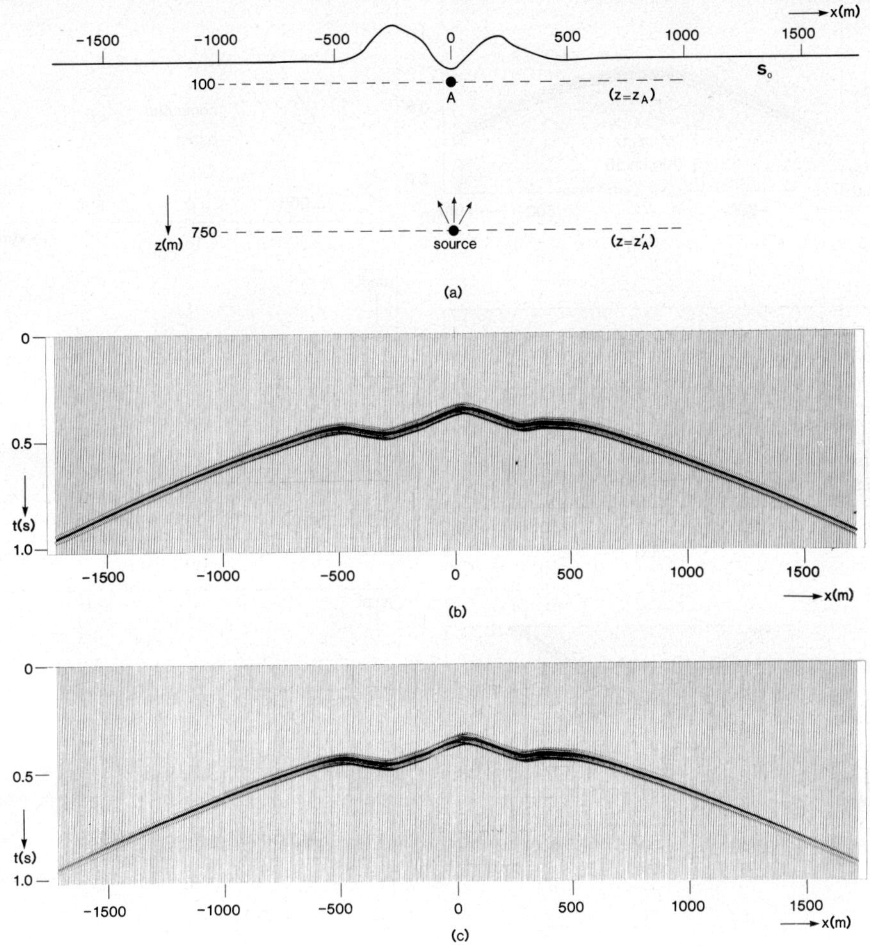

Figure VII-10: a. Homogeneous medium containing a burried dipole source.

b. Pressure field, measured at surface S_o, as a function of space and time.

c. Normal derivative of the data in Figure b.

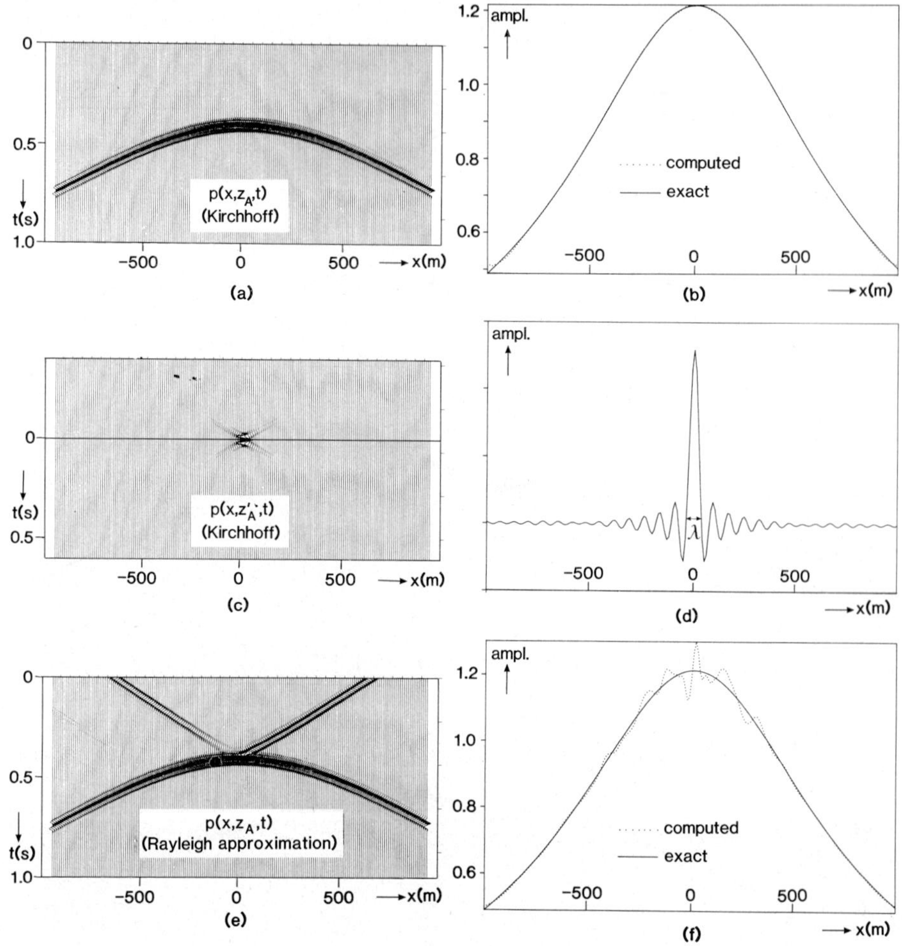

Figure VII-11: *a. Inverse extrapolated data at z_A (Kirchhoff-Helmholtz integral (VII-55))*

b. Maximum amplitude per trace of Figure a.

c. Inverse extrapolated data at z'_A (Kirchhoff-Helmholtz integral (VII-55))

d. Real part of central frequency component (35 Hz) from data in Figure c.

e. Inverse extrapolated data at z_A (Rayleigh approximation)

f. Maximum amplitude per trace of Figure e.

λ. In this example λ equals 2000/35 = 57 m, whereas the width of the mainlobe in Figure VII-11d equals 74 m. The small difference is explained for the greater part by the limited aperture. Finally we carried out inverse wave field extrapolation based upon the Rayleigh approximation. That is, we assumed that only the pressure field $P(\vec{r},\omega)$ at S_0 is available and we approximated equation (VII-55) by replacing $P(\vec{r},\omega)$ by $2P(\vec{r},\omega)$ and by omitting $\nabla P(\vec{r},\omega)$. The result at z_A, transformed back to the time domain, is shown in Figure VII-11e. Note that, apart from the expected hyperbolically shaped dipole response, some significant artifacts are present in these data. The maximum of each trace is shown as a function of the lateral position in Figure VII-11f (dotted line). Note the significant amplitude deviations from the exact, analytically computed response (solid line). Obviously the Rayleigh approximation is not allowed for the configuration of a curved acquisition surface S_0. In the next section we show that the Rayleigh approximation is allowed for a flat surface S_0.

We return to the situation of an inhomogeneous medium and continue our analysis of the error term $\Delta P(\vec{r}_A,\omega)$, as defined by equation (VII-53):

$$\Delta P(\vec{r}_A,\omega) = -2 \iint_{-\infty}^{\infty} \left[\frac{1}{\rho(\vec{r})} \left(\frac{\partial G^+(\vec{r},\vec{r}_A,\omega)}{\partial z} \right)^* P^+(\vec{r},\omega) \right]_{z_1} dxdy. \qquad \text{(VII-56)}$$

The wave field $P^+(\vec{r},\omega)$ at z_1 represents a scattered wave field, hence

$$P^+(\vec{r},\omega) = P_s^+(\vec{r},\omega) \quad \text{at } z=z_1, \qquad \text{(VII-57a)}$$

where the sub-script "s" stands for "scattered", (see Figure VII-12a). The Green's wave field $G^+(\vec{r},\vec{r}_A,\omega)$ at z_1 consists of a term $G_d^+(\vec{r},\vec{r}_A,\omega)$ which propagates directly from $\vec{r}_A$ to z_1 (see Figure VII-12b; the sub-script "d" stands for "direct") and a term $G_s^+(\vec{r},\vec{r}_A,\omega)$ which is scattered by the inhomogeneities above $\vec{r}_A$ before it arrives at z_1 (see Figure VII-12d). Hence

$$G^+(\vec{r},\vec{r}_A,\omega) = G_d^+(\vec{r},\vec{r}_A,\omega) + G_s^+(\vec{r},\vec{r}_A,\omega) \quad \text{at } z=z_1. \qquad \text{(VII-57b)}$$

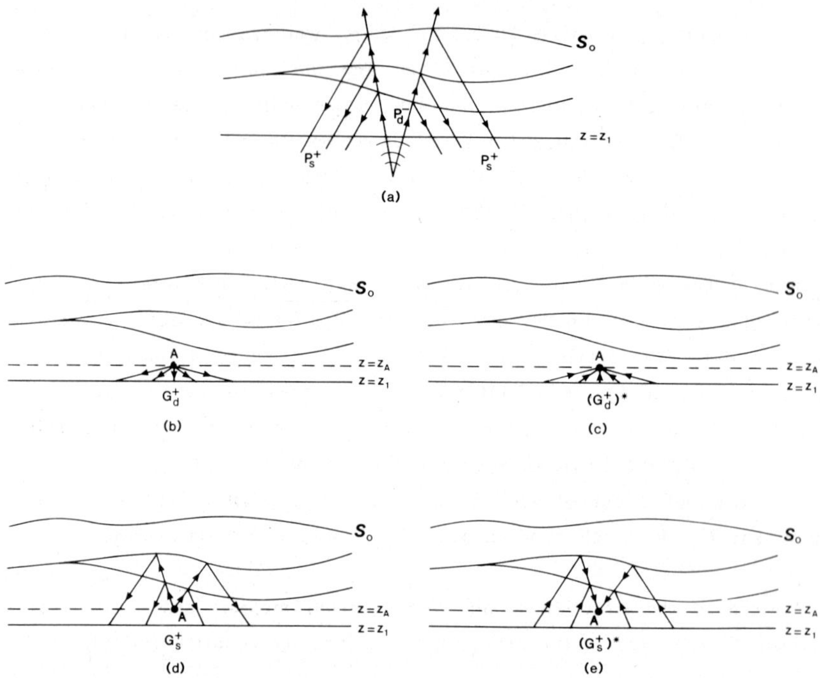

Figure VII-12:

a. *The wave field at z_1 consists of a direct upgoing wave field P_d^- and a scattered downgoing wave field P_s^+.*

b. *G_d^+ at z_1 represents a Green's wave field which propagates directly from $\vec{r}_A$ to z_1.*

c. *$(G_d^+)^*$ at z_1 represents a Green's wave field which propagates directly back from z_1 to $\vec{r}_A$.*

d. *G_s^+ at z_1 represents a Green's wave field which is scattered during propagation from $\vec{r}_A$ to z_1.*

e. *$(G_s^+)^*$ at z_1 represents a Green's wave field which is scattered during back-propagation from z_1 to $\vec{r}_A$.*

Consequently,

$$\left[G^+(\vec{r},\vec{r}_A,\omega)\right]^* = \left[G_d^+(\vec{r},\vec{r}_A,\omega)\right]^* + \left[G_s^+(\vec{r},\vec{r}_A,\omega)\right]^* \quad \text{at } z=z_1. \qquad \text{(VII-57c)}$$

Here $(G_d^+)^*$ propagates directly back from $z=z_1$ and converges to $\vec{r}_A$ from below[1)] (Figure VII-12c). On the other hand, $(G_s^+)^*$ propagates back from $z=z_1$, is scattered by the inhomogeneities above $\vec{r}_A$ and converges to $\vec{r}_A$ from above (Figure VII-12e). With the sub-division made in (VII-57c), we may rewrite (VII-56) as

$$\Delta P(\vec{r}_A,\omega) = \Delta P_1(\vec{r}_A,\omega) + \Delta P_2(\vec{r}_A,\omega), \tag{VII-58a}$$

where

$$\Delta P_1(\vec{r}_A,\omega) = -2\int\int_{-\infty}^{\infty} \left[\frac{1}{\rho(\vec{r})} \left(\frac{\partial G_d^+(\vec{r},\vec{r}_A,\omega)}{\partial z}\right)^* P_s^+(\vec{r},\omega)\right]_{z_1} dxdy \tag{VII-58b}$$

and

$$\Delta P_2(\vec{r}_A,\omega) = -2\int\int_{-\infty}^{\infty} \left[\frac{1}{\rho(\vec{r})} \left(\frac{\partial G_s^+(\vec{r},\vec{r}_A,\omega)}{\partial z}\right)^* P_s^+(\vec{r},\omega)\right]_{z_1} dxdy. \tag{VII-58c}$$

Equation (VII-58b) describes direct back-propagation of the total downgoing wave field P_s^+ at z_1 to $\vec{r}_A$ (see also Figure VII-12c). Hence, ΔP_1 represents the *total downgoing* wave field at $\vec{r}_A$:

$$\Delta P_1(\vec{r}_A,\omega) = P^+(\vec{r}_A,\omega). \tag{VII-59a}$$

Equation (VII-58c) describes back-propagation of the downgoing wave field P_s^+ at z_1, via the scattering medium where the propagation direction changes, to $\vec{r}_A$ (see also Figure VII-12e), so ΔP_2 represents an *upgoing* wave field at $\vec{r}_A$:

$$\Delta P_2(\vec{r}_A,\omega) = \Delta P^-(\vec{r}_A,\omega). \tag{VII-59b}$$

---

1) We will assume that $(G_d^+)^*$ undergoes no scattering in the region between the "reference surface" S_1 at $z=z_1$ and $\vec{r}_A$. This is justified if we let the depth z_1 of the reference surface approach to z_A.

Note that, according to (VII-58c), $\Delta P^-(\vec{r}_A,\omega)$ is proportional to the *product* of the *scattered* wave P_s^+ (Figure VII-12a) and the *scattered* back-propagating Green's wave field $(G_s^+)^*$ (Figure VII-12e). Hence, the magnitude of the error term $\Delta P^-(\vec{r}_A,\omega)$ is proportional to the *squared reflectivity* of the interfaces in the inhomogeneous medium. This result was also found in section VII.2.2 for vertically inhomogeneous media.

Consider again equation (VII-50a) for the *total* wave field at $\vec{r}_A$,

$$P(\vec{r}_A,\omega) = P_0(\vec{r}_A,\omega) + \Delta P(\vec{r}_A,\omega), \tag{VII-60a}$$

with $P_0(\vec{r}_A,\omega)$ and $\Delta P(\vec{r}_A,\omega)$ defined by (VII-50b) and (VII-50c), respectively. On the other hand, define $P(\vec{r}_A,\omega)$ as the sum of a downgoing and an upgoing term, according to

$$P(\vec{r}_A,\omega) \triangleq P^+(\vec{r}_A,\omega) + P^-(\vec{r}_A,\omega). \tag{VII-60b}$$

From the above analysis we thus obtain for the *upgoing* wave field at $\vec{r}_A$

$$P^-(\vec{r}_A,\omega) = P_0^-(\vec{r}_A,\omega) + \Delta P^-(\vec{r}_A,\omega), \tag{VII-60c}$$

with

$$P_0^-(\vec{r}_A,\omega) \triangleq P_0(\vec{r}_A,\omega). \tag{VII-60d}$$

Hence, assuming $\Delta P^-(\vec{r}_A,\omega)$ may be neglected, we obtain for the *upgoing* wave field at $\vec{r}_A$

$$P^-(\vec{r}_A,\omega) \approx P_0^-(\vec{r}_A,\omega), \tag{VII-60e}$$

or, according to (VII-50b),

$$P^-(\vec{r}_A,\omega) \approx \int_{S_0} \frac{1}{\rho(\vec{r})} \left[G^*(\vec{r},\vec{r}_A,\omega)\nabla P(\vec{r},\omega) - \nabla G^*(\vec{r},\vec{r}_A,\omega)P(\vec{r},\omega)\right].\vec{n}\, dS_0. \tag{VII-61}$$

This expression is used by many authors (Schneider, 1978; Berkhout, 1985; Clayton and Stolt, 1981; Castle, 1982; Carter and Frazer, 1984; Wiggins,

1984; Berryhill, 1984). It describes inverse wave field extrapolation (towards the sources) from acquisition surface S_0 to subsurface point $\vec{r}_A$ (Figure VII-8). Our analysis has shown that in equation (VII-61) evanescent waves are neglected and that $\Delta P^-(\vec{r}_A,\omega)$, as defined by (VII-58c) and (VII-59b), is neglected. The magnitude of the latter term is proportional to the squared reflectivity of the interfaces in the inhomogeneous medium. This means not only that internal *multiply reflected waves* are handled erroneously by equation (VII-61), but also that the *primary* wave contribution to $P^-(\vec{r}_A,\omega)$ is not fully correct. The negligence of $\Delta P^-(\vec{r}_A,\omega)$ is justified only when the contrasts in the medium are weak to moderate. In that case equation (VII-61) describes *"true amplitude"* inverse extrapolation of *primary* waves. This is illustrated by an example in section VII.3.6.
When the contrasts in the medium are significant, $\Delta P^-(\vec{r}_A,\omega)$ may not be neglected and should be estimated in an iterative way. This is discussed in chapter IX, section IX.3.

VII.3.4 Acoustic one-way Rayleigh integrals for inverse extrapolation

When the acquisition surface S_0 is a *plane* surface at $z=z_0$, we may rewrite equation (VII-50b) as

$$P_0^-(\vec{r}_A,\omega) = \iint_{-\infty}^{\infty} \frac{1}{\rho} \left[\left(\frac{\partial G}{\partial z}\right)^* P - G^* \frac{\partial P}{\partial z} \right]_{z_0} dxdy. \qquad \text{(VII-62)}$$

For the total wave field at $z=z_0$ we write

$$P(\vec{r},\omega) = P^+(\vec{r},\omega) + P^-(\vec{r},\omega) \qquad \text{at } z=z_0. \qquad \text{(VII-63a)}$$

For the Green's wave field we may choose above $z=z_0$ (outside V) any convenient reference medium. We choose $c(x,y,z<z_0)=c(x,y,z_0)$ and $\rho(x,y,z<z_0)=\rho(x,y,z_0)$. With this choice the Green's wave field at z_0 is purely upgoing (no scattering occurs in the upper half space above z_0), hence,

$$G(\vec{r},\vec{r}_A,\omega) = G^-(\vec{r},\vec{r}_A,\omega) \qquad \text{at } z=z_0. \qquad \text{(VII-63b)}$$

We may thus rewrite equation (VII-62) as

$$P_0^-(\vec{r}_A,\omega) = \iint_{-\infty}^{\infty} \frac{1}{\rho} \left[\left(\frac{\partial G^-}{\partial z}\right)^* (P^+ + P^-) - (G^-)^* \left(\frac{\partial P^+}{\partial z} + \frac{\partial P^-}{\partial z}\right)\right]_{z_0} dxdy. \qquad \text{(VII-64)}$$

In Appendix B, section B.3, we show that this expression may be simplified to

$$P_0^-(\vec{r}_A,\omega) = -2\iint_{-\infty}^{\infty} \left[\frac{1}{\rho(\vec{r})} \left(G^-(\vec{r},\vec{r}_A,\omega)\right)^* \frac{\partial P^-(\vec{r},\omega)}{\partial z}\right]_{z_0} dxdy, \qquad \text{(VII-65a)}$$

or, equivalently,

$$P_0^-(\vec{r}_A,\omega) = 2\iint_{-\infty}^{\infty} \left[\frac{1}{\rho(\vec{r})} \left(\frac{\partial G^-(\vec{r},\vec{r}_A,\omega)}{\partial z}\right)^* P^-(\vec{r},\omega)\right]_{z_0} dxdy. \qquad \text{(VII-65b)}$$

Note that only the wave fields $(G^-)^*$ and P^- which propagate in *opposite* directions through $z=z_0$ contribute to the result $P_0^-(\vec{r}_A,\omega)$, see also Figure VII-13. The underlying assumption is again that evanescent waves may be neglected at z_0.

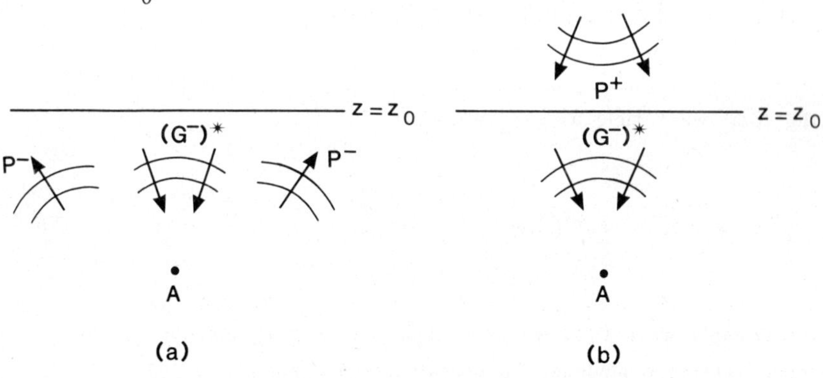

Figure VII-13: Choosing a reflection free upper half-space for G, the Kirchhoff-Helmholtz integral over z_0 (VII-64) consists of terms containing the products of $(G^-)^*$ and P^- at $z=z_0$ (Figure a) and terms containing the products of $(G^-)^*$ and P^+ at $z=z_0$ (Figure b). Only the terms with the ***opposite*** *propagating wave fields at $z=z_0$ (Figure a) contribute to the result $P_0^-(\vec{r}_A,\omega)$.*

Compare equations (VII-65a) and (VII-65b) with equations (V-44a) and (V-44b),

$$P(\vec{r}_A,\omega) = -2\iint_{-\infty}^{\infty}\left[\frac{1}{\rho(\vec{r})} G^-(\vec{r},\vec{r}_A,\omega) \frac{\partial P^+(\vec{r},\omega)}{\partial z}\right]_{z_0} dxdy \tag{VII-66a}$$

and

$$P(\vec{r}_A,\omega) = 2\iint_{-\infty}^{\infty}\left[\frac{1}{\rho(\vec{r})} \frac{\partial G^-(\vec{r},\vec{r}_A,\omega)}{\partial z} P^+(\vec{r},\omega)\right]_{z_0} dxdy. \tag{VII-66b}$$

These are the one-way versions of the acoustic Rayleigh I and Rayleigh II integrals, respectively, for *forward* extrapolation (the sources of P^+ are above z_0, see Figure V-4). Analogously, we will refer to equations (VII-65a) and (VII-65b) as the one-way versions of the acoustic Rayleigh I and Rayleigh II integrals, respectively, for *inverse* extrapolation (the sources of P^- are below z_A, see Figure VII-8). Equations (VII-66a) and (VII-66b) for forward extrapolation yield the *exact total* wave field $P(\vec{r}_A,\omega)$. On the other hand, equations (VII-65a) and (VII-65b) for inverse extrapolation yield $P_0^-(\vec{r}_A,\omega)$, which is an approximate version of the *upgoing* wave field

$$P^-(\vec{r}_A,\omega) = P_0^-(\vec{r}_A,\omega) + \Delta P^-(\vec{r}_A,\omega), \tag{VII-67}$$

see equation (VII-60c). The approximations involve the negligence of evanescent waves and the negligence of the error term $\Delta P^-(\vec{r}_A,\omega)$, which is proportional to (but not restricted to) multiply reflected waves. For the special situation of a homogeneous medium, we may substitute for $(G^-)^*$ the free space solution (V-26b), yielding

$$P^-(\vec{r}_A,\omega) = -\frac{1}{2\pi}\iint_{-\infty}^{\infty}\left[\frac{e^{+jk\Delta r}}{\Delta r} \frac{\partial P^-(\vec{r},\omega)}{\partial z}\right]_{z_0} dxdy, \tag{VII-68a}$$

or, equivalently,

$$P^-(\vec{r}_A,\omega) = \frac{1}{2\pi} \iint_{-\infty}^{\infty} \left[\frac{\partial}{\partial z} \left(\frac{e^{+jk\Delta r}}{\Delta r} \right) P^-(\vec{r},\omega) \right]_{z_0} dxdy. \tag{VII-68b}$$

Here the only approximation concerns the negligence of evanescent waves.

VII.3.5 Matrix formulation of acoustic inverse wave field extrapolation

We return to the inhomogeneous situation. Consider the one-way version of the acoustic Rayleigh II integral (VII-65b) for inverse extrapolation of upgoing waves, which we rewrite in a slightly more general notation according to

$$P^-(x,y,z_1;\omega) \approx$$

$$2\iint_{-\infty}^{\infty} \left[\left(\frac{\partial G^-(x',y',z'=z_0;x,y,z_1;\omega)}{\partial z'} \right)^* \frac{1}{\rho(x',y',z_0)} P^-(x',y',z_0;\omega) \right] dx'dy'. \tag{VII-69a}$$

Note that we replaced P_0^- by P^-, i.e., we neglected the error term ΔP^-, see equation (VII-67). A similar expression can be derived for acoustic inverse extrapolation of downgoing waves, according to

$$P^+(x,y,z_0;\omega) \approx$$

$$-2\iint_{-\infty}^{\infty} \left[\left(\frac{\partial G^+(x',y',z'=z_1;x,y,z_0;\omega)}{\partial z'} \right)^* \frac{1}{\rho(x',y',z_1)} P^+(x',y',z_1;\omega) \right] dx'dy'. \tag{VII-69b}$$

The Green's wave fields in (VII-69a) and (VII-69b) are defined in one and the same reference medium that is reflection free in the upper half-space ($z \le z_0$) as well as in the lower half-space ($z \ge z_1$). Hence, they satisfy the reciprocity relation (V-52),

$$G^-(x_0,y_0,z_0;x_1,y_1,z_1;\omega) = G^+(x_1,y_1,z_1;x_0,y_0,z_0;\omega). \tag{VII-70}$$

This implies that (VII-69a) and (VII-69b) may also be written as

$$P^-(x,y,z_1;\omega) \approx$$

$$2\iint_{-\infty}^{\infty} \left[\left(\frac{\partial G^+(x,y,z_1;x',y',z'=z_0;\omega)}{\partial z'} \right)^* \frac{1}{\rho(x',y',z_0)} \, P^-(x',y',z_0;\omega) \right] dx'dy' \quad \text{(VII-71a)}$$

and

$$P^+(x,y,z_0;\omega) \approx$$

$$-2\iint_{-\infty}^{\infty} \left[\left(\frac{\partial G^-(x,y,z_0;x',y',z'=z_1;\omega)}{\partial z'} \right)^* \frac{1}{\rho(x',y',z_1)} \, P^+(x',y',z_1;\omega) \right] dx'dy'. \quad \text{(VII-71b)}$$

Note the strong resemblance with the acoustic one-way Rayleigh II integrals (VII-43a) and (VII-43b), respectively, for forward extrapolation. In the matrix notation of Appendix A, equations (VII-71a) and (VII-71b) may be replaced by

$$\vec{P}^-(z_1) \approx \langle F^-(z_1,z_0)\rangle \vec{P}^-(z_0) \quad \text{(VII-72a)}$$

and

$$\vec{P}^+(z_0) \approx \langle F^+(z_0,z_1)\rangle \vec{P}^+(z_1), \quad \text{(VII-72b)}$$

respectively, where the approximated inverse wave field extrapolation matrices are defined by

$$\langle \mathbf{F}^-(z_1,z_0)\rangle = 2\left(\frac{\partial \mathbf{G}^+(z_1,z=z_0)}{\partial z} \right)^* \mathbf{M}^{-1}(z_0) \quad \text{(VII-73a)}$$

and

$$\langle \mathbf{F}^+(z_0,z_1)\rangle = -2\left(\frac{\partial \mathbf{G}^-(z_0,z=z_1)}{\partial z} \right)^* \mathbf{M}^{-1}(z_1), \quad \text{(VII-73b)}$$

respectively, with

$$\mathbf{G}^{-}(z_0,z_1) = [\mathbf{G}^{+}(z_1,z_0)]^{T}. \qquad \text{(VII-73c)}$$

For a discussion of the vectors and matrices in (VII-72) and (VII-73) we refer to section V.5.2.

Comparing equations (VII-73a) and (VII-73b) for the inverse wave field extrapolation matrices with equations (VII-45a) and (VII-45b) for the forward wave field extrapolation matrices yields

$$<\mathbf{F}^{-}(z_1,z_0)> = [\mathbf{W}^{+}(z_1,z_0)]^{*} \qquad \text{(VII-74a)}$$

and

$$<\mathbf{F}^{+}(z_0,z_1)> = [\mathbf{W}^{-}(z_0,z_1)]^{*}. \qquad \text{(VII-74b)}$$

VII.3.6 Examples of acoustic inverse wave field extrapolation

We demonstrate the validity of equation (VII-72a),

$$\vec{P}^{-}(z_1) \approx <\mathbf{F}^{-}(z_1,z_0)>\vec{P}^{-}(z_0), \qquad \text{(VII-75a)}$$

with the aid of two numerical examples. Consider the 2-D inhomogeneous acoustic medium, shown in Figure VII-14a. A plane wave source of finite extent is burried in the subsurface at a depth of z_1=600 m. The response at the reflection-free surface z_0=o is shown in Figure VII-14b. This response was computed with a finite difference modeling scheme (Kelly et al., 1976). It represents the acoustic pressure p as a function of the lateral coordinate x and time t. Because the upper half-space $z<z_0$ is homogeneous and acquisition surface z_0 is reflection-free, the recorded pressure represents an upgoing wave field, hence, $p=p^{-}(x,z_0;t)$. By applying a Fourier transform from time (t) to frequency (ω), the data is decomposed into monochromatic wave fields $P^{-}(x,z_0;\omega)$. According to Appendix A, section A.2, a discretized monochromatic wave field may be

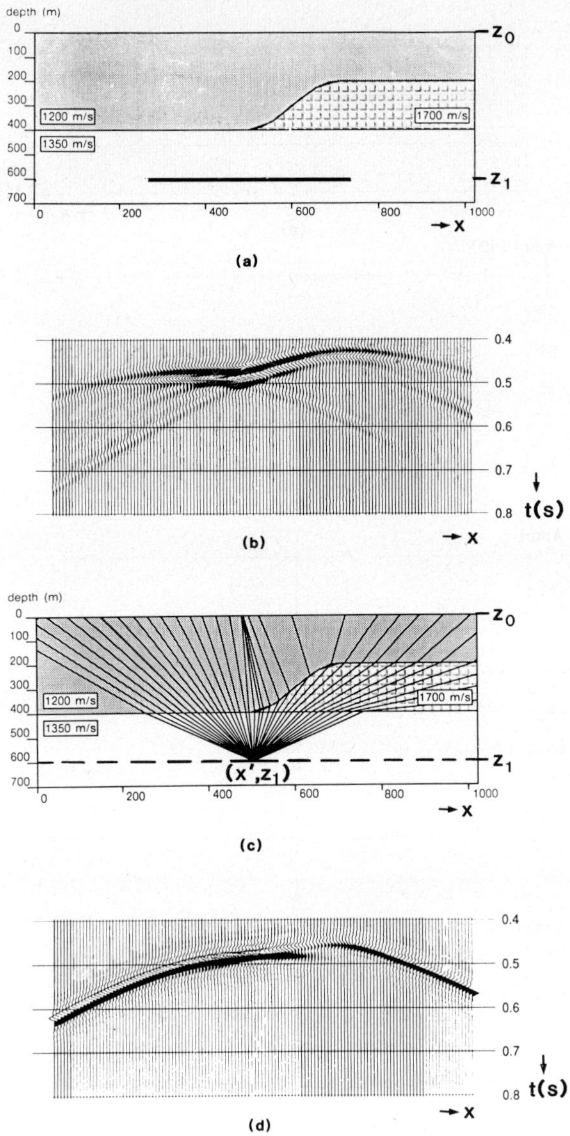

Figure VII-14:

a. 2-D inhomogeneous medium with a burried plane wave source at z_1*=600 m.*

b. Upgoing wave field, registered at z_0*.*

c. Fan of rays, used for Gaussian beam modeling of the Green's wave field.

d. Time-domain representation of a (band-limited) Green's wave field.

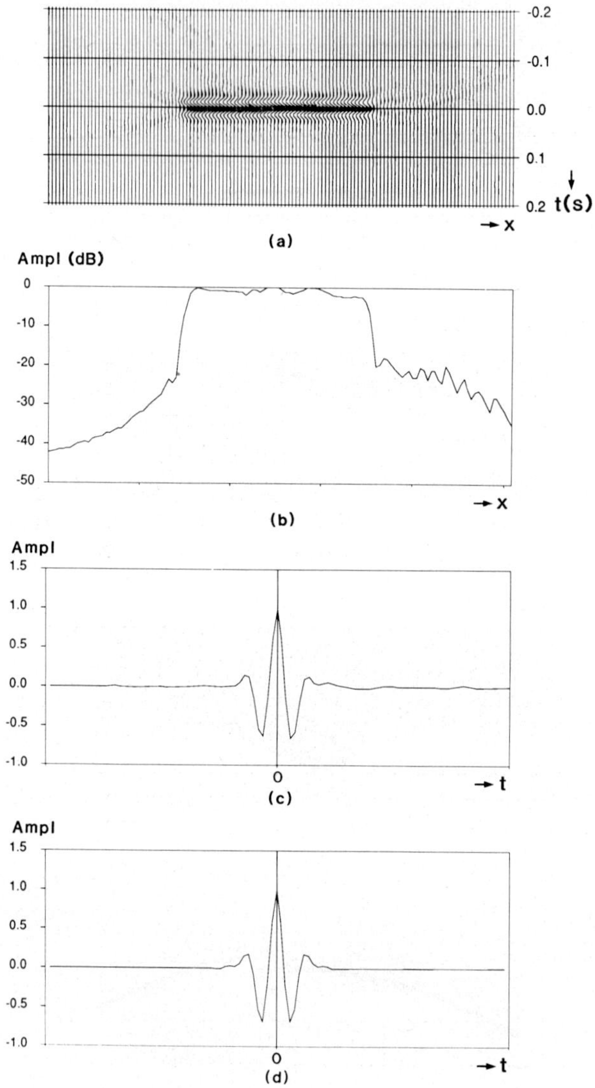

Figure VII-15:

a. Inverse extrapolated data at z_1*=600 m.*

b. Maximum amplitude per trace of Figure a (logarithmic scale).

c. Central trace of Figure a.

d. Wavelet, used for modeling the input data of Figure VII-14b.

represented by a vector $\vec{P}^-(z_0)$. Inverse extrapolation of this upgoing wave field from depth level z_0 to the source depth level z_1 is described by equation (VII-75a), where

$$\langle \mathbf{F}^-(z_1,z_0)\rangle = 2\left(\frac{\partial \mathbf{G}^+(z_1,z=z_0)}{\partial z}\right)^* \mathbf{M}^{-1}(z_0). \qquad \text{(VII-75b)}$$

The numerical modeling of the Green's matrix

$$\mathbf{G}^+(z_1,z_0) = [\mathbf{G}^-(z_0,z_1)]^T \qquad \text{(VII-75c)}$$

was discussed in chapter V (see also Figures VII-14c and VII-14d). By applying (VII-75a) for all frequencies within the seismic band, we obtain a range of monochromatic data vectors $\vec{P}^-(z_1)$. The result, after applying an inverse Fourier transform from ω to t, is shown in Figure VII-15a (Budejicky, 1988). It represents the space-time data $p^-(x,z_1;t)$. Note that the distorting propagation effects of the "overburden" (the medium between z_0 and z_1) have been removed (compare Figure VII-15a with VII-14b).
Figure VII-15b shows the maximum amplitude of each trace as a function of the lateral position x. Note that the amplitude is almost constant along the plane wave source. Also note that some smearing occurs at the edges of the plane wave source. This is due to the negligence of evanescent waves (see section VII.3.3).
Figure VII-15c shows the central trace of the inversely extrapolated data in Figure VII-15a. For comparison, in Figure VII-15d the wavelet is shown that was used for modeling the input data (Figure VII-14b). Apparently, the inverse extrapolation restored the wavelet almost perfectly.

For the following example, consider the 3-D inhomogeneous acoustic medium, shown in Figure VII-16a. A plane wave source of finite extent is burried in the subsurface at a depth of z_1=1000 m. Two cross-sections of the response at the reflection-free surface z_0=o are shown in Figures VII-16b and VII-16c, respectively. Following the same procedure as in the previous example, we obtain the inverse extrapolated data at $z=z_1$, see Figure VII-17 (Kinneging, 1989). Note that the finite plane wave source is correctly positioned and is perfectly aligned at t=o.

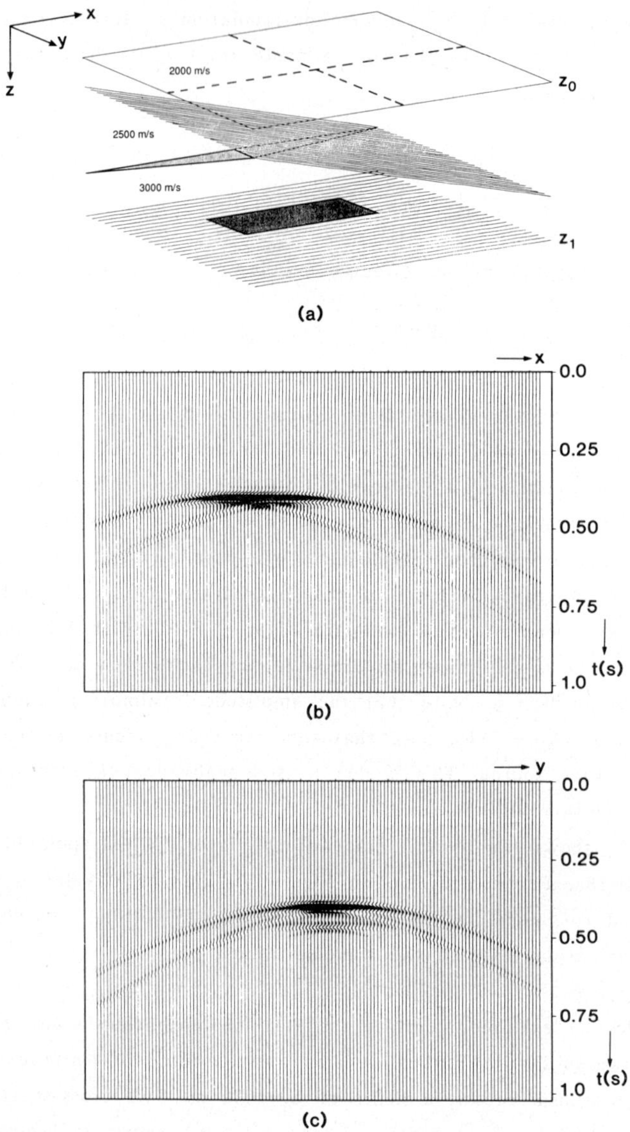

Figure VII-16:

a. 3-D inhomogeneous medium with a burried plane wave source at z_1*=1000 m.*

b. Cross-section for constant y of the upgoing wave field, registered at z_0*=o.*

c. Cross-section for constant x of the upgoing wave field, registered at z_0*=o.*

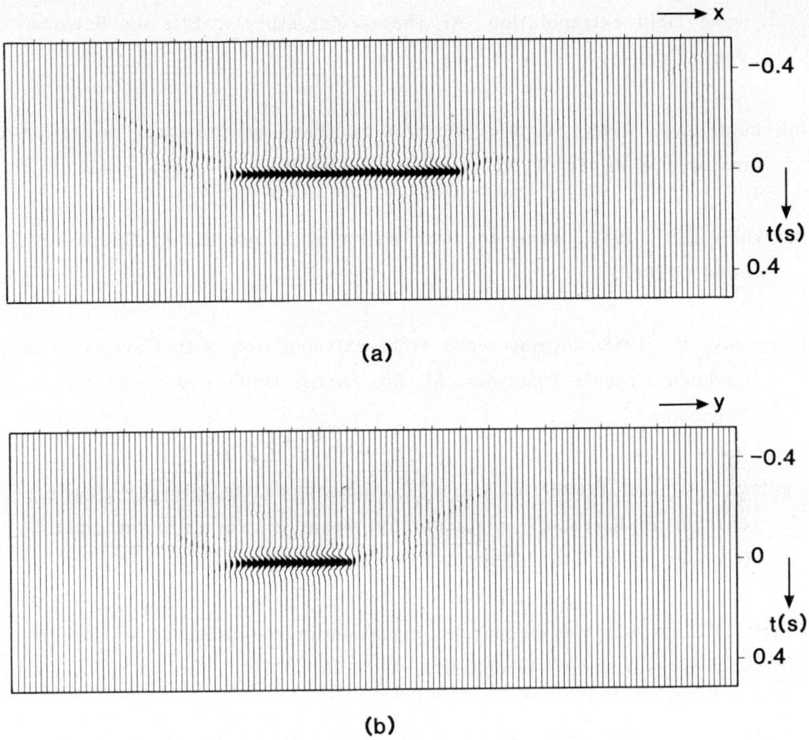

Figure VII-17: Inverse extrapolated data at z_I=1000 m.
a. Cross-section for constant y.
b. Cross-section for constant x.

VII.4 REFERENCES

Abramowitz, M., and Stegun, I.A., 1970, Handbook of mathematical functions: Dover Publications, Inc.

Berkhout, A.J., and Van Wulfften Palthe, D.W., 1979, Migration in terms of spatial deconvolution: Geophysical Prospecting, *27*, 261 - 291.

Berkhout, A.J., 1984, Seismic resolution: resolving power of acoustic echo techniques: Geophysical Press Ltd.

Berkhout, A.J., 1985, Seismic migration: imaging of acoustic energy by wave field extrapolation. A. Theoretical aspects: Elsevier Science Publ. Co., Inc.

Berkhout, A.J., 1986, Seismic inversion in terms of pre-stack migration and multiple elimination: Proceedings of the IEEE, *74*, 415 - 427.

Berryhill, J.R., 1984, Wave equation datuming before stack: Geophysics, *49*, 2064 - 2066.

Budejicky, V., 1988, Inverse wave field extrapolation with Gaussian beam modeled Green's functions: M. Sc. thesis, Delft University of Technology.

Carter, J.A., and Frazer, L.N., 1984, Accommodating lateral velocity changes in Kirchhoff migration by means of Fermat's principle: Geophysics, *49*, 46 - 53.

Castle, R.J., 1982, Wave-equation migration in the presence of lateral velocity variations: Geophysics, *47*, 1001 - 1011.

Clayton, R.W., and Stolt, R.H., 1981, A Born-WKBJ inversion method for acoustic reflection data: Geophysics, *46*, 1559 - 1567.

Cox, H.L.H., Ooms, F.P.J., Wapenaar, C.P.A., and Berkhout, A.J., 1988, Verification of macro subsurface models using a shot record approach: presented at the 58th Annual International SEG Meeting, Anaheim.

Faye, J.P., and Jeannot, J.P., 1986, Pre-stack migration velocities from focussing depth analysis: presented at the 56th Annual International SEG Meeting, Houston.

Kelly, K.R., Ward, R.W., Treitel, S., and Alford, R.M., 1976, Synthetic seismograms: a finite difference approach: Geophysics, *41*, 2 - 27.

Kinneging, N.A., 1989, Three-dimensional redatuming of seismic shot records: D.Sc. thesis, Delft Unviversity of Technology.

Peels, G.L., 1988, True amplitude wave field extrapolation with applications in seismic shot record redatuming: D.Sc. thesis, Delft University of Technology.

Schneider, W.A., 1978, Integral formulation for migration in two and three dimensions: Geophysics, *43*, 49 - 76.

Van der Made, P.M., Van Riel, P., and Berkhout, A.J., 1984, Velocity and subsurface geometry inversion by a parameter estimation in complex inhomogeneous media: presented at the 54th Annual International SEG Meeting, Atlanta.

Wapenaar, C.P.A., Peels, G.L., Budejicky, V., and Berkhout, A.J., 1989, Inverse extrapolation of primary seismic waves: Geophysics, *54*, 853 - 863.

Wiggins, J.W., 1984, Kirchhoff integral extrapolation and migration of non-planar data: Geophysics, *49*, 1239 - 1248.

VIII

ELASTIC INVERSE WAVE FIELD EXTRAPOLATION IN LOW CONTRAST MEDIA

VIII.1 INTRODUCTION

In chapter VI we discussed elastic *forward* wave field extrapolation operators which *simulate* the downward and upward propagation effects of P- and S-waves. These operators were derived from the elastic Kirchhoff-Helmholtz integral (VI-24a) which contains *forward* propagating Green's wave fields. In this chapter we discuss elastic *inverse* wave field extrapolation operators which eliminate the downward and upward propagation effects of P- and S-waves. As in the acoustic case, straightforward inversion of the elastic forward extrapolation operators is unstable. In this chapter we first discuss the elastic version of the matched filter approach to inverse extrapolation through a homogeneous isotropic medium. It is shown that the maximum obtainable spatial resolution from inverse S-wave extrapolation is higher than from inverse P-wave extrapolation (assuming the same frequency).
Next we consider an arbitrarily inhomogeneous anisotropic medium. Our starting point is the elastic Kirchhoff-Helmholtz integral (VI-24b) which contains a backward propagating Green's wave field. We arrive again at inverse extrapolation operators which are the complex conjugate of the forward extrapolation operators and which have restrictions with respect to the maximum obtainable spatial resolution. We analyse the limitations for a medium that contains high contrasts. In chapter X we will modify the elastic inverse extrapolation operators for the latter situation. The operators derived in chapters VIII and X play an essential role in chapter XII, where we discuss an elastic processing scheme for multi-component seismic data.

VIII.2 ELASTIC INVERSE WAVE FIELD EXTRAPOLATION IN HOMOGENEOUS ISOTROPIC MEDIA

For a homogeneous isotropic elastic medium, forward extrapolation of downgoing P-waves may be described by a *spatial convolution* along the x-

and y-axis, according to

$$\Phi^+(x,y,z_1;\omega) = \iint_{-\infty}^{\infty} W^+_{\phi,\phi}(x-x',y-y';z_1,z_0;\omega)\Phi^+(x',y',z_0;\omega)dx'dy', \qquad \text{(VIII-1a)}$$

or, symbolically,

$$\Phi^+(x,y,z_1;\omega) = W^+_{\phi,\phi}(x,y;z_1,z_0;\omega) * \Phi^+(x,y,z_0;\omega), \qquad \text{(VIII-1b)}$$

where

$$W^+_{\phi,\phi}(x,y;z_1,z_0;\omega) = \frac{1}{2\pi}\frac{\partial}{\partial z_0}\left(\frac{e^{-jk_p\Delta r}}{\Delta r}\right), \qquad \text{(VIII-1c)}$$

with

$$\Delta r = \sqrt{x^2 + y^2 + (z_1-z_0)^2} \qquad \text{(VIII-1d)}$$

and

$$k_p = \omega/c_p, \qquad \text{(VIII-1e)}$$

where c_p is the P-wave propagation velocity, see equations (VI-77) to (VI-81). Similarly, the expression for forward extrapolation of downgoing S-waves reads symbolically

$$\Psi^+_\beta(x,y,z_1;\omega) = W^+_{\psi_\beta,\psi_\alpha}(x,y;z_1,z_0;\omega) * \Psi^+_\alpha(x,y,z_0;\omega), \qquad \text{(VIII-2a)}$$

where

$$W^+_{\psi_\beta,\psi_\alpha}(x,y;z_1,z_0;\omega) = \delta_{\alpha\beta}\frac{1}{2\pi}\frac{\partial}{\partial z_0}\left(\frac{e^{-jk_s\Delta r}}{\Delta r}\right), \qquad \text{(VIII-2b)}$$

with

$$k_s = \omega/c_s, \qquad \text{(VIII-2c)}$$

where c_s is the S-wave propagation velocity.
The expressions for forward extrapolation of upgoing P- and S-waves read, respectively,

$$\Phi^-(x,y,z_0;\omega) = W^-_{\phi,\phi}(x,y;z_0,z_1;\omega) * \Phi^-(x,y,z_1;\omega) \tag{VIII-3a}$$

and

$$\Psi^-_\beta(x,y,z_0;\omega) = W^-_{\psi_\beta,\psi_\alpha}(x,y;z_0,z_1;\omega) * \Psi^-_\alpha(x,y,z_1;\omega), \tag{VIII-3b}$$

where

$$W^-_{\phi,\phi}(x,y;z_0,z_1;\omega) = W^+_{\phi,\phi}(x,y;z_1,z_0;\omega) \tag{VIII-3c}$$

and

$$W^-_{\psi_\beta,\psi_\alpha}(x,y;z_0,z_1;\omega) = W^+_{\psi_\beta,\psi_\alpha}(x,y;z_1,z_0;\omega). \tag{VIII-3d}$$

In analogy with section VII.2.1 we describe inverse extrapolation of downgoing and upgoing P- and S-waves as spatial *deconvolution* processes along the x- and y-axis, in the symbolic notation

$$\Phi^+(x,y,z_0;\omega) = F^+_{\phi,\phi}(x,y;z_0,z_1;\omega) * \Phi^+(x,y,z_1;\omega), \tag{VIII-4a}$$

$$\Psi^+_\gamma(x,y,z_0;\omega) = F^+_{\psi_\gamma,\psi_\beta}(x,y;z_0,z_1;\omega) * \Psi^+_\beta(x,y,z_1;\omega), \tag{VIII-4b}$$

$$\Phi^-(x,y,z_1;\omega) = F^-_{\phi,\phi}(x,y;z_1,z_0;\omega) * \Phi^-(x,y,z_0;\omega), \tag{VIII-4c}$$

and

$$\Psi^-_\gamma(x,y,z_1;\omega) = F^-_{\psi_\gamma,\psi_\beta}(x,y;z_1,z_0;\omega) * \Psi^-_\beta(x,y,z_0;\omega), \tag{VIII-4d}$$

where the deconvolution operators F are defined implicitly by

$$F^{+}_{\phi,\phi}(x,y;z_0,z_1;\omega) * W^{+}_{\phi,\phi}(x,y;z_1,z_0;\omega) \hat{=} \delta(x)\delta(y), \tag{VIII-5a}$$

$$F^{+}_{\psi_\gamma,\psi_\beta}(x,y;z_0,z_1;\omega) * W^{+}_{\psi_\beta,\psi_\alpha}(x,y;z_1,z_0;\omega) \hat{=} \delta_{\alpha\gamma}\delta(x)\delta(y), \tag{VIII-5b}$$

$$F^{-}_{\phi,\phi}(x,y;z_1,z_0;\omega) * W^{-}_{\phi,\phi}(x,y;z_0,z_1;\omega) \hat{=} \delta(x)\delta(y) \tag{VIII-5c}$$

and

$$F^{-}_{\psi_\gamma,\psi_\beta}(x,y;z_1,z_0;\omega) * W^{-}_{\psi_\beta,\psi_\alpha}(x,y;z_0,z_1;\omega) \hat{=} \delta_{\alpha\gamma}\delta(x)\delta(y). \tag{VIII-5d}$$

Note that, in analogy with (VIII-3c) and (VIII-3d),

$$F^{-}_{\phi,\phi}(x,y;z_1,z_0;\omega) = F^{+}_{\phi,\phi}(x,y;z_0,z_1;\omega) \tag{VIII-5e}$$

and

$$F^{-}_{\psi_\gamma,\psi_\beta}(x,y;z_1,z_0;\omega) = F^{+}_{\psi_\gamma,\psi_\beta}(x,y;z_0,z_1;\omega). \tag{VIII-5f}$$

Expressions (VIII-1) to (VIII-5) for the elastic situation show a high degree of similarity with expressions (VII-2) to (VII-5) for the acoustic situation. Therefore we may follow the same approach as in section VII.2.1 to find explicit expressions for the deconvolution operators F. In analogy with the acoustic *matched inverse* operators (VII-19a) and (VII-19b) we define the elastic matched inverse operators by

$$\langle F^{+}_{\phi,\phi}(x,y;z_0,z_1;\omega)\rangle \hat{=} \left[W^{+}_{\phi,\phi}(x,y;z_1,z_0;\omega)\right]^{*}, \tag{VIII-6a}$$

$$\langle F^{+}_{\psi_\gamma,\psi_\beta}(x,y;z_0,z_1;\omega)\rangle \hat{=} \delta_{\alpha\gamma}\left[W^{+}_{\psi_\beta,\psi_\alpha}(x,y;z_1,z_0;\omega)\right]^{*}, \tag{VIII-6b}$$

$$\langle F^{-}_{\phi,\phi}(x,y;z_1,z_0;\omega)\rangle \hat{=} \left[W^{-}_{\phi,\phi}(x,y;z_0,z_1;\omega)\right]^{*} \tag{VIII-6c}$$

and

$$<F^-_{\psi_\gamma,\psi_\beta}(x,y;z_1,z_0;\omega)> \hat{=} \delta_{\alpha\gamma}\left[W^-_{\psi_\beta,\psi_\alpha}(x,y;z_0,z_1;\omega)\right]^* . \tag{VIII-6d}$$

These are *stable spatially band-limited* approximations of the exact inverse operators as defined in (VIII-5). We speak of spatially band-limited approximations because in the wavenumber-frequency domain the inverse operators are exact for the propagating wavenumber area only. For the inverse P-wave extrapolation operators (VIII-6a) and (VIII-6c) the propagating wavenumber area is defined by

$$k_x^2 + k_y^2 \leq k_p^2 = \omega^2/c_p^2; \tag{VIII-7a}$$

for the inverse S-wave extrapolation operators (VIII-6b) and (VIII-6d) the propagating wavenumber area is defined by

$$k_x^2 + k_y^2 \leq k_s^2 = \omega^2/c_s^2. \tag{VIII-7b}$$

Since $c_s<c_p$, the spatial band-limitation for the inverse S-wave operators is less restrictive than for the inverse P-wave operators. In section VII.2.1 we have seen that the spatial band-limitation imposes a restriction to the maximum obtainable spatial resolution. The width of the main lobe of the acoustic inversion result in Figure VII-6 equals approximately $6\lambda/5$, where $\lambda=2\pi c/\omega$ represents the wavelength for acoustic waves.
Accordingly, the spatial resolution of elastic inverse P-wave extrapolation is determined by $6\lambda_p/5$, whereas the spatial resolution of elastic inverse S-wave extrapolation is determined by $6\lambda_s/5$, where

$$\lambda_p = 2\pi c_p/\omega \tag{VIII-8a}$$

and

$$\lambda_s = 2\pi c_s/\omega. \tag{VIII-8b}$$

Note that

$$\lambda_s < \lambda_p, \tag{VIII-8c}$$

which means that, for a given frequency ω, inverse S-wave extrapolation yields a better spatial resolution than inverse P-wave extrapolation. This conclusion should be interpreted with care. In practical seismic data the *temporal* bandwidth of the S-wave registrations is generally much smaller than the temporal bandwidth of the P-wave registrations (Garotta, 1987). Therefore in practice the spatial resolution obtained from S-data is not significantly better than that obtained from P-data.

VIII.3 ELASTIC INVERSE WAVE FIELD EXTRAPOLATION IN ARBITRARILY INHOMOGENEOUS ANISOTROPIC MEDIA

VIII.3.1 Introduction

For an arbitrarily inhomogeneous anisotropic elastic medium, forward extrapolation of P- and S-waves may be described by the elastic one-way Rayleigh II integrals, in the matrix notation represented by

$$\vec{\underset{\sim}{D}}^+(z_1) = \underset{\sim}{W}^+(z_1,z_0)\vec{\underset{\sim}{D}}^+(z_0) \tag{VIII-9}$$

and

$$\vec{\underset{\sim}{D}}^-(z_0) = \underset{\sim}{W}^-(z_0,z_1)\vec{\underset{\sim}{D}}^-(z_1), \tag{VIII-10}$$

where the *multi-component* data vectors are defined as

$$\vec{\underset{\sim}{D}}^{\pm}(z_0) = \begin{bmatrix} \vec{\Phi}^{\pm}(z_0) \\ \vec{\Psi}_x^{\pm}(z_0) \\ \vec{\Psi}_y^{\pm}(z_0) \end{bmatrix} \quad \text{and} \quad \vec{\underset{\sim}{D}}^{\pm}(z_1) = \begin{bmatrix} \vec{\Phi}^{\pm}(z_1) \\ \vec{\Psi}_x^{\pm}(z_1) \\ \vec{\Psi}_y^{\pm}(z_1) \end{bmatrix} \tag{VIII-11}$$

and where the *multi-component* extrapolation operators are defined as

$$\underset{\sim}{W}^{+}(z_1,z_o) = \begin{bmatrix} W^{+}_{\phi,\phi}(z_1,z_o) & W^{+}_{\phi,\psi_x}(z_1,z_o) & W^{+}_{\phi,\psi_y}(z_1,z_o) \\ W^{+}_{\psi_x,\phi}(z_1,z_o) & W^{+}_{\psi_x,\psi_x}(z_1,z_o) & W^{+}_{\psi_x,\psi_y}(z_1,z_o) \\ W^{+}_{\psi_y,\phi}(z_1,z_o) & W^{+}_{\psi_y,\psi_x}(z_1,z_o) & W^{+}_{\psi_y,\psi_y}(z_1,z_o) \end{bmatrix} \quad \text{(VIII-12a)}$$

and

$$\underset{\sim}{W}^{-}(z_o,z_1) = \begin{bmatrix} W^{-}_{\phi,\phi}(z_o,z_1) & W^{-}_{\phi,\psi_x}(z_o,z_1) & W^{-}_{\phi,\psi_y}(z_o,z_1) \\ W^{-}_{\psi_x,\phi}(z_o,z_1) & W^{-}_{\psi_x,\psi_x}(z_o,z_1) & W^{-}_{\psi_x,\psi_y}(z_o,z_1) \\ W^{-}_{\psi_y,\phi}(z_o,z_1) & W^{-}_{\psi_y,\psi_x}(z_o,z_1) & W^{-}_{\psi_y,\psi_y}(z_o,z_1) \end{bmatrix}, \quad \text{(VIII-12b)}$$

see section VI.5.2. In this matrix notation, the inverse problem can be formulated as follows

$$\vec{\underset{\sim}{D}}^{+}(z_o) = \underset{\sim}{F}^{+}(z_o,z_1)\vec{\underset{\sim}{D}}^{+}(z_1) \quad \text{(VIII-13)}$$

and

$$\vec{\underset{\sim}{D}}^{-}(z_1) = \underset{\sim}{F}^{-}(z_1,z_o)\vec{\underset{\sim}{D}}^{-}(z_o), \quad \text{(VIII-14)}$$

where the inverse extrapolation operators are defined as

$$\underset{\sim}{F}^{+}(z_o,z_1) \triangleq [\underset{\sim}{W}^{+}(z_1,z_o)]^{-1} \quad \text{(VIII-15a)}$$

and

$$\underset{\sim}{F}^{-}(z_1,z_0) \hat{=} [\underset{\sim}{W}^{-}(z_0,z_1)]^{-1}. \qquad \text{(VIII-15b)}$$

Inversion according to equation (VIII-15) is unstable. Therefore, rather than trying to invert the forward problem, in this section we discuss the inverse extrapolation problem step by step, using the elastic Kirchhoff-Helmholtz integral as the starting point (Wapenaar and Haimé, 1989). After a number of necessary approximations we finally arrive at stable, spatially band-limited inverse extrapolation operators which read in the matrix notation

$$<\underset{\sim}{F}^{+}(z_0,z_1)> = [\underset{\sim}{W}^{-}(z_0,z_1)]^{*} \qquad \text{(VIII-16a)}$$

and

$$<\underset{\sim}{F}^{-}(z_1,z_0)> = [\underset{\sim}{W}^{+}(z_1,z_0)]^{*}. \qquad \text{(VIII-16b)}$$

VIII.3.2 Elastic Kirchhoff-Helmholtz integrals for inverse extrapolation

In section VI.3.2 we have derived the following two equivalent versions of the elastic Kirchhoff-Helmholtz integral,

$$V_m(\vec{r}_A,\omega) = -\oint_S [\boldsymbol{\Theta}_m(\vec{r},\vec{r}_A,\omega)\vec{V}(\vec{r},\omega) - \tau(\vec{r},\omega)\vec{G}_m(\vec{r},\vec{r}_A,\omega)].\vec{n}\,dS \qquad \text{(VIII-17a)}$$

and

$$V_m(\vec{r}_A,\omega) = -\oint_S [\boldsymbol{\Theta}^{*}_m(\vec{r},\vec{r}_A,\omega)\vec{V}(\vec{r},\omega) + \tau(\vec{r},\omega)\vec{G}^{*}_m(\vec{r},\vec{r}_A,\omega)].\vec{n}\,dS. \qquad \text{(VIII-17b)}$$

These expressions describe the elastic wave field at $\vec{r}_A$ in a source-free volume V in terms of the elastic wave field on surface S, enclosing V (Figure VI-1). In (VIII-17a), $\vec{G}_m$ and $\boldsymbol{\Theta}_m$ represent the velocity and stress of a *forward* propagating Green's wave field; in (VIII-17b), $\vec{G}^{*}_m$ and $\boldsymbol{\Theta}^{*}_m$ represent the velocity and stress of a *backward* propagating Green's wave field.

In chapter VI we used the Kirchoff-Helmholtz integral (VIII-17a) with the forward propagating Green's wave field as a starting point for deriving forward wave field extrapolation operators. Here we use the Kirchhoff-Helmholtz integral (VIII-17b) with the backward propagating Green's wave field as a starting point for deriving inverse wave field extrapolation operators. In analogy with section VI.3.4 we modify equation (VIII-17b) according to

$$\Omega(\vec{r}_A,\omega) = -\oint_S [\boldsymbol{\Theta}^*_\Omega(\vec{r},\vec{r}_A,\omega)\vec{V}(\vec{r},\omega) + \tau(\vec{r},\omega)\vec{G}^*_\Omega(\vec{r},\vec{r}_A,\omega)]\,.\vec{n}\,dS, \qquad \text{(VIII-18)}$$

see Figure VIII-1.

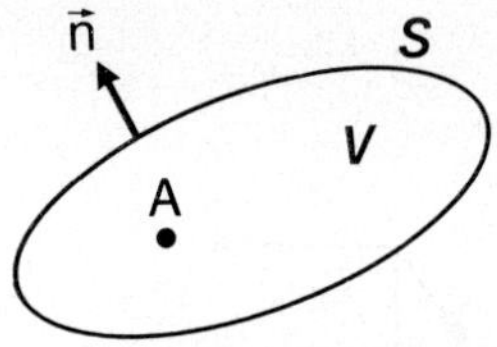

$$\Omega(\vec{r}_A,\omega) = -\oint_S [\boldsymbol{\Theta}^*_\Omega\vec{V}+\tau\vec{G}^*_\Omega]\,.\vec{n}\,dS$$

Figure VIII-1: Elastic Kirchhoff-Helmholtz integral. The Green's wave field $(\vec{G}_\Omega,\boldsymbol{\Theta}_\Omega)$ *may be excited either by an impulsive force, an impulsive P-wave source or an impulsive S-wave source at* $\vec{r}_A$. *Accordingly,* $\Omega(\vec{r}_A,\omega)$ *may represent either the velocity, the P-wave potential or the S-wave potential at* $\vec{r}_A$.

1. If we choose for the Green's wave field an impulsive force in the m-direction at $\vec{r}_A$(i.e., $\vec{G}_\Omega=\vec{G}_m$; $\boldsymbol{\Theta}_\Omega=\boldsymbol{\Theta}_m$), then $\Omega(\vec{r}_A,\omega)$ represents the m-component of the velocity $\vec{V}$ at $\vec{r}_A$.

2. If we choose for the Green's wave field an impulsive P-wave source at $\vec{r}_A$ (i.e., $\vec{G}_\Omega=\vec{G}_\phi$;$\boldsymbol{\Theta}_\Omega=\boldsymbol{\Theta}_\phi$), then $\Omega(\vec{r}_A,\omega)$ represents the scaled P-wave potential $j\omega\Phi$ at $\vec{r}_A$.

3. If we choose for the Green's wave field an impulsive S_h-wave source at $\vec{r}_A$ (i.e., $\vec{G}_\Omega=\vec{G}_{\psi_h}$; $\mathbf{\Theta}_\Omega=\mathbf{\Theta}_{\psi_h}$), then $\Omega(\vec{r}_A,\omega)$ represents the h-component of the scaled S-wave potential $j\omega\vec{\Psi}$ at $\vec{r}_A$.

For the half-space geometry of Figure VI-4, the backward propagating Green's wave field does not satisfy Sommerfeld's radiation condition on S_1. Therefore, consider the modified configuration shown in Figure VIII-2. We assume that an elastic wave field, radiated by (secondary) sources in the subsurface, has been measured on "acquisition surface" S_0. Our aim is to find an expression which describes the elastic wave field at $\vec{r}_A$ in the subsurface in terms of the elastic wave field on S_0. We constructed a surface S enclosing a volume V such that $\vec{r}_A$ lies in V and such that V is source-free. Closed surface S consists of "acquisition surface" S_0, a plane horizontal "reference surface" S_1 at $z=z_1$ (between $\vec{r}_A$ and the sources of the elastic wave field) and a cylindrical surface S_2 with a vertical axis through $\vec{r}_A$ and radius r.

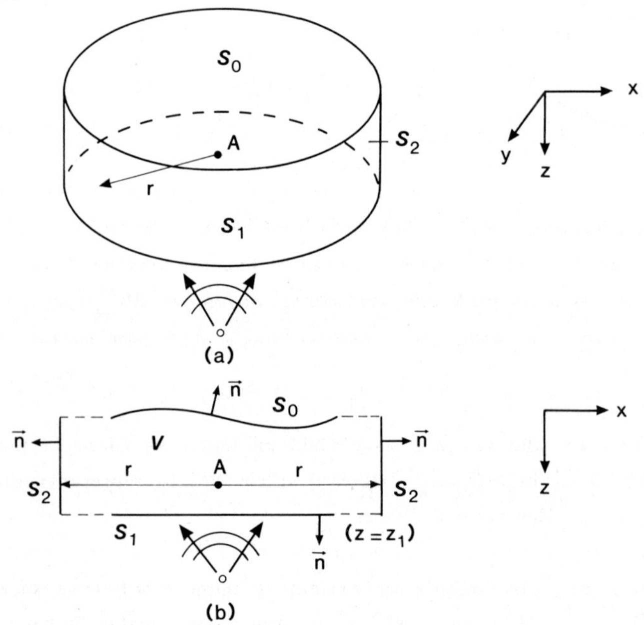

Figure VIII-2: Modified geometry for the Kirchhoff-Helmholtz integral (VIII-18) with the backward propagating Green's wave field. Under certain conditions (discussed in the text) the contribution of this integral over S_1 and S_2 can be neglected.

a. Perspective view. b. Cross section for y=0.

For this configuration we analyse Kirchhoff-Helmholtz integral (VIII-18), with the backward propagating Green's wave field. The contribution of this integral over S_2 vanishes when r goes to infinity (the cylindrical surface is proportional to r, the integrand is proportional to $1/r^2$). So for the geometry of Figure VIII-2, Kirchhoff-Helmholtz integral (VIII-18) may be replaced by

$$\Omega(\vec{r}_A,\omega) = \Omega_0(\vec{r}_A,\omega) + \Delta\Omega(\vec{r}_A,\omega), \qquad \text{(VIII-19a)}$$

where

$$\Omega_0(\vec{r}_A,\omega) \triangleq -\int_{S_0} [\boldsymbol{\Theta}^*_\Omega\vec{V} + \boldsymbol{\tau}\vec{G}^*_\Omega].\vec{n}\,dS_0 \qquad \text{(VIII-19b)}$$

and

$$\Delta\Omega(\vec{r}_A,\omega) \triangleq -\int_{S_1} [\boldsymbol{\Theta}^*_\Omega\vec{V} + \boldsymbol{\tau}\vec{G}^*_\Omega].\vec{n}\,dS_1. \qquad \text{(VIII-19c)}$$

When $\Delta\Omega(\vec{r}_A,\omega)$, as defined in (VIII-19c), may be neglected, then equation (VIII-19b) describes *inverse* wave field extrapolation (towards the sources) from acquisition surface S_0 to subsurface point $\vec{r}_A$. In the next section we investigate under which conditions $\Delta\Omega(\vec{r}_A,\omega)$ may be neglected.

VIII.3.3 Error analysis

At $z=z_1$ (the depth of "reference surface" S_1), the elastic wave field consists of upgoing waves (including higher order terms), related to the sources below z_1, and downgoing waves (including higher order terms), caused by scattering above z_1, hence

$$\vec{V}(\vec{r},\omega) = \vec{V}^+(\vec{r},\omega) + \vec{V}^-(\vec{r},\omega) \qquad \text{at } z=z_1 \qquad \text{(VIII-20a)}$$

and

$$\vec{\tau}_z(\vec{r},\omega) = \vec{\tau}_z^{\,+}(\vec{r},\omega) + \vec{\tau}_z^{\,-}(\vec{r},\omega) \qquad \text{at } z=z_1 \tag{VIII-20b}$$

(the traction vector $\vec{\tau}_z$ is the third column of stress tensor $\boldsymbol{\tau}$). For the Green's wave field we may choose below $z=z_1$, (outside V) any convenient reference medium. We choose $c_{ijk\ell}(x,y,z>z_1)=c_{ijk\ell}(x,y,z_1)$ and $\rho(x,y,z>z_1)=\rho(x,y,z_1)$. With this choice the Green's wave field at z_1 is purely downgoing (the Green's source is situated at $\vec{r}_A$ above z_1; the half-space below z_1 is reflection free), hence

$$\vec{G}_\Omega(\vec{r},\vec{r}_A,\omega) = \vec{G}_\Omega^{\,+}(\vec{r},\vec{r}_A,\omega) \qquad \text{at } z=z_1 \tag{VIII-21a}$$

and

$$\vec{\Theta}_{z,\Omega}(\vec{r},\vec{r}_A,\omega) = \vec{\Theta}_{z,\Omega}^{\,+}(\vec{r},\vec{r}_A,\omega) \qquad \text{at } z=z_1 \tag{VIII-21b}$$

(the Green's traction vector $\vec{\Theta}_{z,\Omega}$ is the third column of the Green's stress tensor $\boldsymbol{\Theta}_\Omega$). We may thus rewrite equation (VIII-19c) as

$$\Delta\Omega(\vec{r}_A,\omega) = -\int\!\!\int_{-\infty}^{\infty} \left[(\vec{\Theta}_{z,\Omega}^{\,+})^* \cdot (\vec{V}^+ + \vec{V}^-) + (\vec{G}_\Omega^{\,+})^* \cdot (\vec{\tau}_z^{\,+} + \vec{\tau}_z^{\,-})\right]_{z_1} dxdy. \tag{VIII-22}$$

In Appendix C, section C.3, we analyse the interactions between one-way elastic wave fields and one-way backward propagating Green's wave fields, assuming for simplicity that the medium is locally homogeneous and isotropic at z_1. The starting point of this analysis is equation (C-22), which is identical to equation (VIII-22) if we choose $\vec{G}_\Omega^{\,-}=\vec{\Theta}_{z,\Omega}^{\,-}=\vec{o}$ and if we replace Ω_o by $-\Delta\Omega$ and z_o by z_1. Hence, if we make the same alterations in the final result, equation (C-28b), we obtain

$$\Delta\Omega(\vec{r}_A,\omega) \approx \frac{2}{j\omega} \int\!\!\int_{-\infty}^{\infty} \frac{1}{\rho(z_1)} \left[\left(\frac{\partial \Gamma_{\phi,\Omega}^+}{\partial z}\right)^* \Phi^+ + \left(\frac{\partial \vec{\Gamma}_{\psi,\Omega}^{\,+}}{\partial z}\right)^* \cdot \vec{\Psi}^+ \right]_{z_1} dxdy, \tag{VIII-23}$$

where Φ^+ and $\vec{\Psi}^+$ are the P- and S-wave potentials for the velocity $\vec{V}^+$ at z_1 and where $\Gamma^+_{\phi,\Omega}$ and $\vec{\Gamma}^+_{\psi,\Omega}$ are the P- and S-wave potentials for the Green's velocity $\vec{G}^+_\Omega$ at z_1, see equation (C-3) and Figure C-1. Apparently, only the wave fields which propagate in *opposite* directions through $z=z_1$ contribute to the error term $\Delta\Omega(\vec{r}_A,\omega)$ (see Figure VII-9b for the acoustic equivalence). The underlying assumption is that *evanescent waves* are negligible at $z=z_1$. This assumption is satisfied when the source of the elastic wave field and the source (at $\vec{r}_A$) of the Green's wave field are not both in the direct vicinity of the "reference surface" S_1, or in other words, when $\vec{r}_A$ is not too close to the source of the elastic wave field. In the following we replace "≈" by "=" whenever the only approximation concerns the negligence of evanescent waves.

Let us first discuss the consequence of this important result for the situation of a homogeneous medium. Then no scattering occurs, hence

$$\Phi^+(\vec{r},\omega) = 0 \qquad \text{at } z=z_1 \tag{VIII-24a}$$

and

$$\vec{\Psi}^+(\vec{r},\omega) = \vec{0} \qquad \text{at } z=z_1 \tag{VIII-24b}$$

and, consequently,

$$\Delta\Omega(\vec{r}_A,\omega) = 0. \tag{VIII-24c}$$

With this result we may rewrite equation (VIII-19) as

$$\Omega(\vec{r}_A,\omega) = \Omega_0(\vec{r}_A,\omega) =$$

$$-\int_{S_0} [\boldsymbol{\Theta}^*_\Omega(\vec{r},\vec{r}_A,\omega)\vec{V}(\vec{r},\omega) + \boldsymbol{\tau}(\vec{r},\omega)\vec{G}^*_\Omega(\vec{r},\vec{r}_A,\omega)].\vec{n}\,dS_0. \tag{VIII-25}$$

This Kirchhoff-Helmholtz integral describes *inverse* wave field extrapolation from acquisition surface S_0 to subsurface point $\vec{r}_A$. The only approximation is a spatial bandlimitation (the negligence of evanescent waves at z_1). This imposes a restriction to the maximum obtainable spatial resolution, see also section VII.2.

For the situation of an inhomogeneous medium, a similar analysis as presented in section VII.3.3 leads to the conclusion that we may write for the *upgoing* wave field at $\vec{r}_A$:

$$\Omega^-(\vec{r}_A,\omega) = \Omega_0^-(\vec{r}_A,\omega) + \Delta\Omega^-(\vec{r}_A,\omega), \tag{VIII-26a}$$

where

$$\Omega_0^-(\vec{r}_A,\omega) \triangleq \Omega_0(\vec{r}_A,\omega) = -\int_{S_0} [\boldsymbol{\Theta}_\Omega^* \vec{V} + \boldsymbol{\tau}\vec{G}_\Omega^*].\vec{n}\,dS_0 \tag{VIII-26b}$$

and

$$\Delta\Omega^-(\vec{r}_A,\omega) = \frac{2}{j\omega}\int\!\!\int_{-\infty}^{\infty} \frac{1}{\rho(z_1)} \left[\left(\frac{\partial\Gamma^+_{\phi,\Omega}}{\partial z}\right)^*_s \Phi^+_s + \left(\frac{\partial\vec{\Gamma}^+_{\psi,\Omega}}{\partial z}\right)^*_s.\vec{\Psi}^+_s \right]_{z_1} dxdy, \tag{VIII-26c}$$

the sub-script "s" denoting that only scattered waves are considered (see Figures VII-12a and VII-12e for the acoustic equivalence). According to (VIII-26c), $\Delta\Omega^-(\vec{r}_A,\omega)$ is proportional to *products* of *scattered* wave fields. Hence, the magnitude of the error term $\Delta\Omega^-(\vec{r}_A,\omega)$ is proportional to the *squared reflectivity* of the interfaces in the inhomogeneous medium.

Neglecting $\Delta\Omega^-(\vec{r}_A,\omega)$ means not only that *multiply reflected waves* are handled erroneously by equation (VIII-26), but also that the *primary* wave contribution to $\Omega^-(\vec{r}_A,\omega)$ is not fully correct. The negligence of $\Delta\Omega^-(\vec{r}_A,\omega)$ is justified only when the contrasts in the medium are weak to moderate. In that case equation (VIII-26b) describes *'true amplitude'* inverse extrapolation of *primary* waves. This is illustrated by an example in section VIII.3.6. When the contrasts in the medium are significant, $\Delta\Omega^-(\vec{r}_A,\omega)$ may not be neglected and should be estimated in an iterative way. This is discussed in chapter X, section X.3.

VIII.3.4 Elastic one-way Rayleigh integrals for inverse extrapolation

When the acquisition surface S_o is a plane surface at $z=z_o$, we may rewrite equation (VIII-19b) as

$$\Omega_o^-(\vec{r}_A,\omega) = \iint_{-\infty}^{\infty} [\vec{\Theta}_{z,\Omega}^* \cdot \vec{V} + \vec{G}_\Omega^* \cdot \vec{\tau}_z]_{z_o} \, dxdy. \qquad \text{(VIII-27)}$$

For the total wave field at $z=z_o$ we write

$$\vec{V}(\vec{r},\omega) = \vec{V}^+(\vec{r},\omega) + \vec{V}^-(\vec{r},\omega) \qquad \text{at } z=z_o \qquad \text{(VIII-28a)}$$

and

$$\vec{\tau}_z(\vec{r},\omega) = \vec{\tau}_z^{\,+}(\vec{r},\omega) + \vec{\tau}_z^{\,-}(\vec{r},\omega) \qquad \text{at } z=z_o. \qquad \text{(VIII-28b)}$$

For the Green's wave field we may choose above $z=z_o$ (outside V) any convenient reference medium. We choose $c_{ijk\ell}(x,y,z<z_o)=c_{ijk\ell}(x,y,z_o)$ and $\rho(x,y,z<z_o)=\rho(x,y,z_o)$. With this choice the Green's wave field at z_o is purely upgoing (no scattering occurs in the upper half-space above z_o), hence,

$$\vec{G}_\Omega(\vec{r},\vec{r}_A,\omega) = \vec{G}_\Omega^-(\vec{r},\vec{r}_A,\omega) \qquad \text{at } z=z_o \qquad \text{(VIII-29a)}$$

and

$$\vec{\Theta}_{z,\Omega}(\vec{r},\vec{r}_A,\omega) = \vec{\Theta}_{z,\Omega}^-(\vec{r},\vec{r}_A,\omega) \qquad \text{at } z=z_o. \qquad \text{(VIII-29b)}$$

We may thus rewrite equation (VIII-27) as

$$\Omega_o^-(\vec{r}_A,\omega) = \iint_{-\infty}^{\infty} [(\vec{\Theta}_{z,\Omega}^-)^* . (\vec{V}^+ + \vec{V}^-) + (\vec{G}_\Omega^-)^* . (\vec{\tau}_z^{\,+} + \vec{\tau}_z^{\,-})]_{z_o} \, dxdy. \qquad \text{(VIII-30)}$$

Assuming for simplicity that the medium is locally homogeneous and isotropic at z_0, we show in Appendix C, section C.3, that (VIII-30) may be simplified to

$$\Omega_0^-(\vec{r}_A,\omega) = \frac{2}{j\omega} \iint_{-\infty}^{\infty} \frac{1}{\rho(z_0)} \left[\left(\Gamma_{\phi,\Omega}^-\right)^* \frac{\partial \Phi^-}{\partial z} + \left(\vec{\Gamma}_{\psi,\Omega}^-\right)^* \cdot \frac{\partial \vec{\Psi}^-}{\partial z} \right]_{z_0} dxdy, \qquad \text{(VIII-31a)}$$

or, equivalently, to

$$\Omega_0^-(\vec{r}_A,\omega) = \frac{-2}{j\omega} \iint_{-\infty}^{\infty} \frac{1}{\rho(z_0)} \left[\left(\frac{\partial \Gamma_{\phi,\Omega}^-}{\partial z}\right)^* \Phi^- + \left(\frac{\partial \vec{\Gamma}_{\psi,\Omega}^-}{\partial z}\right)^* \cdot \vec{\Psi}^- \right]_{z_0} dxdy, \qquad \text{(VIII-31b)}$$

where Φ^- and $\vec{\Psi}^-$ are the P- and S-wave potentials for the velocity $\vec{V}^-$ at z_0 and where $\Gamma_{\phi,\Omega}^-$ and $\vec{\Gamma}_{\psi,\Omega}^-$ are the P- and S-wave potentionals for the Green's velocity $\vec{G}_\Omega^-$ at z_0, see equation (C-3) and Figure C-1. Again, only the wave fields which propagate in *opposite* directions through $z=z_0$ contribute to the result $\Omega_0^-(\vec{r}_A,\omega)$ (see Figure VII-13a for the acoustic equivalence). The underlying assumption is again that evanescent waves may be neglected at z_0.

Compare equations (VIII-31a) and (VIII-31b) with equations (VI-66a) and (VI-66b),

$$\Omega(\vec{r}_A,\omega) = \frac{2}{j\omega} \iint_{-\infty}^{\infty} \frac{1}{\rho(z_0)} \left[\Gamma_{\phi,\Omega}^- \frac{\partial \Phi^+}{\partial z} + \vec{\Gamma}_{\psi,\Omega}^- \cdot \frac{\partial \vec{\Psi}^+}{\partial z} \right]_{z_0} dxdy \qquad \text{(VIII-32a)}$$

and

$$\Omega(\vec{r}_A,\omega) = \frac{-2}{j\omega} \iint_{-\infty}^{\infty} \frac{1}{\rho(z_0)} \left[\frac{\partial \Gamma_{\phi,\Omega}^-}{\partial z} \Phi^+ + \frac{\partial \vec{\Gamma}_{\psi,\Omega}^-}{\partial z} \cdot \vec{\Psi}^+ \right]_{z_0} dxdy. \qquad \text{(VIII-32b)}$$

These are the one-way versions of the elastic Rayleigh I and Rayleigh II integrals, respectively, for *forward* extrapolation (the sources of Φ^+ and $\vec{\Psi}^+$

are above z_0, see Figure VI-4). Analogously, we will refer to equations (VIII-31a) and (VIII-31b) as the one-way versions of the elastic Rayleigh I and Rayleigh II integrals, respectively, for *inverse* extrapolation (the sources of Φ^- and $\vec{\Psi}^-$ are below z_A, see Figure VIII-2). Equations (VIII-32a) and (VIII-32b) for forward extrapolation yield the *exact total* wave field $\Omega(\vec{r}_A,\omega)$. On the other hand, equations (VIII-31a) and (VIII-31b) for inverse extrapolation yield $\Omega_0^-(\vec{r}_A,\omega)$, which is an approximate version of the *upgoing* wave field

$$\Omega^-(\vec{r}_A,\omega) = \Omega_0^-(\vec{r}_A,\omega) + \Delta\Omega^-(\vec{r}_A,\omega), \tag{VIII-33}$$

see equation (VIII-26a). The approximations concern the negligence of evanescent waves and the negligence of the error term $\Delta\Omega^-(\vec{r}_A,\omega)$, which is proportional to (but not restricted to) multiply reflected waves.

Depending on the choice of the source for the Green's wave fields, Ω_0^- in equations (VIII-31a) and (VIII-31b) may represent either V_m^- for m=1, 2, 3 or $j\omega\Phi^-$, or $j\omega\Psi_h^-$ for h=1, 2, 3, see section VIII.3.2. We consider the latter two cases. If the Green's wave fields have an impulsive P-wave source at $\vec{r}_A$, then Ω_0^- represents the scaled P-wave potential $j\omega\Phi^-$ at $\vec{r}_A$, hence

$$\Phi^-(\vec{r}_A,\omega) \approx \frac{-2}{\omega^2}\iint_{-\infty}^{\infty}\frac{1}{\rho(z_0)}\left[(\Gamma^-_{\phi,\phi})^*\frac{\partial\Phi^-}{\partial z} + (\vec{\Gamma}^-_{\psi,\phi})^*\cdot\frac{\partial\vec{\Psi}^-}{\partial z}\right]_{z_0} dxdy, \tag{VIII-34a}$$

or, alternatively,

$$\Phi^-(\vec{r}_A,\omega) \approx \frac{2}{\omega^2}\iint_{-\infty}^{\infty}\frac{1}{\rho(z_0)}\left[\left(\frac{\partial\Gamma^-_{\phi,\phi}}{\partial z}\right)^*\Phi^- + \left(\frac{\partial\vec{\Gamma}^-_{\psi,\phi}}{\partial z}\right)^*\cdot\vec{\Psi}^-\right]_{z_0} dxdy. \tag{VIII-34b}$$

On the other hand, if the Green's wave fields have an impulsive S_h-wave source at $\vec{r}_A$, then Ω_0^- represents the h-component of the scaled S-wave potential $j\omega\vec{\Psi}^-$ at $\vec{r}_A$, hence

$$\Psi_h^-(\vec{r}_A,\omega) \approx \frac{-2}{\omega^2} \iint_{-\infty}^{\infty} \frac{1}{\rho(z_0)} \left[\left(\Gamma^-_{\phi,\psi_h}\right)^* \frac{\partial \Phi^-}{\partial z} + \left(\vec{\Gamma}^-_{\psi,\psi_h}\right)^* \cdot \frac{\partial \vec{\Psi}^-}{\partial z} \right]_{z_0} dxdy, \quad \text{(VIII-35a)}$$

or, alternatively,

$$\Psi_h^-(\vec{r}_A,\omega) \approx \frac{2}{\omega^2} \iint_{-\infty}^{\infty} \frac{1}{\rho(z_0)} \left[\left(\frac{\partial \Gamma^-_{\phi,\psi_h}}{\partial z}\right)^* \Phi^- + \left(\frac{\partial \vec{\Gamma}^-_{\psi,\psi_h}}{\partial z}\right)^* \cdot \vec{\Psi}^- \right]_{z_0} dxdy. \quad \text{(VIII-35b)}$$

For the special situation of a homogeneous isotropic medium, we may substitute the free space solutions (VI-44a) and (VI-44b) for the Green's functions $\Gamma^-_{\phi,\phi}$ and $\Gamma^-_{\psi_k,\psi_h}$ (bear in mind that $\Gamma^-_{\psi_k,\psi_h}$ denotes the k-component of $\vec{\Gamma}^-_{\psi,\psi_h}$). Hence, for this situation equations (VIII-34a) and (VIII-34b) read

$$\Phi^-(\vec{r}_A,\omega) = \frac{-1}{2\pi} \iint_{-\infty}^{\infty} \left[\frac{e^{+jk_p\Delta r}}{\Delta r} \frac{\partial \Phi^-(\vec{r},\omega)}{\partial z} \right]_{z_0} dxdy, \quad \text{(VIII-36a)}$$

or

$$\Phi^-(\vec{r}_A,\omega) = \frac{1}{2\pi} \iint_{-\infty}^{\infty} \left[\frac{\partial}{\partial z} \left(\frac{e^{+jk_p\Delta r}}{\Delta r}\right) \Phi^-(\vec{r},\omega) \right]_{z_0} dxdy, \quad \text{(VIII-36b)}$$

respectively, with

$$k_p = \omega/c_p \quad \text{(VIII-36c)}$$

and

$$\Delta r = |\vec{r}-\vec{r}_A|. \quad \text{(VIII-36d)}$$

For the same situation equations (VIII-35a) and (VIII-35b) read

$$\Psi_h^-(\vec{r}_A,\omega) = \frac{-1}{2\pi}\iint_{-\infty}^{\infty}\left[\left((\delta_{kh}+\frac{1}{k_s^2}\partial_k\partial_h)\frac{e^{+jk_s\Delta r}}{\Delta r}\right)\frac{\partial\Psi_k^-(\vec{r},\omega)}{\partial z}\right]_{z_0} dxdy, \qquad \text{(VIII-37a)}$$

or

$$\Psi_h^-(\vec{r}_A,\omega) = \frac{1}{2\pi}\iint_{-\infty}^{\infty}\left[\frac{\partial}{\partial z}\left((\delta_{kh}+\frac{1}{k_s^2}\partial_k\partial_h)\frac{e^{+jk_s\Delta r}}{\Delta r}\right)\Psi_k^-(\vec{r},\omega)\right]_{z_0} dxdy, \qquad \text{(VIII-37b)}$$

respectively, with

$$k_s = \omega/c_s. \qquad \text{(VIII-37c)}$$

The only approximation in (VIII-36) and (VIII-37) concerns the negligence of evanescent waves.

VIII.3.5 Matrix formulation of elastic inverse wave field extrapolation

We return to the inhomogeneous anisotropic situation. Consider the one-way version of the elastic Rayleigh II integral (VIII-31b) for inverse extrapolation of upgoing waves. Following the same procedure as in section VII.3.5, this expression may be rewritten in the matrix notation of Appendix A, yielding

$$\frac{1}{j\omega}\vec{\Omega}^-(z_1) \approx \langle \mathbf{F}^-_{\Omega,\phi}(z_1,z_0)\rangle \vec{\Phi}^-(z_0) + \langle \mathbf{F}^-_{\Omega,\psi_\alpha}(z_1,z_0)\rangle \vec{\Psi}^-_\alpha(z_0). \qquad \text{(VIII-38a)}$$

Similarly, for inverse extrapolation of downgoing waves we obtain in the matrix notation

$$\frac{1}{j\omega}\vec{\Omega}^{+}(z_o) \approx \langle \mathbf{F}^{+}_{\Omega,\phi}(z_o,z_1)\rangle \vec{\Phi}^{+}(z_1) + \langle \mathbf{F}^{+}_{\Omega,\psi_\alpha}(z_o,z_1)\rangle \vec{\Psi}^{+}_{\alpha}(z_1). \tag{VIII-38b}$$

We neglected the error terms $\Delta\Omega^{\pm}$. The approximated inverse wave field extrapolation matrices are defined by

$$\langle \mathbf{F}^{-}_{\Omega,\phi}(z_1,z_o)\rangle = \frac{2}{\omega^2}\left(\frac{\partial \Gamma^{+}_{\Omega,\phi}(z_1,z=z_o)}{\partial z}\right)^{*} \mathbf{M}^{-1}(z_o), \tag{VIII-39a}$$

$$\langle \mathbf{F}^{-}_{\Omega,\psi_\alpha}(z_1,z_o)\rangle = \frac{2}{\omega^2}\left(\mathcal{L}^{+}_{\Omega,\psi_\alpha}(z_1,z_o)\right)^{*} \mathbf{M}^{-1}(z_o), \tag{VIII-39b}$$

$$\langle \mathbf{F}^{+}_{\Omega,\phi}(z_o,z_1)\rangle = \frac{-2}{\omega^2}\left(\frac{\partial \Gamma^{-}_{\Omega,\phi}(z_o,z=z_1)}{\partial z}\right)^{*} \mathbf{M}^{-1}(z_1) \tag{VIII-39c}$$

and

$$\langle \mathbf{F}^{+}_{\Omega,\psi_\alpha}(z_o,z_1)\rangle = \frac{-2}{\omega^2}\left(\mathcal{L}^{-}_{\Omega,\psi_\alpha}(z_o,z_1)\right)^{*} \mathbf{M}^{-1}(z_1). \tag{VIII-39d}$$

For a discussion of the vectors and matrices in (VIII-38) and (VIII-39) we refer to section VI.5.2. Comparing equations (VIII-39a) to (VIII-39d) for the inverse wave field extrapolation matrices with equations (VI-86b), (VI-86c), (VI-87b) and (VI-87c) for the forward wave field extrapolation matrices yields

$$\langle \mathbf{F}^{-}_{\Omega,\phi}(z_1,z_o)\rangle = [\mathbf{W}^{+}_{\Omega,\phi}(z_1,z_o)]^{*}, \tag{VIII-40a}$$

$$\langle \mathbf{F}^{-}_{\Omega,\psi_\alpha}(z_1,z_o)\rangle = [\mathbf{W}^{+}_{\Omega,\psi_\alpha}(z_1,z_o)]^{*}, \tag{VIII-40b}$$

$$\langle \mathbf{F}^{+}_{\Omega,\phi}(z_o,z_1)\rangle = [\mathbf{W}^{-}_{\Omega,\phi}(z_o,z_1)]^{*} \tag{VIII-40c}$$

and

$$\langle \mathbf{F}^{+}_{\Omega,\psi_\alpha}(z_o,z_1)\rangle = [\mathbf{W}^{-}_{\Omega,\psi_\alpha}(z_o,z_1)]^{*}. \tag{VIII-40d}$$

Note that equations (VIII-38a) and (VIII-38b) for elastic inverse extrapolation may be elegantly rewritten as

$$\vec{\underset{\sim}{D}}^{-}(z_1) \approx \langle \underset{\sim}{F}^{-}(z_1,z_0)\rangle \vec{\underset{\sim}{D}}^{-}(z_0) \tag{VIII-41a}$$

and

$$\vec{\underset{\sim}{D}}^{+}(z_0) \approx \langle \underset{\sim}{F}^{+}(z_0,z_1)\rangle \vec{\underset{\sim}{D}}^{+}(z_1), \tag{VIII-41b}$$

respectively, where the multi-component data vectors $\vec{\underset{\sim}{D}}^{\pm}$ are defined by equation (VIII-11) and where the multi-component inverse extrapolation operators are defined as

$$\langle \underset{\sim}{F}^{-}(z_1,z_0)\rangle = \begin{bmatrix} \langle F^{-}_{\phi,\phi}(z_1,z_0)\rangle & \langle F^{-}_{\phi,\psi_x}(z_1,z_0)\rangle & \langle F^{-}_{\phi,\psi_y}(z_1,z_0)\rangle \\ \langle F^{-}_{\psi_x,\phi}(z_1,z_0)\rangle & \langle F^{-}_{\psi_x,\psi_x}(z_1,z_0)\rangle & \langle F^{-}_{\psi_x,\psi_y}(z_1,z_0)\rangle \\ \langle F^{-}_{\psi_y,\phi}(z_1,z_0)\rangle & \langle F^{-}_{\psi_y,\psi_x}(z_1,z_0)\rangle & \langle F^{-}_{\psi_y,\psi_y}(z_1,z_0)\rangle \end{bmatrix} \tag{VIII-42a}$$

and

$$\langle \underset{\sim}{F}^{+}(z_0,z_1)\rangle = \begin{bmatrix} \langle F^{+}_{\phi,\phi}(z_0,z_1)\rangle & \langle F^{+}_{\phi,\psi_x}(z_0,z_1)\rangle & \langle F^{+}_{\phi,\psi_y}(z_0,z_1)\rangle \\ \langle F^{+}_{\psi_x,\phi}(z_0,z_1)\rangle & \langle F^{+}_{\psi_x,\psi_x}(z_0,z_1)\rangle & \langle F^{+}_{\psi_x,\psi_y}(z_0,z_1)\rangle \\ \langle F^{+}_{\psi_y,\phi}(z_0,z_1)\rangle & \langle F^{+}_{\psi_y,\psi_x}(z_0,z_1)\rangle & \langle F^{+}_{\psi_y,\psi_y}(z_0,z_1)\rangle \end{bmatrix}. \tag{VIII-42b}$$

From equations (VIII-40) and (VIII-42) we easily find

$$\langle \underset{\sim}{F}^{-}(z_1,z_0)\rangle = [\underset{\sim}{W}^{+}(z_1,z_0)]^{*} \tag{VIII-43a}$$

and

$$\langle \underset{\sim}{F}^{+}(z_0,z_1)\rangle = [\underset{\sim}{W}^{-}(z_0,z_1)]^{*}, \tag{VIII-43b}$$

where the multi-component extrapolation operators $\underset{\sim}{W}^{+}(z_1,z_0)$ and $\underset{\sim}{W}^{-}(z_0,z_1)$ are defined by (VIII-12a) and (VIII-12b), respectively.

When converted waves during propagation may be neglected, then the expressions for the multi-component inverse extrapolation operators simplify to

$$<\underset{\sim}{F}^-(z_1,z_0)> \approx \begin{bmatrix} <F^-_{\phi,\phi}(z_1,z_0)> & 0 & 0 \\ 0 & <F^-_{\psi_x,\psi_x}(z_1,z_0)> & 0 \\ 0 & 0 & <F^-_{\psi_y,\psi_y}(z_1,z_0)> \end{bmatrix} \tag{VIII-44a}$$

and

$$<\underset{\sim}{F}^+(z_0,z_1)> \approx \begin{bmatrix} <F^+_{\phi,\phi}(z_0,z_1)> & 0 & 0 \\ 0 & <F^+_{\psi_x,\psi_x}(z_0,z_1)> & 0 \\ 0 & 0 & <F^+_{\psi_y,\psi_y}(z_0,z_1)> \end{bmatrix} . \tag{VIII-44b}$$

Substitution of (VIII-44a) into (VIII-41a) yields

$$\vec{\Phi}^-(z_1) \approx <F^-_{\phi,\phi}(z_1,z_0)>\vec{\Phi}^-(z_0) \tag{VIII-45a}$$

for inverse extrapolation of upgoing P-waves and

$$\vec{\Psi}^-_x(z_1) \approx <F^-_{\psi_x,\psi_x}(z_1,z_0)>\vec{\Psi}^-_x(z_0), \tag{VIII-45b}$$

$$\vec{\Psi}^-_y(z_1) \approx <F^-_{\psi_y,\psi_y}(z_1,z_0)>\vec{\Psi}^-_y(z_0) \tag{VIII-45c}$$

for inverse extrapolation of upgoing S-waves.
Note the high degree of similarity with the acoustic algorithm (VII-72a) for inverse extrapolation of upgoing waves

$$\vec{P}^-(z_1) \approx <F^-(z_1,z_0)>\vec{P}^-(z_0).$$

Similarly, substitution of (VIII-44b) into (VIII-41b) yields

$$\vec{\Phi}^+(z_0) \approx <F^+_{\phi,\phi}(z_0,z_1)>\vec{\Phi}^+(z_1) \tag{VIII-46a}$$

for inverse extrapolation of downgoing P-waves and

$$\vec{\Psi}^{+}_{x}(z_0) \approx \langle F^{+}_{\psi_x,\psi_x}(z_0,z_1)\rangle \vec{\Psi}^{+}_{x}(z_1), \tag{VIII-46b}$$

$$\vec{\Psi}^{+}_{y}(z_0) \approx \langle F^{+}_{\psi_y,\psi_y}(z_0,z_1)\rangle \vec{\Psi}^{+}_{y}(z_1), \tag{VIII-46c}$$

for inverse extrapolation of downgoing S-waves. Note the high degree of similarity with the acoustic algorithm (VII-72b) for inverse extrapolation of downgoing waves

$$\vec{P}^{+}(z_0) \approx \langle F^{+}(z_0,z_1)\rangle \vec{P}^{+}(z_1).$$

For practical applications in low contrast media, the simplified expressions (VIII-45) and (VIII-46) are preferred above the more correct expressions (VIII-38a) and (VIII-38b) for the following reason. The extrapolation operators $\langle F \rangle$ are based on Green's wave fields Γ which are modeled in a geologically oriented reference medium (the macro subsurface model), see also section VI.5.2. The most important parameters in such a reference medium are the P- and S-wave propagation velocities. In practice a reference medium contains errors. When these errors are small, then the simplified expressions (VIII-45) and (VIII-46) may still yield reasonably accurate results. For this situation the accuracy will not improve, however, by taking into account the converted waves, as in (VIII-38a) and (VIII-38b), because here two wave fields are superimposed which may be mutually shifted because they are based on the (slightly erroneous) P-wave velocity and the (slightly erroneous) S-wave velocity, respectively.

In conclusion, when the contrasts in the medium are weak to moderate, wave conversion during propagation plays a minor role, so the simplified expressions (VIII-45) and (VIII-46) yield sufficient accuracy, even in the presence of small errors in the macro subsurface model. When the contrasts in the medium are significant then we should *not* use the full expressions (VIII-38a) and (VIII-38b) for the following two reasons:

1. The error term $\Delta\Omega^{-}$, as defined in (VIII-26c), is not negligible.
2. The two terms in (VIII-38a) and (VIII-38b) are mutually shifted in the presence of small errors in the macro subsurface model.

In chapter X, section X.3, we discuss an iterative scheme for designing inverse wave field extrapolation operators which tackles both problems.

VIII.3.6 Examples of elastic inverse wave field extrapolation

We demonstrate the validity of equations (VIII-45a) and (VIII-45c) with the aid of two numerical examples (Haimé, 1987). Consider the 2-D inhomogeneous elastic medium, shown in Figure VIII-3a. A plane P-wave source of finite extent is burried in the subsurface at a depth of z_2=2000m. The response at the reflection-free surface z_0=0m is shown in Figures VIII-3b, c. This response was computed with a finite difference modeling scheme. It represents the vertical (b) and horizontal (c) components of the partical velocity $\vec{v}$ as a function of the lateral coordinate x and time t. Because the upper half-space $z<z_0$ is homogeneous and acquisition surface z_0 is reflection-free, the recorded velocity represents an upgoing wave field, hence, $\vec{v}=\vec{v}^-(x,z_0;t)$.

In seismic practice the vertical motions are often associated to P-waves whereas the horizontal motions are often associated to SV-waves. However, due to the complex overburden in this example the vertical velocity data also contain SV-wave contributions and the horizontal velocity data also contain P-wave contributions. Therefore we prefer to speak of *pseudo* P-data and *pseudo* SV-data, respectively. Figures VIII-3d and VIII-3e show the *true* P-data and the *true* SV-data in terms of the potentials $\phi^-(x,z_0;t)$ and $\psi_y^-(x,z_0;t)$, respectively. They are related to the velocity data according to equations (II-31c) and (II-31d), respectively.

By applying a Fourier transform from time (t) to frequency (ω), the P-wave potential $\phi^-(x,z_0;t)$ is decomposed into monochromatic wave fields $\Phi^-(x,z_0;\omega)$. According to Appendix A, section A.2, a discretized monochromatic wave field may be represented by a vector $\vec{\Phi}^-(z_0)$. Inverse extrapolation of this upgoing wave field from depth level z_0=0m to depth level z_1=1200m is described by equation (VIII-45a), where

$$<\mathbf{F}^-_{\phi,\phi}(z_1,z_0)> = \frac{2}{\omega^2}\left(\frac{\partial\mathbf{\Gamma}^+_{\phi,\phi}(z_1,z=z_0)}{\partial z}\right)^* \mathbf{M}^{-1}(z_0). \qquad \text{(VIII-47a)}$$

The numerical modeling of the Green's matrix

$$\Gamma^{+}_{\phi,\phi}(z_1,z_0) = [\Gamma^{-}_{\phi,\phi}(z_0,z_1)]^T \qquad \text{(VIII-47b)}$$

was discussed in chapter VI (see Figure VI-6d). By applying (VIII-45a) for all frequencies within the seismic band, we obtain a range of monochromatic data vectors $\vec{\Phi}^{-}(z_1)$. The result, after applying an inverse Fourier transform from ω to t, is shown in Figure VIII-4a. It represents the space-time data $\phi^{-}(x,z_1;t)$. Note that the distorting propagation effects of the overburden have been removed. For comparison, Figure VIII-4b shows the exact upgoing P-wave potential at depth level z_1. The inverse extrapolated data (a) and the direct modeled data (b) show a very good agreement. This is also illustrated in Figure VIII-4c where the amplitude cross-sections are compared. Apparently, the underlying assumption of equation (VIII-47a) (weak to moderate contrasts) was satisfied for this example.

For the next example, we consider the same medium, this time with a plane SV-wave source burried in the subsurface at a depth of z_2=2000 m, see Figure VIII-5a. The response in terms of the vertical and horizontal velocity components is shown in Figures VIII-5b and VIII-5c, respectively; the response in terms of the P-wave and SV-wave potentials is shown in Figures VIII-5d and VIII-5e, respectively. The Fourier transformed SV-wave potential is respresented by a vector $\vec{\Psi}^{-}_y(z_0)$. Inverse extrapolation from z_0 to z_1 is described by equation (VIII-45c), where

$$\langle \mathbf{F}^{-}_{\psi_y,\psi_y}(z_1,z_0)\rangle = \frac{2}{\omega^2}\left(\mathscr{L}^{+}_{\psi_y,\psi_y}(z_1,z_0)\right)^{*} \mathbf{M}^{-1}(z_0). \qquad \text{(VIII-48)}$$

The numerical modeling of the Green's matrix was discussed in chapter VI (see Figure VI-7e).

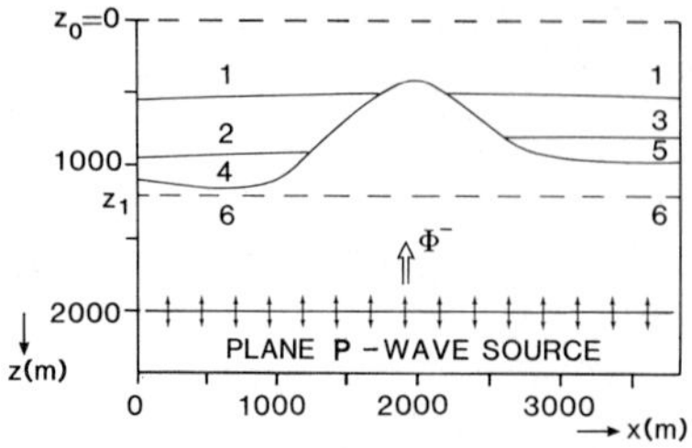

layer nr	velocities	layer nr	velocities
1	Cp=2600 m/s Cs=1500 m/s	4	Cp=2800 m/s Cs=1600 m/s
2	Cp=3200 m/s Cs=1900 m/s	5	Cp=3000 m/s Cs=1800 m/s
3	Cp=3200 m/s Cs=1900 m/s	6	Cp=3600 m/s Cs=2078 m/s

(a)

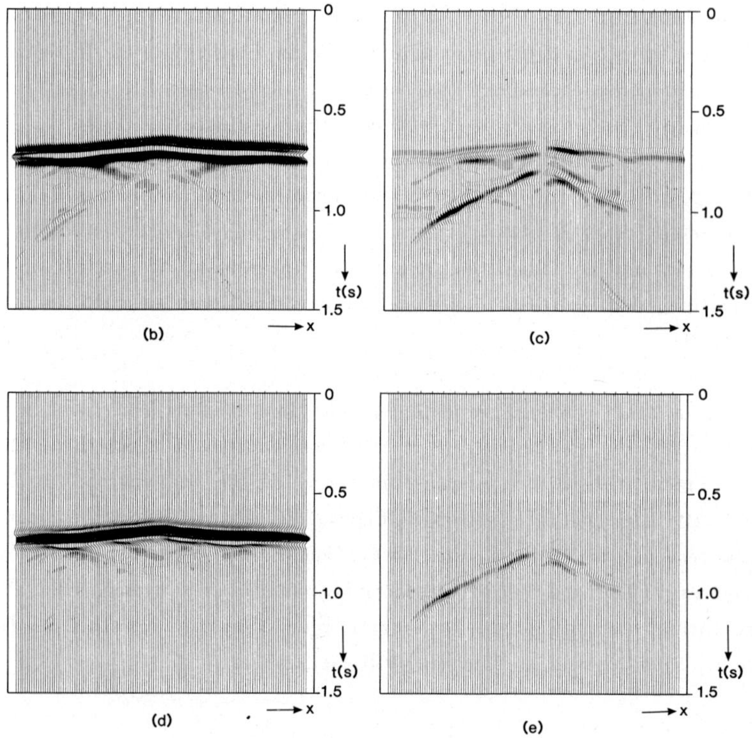

Figure VIII-3: *a. Inhomogeneous elastic medium with a burried plane P-wave source at z_2=2000 m.*

b. Vertical component of the velocity, registered at z_o (pseudo P-data).

c. Horizontal component of the velocity, registered at z_o (pseudo SV-data).

d. Upgoing P-wave potential at z_o (true P-data).

e. Upgoing SV-wave potential at z_o (true SV-data).

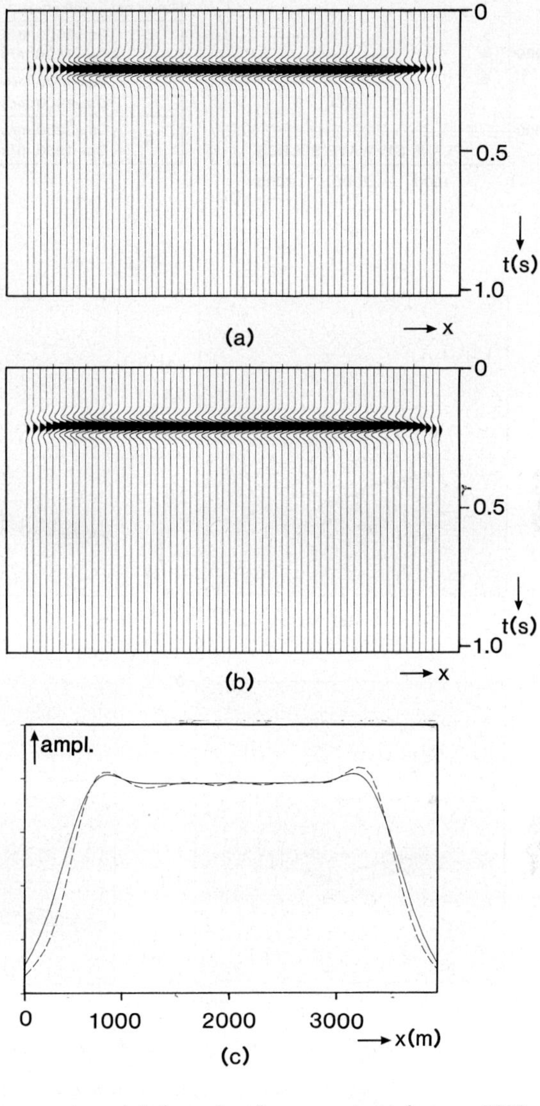

Figure VIII-4: *a. Inverse extrapolated upgoing P-wave potential at* z_1*=1200 m.*

b. Exact upgoing P-wave potential at z_1*=1200 m.*

c. Maximum amplitude per trace of Figure a (dotted line) and b (solid line).

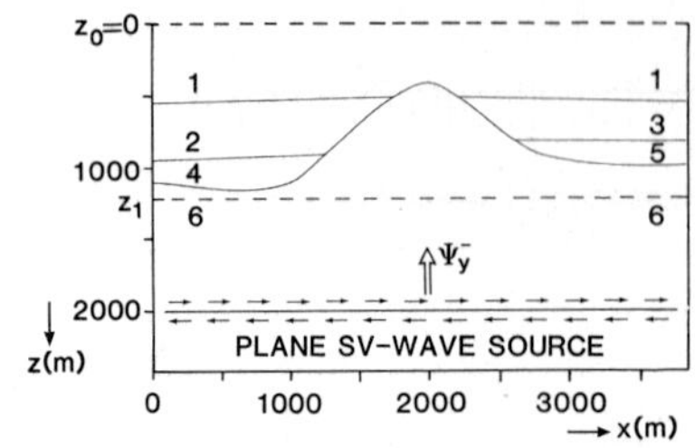

layer nr	velocities	layer nr	velocities
1	Cp=2600 m/s Cs=1500 m/s	4	Cp=2800 m/s Cs=1600 m/s
2	Cp=3200 m/s Cs=1900 m/s	5	Cp=3000 m/s Cs=1800 m/s
3	Cp=3200 m/s Cs=1900 m/s	6	Cp=3600 m/s Cs=2078 m/s

(a)

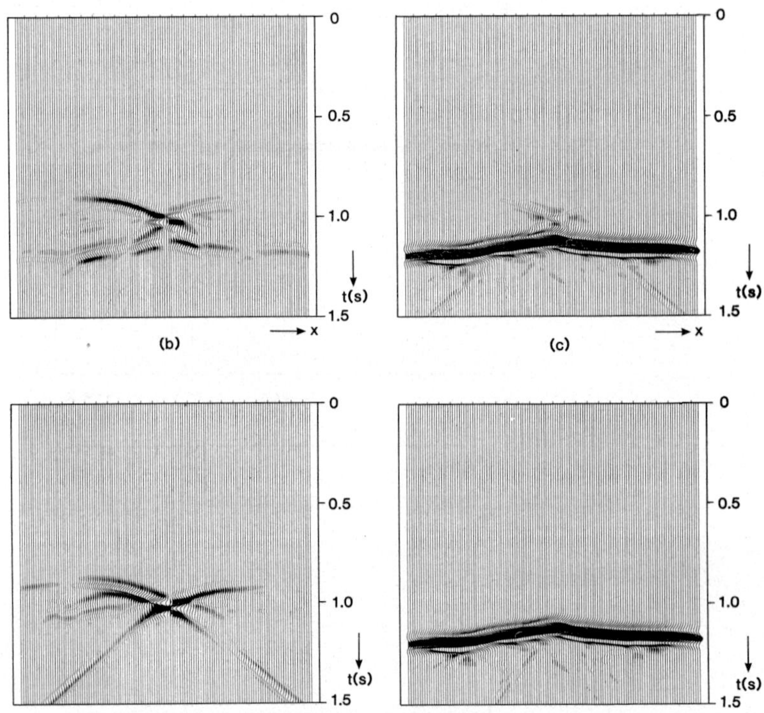

(b) (c)

(d) (e)

Figure VIII-5: *a. Inhomogeneous elastic medium with a burried plane SV-wave source at z_2=2000 m.*

b. Vertical component of the velocity, registered at z_o (pseudo P-data).

c. Horizontal component of the velocity, registered at z_o (pseudo SV-data).

d. Upgoing P-wave potential at z_o (true P-data).

e. Upgoing SV-wave potential at z_o (true SV-data).

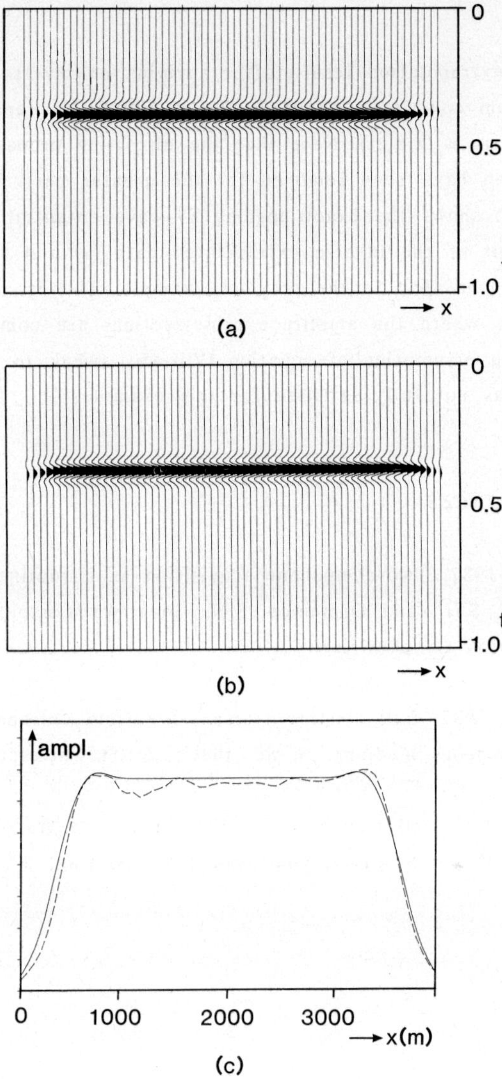

Figure VIII-6: *a. Inverse extrapolated upgoing SV-wave potential at z_1=1200 m.*
b. Exact upgoing SV-wave potential at z_1=1200 m.
c. Maximum amplitude per trace of Figure a (dotted line) and b (solid line).

The inverse extrapolation result, after applying an inverse Fourier transform from ω to t, is shown in Figure VIII-6a. It represents the space-time data $\psi_y^-(x,z_1;t)$. Note that the distorting propagation effects of the overburden have been removed for the greater part. For comparison, Figure VIII-6b shows the exact upgoing SV-wave potential at depth level z_1. The agreement of the inverse extrapolated data (a) and the direct modeled data is less good than in the previous example. This is also illustrated in Figure VIII-6c where the amplitude cross-sections are compared. Apparently, the underlying assumption of equation (VIII-45c) (weak to moderate contrasts), was not fully satisfied for this example.

VIII.4 REFERENCES

Garotta, R., 1987, Two-component acquisition as a routine procedure, *in* Danbom, S.H., and Domenico, S.N., Shear-wave exploration: Society of Exploration Geophysicists.

Haimé, G.C., 1987, Full elastic inverse wave field extrapolation through an inhomogeneous medium: M. Sc. thesis, Delft University of Technology.

Wapenaar, C.P.A., and Haimé, G.C., 1989, Elastic extrapolation of primary seismic P and S waves: Geophysical Prospecting, *37*, in press.

IX

ACOUSTIC INVERSE WAVE FIELD EXTRAPOLATION IN HIGH CONTRAST MEDIA

IX.1. INTRODUCTION

For the configuration of Figure IX-1, inverse extrapolation of the upgoing wave field P^- from acquisition surface $z=z_0$ to subsurface point $\vec{r}_A$ is described by

$$P^-(\vec{r}_A,\omega) = P_0^-(\vec{r}_A,\omega) + \Delta P^-(\vec{r}_A,\omega), \tag{IX-1a}$$

where

$$P_0^-(\vec{r}_A,\omega) = 2\int\int_{-\infty}^{\infty}\left[\frac{1}{\rho(\vec{r})}\left(\frac{\partial G^-(\vec{r},\vec{r}_A,\omega)}{\partial z}\right)^* P^-(\vec{r},\omega)\right]_{z_0} dxdy \tag{IX-1b}$$

and

$$\Delta P^-(\vec{r}_A,\omega) = -2\int\int_{-\infty}^{\infty}\left[\frac{1}{\rho(\vec{r})}\left(\frac{\partial G_s^+(\vec{r},\vec{r}_A,\omega)}{\partial z}\right)^* P_s^+(\vec{r},\omega)\right]_{z_1} dxdy, \tag{IX-1c}$$

see section VII.3. The only approximation concerns the negligence of

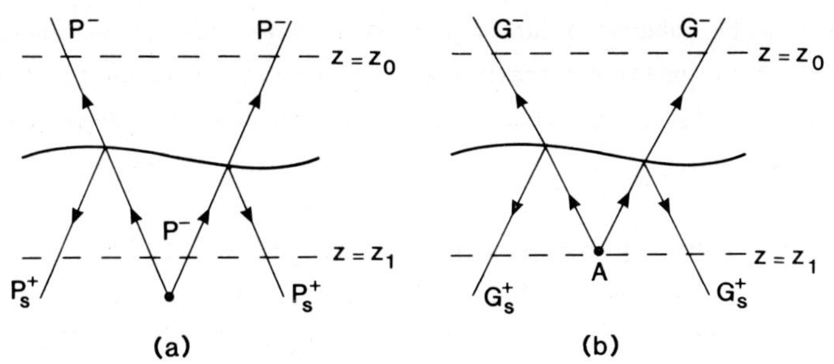

Figure IX-1: The error in acoustic inverse wave field extrapolation is proportional to the product of the scattered wave field P_s^+ (Figure a) and the scattered Green's wave field G_s^+ (Figure b).

evanescent waves at $z=z_0$ and $z=z_1$. However, in practice no measurements of $P_s^+(\vec{r},\omega)$ are available at $z=z_1$, so the term $\Delta P^-(\vec{r}_A,\omega)$ is neglected, which means that equation (IX-1) is approximated by

$$P^-(\vec{r}_A,\omega) \approx 2\int\int_{-\infty}^{\infty} \left[\frac{1}{\rho(\vec{r})} \left(\frac{\partial G^-(\vec{r},\vec{r}_A,\omega)}{\partial z} \right)^* P^-(\vec{r},\omega) \right]_{z_0} dxdy. \tag{IX-2}$$

In section VII.3.3 we showed that this approximation is justified when the contrasts in the medium are low: $\Delta P^-(\vec{r}_A,\omega)$ is proportional to the *product* of the *scattered* wave field P_s^+ at $z=z_1$ (Figure IX-1a) and the *scattered* Green's wave field G_s^+ at $z=z_1$ (Figure IX-1b).

The subject of this chapter is acoustic inverse wave field extrapolation in high contrast media, where equation (IX-2) breaks down. Looking at the "error term" $\Delta P^-(\vec{r}_A,\omega)$, as defined by (IX-1c), we see that there are two possible paths to be followed:

1. By choosing a non-scattering reference medium for the Green's wave field, G_s^+ will be zero and, consequently, $\Delta P^-(\vec{r}_A,\omega)$ will vanish.
   Of course we cannot choose just any reference medium for the Green's wave field. However, by applying the inverse wave field extrapolation *recursively* from layer interface to layer interface (taking into account the boundary conditions at the layer interfaces), the total reference medium may be subdivided into a number of reference layers, each of which may be (almost) non-scattering. The *recursive* approach to acoustic inverse wave field extrapolation in high contrast media is discussed in section IX.2.

2. On the other hand, for non-recursive applications, we should derive an (approximate) expression for $P_s^+(\vec{r},\omega)$ at $z=z_1$. In this way $\Delta P^-(\vec{r}_A,\omega)$ can be estimated and, subsequently, added to $P_0^-(\vec{r}_A,\omega)$. In section IX.3 we discuss an *iterative* approach to acoustic inverse wave field extrapolation in high contrast media, which is based on iteratively estimating the "error term" $\Delta P^-(\vec{r}_A,\omega)$.

IX.2 RECURSIVE ACOUSTIC INVERSE WAVE FIELD EXTRAPOLATION

IX.2.1 Principle

Consider the i'th layer between the arbitrarily curved surfaces S_{i-1} and S_i in an inhomogeneous acoustic medium (Figure IX-2a). The medium parameters in this layer read $c_i(\vec{r})$ and $\rho_i(\vec{r})$. We will assume that the contrasts within the layer are weak. However, the contrasts at surfaces S_{i-1} and S_i may be arbitrarily high.

Volume V_i for the Kirchhoff-Helmholtz integral is bounded by surfaces S_{i-1} and S_i. For the Green's wave field G we may choose outside V_i any convenient reference medium. We choose this reference medium such that S_{i-1} and S_i are non-reflecting for G (Figure IX-2b). In the special situation of a homogeneous layer, the reference medium for G would be homogeneous throughout, so G would be just the free space Green's function.

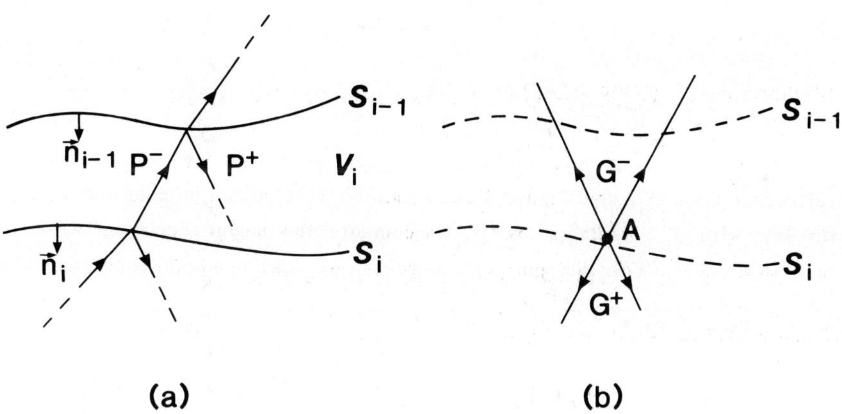

Figure IX-2: a. Layer i in an inhomogeneous acoustic medium.
b. Non-reflecting reference medium for the Green's wave field.

According to chapter VII, inverse wave field extrapolation from S_{i-1} to any point $\vec{r}_A$ in V_i is described by

$$P^-(\vec{r}_A,\omega) =$$

$$\int_{S_{i-1}} \frac{-1}{\rho_i(\vec{r})} \left[G^*(\vec{r},\vec{r}_A,\omega) \frac{\partial P(\vec{r},\omega)}{\partial n_{i-1}} - \frac{\partial G^*(\vec{r},\vec{r}_A,\omega)}{\partial n_{i-1}} P(\vec{r},\omega) \right] dS_{i-1},^{1)} \qquad \text{(IX-3a)}$$

where $\partial/\partial n_{i-1}$ stands for $\vec{n}_{i-1}.\nabla$ and $\vec{n}_{i-1}$ being the downward pointing normal vector on surface S_{i-1}. In (IX-3a), $P(\vec{r},\omega)$ represents the *total* acoustic wave field at $\vec{r}$ on S_{i-1}, whereas $P^-(\vec{r}_A,\omega)$ represents the *upgoing* acoustic wave field at $\vec{r}_A$ below S_{i-1}. In the following we take $\vec{r}_A$ just above surface S_i, see Figure IX-2b. $P^-(\vec{r}_A,\omega)$ thus represents the upgoing wave field just above S_i. For the *primary* wave, this upgoing term represents the total wave field just above S_i, see Figure IX-2a. Hence, we can easily find the total wave field just below S_i by applying the following boundary condition for the acoustic pressure:

$$\lim_{\vec{r}_A \uparrow S_i} [P(\vec{r}_A,\omega)] = \lim_{\vec{r}_A \downarrow S_i} [P^-(\vec{r}_A,\omega)]. \qquad \text{(IX-3b)}$$

This total wave field just below S_i (the upgoing primary wave plus its reflection from S_i, see Figure IX-2a) can be used again in equation (IX-3a), with S_{i-1} replaced by S_i, to compute the upgoing primary wave just above S_{i+1}. For this purpose, however, we also need an expression for $\partial P(\vec{r},\omega)/\partial n_i$. From (IX-3a) we obtain

$$\frac{\partial P^-(\vec{r}_A,\omega)}{\partial n_{i,A}} = \qquad \text{(IX-4a)}$$

$$\int_{S_{i-1}} \frac{-1}{\rho_i(\vec{r})} \left[\frac{\partial G^*(\vec{r},\vec{r}_A,\omega)}{\partial n_{i,A}} \frac{\partial P(\vec{r},\omega)}{\partial n_{i-1}} - \frac{\partial^2 G^*(\vec{r},\vec{r}_A,\omega)}{\partial n_{i-1} \partial n_{i,A}} P(\vec{r},\omega) \right] dS_{i-1},$$

---

1) Here the summation convention does not apply to the layer index i.

where $\partial/\partial n_{i,A}$ stands for $\vec{n}_i.\nabla_A$, $\vec{n}_i$ being the downward pointing normal vector on surface S_i and ∇_A being the gradient at $\vec{r}_A$. In analogy with (IX-3b), the boundary condition for the particle velocity at S_i reads

$$\lim_{\vec{r}_A \uparrow S_i} \left[\frac{1}{\rho_{i+1}(\vec{r}_A)} \frac{\partial P(\vec{r}_A,\omega)}{\partial n_{i,A}} \right] = \lim_{\vec{r}_A \downarrow S_i} \left[\frac{1}{\rho_i(\vec{r}_A)} \frac{\partial P^-(\vec{r}_A,\omega)}{\partial n_{i,A}} \right] . \tag{IX-4b}$$

Equations (IX-3) and (IX-4) together can be used in a *recursive* mode to extrapolate the primary upgoing wave field inversely from layer interface to layer interface.

Generally, in recursive one-way *forward* wave field extrapolation schemes, significant additional effort is required at the layer boundaries in order to take transmission effects into account. Here we have presented a recursive scheme for *inverse* wave field extrapolation, in which transmission effects are simply taken into account by (IX-3b) and (IX-4b). For the situation of homogeneous layers, the free space Green's wave field (V-26) may be substituted into (IX-3a) and (IX-4a). In this case the only approximation in the recursive scheme concerns the negligence of internal multiples and evanescent waves, irrespective of the magnitude of the contrasts and the shape of the interfaces.

IX.2.2 Example

We illustrate the validity of the recursive approach with the aid of a simple example. We consider two homogeneous half-spaces separated by an interface at z_1=500 m, see Figure IX-3a. The medium parameters of the upper half-space read c_1=1500 m/s and ρ_1=1000 kg/m^3; the medium parameters of the lower half-space read c_2=3000 m/s and ρ_2=2000 kg/m^3. Note the significant contrast at z_1. A 2-D acoustic wave field is radiated by a source in the lower half-space at (x=o, z_3=1500 m). The upgoing wave field $p^-(x,z_o,t)$ for z_o=o m is shown as a function of x and t in Figure IX-3b; the upgoing wave field $p^-(x,z_2,t)$ for z_2=1000 m is shown in Figure IX-3c.

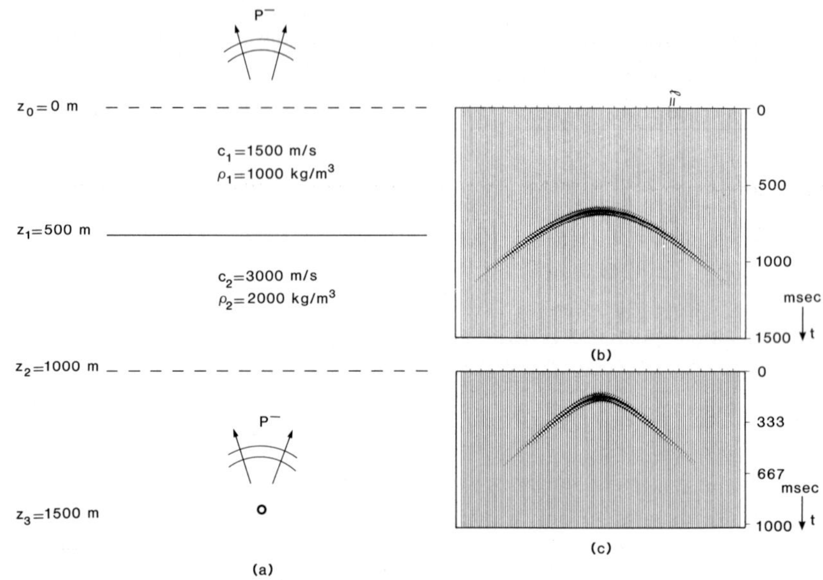

figure IX-3: a. Inhomogeneous acoustic medium with a high contrast at z_1.
b. Upgoing wave field at z_0.
c. Upgoing wave field at z_2.

In the wavenumber-frequency domain, these wave fields are related via a forward wave field extrapolation operator, according to

$$\tilde{P}^-(z_0) = \tilde{W}^-(z_0,z_2)\tilde{P}^-(z_2), \tag{IX-5a}$$

where

$$\tilde{W}^-(z_0,z_2) = \tilde{W}^-(z_0,z_1)\tilde{T}^-(z_1)\tilde{W}^-(z_1,z_2), \tag{IX-5b}$$

with

$$\tilde{W}^-(z_0,z_1) = \exp(-jk_{z,1}|z_1-z_0|), \tag{IX-5c}$$

$$\tilde{W}^-(z_1,z_2) = \exp(-jk_{z,2}|z_2-z_1|), \tag{IX-5d}$$

$$\tilde{T}^-(z_1) = 1-\tilde{R}^+(z_1) \tag{IX-5e}$$

and

$$\tilde{R}^+(z_1) = \frac{\rho_2 k_{z,1} - \rho_1 k_{z,2}}{\rho_2 k_{z,1} + \rho_1 k_{z,2}}, \tag{IX-5f}$$

see section VII.2.2. Before we apply the recursive approach, we show that the modified matched filter approach fails for this configuration. Following section VII.2.2, inverse extrapolation according to the modified matched filter approach reads

$$<\tilde{P}^-(z_2)> = <\tilde{F}^-(z_2,z_o)>\tilde{P}^-(z_o), \tag{IX-6a}$$

where

$$<\tilde{F}^-(z_2,z_o)> \triangleq [\tilde{W}^+(z_2,z_o)]^*, \tag{IX-6b}$$

or

$$<\tilde{F}^-(z_2,z_o)> \triangleq [\tilde{W}^+(z_2,z_1)\tilde{T}^+(z_1)\tilde{W}^+(z_1,z_o)]^*, \tag{IX-6c}$$

with

$$\tilde{W}^+(z_1,z_o) = \tilde{W}^-(z_o,z_1), \tag{IX-6d}$$

$$\tilde{W}^+(z_2,z_1) = \tilde{W}^-(z_1,z_2) \tag{IX-6e}$$

and

$$\tilde{T}^+(z_1) = 1 + \tilde{R}^+(z_1). \tag{IX-6f}$$

Note that for propagating waves, $<\tilde{F}^-(z_2,z_o)>$ is related to the forward operator $\tilde{W}^-(z_o,z_2)$, according to

$$\langle \tilde{F}^-(z_2,z_o)\rangle = [1-(\tilde{R}^+(z_1))^2]\,[\tilde{W}^-(z_o,z_2)]^{-1}. \tag{IX-6g}$$

The result of (IX-6a), transformed back to the space-time domain, is shown in Figure IX-4a. It represents the upgoing wave field $\langle p^-(x,z_2,t)\rangle$ for z_2=1000 m as a function of x and t. The maximum amplitude of each trace is shown as a function of x in Figure IX-4b (dotted line). Note the important mismatch with the exact result (solid line). This could be expected because for high $\tilde{R}^+(z_1)$, operator $\langle \tilde{F}^-(z_2,z_o)\rangle$ deviates significantly from $[\tilde{W}^-(z_o,z_2)]^{-1}$, see equation (IX-6g).

Next we analyse the recursive approach, as discussed in the previous section. For the laterally invariant configuration of Figure IX-3a, the integrals in equations (IX-3a) and (IX-4a) represent spatial convolutions, which correspond to multiplications in the wavenumber-frequency domain. For the first inverse extrapolation step from z_o to z_1 we rewrite (IX-3a) and (IX-4a) as

$$\tilde{P}^-(z_1) = \frac{-1}{\rho_1}\left[\tilde{G}^*(z_o,z_1)\,\frac{\partial \tilde{P}^-(z_o)}{\partial z_o} - \frac{\partial \tilde{G}^*(z_o,z_1)}{\partial z_o}\,\tilde{P}^-(z_o)\right] \tag{IX-7a}$$

and

$$\frac{\partial \tilde{P}^-(z_1)}{\partial z_1} = \frac{-1}{\rho_1}\left[\frac{\partial \tilde{G}^*(z_o,z_1)}{\partial z_1}\,\frac{\partial \tilde{P}^-(z_o)}{\partial z_o} - \frac{\partial^2 \tilde{G}^*(z_o,z_1)}{\partial z_o \partial z_1}\,\tilde{P}^-(z_o)\right], \tag{IX-7b}$$

respectively. Note that the total wave field at z_o is equal to the upgoing wave field $\tilde{P}^-(z_o)$, see Figure IX-3a. It satisfies one-way wave equation (III-45b):

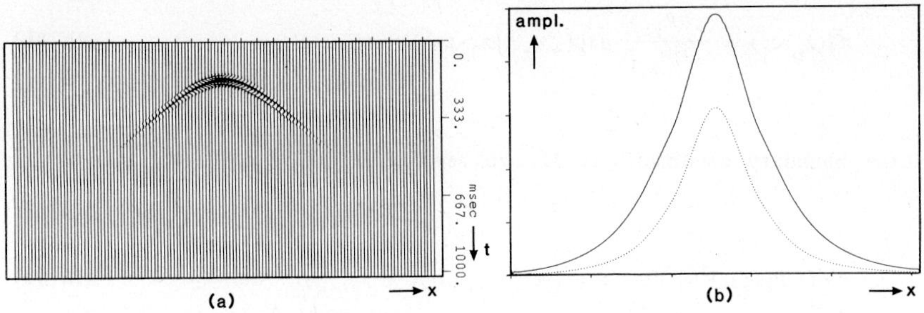

Figure IX-4: Inverse extrapolation according to the modified matched filter approach.
a. Upgoing wave field at z_2.
b. Maximum amplitude per trace (dotted line) compared with the exact result (solid line).

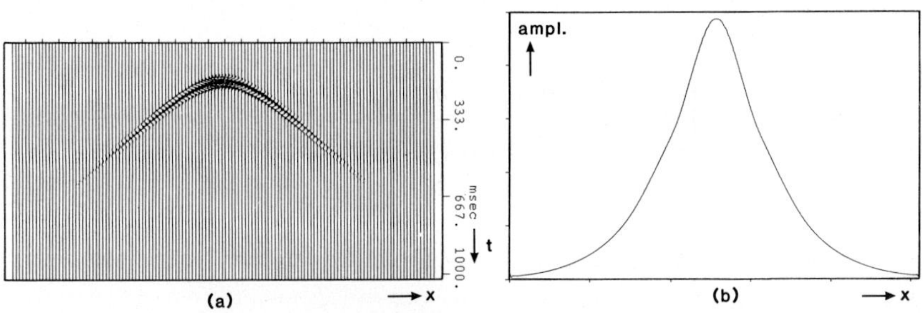

Figure IX-5: Inverse extrapolation according to the recursive approach.
a. Upgoing wave field at z_2.
b. Maximum amplitude per trace (dotted line) compared with the exact result (solid line).
(The dotted line is hidden by the solid line).

$$\frac{\partial \tilde{P}^-(z_0)}{\partial z_0} = jk_{z,1}\tilde{P}^-(z_0). \tag{IX-8a}$$

The Green's wave field $\tilde{G}(z_0,z_1)$ is given by the free space solution (V-27a):

$$\tilde{G}(z_o,z_1) = \frac{\rho_1}{2jk_{z,1}} \exp(-jk_{z,1}|z_1-z_o|). \tag{IX-8b}$$

For the boundary conditions at z_1, we rewrite (IV-3b) and (IV-4b) as

$$\lim_{\epsilon \to o} \tilde{P}(z_1+\epsilon) = \lim_{\epsilon \to o} \tilde{P}^-(z_1-\epsilon) \tag{IX-9a}$$

and

$$\lim_{\epsilon \to o} \frac{1}{\rho_2} \frac{\partial \tilde{P}(z_1+\epsilon)}{\partial z_1} = \lim_{\epsilon \to o} \frac{1}{\rho_1} \frac{\partial \tilde{P}^-(z_1-\epsilon)}{\partial z_1}, \tag{IX-9b}$$

respectively. For the second inverse extrapolation step from z_1 to z_2 we rewrite (IX-3a) as

$$\tilde{P}^-(z_2) = \frac{-1}{\rho_2} \left[\tilde{G}^*(z_1,z_2) \frac{\partial \tilde{P}(z_1)}{\partial z_1} - \frac{\partial \tilde{G}^*(z_1,z_2)}{\partial z_1} \tilde{P}(z_1) \right], \tag{IX-10}$$

where the Green's wave field $\tilde{G}(z_1,z_2)$ is given by the free space solution:

$$\tilde{G}(z_1,z_2) = \frac{\rho_2}{2jk_{z,2}} \exp(-jk_{z,2}|z_2-z_1|). \tag{IX-11}$$

The result of (IX-10), transformed back to the space-time domain, is shown in Figure IX-5a. It represents the upgoing wave field $p^-(x,z_2,t)$ for z_2=1000 m as a function of x and t. The maximum of each trace is shown as a function of x in Figure IX-5b (dotted line). Note the perfect match with the exact result (solid line). This very good result is explained analytically as follows. Substitution of (IX-8) into (IX-7) yields for propagating waves ($k_{z,1}$ is real)

$$\tilde{P}^-(z_1) = \exp(+jk_{z,1}|z_1-z_0|)\tilde{P}^-(z_0) \tag{IX-12a}$$

and

$$\frac{\partial\tilde{P}^-(z_1)}{\partial z_1} = jk_{z,1}\exp(+jk_{z,1}|z_1-z_0|)\tilde{P}^-(z_0). \tag{IX-12b}$$

Applying boundary conditions (IX-9a) and (IX-9b) and substituting the results together with (IX-11) into (IX-10) yields for propagating waves ($k_{z,2}$ is real)

$$\tilde{P}^-(z_2) = \exp(+jk_{z,2}|z_2-z_1|)\left(\frac{\rho_2 k_{z,1}+\rho_1 k_{z,2}}{2\rho_1 k_{z,2}}\right)\exp(+jk_{z,1}|z_1-z_0|)\tilde{P}^-(z_0), \tag{IX-13a}$$

or

$$\tilde{P}^-(z_2) = \tilde{F}^-(z_2,z_0)\tilde{P}^-(z_0), \tag{IX-13b}$$

where

$$\tilde{F}^-(z_2,z_0) = [\tilde{W}^-(z_0,z_1)\tilde{T}^-(z_1)\tilde{W}^-(z_1,z_2)]^{-1} = [\tilde{W}^-(z_0,z_2)]^{-1}. \tag{IX-13c}$$

Equation (IX-13) confirms that inverse wave field extrapolation according to the recursive approach yields a result which is *exact* for the primary waves in the propagating wavenumber area. Hence, even in the presence of significant contrasts, the only approximation concerns the negligence of internal multiples and evanescent waves. Bear in mind that we presented this analysis in the wavenumber domain only as an illustration. In practice, (recursive) inverse wave field extrapolation is entirely carried out in the space domain.

IX.3 ITERATIVE ACOUSTIC INVERSE WAVE FIELD EXTRAPOLATION

IX.3.1 Principle

We consider again the non-recursive expression (IX-1) for acoustic inverse extrapolation of the upgoing wave field P^- from acquisition surface $z=z_0$ to subsurface point $\vec{r}_A$ (Figure IX-1):

$$P^-(\vec{r}_A,\omega) = P_0^-(\vec{r}_A,\omega) + \Delta P^-(\vec{r}_A,\omega), \tag{IX-14a}$$

where

$$P_0^-(\vec{r}_A,\omega) = 2\iint_{-\infty}^{\infty}\left[\frac{1}{\rho(\vec{r})}\left(\frac{\partial G^-(\vec{r},\vec{r}_A,\omega)}{\partial z}\right)^* P^-(\vec{r},\omega)\right]_{z_0} dxdy \tag{IX-14b}$$

and

$$\Delta P^-(\vec{r}_A,\omega) = -2\iint_{-\infty}^{\infty}\left[\frac{1}{\rho(\vec{r})}\left(\frac{\partial G_s^+(\vec{r},\vec{r}_A,\omega)}{\partial z}\right)^* P_s^+(\vec{r},\omega)\right]_{z_1} dxdy. \tag{IX-14c}$$

Since $P_s^+(\vec{r},\omega)$ at z_1 is unknown, $\Delta P^-(\vec{r}_A,\omega)$ cannot be computed directly. When the contrasts in the medium are significant, $\Delta P^-(\vec{r}_A,\omega)$ may not just be neglected. In this section we discuss how $\Delta P^-(\vec{r}_A,\omega)$ can be computed iteratively. For this purpose we choose the "reference surface" $z=z_1$ just below $\vec{r}_A$ (see Figure IX-1b).

The first step is described by equation (IX-14b), which is illustrated in Figure IX-6a. Here $P^-(\vec{r},\omega)$ for $\vec{r}=(x,y,z=z_0)$ represents the upgoing wave field at $z=z_0$. $G^-(\vec{r},\vec{r}_A,\omega)$ for $\vec{r}=(x,y,z=z_0)$ represents an upgoing Green's wave field at $z=z_0$, related to a source at $\vec{r}_A=(x_A,y_A,z_A)$. Hence, $\left(G^-(\vec{r},\vec{r}_A,\omega)\right)^*$ for $\vec{r}=(x,y,z=z_0)$ represents a downgoing Green's wave field at $z=z_0$ which propagates backward to $\vec{r}_A$.
Applying this backward propagating Green's wave field to the upgoing wave field at $z=z_0$ according to equation (IX-14b), yields the "zeroth order" estimate $P_0^-(\vec{r}_A,\omega)$ of the true upgoing wave field $P^-(\vec{r}_A,\omega)$.

The second step is described by

$$P_{s,1}^+(\vec{r}_A,\omega) = -2\iint_{-\infty}^{\infty}\left[\frac{1}{\rho(\vec{r})}\frac{\partial G_s^+(\vec{r}_A,\vec{r},\omega)}{\partial z} P_0^-(\vec{r},\omega)\right]_{z_1} dxdy, \tag{IX-15}$$

which is illustrated in Figure IX-6b.[1)] Here $G_s^+(\vec{r}_A,\vec{r},\omega)$ represents a scattered downgoing Green's wave field at $\vec{r}_A$, related to sources at

---

[1)] A similar situation was discussed in section V.5.1, see equation (V-50b) and Figure V-11b.

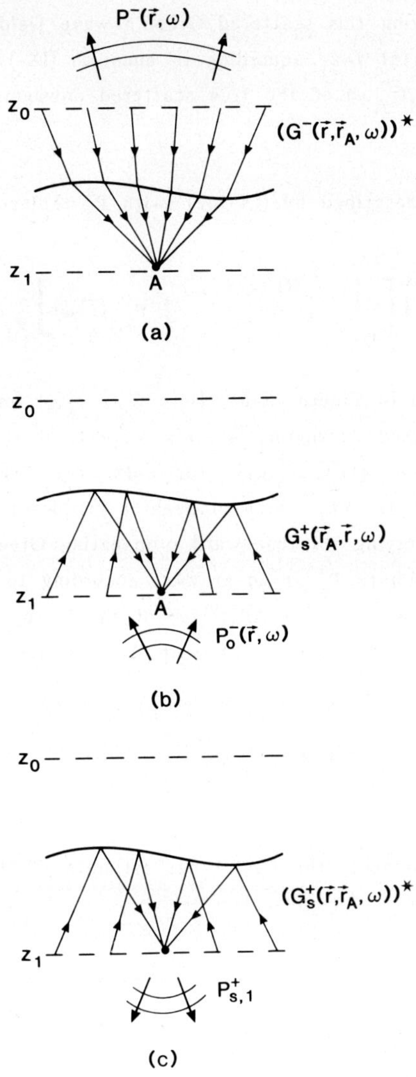

Figure IX-6: Iterative acoustic inverse wave field extrapolation.
a. Back-propagation of the upgoing wave from z_o to $\vec{r}_A$.
b. Simulation of the scattered downgoing wave at z_1 from the upgoing wave at z_1.
c. Back-propagation of the simulated scattered downgoing wave from z_1 to $\vec{r}_A$.
The latter two steps are applied iteratively.

$\vec{r}'=(x,y,z=z_1)$. Applying this scattered Green's wave field to the upgoing wave field $P_0^-(\vec{r}',\omega)$ at $z=z_1$ according to equation (IX-15), yields the first order estimate $P_{s,1}^+(\vec{r}_A,\omega)$ of the true scattered downgoing wave field $P_s^+(\vec{r}_A,\omega)$.

The third step is described by (IX-14c) (with P_s^+ replaced by $P_{s,1}^+$), hence

$$\Delta P_1^-(\vec{r}_A,\omega) = -2\iint_{-\infty}^{\infty}\left[\frac{1}{\rho(\vec{r}')}\left(\frac{\partial G_s^+(\vec{r}',\vec{r}_A,\omega)}{\partial z}\right)^* P_{s,1}^+(\vec{r}',\omega)\right]_{z_1} dxdy, \tag{IX-16}$$

which is illustrated in Figure IX-6c. Here $G_s^+(\vec{r}',\vec{r}_A,\omega)$ for $\vec{r}'=(x,y,z=z_1)$ represents a scattered downgoing Green's wave field at $z=z_1$, related to a source at $\vec{r}_A$. Hence, $\left(G_s^+(\vec{r}',\vec{r}_A,\omega)\right)^*$ for $\vec{r}'=(x,y,z=z_1)$ represents an upgoing Green's wave field at $z=z_1$ which propagates backward via the scattering medium to $\vec{r}_A$. Applying this backward propagating Green's wave field to the first order estimate $P_{s,1}^+(\vec{r}',\omega)$ at $z=z_1$ according to equation (IX-16), yields the first order estimate $\Delta P_1^-(\vec{r}_A,\omega)$ of the "error term" $\Delta P^-(\vec{r}_A,\omega)$. Substituting this result into (IX-14a) (with ΔP^- replaced by ΔP_1^-), yields

$$P_1^-(\vec{r}_A,\omega) = P_0^-(\vec{r}_A,\omega) + \Delta P_1^-(\vec{r}_A,\omega), \tag{IX-17}$$

where $P_1^-(\vec{r}_A,\omega)$ represents the first order estimate of the true upgoing wave field $P^-(\vec{r}_A,\omega)$.

The essence of this scheme is that the unknown scattered downgoing wave P_s^+ at z_1 is simulated by *forward modeling* in the second step. Since the "input" P_0^- for this forward modeling is not exact, the second and third step should preferably be carried out iteratively. In this case, iteration ℓ (with $\ell \geq 1$) reads

$$P_{s,\ell}^+(\vec{r}_A,\omega) = -2\iint_{-\infty}^{\infty}\left[\frac{1}{\rho(\vec{r}')}\frac{\partial G_s^+(\vec{r}_A,\vec{r}',\omega)}{\partial z} P_{\ell-1}^-(\vec{r}',\omega)\right]_{z_1} dxdy, \tag{IX-18a}$$

$$\Delta P_\ell^-(\vec{r}_A,\omega) = -2\int\int_{-\infty}^{\infty}\left[\frac{1}{\rho(\vec{r})}\left(\frac{\partial G_s^+(\vec{r},\vec{r}_A,\omega)}{\partial z}\right)^* P_{s,\ell}^+(\vec{r},\omega)\right]_{z_1} dxdy \qquad \text{(IX-18b)}$$

and

$$P_\ell^-(\vec{r}_A,\omega) = P_o^-(\vec{r}_A,\omega) + \Delta P_\ell^-(\vec{r}_A,\omega), \qquad \text{(IX-18c)}$$

with $P_o^-(\vec{r}_A,\omega)$ being given by equation (IX-14b).

Note that in this iterative scheme the computations must be carried out for all $\vec{r}_A$ just above $z=z_1$. Therefore the matrix notation is more appropriate. In the matrix notation of Appendix A, expressions (IX-18a), (IX-18b) and (IX-18c) may be replaced by

$$\vec{P}_{s,\ell}^{+}(z_1) = \mathbf{W}_s^+(z_1,z_1)\vec{P}_{\ell-1}^{-}(z_1), \qquad \text{(IX-19a)}$$

$$\Delta\vec{P}_\ell^{-}(z_1) = [\mathbf{W}_s^+(z_1,z_1)]^*\vec{P}_{s,\ell}^{+}(z_1) \qquad \text{(IX-19b)}$$

and

$$\vec{P}_\ell^{-}(z_1) = \vec{P}_o^{-}(z_1) + \Delta\vec{P}_\ell^{-}(z_1), \qquad \text{(IX-19c)}$$

where $\vec{P}_o^-(z_1)$ is defined by expression (IX-14b) in the matrix notation:

$$\vec{P}_o^{-}(z_1) = \langle\mathbf{F}^-(z_1,z_o)\rangle\vec{P}^{-}(z_o). \qquad \text{(IX-20)}$$

In expressions (IX-19a) and (IX-19b), the forward extrapolation operator $\mathbf{W}_s^+(z_1,z_1)$ is defined as

$$\mathbf{W}_s^+(z_1,z_1) = -2\,\frac{\partial \mathbf{G}_s^+(z_1,z=z_1)}{\partial z}\,\mathbf{M}^{-1}(z_1), \qquad \text{(IX-21a)}$$

where the columns of $\mathbf{G}_s^+(z_1,z_1)$ contain the discretized versions of $G_s^+(\vec{r}_A,\vec{r},\omega)=G_s^+(\vec{r},\vec{r}_A,\omega)$ for $z=z_A=z_1$; in expression (IX-20), the modified matched inverse operator $<\mathbf{F}^-(z_1,z_0)>$ is defined as

$$<\mathbf{F}^-(z_1,z_0)> = 2\left(\frac{\partial \mathbf{G}^+(z_1,z=z_0)}{\partial z}\right)^* \mathbf{M}^{-1}(z_0), \tag{IX-21b}$$

where the columns of $\mathbf{G}^+(z_1,z_0)$ contain the discretized versions of $G^+(\vec{r}_A,\vec{r},\omega)=G^-(\vec{r},\vec{r}_A,\omega)$ for $z=z_0$ and $z_A=z_1$. For a further discussion of the matrices and vectors in (IX-19), (IX-20) and (IX-21) we refer to sections V.5.2 and VII.3.5.

Expressions (IX-19a), (IX-19b) and (IX-19c) can be combined into one expression, according to

$$\vec{P}_\ell^-(z_1) = \vec{P}_0^-(z_1) + \Xi(z_1)\vec{P}_{\ell-1}^-(z_1), \tag{IX-22a}$$

where

$$\Xi(z_1) \triangleq [\mathbf{W}_s^+(z_1,z_1)]^* \mathbf{W}_s^+(z_1,z_1). \tag{IX-22b}$$

Upon substitution of (IX-20), equation (IX-22a) can be rewritten as

$$\vec{P}_\ell^-(z_1) = \mathbf{F}_\ell^-(z_1,z_0)\vec{P}^-(z_0), \tag{IX-23a}$$

where the ℓ'th order estimate $\mathbf{F}_\ell^-(z_1,z_0)$ of the inverse operator $\mathbf{F}^-(z_1,z_0)$ reads

$$\mathbf{F}_\ell^-(z_1,z_0) = \sum_{m=0}^{\ell} \left(\Xi(z_1)\right)^m <\mathbf{F}^-(z_1,z_0)>. \tag{IX-23b}$$

Equation (IX-23a) elegantly describes inverse extrapolation of the upgoing wave field $\vec{P}^-$ at depth level z_0, yielding the ℓ'th order estimate of the upgoing wave field $\vec{P}^-$ at depth level z_1. Equation (IX-23b) relates the ℓ'th order estimate of the inverse operator $\mathbf{F}^-(z_1,z_0)$ to the modified matched inverse operator $<\mathbf{F}^-(z_1,z_o)>$. Note that

$$\mathbf{F}_o^-(z_1,z_o) = <\mathbf{F}^-(z_1,z_o)>, \tag{IX-23c}$$

which means that for ℓ=o the iterative approach to inverse wave field extrapolation degrades to the modified matched inverse approach, which was discussed in chapter VII. Finally, for inverse extrapolation of downgoing waves the corresponding expressions read

$$\vec{\mathbf{P}}_\ell^+(z_o) = \mathbf{F}_\ell^+(z_o,z_1)\vec{\mathbf{P}}^+(z_1), \tag{IX-24a}$$

where

$$\mathbf{F}_\ell^+(z_o,z_1) = \sum_{m=o}^{\ell} \left(\Xi(z_o)\right)^m <\mathbf{F}^+(z_o,z_1)>, \tag{IX-24b}$$

with

$$\Xi(z_o) \triangleq [\mathbf{W}_s^-(z_o,z_o)]^* \mathbf{W}_s^-(z_o,z_o). \tag{IX-24c}$$

Matrices $\mathbf{W}_s^-(z_o,z_o)$ and $<\mathbf{F}^+(z_o,z_1)>$ are defined in a similar way as above. In the next section we show for some simplified situations that for $\ell\to\infty$ the iterative approach converges to exact inverse wave field extrapolation for the propagating wavenumber area.

IX.3.2 Examples

We illustrate the validity of the iterative approach with the aid of some simple examples. We consider again the situation depicted in Figure IX-3. For the the laterally invariant configuration of Figure IX-3a, we replace (IX-23a) and (IX-23b) by the following expressions in the wavenumber-frequency domain:

$$\tilde{P}_\ell^-(z_2) = \tilde{F}_\ell^-(z_2,z_o)\tilde{P}^-(z_o), \tag{IX-25a}$$

where

$$\tilde{F}_\ell^-(z_2,z_o) = \sum_{m=o}^{\ell} \left(\tilde{\Xi}(z_2)\right)^m \langle\tilde{F}^-(z_2,z_o)\rangle, \tag{IX-25b}$$

with

$$\tilde{\Xi}(z_2) = [\tilde{W}_s^+(z_2,z_2)]^* \tilde{W}_s^+(z_2,z_2). \tag{IX-25c}$$

The forward extrapolation operator $\tilde{W}_s^+(z_2,z_2)$ describes upward propagation from z_2 to z_1, scattering at z_1 and downward propagation from z_1 to z_2, hence

$$\tilde{W}_s^+(z_2,z_2) = \tilde{W}^+(z_2,z_1)\tilde{R}^-(z_1)\tilde{W}^-(z_1,z_2), \tag{IX-26a}$$

where

$$\tilde{R}^-(z_1) = -\tilde{R}^+(z_1), \tag{IX-26b}$$

with $\tilde{W}^+(z_2,z_1)$, $\tilde{W}^-(z_1,z_2)$ and $\tilde{R}^+(z_1)$ defined in section IX.2.2. Substitution into (IX-25) yields for the propagating wavenumber area

$$\tilde{F}_\ell^-(z_2,z_o) = \sum_{m=o}^{\ell} \left(\tilde{R}^+(z_1)\right)^{2m} \langle\tilde{F}^-(z_2,z_o)\rangle. \tag{IX-27a}$$

In section IX.2.2 we found that for propagating waves the modified matched inverse operator is related to the exact inverse operator, according to

$$\langle\tilde{F}^-(z_2,z_o)\rangle = \left[1-\left(\tilde{R}^+(z_1)\right)^2\right]\tilde{F}^-(z_2,z_o), \tag{IX-27b}$$

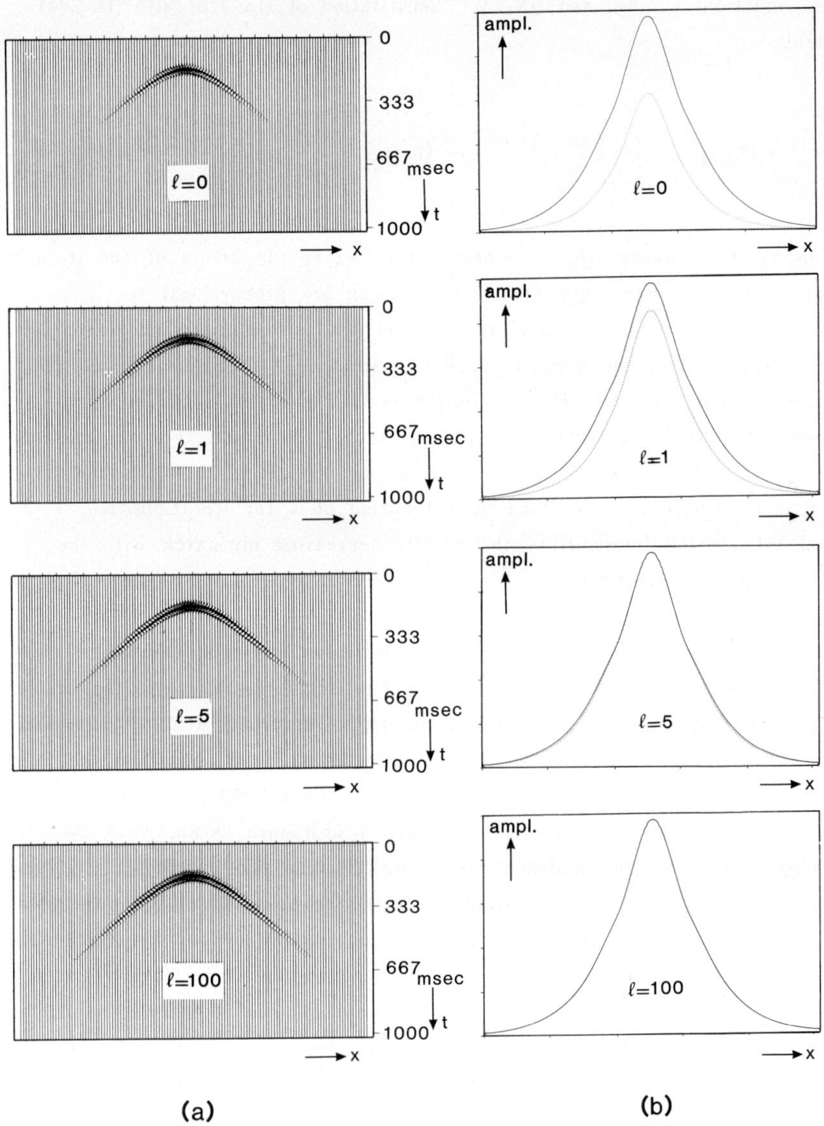

Figure IX-7: Inverse extrapolation of the data in Figure IX-3b, according to the iterative approach.

a. Upgoing wave field at z_2 for iterations 0, 1, 5 and 100.

b. Maximum amplitude per trace (dotted lines) compared with the exact result (solid lines) for iterations 0, 1, 5 and 100.

see equations (IX-6g) and (IX-13c). Substitution of (IX-27b) into (IX-27a) yields

$$\tilde{F}_\ell^-(z_2,z_0) = \left[1-\left(\tilde{R}^+(z_1)\right)^{2\ell+2}\right]\tilde{F}^-(z_2,z_0). \tag{IX-28}$$

This result expresses that for propagating waves the errors of the iterative approach to inverse wave field extrapolation are proportional to $\left(\tilde{R}^+(z_1)\right)^{2\ell+2}$. For pre-critical propagation (i.e., for propagating waves in both half-spaces), the modulus of $\tilde{R}^+(z_1)$ is smaller than one, hence, the errors vanish for $\ell\to\infty$. This is illustrated in Figure IX-7. Figure IX-7a shows the inverse extrapolated upgoing wave field $p_\ell^-(x,z_2,t)$ as a function of x and t for iterations ℓ=0, 1, 5 and 100. Figure IX-7b shows the maximum amplitude per trace as a function of x for iterations ℓ=0, 1, 5 and 100 (dotted lines). Note the rapidly decreasing mismatch with the exact result (solid lines).

In the next example we show that the iterative approach takes also care of internal multiple reflections. Consider the configuration depicted in Figure IX-8a. The medium contains significant contrasts at z_1=300 m and z_2=800 m. A 2-D acoustic wave field is radiated by a source in the lower half-space at (x=o, z_4=1800 m). The upgoing wave field $p^-(x,z_0,t)$ for z_0=0 m is shown as a function of x and t in Figure IX-8b. Note the events related to the multiple reflections between the contrasts at z_1 and z_2. The exact upgoing wave field $p^-(x,z_3,t)$ for z_3=1200 m is shown in Figure IX-8c. Note that this wave field contains no multiple reflections. Figure IX-9a shows the inverse extrapolated upgoing wave field $p_\ell^-(x,z_3,t)$ for iterations ℓ=0, 1, 2 and 5. Note that the number of internal multiples in the extrapolation result decreases when ℓ increases. Finally, Figure IX-9b shows the maximum amplitude per trace for iterations ℓ=0, 1, 2 and 5 (dotted lines). Note again the rapidly decreasing mismatch with the exact result (solid lines). The small deviations at the edges are due to the negligence of evanescent waves.

Internal multiples can only be handled correctly when the macro subsurface model is accurately known (in the example we used the exact subsurface

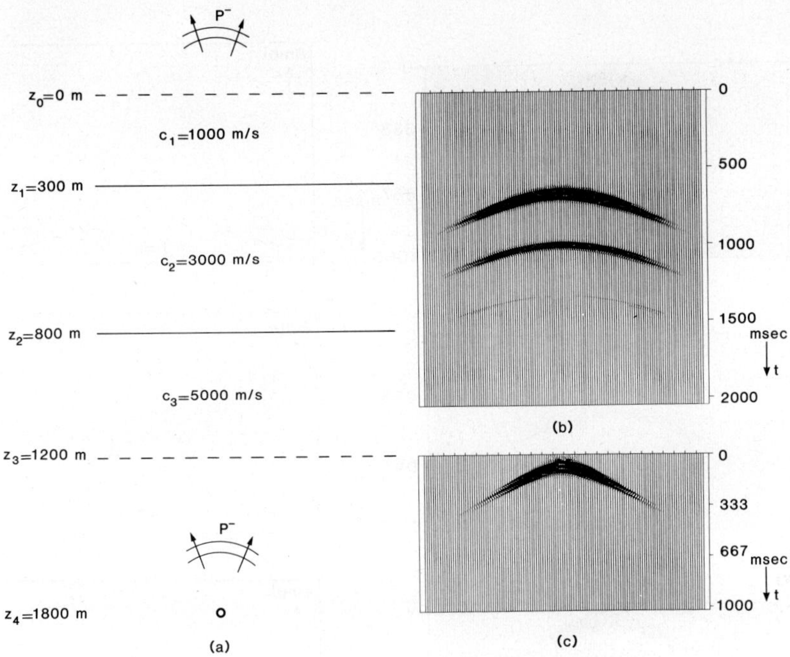

Figure IX-8: a. Inhomogeneous medium with high contrasts at z_1 *and* z_2.
b. Upgoing wave field at z_o, *containing internal multiple reflections.*
c. Upgoing wave field at z_3.

model). Even small errors in the macro subsurface model may already give rise to significant artifacts. These *additional* artifacts can only be avoided at the cost of the correct multiple handling of the iterative operator. Finally, in practical situations with high contrasts, one iteration (ℓ=1) will generally be sufficient. Higher order iterations will hardly improve the result, particularly when the macro subsurface model is slightly in error.

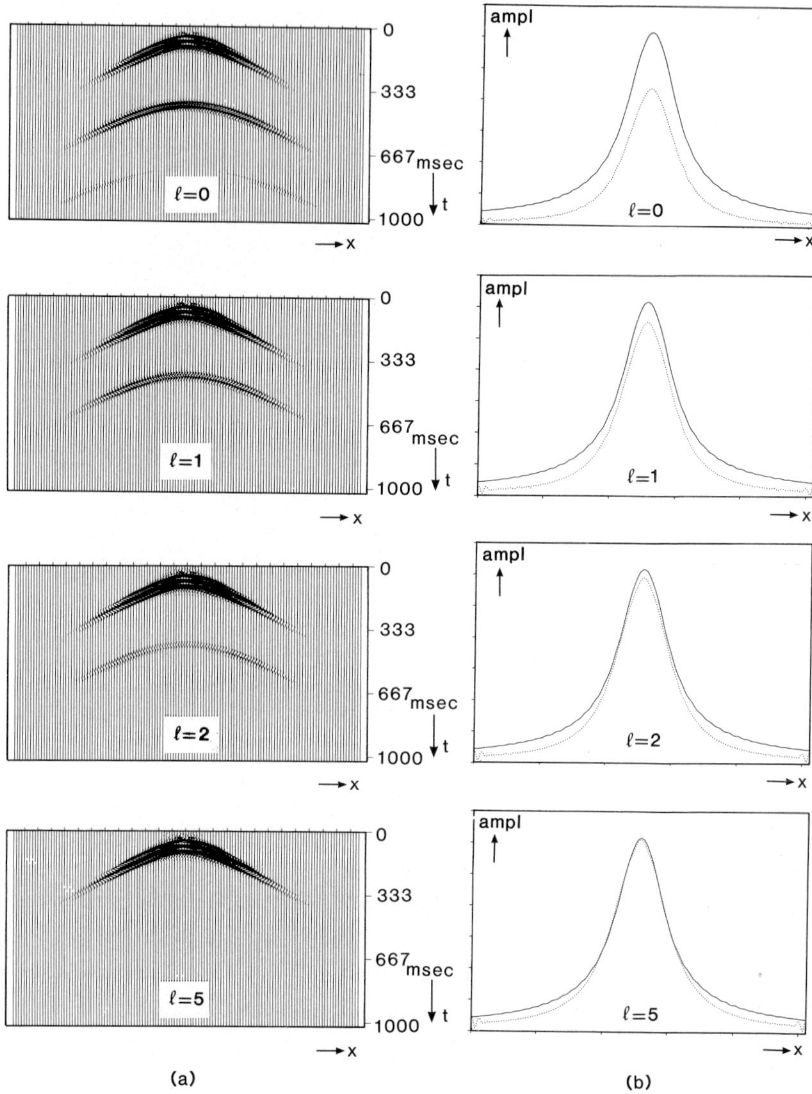

Figure IX-9: Inverse extrapolation of the data in Figure IX-8b, according to the iterative approach.

a. Upgoing wave field at z_3 for iterations 0, 1, 2 and 5.

b. Maximum amplitude per trace (dotted lines) compared with the exact result (solid lines) for iterations 0, 1, 2 and 5.

X

ELASTIC INVERSE WAVE FIELD EXTRAPOLATION IN HIGH CONTRAST MEDIA

X.1. INTRODUCTION

For the configuration of Figure X-1, inverse extrapolation of the upgoing P- and S-waves (Φ^- and $\vec{\Psi}^-$, respectively) from acquisition surface $z=z_0$ to subsurface point $\vec{r}_A$ is described by

$$\Omega^-(\vec{r}_A,\omega) = \Omega_0^-(\vec{r}_A,\omega) + \Delta\Omega^-(\vec{r}_A,\omega), \tag{X-1a}$$

where

$$\Omega_0^-(\vec{r}_A,\omega) = \frac{-2}{j\omega}\iint_{-\infty}^{\infty} \frac{1}{\rho(z_0)} \left[\left(\frac{\partial\Gamma^-_{\phi,\Omega}}{\partial z}\right)^* \Phi^- + \left(\frac{\partial\vec{\Gamma}^-_{\psi,\Omega}}{\partial z}\right)^* \cdot \vec{\Psi}^- \right]_{z_0} dxdy \tag{X-1b}$$

and

$$\Delta\Omega^-(\vec{r}_A,\omega) = \frac{2}{j\omega}\iint_{-\infty}^{\infty} \frac{1}{\rho(z_1)} \left[\left(\frac{\partial\Gamma^+_{\phi,\Omega}}{\partial z}\right)_s^* \Phi_s^+ + \left(\frac{\partial\vec{\Gamma}^+_{\psi,\Omega}}{\partial z}\right)_s^* \cdot \vec{\Psi}_s^+ \right]_{z_1} dxdy, \tag{X-1c}$$

see section VIII.3. Depending on the choice of the source for the Green's wave fields, Ω^- may represent either V_m^- for m=1, 2, 3 or $j\omega\Phi^-$, or $j\omega\Psi_h^-$ for h=1, 2, 3. The only approximation in (X-1) concerns the negligence of evanescent waves at $z=z_0$ and $z=z_1$. However, in practice no measurements of Φ_s^+ and $\vec{\Psi}_s^+$ are available at $z=z_1$, so the term $\Delta\Omega^-(\vec{r}_A,\omega)$ is neglected, which means that equation (X-1) is approximated by

$$\Omega^-(\vec{r}_A,\omega) \approx \frac{-2}{j\omega}\iint_{-\infty}^{\infty} \frac{1}{\rho(z_0)}\left[\left(\frac{\partial\Gamma^-_{\phi,\Omega}}{\partial z}\right)^{*}\Phi^- + \left(\frac{\partial\vec{\Gamma}^-_{\psi,\Omega}}{\partial z}\right)^{*}\cdot\vec{\Psi}^-\right]_{z_0} dxdy. \quad \text{(X-2)}$$

In section VIII.3.3 we showed that this approximation is justified when the contrasts in the medium are low: $\Delta\Omega^-(\vec{r}_A,\omega)$ is proportional to the *product* of scattered wave fields at $z=z_1$ (Figure X-1).

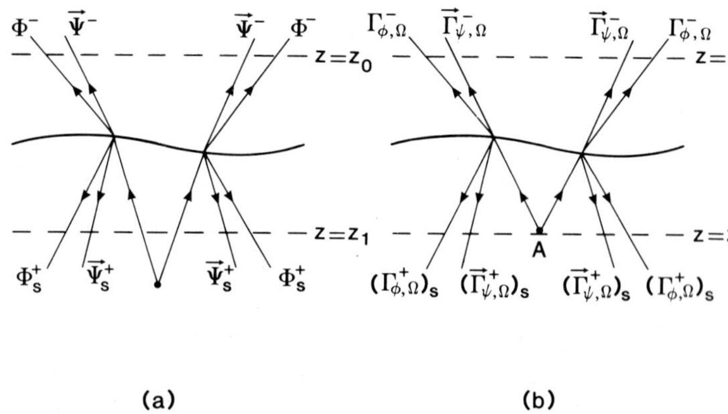

Figure X-1: The error in elastic inverse wave field extrapolation is proportional to the product of scattered elastic wave fields (Figure a) and scattered Green's wave fields (Figure b).

The subject of this chapter is elastic inverse wave field extrapolation in high contrast media, where equation (X-2) breaks down. Similar as in chapter IX, we discuss a *recursive* and an *iterative* approach.

X.2 RECURSIVE ELASTIC INVERSE WAVE FIELD EXTRAPOLATION

Consider the i'th layer between arbitrarily curved surfaces S_{i-1} and S_i in an inhomogeneous anisotropic elastic medium (see Figure IX-2a for the acoustic equivalence). The medium parameters in this layer read $[c_{k\ell mn}(\vec{r})]_i$ and $[\rho(\vec{r})]_i$. We will assume that the contrasts within the layer are weak. However, the contrasts at surfaces S_{i-1} and S_i may be arbitrarily high.

Volume V_i for the Kirchhoff-Helmholtz integral is bounded by surfaces S_{i-1} and S_i. For the Green's wave field we may choose outside V_i any convenient reference medium. We choose this reference medium such that S_{i-1} and S_i are non-reflecting (see Figure IX-2b for the acoustic equivalence). In the special situation of a homogeneous layer, the reference medium would be homogeneous throughout.

According to chapter VIII, inverse wave field extrapolation from S_{i-1} to any point $\vec{r}_A$ in V_i is described by

$$V_m^-(\vec{r}_A,\omega) = \int_{S_{i-1}} [\boldsymbol{\Theta}_m^*(\vec{r},\vec{r}_A,\omega)\vec{V}(\vec{r},\omega) + \boldsymbol{\tau}(\vec{r},\omega)\vec{G}_m^*(\vec{r},\vec{r}_A,\omega)]\,.\vec{n}_{i-1}dS_{i-1},^{1)} \tag{X-3a}$$

where $\vec{n}_{i-1}$ is the downward pointing normal vector on surface S_{i-1}. $\vec{V}(\vec{r},\omega)$ and $\boldsymbol{\tau}(\vec{r},\omega)$ represent the *total* elastic wave field at $\vec{r}$ on S_{i-1}, whereas $V_m^-(\vec{r}_A,\omega)$ for m=1, 2, 3 represents the components of the *upgoing* elastic velocity wave field $\vec{V}^-(\vec{r}_A,\omega)$ at $\vec{r}_A$ below S_{i-1}. In the following we take $\vec{r}_A$ just above surface S_i (see Figure IX-2b for the acoustic equivalence). Hence, $\vec{V}^-(\vec{r}_A,\omega)$ now represents the upgoing wave field just above S_i. For the *primary* wave, this upgoing term represents the total wave field just above S_i (see Figure IX-2a for the acoustic equivalence). Hence, we can easily find the total wave field just below S_i by applying the following boundary condition for the particle velocity:

$$\lim_{\vec{r}_A \uparrow S_i} [\vec{V}(\vec{r}_A,\omega)] = \lim_{\vec{r}_A \downarrow S_i} [\vec{V}^-(\vec{r}_A,\omega)]. \tag{X-3b}$$

This total wave field just below S_i (the upgoing primary wave plus its reflection from S_i, see Figure IX-2a for the acoustic equivalence) can be used again in equation (X-3a), with S_{i-1} replaced by S_i, to compute the upgoing primary wave just above S_{i+1}. For this purpose, however, we also need an expression for $\boldsymbol{\tau}(\vec{r},\omega)$.

---

1) Here the summation convention does not apply to the layer index i.

Consider the stress-velocity relation (II-21), which reads in the space-frequency domain

$$j\omega\tau_{k\ell} = c_{k\ell mn}\partial_n V_m .$$

Substitution of (X-3a) yields

$$\tau^-_{k\ell}(\vec{r}_A,\omega) = \frac{1}{j\omega}\left[c_{k\ell mn}(\vec{r}_A)\right] \int_{i\,S_{i-1}} \left[\partial_{n_A} \mathbf{\Theta}^*_m(\vec{r}\,',\vec{r}_A,\omega)\vec{V}(\vec{r}\,',\omega) + \boldsymbol{\tau}(\vec{r}\,',\omega)\partial_{n_A} \vec{G}^*_m(\vec{r}\,',\vec{r}_A,\omega)\right].\vec{n}\,'_{i-1} dS_{i-1}, \qquad \text{(X-4a)}$$

where ∂_{n_A} for n=1, 2, 3 denotes differentiation with respect to the Green's source point coordinates x_A, y_A, z_A, respectively, and where $\tau^-_{k\ell}(\vec{r}_A,\omega)$ for k=1, 2, 3 and ℓ=1, 2, 3 represents the components of the upgoing elastic stress wave field $\boldsymbol{\tau}^-(\vec{r}_A,\omega)$. In analogy with (X-3b), the boundary condition for the traction at S_i reads

$$\lim_{\vec{r}_A \uparrow S_i} \left[\boldsymbol{\tau}(\vec{r}_A,\omega)\vec{n}_i\right] = \lim_{\vec{r}_A \downarrow S_i} \left[\boldsymbol{\tau}^-(\vec{r}_A,\omega)\vec{n}_i\right], \qquad \text{(X-4b)}$$

where $\vec{n}_i$ is the downward pointing normal vector on surface S_i.

Equations (X-3) and (X-4) together can be used in a *recursive* mode to extrapolate the primary upgoing wave field inversely from layer interface to layer interface.

Generally, in recursive one-way *forward* wave field extrapolation schemes, significant additional effort is required at the layer boundaries in order to take transmission effects into account. Here we have presented a recursive scheme for *inverse* wave field extrapolation in which transmission effects are simply taken into account by (X-3b) and (X-4b). However, this recursive scheme is very sensitive to errors in the ratio of the P- and S-wave propagation velocities in each layer.

For practical applications of elastic inverse wave field extrapolation in high contrast media we prefer the iterative approach which is discussed in the next section.

X.3 ITERATIVE ELASTIC INVERSE WAVE FIELD EXTRAPOLATION

X.3.1 Principle

The derivation of the matrix operators for iterative *elastic* inverse wave field extrapolation is very similar to the derivation for the acoustic situation, as presented in section IX.3.1. We skip the derivation and present the results only. In analogy with (IX-22a), iterative inverse extrapolation of upgoing P-waves from z_0 to z_1 reads in the matrix notation of Appendix A

$$\left(\vec{\Phi}^-(z_1)\right)_\ell = \left(\vec{\Phi}^-(z_1)\right)_0 + \Xi_{\phi,\phi}(z_1)\left(\vec{\Phi}^-(z_1)\right)_{\ell-1}, \tag{X-5a}$$

where, in analogy with (VIII-45a), the "zeroth order" estimate reads

$$\left(\vec{\Phi}^-(z_1)\right)_0 = \langle \mathbf{F}^-_{\phi,\phi}(z_1,z_0)\rangle \vec{\Phi}^-(z_0), \tag{X-5b}$$

and where, in analogy with (IX-22b)

$$\Xi_{\phi,\phi}(z_1) = [\mathbf{W}^+_{\phi,\phi}(z_1,z_1)]^*_s[\mathbf{W}^+_{\phi,\phi}(z_1,z_1)]_s + [\mathbf{W}^+_{\phi,\psi_\alpha}(z_1,z_1)]^*_s[\mathbf{W}^+_{\psi_\alpha,\phi}(z_1,z_1)]_s$$

$$+ [\mathbf{W}^+_{\phi,\psi_\alpha}(z_1,z_0)]^*[\mathbf{W}^-_{\psi_\alpha,\phi}(z_0,z_1)]. \tag{X-5c}$$

In (X-5b), $<\mathbf{F}^{-}_{\phi,\phi}(z_1,z_o)>$ represents the modified matched inverse operator as discussed in section VIII.3.5. In (X-5c), the matrices $[\mathbf{W}^{+}_{\phi,\phi}(z_1,z_1)]_s$, $[\mathbf{W}^{+}_{\phi,\psi_\alpha}(z_1,z_1)]_s$ and $[\mathbf{W}^{+}_{\psi_\alpha,\phi}(z_1,z_1)]_s$ describe forward extrapolation from z_1 to z_1 via the *scattering* medium above z_1 (see Figure IX-6b for the acoustic equivalence); the sub-scripts ϕ an ψ_α refer to P-waves and S_α-waves[1)], respectively. In the iterative scheme these matrices take care of the scattered downgoing P- and S-waves at depth level z_1. Finally, in (X-5c) matrices $\mathbf{W}^{+}_{\phi,\psi_\alpha}(z_1,z_o)$ and $\mathbf{W}^{-}_{\psi_\alpha,\phi}(z_o,z_1)$ are forward extrapolation operators, as defined in section VI.5.2. In the iterative scheme these matrices take care of the converted upgoing S-waves at depth level z_o, which were neglected for practical reasons in equation (VIII-45a), see the discussion in section VIII.3.5.

Upon substitution of (X-5b), equation (X-5a) can be rewritten as

$$\left(\vec{\Phi}^{-}(z_1)\right)_\ell = \left(\mathbf{F}^{-}_{\phi,\phi}(z_1,z_o)\right)_\ell \vec{\Phi}^{-}(z_o), \tag{X-6a}$$

where the ℓ'th order inverse P-wave extrapolation operator reads

$$\left(\mathbf{F}^{-}_{\phi,\phi}(z_1,z_o)\right)_\ell = \sum_{m=o}^{\ell} \left(\Xi_{\phi,\phi}(z_1)\right)^m <\mathbf{F}^{-}_{\phi,\phi}(z_1,z_o)>. \tag{X-6b}$$

Equation (X-6a) elegantly describes inverse extrapolation of the upgoing wave field $\vec{\Phi}^{-}$ at depth level z_o, yielding the ℓ'th order estimate of the upgoing wave field $\vec{\Phi}^{-}$ at depth level z_1. Equation (X-6b) relates the ℓ'th order estimate of the inverse P-wave extrapolation operator to the modified matched inverse operator $<\mathbf{F}^{-}_{\phi,\phi}(z_1,z_o)>$.

The expressions for iterative inverse extrapolation of upgoing S_y-waves from z_o to z_1 read

---

1) S_α-waves for $\alpha=1$, 2 are polarized in the plane perpendicular to the x- or y-axis, respectively, see section II.3.2.

$$\left(\vec{\Psi}_y^-(z_1)\right)_\ell = \left(\vec{\Psi}_y^-(z_1)\right)_o + \Xi_{\psi_y,\psi_y}(z_1)\left(\vec{\Psi}_y^-(z_1)\right)_{\ell-1}, \tag{X-7a}$$

where, in analogy with (VIII-45c), the "zeroth order" estimate reads

$$\left(\vec{\Psi}_y^-(z_1)\right)_o = \langle F^-_{\psi_y,\psi_y}(z_1,z_o)\rangle \vec{\Psi}_y^-(z_o), \tag{X-7b}$$

and where

$$\Xi_{\psi_y,\psi_y}(z_1) = [W^+_{\psi_y,\phi}(z_1,z_1)]_s^* [W^+_{\phi,\psi_y}(z_1,z_1)]_s + [W^+_{\psi_y,\psi_\alpha}(z_1,z_1)]_s^* [W^+_{\psi_\alpha,\psi_y}(z_1,z_1)]_s$$
$$+ [W^+_{\psi_y,\phi}(z_1,z_o)]^* [W^-_{\phi,\psi_y}(z_o,z_1)] + [W^+_{\psi_y,\psi_x}(z_1,z_o)]^* [W^-_{\psi_x,\psi_y}(z_o,z_1)]. \tag{X-7c}$$

In (X-7b), $\langle F^-_{\psi_y,\psi_y}(z_1,z_o)\rangle$ represents the matched inverse operator as discussed in section VIII.3.5. In (X-7c), the forward extrapolation matrices are defined similarly as in (X-5c).

Upon substitution of (X-7b), equation (X-7a) can be rewritten as

$$\left(\vec{\Psi}_y^-(z_1)\right)_\ell = \left(F^-_{\psi_y,\psi_y}(z_1,z_o)\right)_\ell \vec{\Psi}_y^-(z_o), \tag{X-8a}$$

where the ℓ'th order inverse S_y-wave extrapolation operators reads

$$\left(F^-_{\psi_y,\psi_y}(z_1,z_o)\right)_\ell = \sum_{m=o}^{\ell} \left(\Xi_{\psi_y,\psi_y}(z_1)\right)^m \langle F^-_{\psi_y,\psi_y}(z_1,z_o)\rangle. \tag{X-8b}$$

Equation (X-8a) elegantly describes inverse extrapolation of the upgoing wave field $\vec{\Psi}_y^-$ at depth level z_o, yielding the ℓ'th order estimate of the upgoing wave field $\vec{\Psi}_y^-$ at depth level z_1. Equation (X-8b) relates the ℓ'th order estimate of the inverse S_y-wave extrapolation operator to the modified matched inverse operator $\langle F^-_{\psi_y,\psi_y}(z_1,z_o)\rangle$.

The expressions for iterative inverse extrapolation of upgoing S_x-waves from z_o to z_1 are obtained by interchanging the indices x and y in equations (X-7) and (X-8). Finally, similar expressions can be found for iterative inverse extrapolation of *downgoing* P- and S-waves from z_1 to z_o. This is left to the reader.

X.3.2 Examples

We illustrate the validity of the iterative approach with the aid of some simple examples. We consider two homogeneous isotropic half-spaces separated by an interface at z_1=600 m, see Figure X-2a. The medium parameters of the upper half-space read $c_{p,1}$=2600 m/s, $c_{s,1}$=1500 m/s and ρ_1=1000 kg/m^3; the medium parameters of the lower half-space read $c_{p,2}$=3800 m/s, $c_{s,2}$=2800 m/s and ρ_2=1000 kg/m^3. Note the significant contrast at z_1. A 2-D elastic wave field is radiated by a P-wave source in the lower half-space at (x=o, z_3=1600 m). The upgoing P-wave field $\phi^-(x,z_o,t)$ for z_o=o m is shown as a function of x and t in Figure X-2b. Inverse extrapolation to depth level z_2=700 m is carried out according to equation (X-6). The amplitude cross sections of $\left(\phi^-(x,z_2,t)\right)_\ell$ for iterations ℓ=0, 1, 2 and 5 are shown in Figure X-3 (dotted lines). Note the rapidly decreasing mismatch with the exact result (solid lines).

In Figure X-4a we consider the same subsurface configuration, this time with an SV-wave[1)] source in the lower half-space at (x=o, z_3=1600 m). The upgoing SV-wave field $\psi_y^-(x,z_o,t)$ for z_o=o m is shown as a function of x and t in Figure X-4b. Inverse extrapolation to depth level z_2= 700 m is carried out according to equation (X-8). The amplitude cross-sections of $\left(\psi_y^-(x,z_2,t)\right)_\ell$ for iterations ℓ=0, 1, 2 and 5 are shown in Figure X-5 (dotted lines). Again, note the rapidly decreasing mismatch with the exact result (solid lines).

---

1) In the 2-D situation S_y-waves are equivalent with SV-waves.

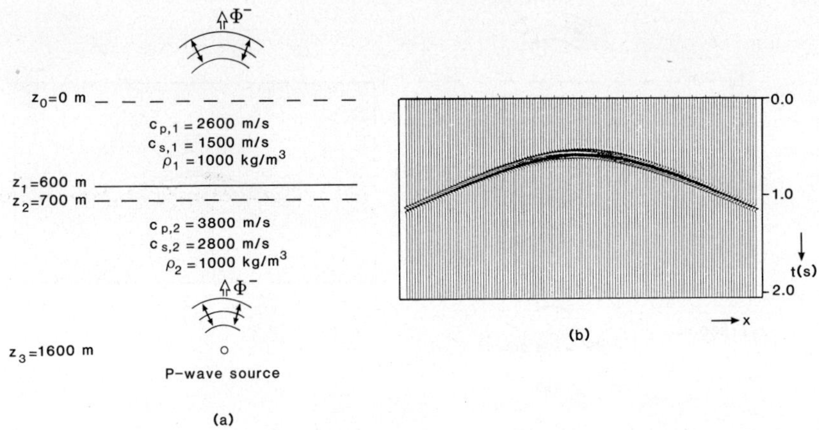

Figure X-2: a. Inhomogeneous elastic medium with a high contrast at z_1 *and a burried P-wave source at* z_3.
b. Upgoing P-wave field at z_0.

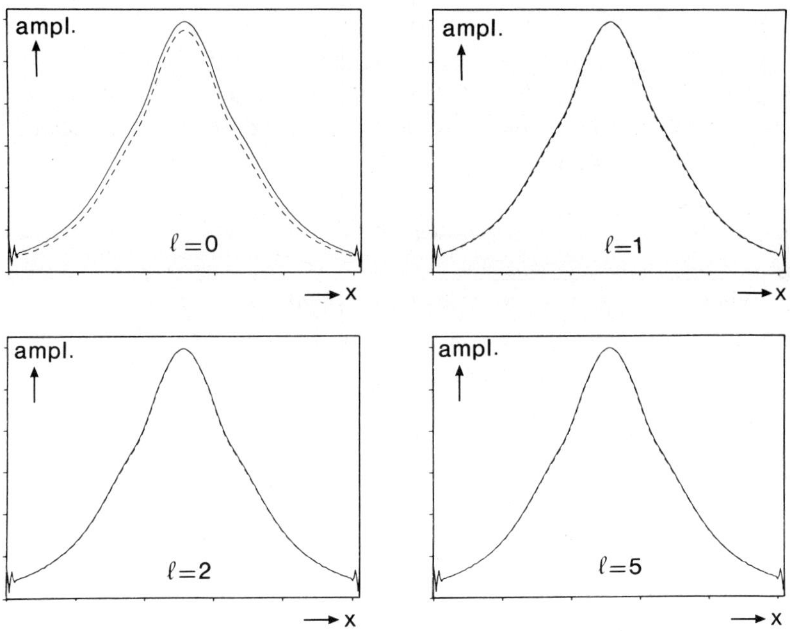

Figure X-3: Amplitude cross-sections of the inverse extrapolated upgoing P-wave at z_2 *(dotted lines), compared with the exact result (solid lines) for iterations* ℓ*=0, 1, 2 and 5.*

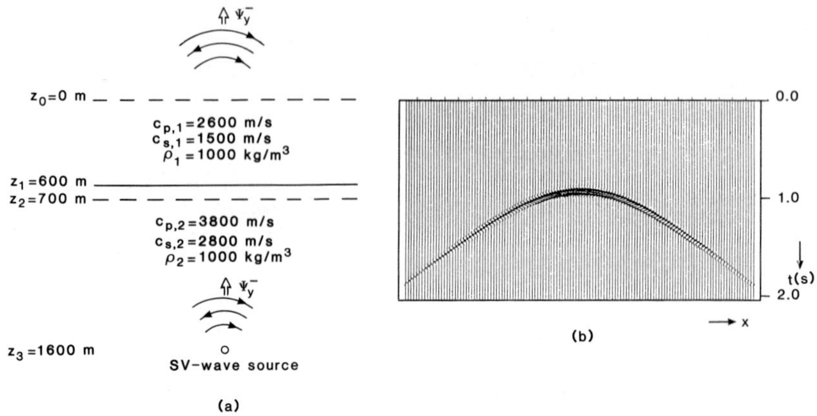

Figure X-4: a. Inhomogeneous elastic medium with a high contrast at z_1 and a burried SV-wave source at z_3.
b. Upgoing SV-wave field at z_0.

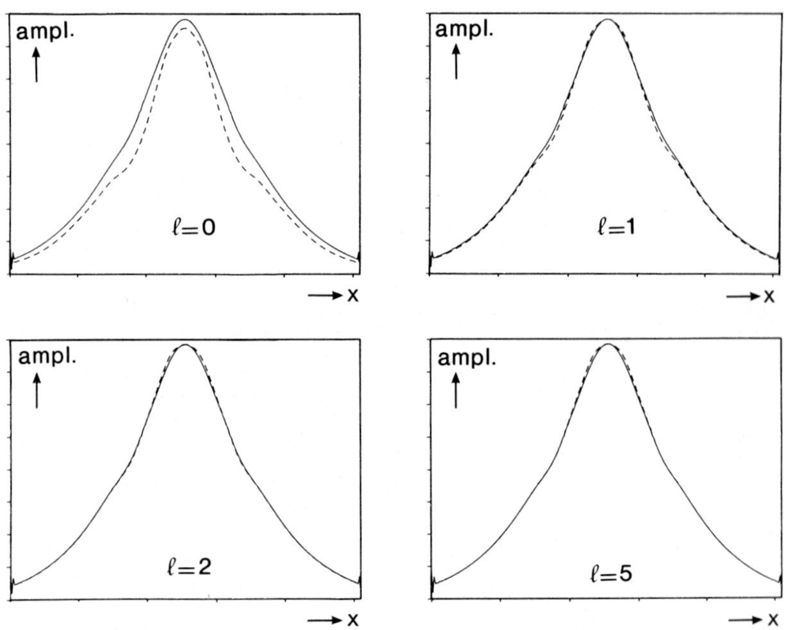

Figure X-5: Amplitude cross-sections of the inverse extrapolated upgoing SV-wave at z_2 (dotted lines), compared with the exact result (solid lines) for iterations ℓ=0, 1, 2 and 5.

XI

ACOUSTIC REDATUMING OF SINGLE-COMPONENT SEISMIC DATA

XI.1 INTRODUCTION

As we explained in the introduction, redatuming is the computational process which transforms seismic surface data in such a way as if they were recorded at a "datum plane" in the subsurface. This process is essentially based on the elimination of wave propagation effects (down *and* up) between the surface and the new datum in the subsurface. In chapters VII and IX we have discussed acoustic inverse wave field extrapolation operators which eliminate the propagation effects from primary *one-way* (i.e., downgoing or upgoing) acoustic wave fields. These operators play a central role in the redatuming scheme. However, they cannot be applied directly to the seismic data because in seismic data acquisition *two-way* wave fields (including multiple reflections related to the earth's surface) are recorded. Obviously, pre-processing is required at the acquisition surface in order to decompose the recorded two-way wave fields into one-way wave fields and to remove the multiple reflections related to the earth's surface.

Before we discuss the actual acoustic processing scheme, we present a forward model for single-component seismic data. This forward model is built up step by step. Following Berkhout (1985), first we discuss the relationship between the reflectivity properties of the subsurface and the primary *one-way* seismic response at the surface. Next we include the multiple reflections related to the earth's surface. Finally we include the acquisition properties, i.e., we transform the one-way seismic response into a two-way seismic response. It is important to bear in mind that the description of this forward model is not a proposal for a numerical modeling scheme for single-component seismic data. It only serves as an introduction to the acoustic processing scheme which is discussed in sections XI.3 and XI.4.

XI.2 FORWARD MODEL FOR SINGLE-COMPONENT SEISMIC DATA

First we derive an expression for the one-way seismic response at z_0 of a single interface at $z_1 > z_0$ between two homogeneous acoustic half-spaces, see Figure XI-1. A downgoing acoustic wave field is incident from the upper half-space $z < z_0$. In the space-frequency domain, the acoustic pressure of this downgoing wave field at z_0 reads $P^+(x,y,z_0;\omega)$. In the following we will make extensively use of the matrix notation of Appendix A. In this notation, $P^+(x,y,z_0;\omega)$ is replaced by the vector $\vec{P}^+(z_0)$. The relationship between the downgoing wave fields at z_0 and z_1 is described by the forward extrapolation matrix $\mathbf{W}^+(z_1,z_0)$, according to

$$\vec{P}^+(z_1) = \mathbf{W}^+(z_1,z_0)\vec{P}^+(z_0), \tag{XI-1a}$$

see section III.3.2. The relationship between the downgoing and upgoing wave fields at z_1 is described by the reflection matrix $\mathbf{R}^+(z_1)$, according to

$$\vec{P}^-(z_1) = \mathbf{R}^+(z_1)\vec{P}^+(z_1), \tag{XI-1b}$$

see section III.3.3. Finally, the relationship between the upgoing wave fields at z_1 and z_0 is described by the forward extrapolation matrix $\mathbf{W}^-(z_0,z_1)$, according to

$$\vec{P}^-(z_0) = \mathbf{W}^-(z_0,z_1)\vec{P}^-(z_1), \tag{XI-1c}$$

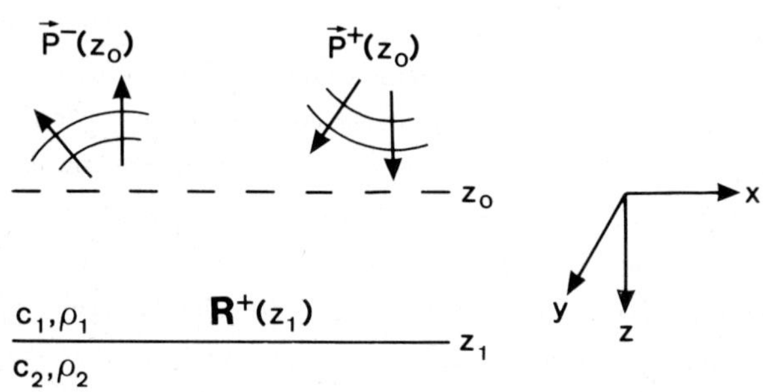

Figure XI-1: Two homogeneous acoustic half-spaces, separated by an interface at z_1.

see again section III.3.2. The combination of equations (XI-1a), (XI-1b) and (XI-1c) yields the following relationship between the downgoing and upgoing wave fields at z_0:

$$\vec{P}^-(z_0) = \mathbf{W}^-(z_0,z_1)\mathbf{R}^+(z_1)\mathbf{W}^+(z_1,z_0)\vec{P}^+(z_0), \tag{XI-2}$$

or

$$\vec{P}^-(z_0) = \mathbf{X}(z_0,z_0)\vec{P}^+(z_0), \tag{XI-3a}$$

where

$$\mathbf{X}(z_0,z_0) = \mathbf{W}^-(z_0,z_1)\mathbf{R}^+(z_1)\mathbf{W}^+(z_1,z_0), \tag{XI-3b}$$

see Figure XI-2. Matrix $\mathbf{X}(z_0,z_0)$, as defined by equation (XI-3b), is the *one-way response matrix.* For the subsurface configuration of Figure XI-1, equations (XI-3a) and (XI-3b) are exact.

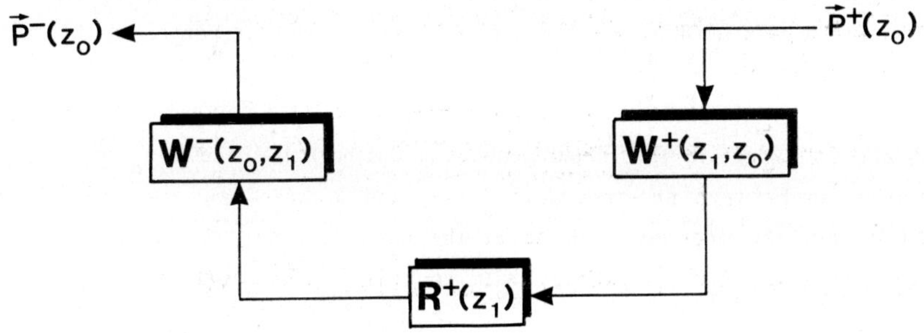

Figure XI-2: Diagram, showing the one-way response at z_0 of a reflector at z_1.

Let us now consider an arbitrarily inhomogeneous acoustic half-space $z>z_0$ below a reflection free surface z_0. Again we assume that a downgoing acoustic wave field $\vec{P}^+(z_0)$ is incident from the upper half space $z<z_0$. We investigate the following forward model for the upgoing wave field $\vec{P}^-(z_0)$:

$$\vec{P}^{-}(z_0) = \mathbf{X}(z_0,z_0)\vec{P}^{+}(z_0), \tag{XI-4a}$$

where

$$\mathbf{X}(z_0,z_0) = \int_{z_0}^{\infty} \mathbf{W}^{-}(z_0,z)\mathbb{R}^{+}(z)\mathbf{W}^{+}(z,z_0)dz. \tag{XI-4b}$$

For the operators $\mathbf{W}^{+}(z,z_0)$ and $\mathbf{W}^{-}(z_0,z)$ we refer to section V.5.2. Note that the forward model of equation (XI-3) can be seen as a special case of equation (XI-4) when we define

$$\mathbb{R}^{+}(z) = \mathbf{R}^{+}(z)\delta(z-z_1). \tag{XI-5}$$

For the general inhomogeneous situation, the one-way response matrix (XI-4b) is not exact. The main approximation is the negligence of internal multiple reflections. Furthermore, the definition of $\mathbb{R}^{+}(z)$ is not straightforward and therefore it is not recommended to use equation (XI-4b) in numerical forward modeling. Equation (XI-4b) is very useful, however, to derive the relationship between the primary one-way responses at different depth levels.

In the following, in the inhomogeneous acoustic half-space $z>z_0$ we distinguish between an *overburden* $z_0<z\leq z_t$ and a *target zone* $z>z_t$, see Figure XI-3a. For the one-way response at the upper boundary z_t of the target zone, we may write in analogy with equations (XI-4a) and (XI-4b),

$$\vec{P}^{-}(z_t) = \mathbf{X}(z_t,z_t)\vec{P}^{+}(z_t), \tag{XI-6a}$$

where

$$\mathbf{X}(z_t,z_t) = \int_{z_t}^{\infty} \mathbf{W}^{-}(z_t,z)\mathbb{R}^{+}(z)\mathbf{W}^{+}(z,z_t)dz, \tag{XI-6b}$$

see Figure XI-3b. Using the following properties of the one-way extrapolation matrices,

$$\mathbf{W}^+(z,z_o) = \mathbf{W}^+(z,z_t)\mathbf{W}^+(z_t,z_o) \qquad \text{for } z_o<z_t<z \tag{XI-7a}$$

and

$$\mathbf{W}^-(z_o,z) = \mathbf{W}^-(z_o,z_t)\mathbf{W}^-(z_t,z) \qquad \text{for } z_o<z_t<z, \tag{XI-7b}$$

equation (XI-4b) can now be rewritten as

$$\mathrm{X}(z_o,z_o) = \int_{z_o}^{z_t} \mathbf{W}^-(z_o,z)\mathbb{R}^+(z)\mathbf{W}^+(z,z_o)dz$$

$$+ \mathbf{W}^-(z_o,z_t)\mathrm{X}(z_t,z_t)\mathbf{W}^+(z_t,z_o). \tag{XI-8}$$

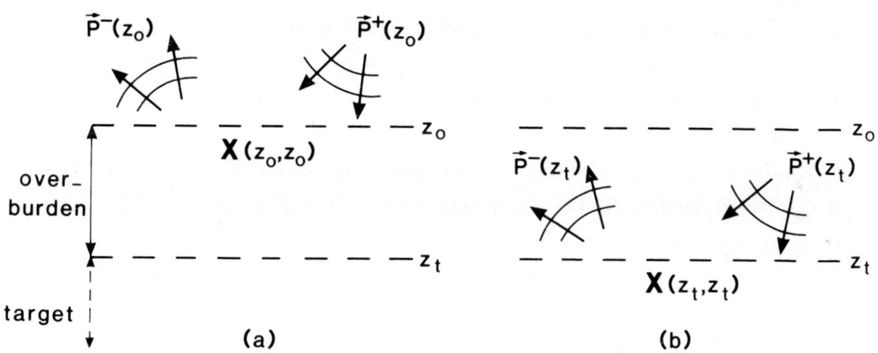

Figure XI-3: a. The primary one-way response matrix $X(z_o,z_o)$ describes the relationship between primary downgoing and upgoing wave fields at depth level z_o.

b. The primary one-way response matrix $X(z_t,z_t)$ describes the relationship between primary downgoing and upgoing wave fields at depth level z_t.

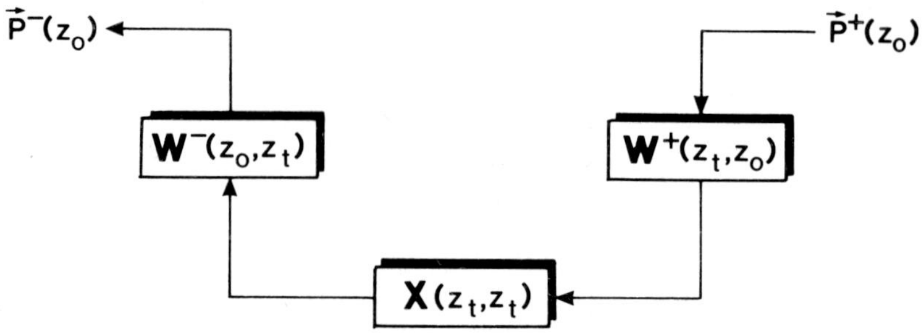

Figure XI-4: Diagram, showing the primary one-way response at z_o of a target zone below z_t.
(The response of the overburden is ignored).

The first term in the right-hand side describes the primary one-way response at z_o of the overburden; the second term in the right-hand side describes the primary one-way response at z_o of the target zone, see also Figure XI-4. In section XI.4, equation (XI-8) will be used as the starting point for the derivation of an acoustic redatuming scheme.

So far we assumed that the surface z_o is reflection free. In practical seismic situations, however, surface z_o represents the earth's *free surface* which is a perfect reflector for the upcoming waves $\vec{P}^-(z_o)$. Therefore in the forward model (XI-4a) we should write for the total downgoing wave field at z_o,

$$\vec{P}^+(z_o) = \vec{P}_r^+(z_o) + \vec{P}_s^+(z_o). \tag{XI-9a}$$

Here $\vec{P}_r^+(z_o)$ is the reflected upgoing wave field at z_o, according to

$$\vec{P}_r^+(z_o) = R_{fr}^-(z_o)\vec{P}^-(z_o), \tag{XI-9b}$$

where $R_{fr}^-(z_o)$ describes the reflectivity of the earth's free surface for upgoing waves. Note that for an acoustic free surface[1)]

$$R^-_{fr}(z_0) = -I. \tag{XI-9c}$$

In equation (XI-9a), $\vec{P}^+_s(z_0)$ is the downgoing source wave field at z_0. The relationship between $\vec{P}^+_s(z_0)$ and the seismic sources at or below z_0 is discussed later on. Upon substitution of (XI-9a) and (XI-9b) into the forward model (XI-4a) we obtain the following implicit expression for the upgoing wave field $\vec{P}^-(z_0)$:

$$\vec{P}^-(z_0) = X(z_0,z_0)\left[R^-_{fr}(z_0)\vec{P}^-(z_0) + \vec{P}^+_s(z_0)\right], \tag{XI-10}$$

see Figure XI-5. This expression can be rewritten explicitly, according to

$$\vec{P}^-(z_0) = X_{fr}(z_0,z_0)\vec{P}^+_s(z_0), \tag{XI-11a}$$

where the free surface one-way response matrix $X_{fr}(z_0,z_0)$ is defined as

$$X_{fr}(z_0,z_0) = \left[I - X(z_0,z_0)R^-_{fr}(z_0)\right]^{-1}X(z_0,z_0), \tag{XI-11b}$$

or, rewriting the inverse matrix as a series expansion,

$$X_{fr}(z_0,z_0) = \left[I + \sum_{m=1}^{\infty}\left(X(z_0,z_0)R^-_{fr}(z_0)\right)^m\right]X(z_0,z_0). \tag{XI-11c}$$

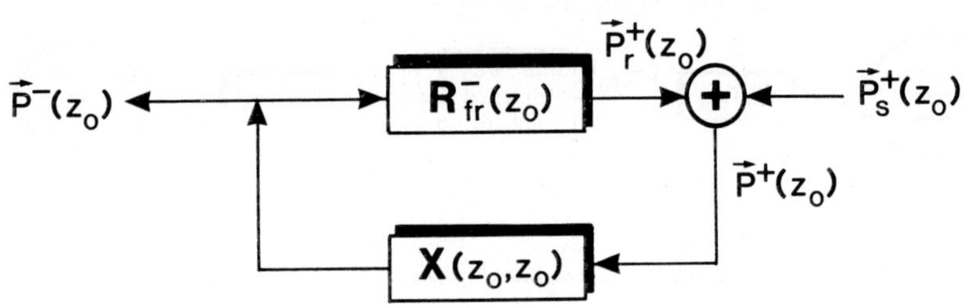

Figure XI-5: One-way forward model, including surface related multiple reflections.

---

1) Choose $H^{-1}_{1,u}M_u$=O in equation (III-89a).

The latter expression clearly shows that the free surface generates an infinite number of *multiple reflections.*

Next we discuss the relationship between the one-way wave fields in the forward model (XI-11) and the two-way seismic data, see Figure XI-6. First we derive the relationship between the two-way seismic source and the downgoing source wave field $\vec{P}_s^+(z_0)$. For a pressure source $\vec{P}_s(z_0)$[1] at the free surface (Figure XI-7a) this relationship is very simple. Since for this situation the upgoing source wave field must be zero we may write

$$\vec{P}_s(z_0) = \vec{P}_s^+(z_0), \tag{XI-12a}$$

or

$$\vec{P}_s^+(z_0) \hat{=} C_s^{-1}(z_0)\vec{P}_s(z_0), \tag{XI-12b}$$

where the *source-decomposition* operator $C_s^{-1}(z_0)$ simply reads

$$C_s^{-1}(z_0) = I. \tag{XI-12c}$$

For a burried source (Figure XI-7b) the situation is more complicated. We consider a volume injection source at depth level z_s, in the vector notation represented by

$$\vec{I}_v(z) = \vec{I}_{v,0}(z_s)\delta(z-z_s).^{2)} \tag{XI-13a}$$

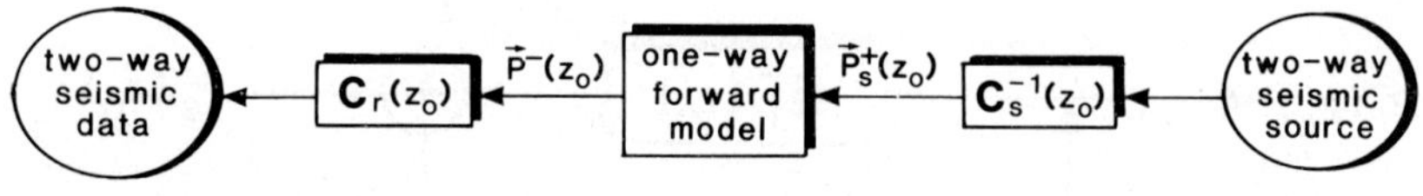

Figure XI-6: Diagram, showing the relationship between the one-way forward model (Figure XI-5) and the two-way seismic data.
*$C_s^{-1}(z_0)$: **decomposition** operator for the source wave field.*
*$C_r(z_0)$: **composition** operator for the received wave field.*

1) For a *point* pressure source, vector $\vec{P}_s(z_0)$ contains only one non-zero element, its value representing the source signature $S(\omega)$.

2) For a *point* source of volume injection, vector $\vec{I}_{v,0}(z_s)$ contains only one non-zero element, its value representing the source signature $S(\omega)$.

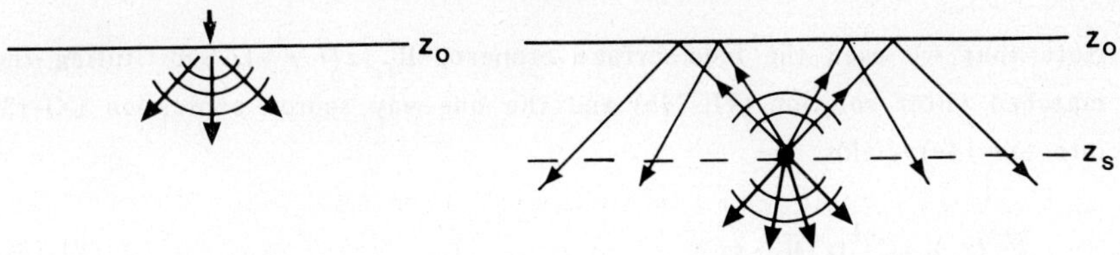

Figure XI-7: a. Pressure source (vibrator) at the free surface.
b. Volume injection source (airgun) below the free surface.

According to equation (III-79c) the one-way representation of this source reads

$$\vec{S}^{\pm}(z) = \vec{S}_0^{\pm}(z_s)\delta(z-z_s), \tag{XI-13b}$$

where

$$\vec{S}_0^{\pm}(z_s) = \frac{1}{2}\, j\omega^2 \mathbf{H}_1^{-1}(z_s)\mathbf{M}(z_s)\vec{I}_{v,0}(z_s), \tag{XI-13c}$$

with matrices $\mathbf{H}_1$ and $\mathbf{M}$ as defined in section III.3.2. For the total downgoing source wave field at z_s we write

$$\vec{P}_s^{+}(z_s) = \vec{S}_0^{+}(z_s) + \mathbf{W}^{+}(z_s,z_0)\mathbf{R}_{fr}^{-}(z_0)\mathbf{W}^{-}(z_0,z_s)\vec{S}_0^{-}(z_s). \tag{XI-14a}$$

The second term in the right-hand side represents the "ghost" wave field related to the free surface, see Figure XI-7b. We obtain an expression for the effective downgoing source wave field at z_0 by applying *inverse* extrapolation to $\vec{P}_s^{+}(z_s)$, according to

$$\vec{P}_s^{+}(z_0) = \mathbf{F}^{+}(z_0,z_s)\vec{P}_s^{+}(z_s), \tag{XI-14b}$$

or

$$\vec{P}_s^{+}(z_0) = \mathbf{F}^{+}(z_0,z_s)\vec{S}_0^{+}(z_s) - \mathbf{W}^{-}(z_0,z_s)\vec{S}_0^{-}(z_s), \tag{XI-14c}$$

where

$$\mathbf{F}^{+}(z_0,z_s) \triangleq [\mathbf{W}^{+}(z_s,z_0)]^{-1}. \tag{XI-14d}$$

Note that we used the free surface property $\mathbf{R}^{-}_{fr}(z_0) = -\mathbf{I}$. Substituting the matched filter solution (VII-74b) and the one-way source expression (XI-13c) into (XI-14c) yields

$$\vec{P}^{+}_{s}(z_0) = \mathbf{C}^{-1}_{s}(z_0)\vec{I}_{v,o}(z_s), \tag{XI-15a}$$

where the source-decomposition operator $\mathbf{C}^{-1}_{s}(z_0)$ reads

$$\mathbf{C}^{-1}_{s}(z_0) = \frac{1}{2} j\omega^2 \left([\mathbf{W}^{-}(z_0,z_s)]^{*} - \mathbf{W}^{-}(z_0,z_s) \right) \mathbf{H}^{-1}_{1}(z_s)\mathbf{M}(z_s), \tag{XI-15b}$$

or, upon substitution of the Taylor series expansion (III-81b),

$$\mathbf{C}^{-1}_{s}(z_0) = -(z_s-z_0)\omega^2 \left[\mathbf{I} + \sum_{m=1}^{\infty} (-1)^m \frac{(z_s-z_0)^{2m}}{(2m+1)!} \mathbf{H}_{2m}(z_s) \right] \mathbf{M}(z_s). \tag{XI-15c}$$

When $|z_s-z_0|$ is small in comparison with the half wavelength, this expression simplifies to

$$\mathbf{C}^{-1}_{s}(z_0) \approx -(z_s-z_0)\omega^2 \mathbf{M}(z_s). \tag{XI-15d}$$

Hence, for small $|z_s-z_0|$, the "source-decomposition" degenerates to frequency- and space-dependent scaling as $\mathbf{M}(z_s)$ is a diagonal matrix, containing the discretized mass density ρ at depth level z_s.

Next we derive expressions for the *receiver-composition* operator $\mathbf{C}_r(z_0)$. According to equations (III-76) and (III-77a), the relationship between two-way and one-way acoustic wave fields reads

$$\begin{bmatrix} \vec{P}(z) \\ \vec{V}_z(z) \end{bmatrix} = \begin{bmatrix} \mathbf{I} & \mathbf{I} \\ \frac{1}{\omega}\mathbf{M}^{-1}(z)\mathbf{H}_1(z) & \frac{-1}{\omega}\mathbf{M}^{-1}(z)\mathbf{H}_1(z) \end{bmatrix} \begin{bmatrix} \vec{P}^{+}(z) \\ \vec{P}^{-}(z) \end{bmatrix}, \tag{XI-16a}$$

or

$$\vec{P}(z) = \vec{P}^{+}(z) + \vec{P}^{-}(z) \tag{XI-16b}$$

and

$$\vec{V}_z(z) = \frac{1}{\omega} \mathbf{M}^{-1}(z)\mathbf{H}_1(z) [\vec{P}^{+}(z) - \vec{P}^{-}(z)]. \tag{XI-16c}$$

For *velocity* receivers (geophones) at the free surface (Figure XI-8a) we obtain upon substitution of (XI-9b) into (XI-16c)[1)]

$$\vec{V}_z(z_0) = \frac{1}{\omega} \mathbf{M}^{-1}(z_0)\mathbf{H}_1(z_0)\left[\mathbf{R}^-_{fr}(z_0)\vec{P}^-(z_0) - \vec{P}^-(z_0)\right], \tag{XI-17a}$$

or

$$\vec{V}_z(z_0) \triangleq C_r(z_0)\vec{P}^-(z_0), \tag{XI-17b}$$

where the receiver-composition operator $C_r(z_0)$ reads

$$C_r(z_0) = -\frac{2}{\omega} \mathbf{M}^{-1}(z_0)\mathbf{H}_1(z_0). \tag{XI-17c}$$

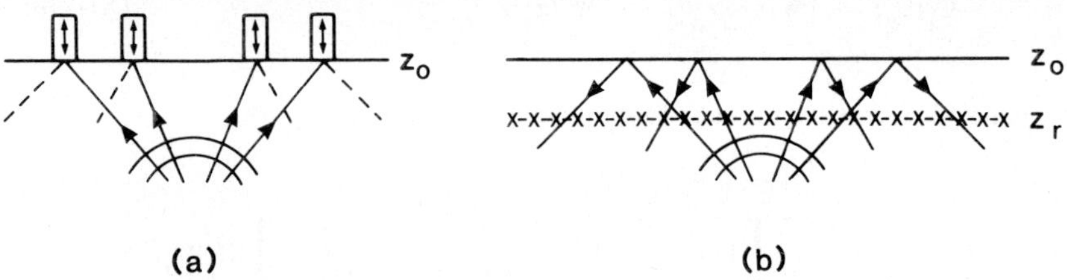

Figure XI-8: a. Velocity receivers (geophones) at the free surface.
b. Pressure receivers (hydrophones) below the free surface.

For *pressure* receivers (hydrophones) below the free surface (Figure XI-8b) we obtain from (XI-16b)

$$\vec{P}(z_r) = \mathbf{W}^+(z_r,z_0)\mathbf{R}^-_{fr}(z_0)\mathbf{W}^-(z_0,z_r)\vec{P}^-(z_r) + \vec{P}^-(z_r). \tag{XI-18a}$$

The first term in the right-hand side represents the "ghost" wave field related to the free surface. The upgoing wave field $\vec{P}^-(z_r)$ is related to $\vec{P}^-(z_0)$ via an *inverse* extrapolation operator, according to

---

1) We ignore the direct source wave and we use the free surface property $\mathbf{R}^-_{fr}(z_0)=-\mathbf{I}$.

$$\vec{P}^-(z_r) = \mathbf{F}^-(z_r,z_o)\vec{P}^-(z_o), \tag{XI-18b}$$

where

$$\mathbf{F}^-(z_r,z_o) \triangleq [\mathbf{W}^-(z_o,z_r)]^{-1}. \tag{XI-18c}$$

Substitution of (XI-18b) into (XI-18a) yields

$$\vec{P}(z_r) = [-\mathbf{W}^+(z_r,z_o) + \mathbf{F}^-(z_r,z_o)]\vec{P}^-(z_o), \tag{XI-18d}$$

or, using the matched filter solution (VII-74a),

$$\vec{P}(z_r) = \mathbf{C}_r(z_o)\vec{P}^-(z_o), \tag{XI-19a}$$

where the receiver-composition operator $\mathbf{C}_r(z_o)$ reads

$$\mathbf{C}_r(z_o) = [\mathbf{W}^+(z_r,z_o)]^* - \mathbf{W}^+(z_r,z_o), \tag{XI-19b}$$

or, upon substitution of the Taylor series expansion (III-81b),

$$\mathbf{C}_r(z_o) = 2(z_r-z_o)\left[\mathbf{I} + \sum_{m=1}^{\infty}(-1)^m \frac{(z_r-z_o)^{2m}}{(2m+1)!}\mathbf{H}_{2m}(z_o)\right]j\mathbf{H}_1(z_o). \tag{XI-19c}$$

When $|z_r-z_o|$ is small in comparison with the half wavelength, this expression simplifies to

$$\mathbf{C}_r(z_o) \approx 2(z_r-z_o)j\mathbf{H}_1(z_o). \tag{XI-19d}$$

Hence, for small $|z_r-z_o|$, the "receiver-composition" degenerates to taking the scaled vertical derivative of the upgoing wave field at z_o (see also equation (III-80b)).

We summarize our expressions for the forward model for single-component seismic data.

We introduce a source vector $\vec{S}(z_o)$ which may represent either a pressure source $\vec{P}_s(z_o)$ at the free surface (Figure XI-7a), or a volume injection source $\vec{I}_{v,o}(z_s)$ below the free surface (Figure XI-7b). We introduce a data vector $\vec{P}(z_0)$ which may represent either the velocity $\vec{V}_z(z_0)$ at the free surface (Figure XI-8a), or the pressure $\vec{P}(z_r)$ below the free surface (Figure XI-8b). The data vector $\vec{P}(z_0)$ is related to the source vector $\vec{S}(z_0)$ according to

$$\vec{P}(z_0) = C_r(z_0)X_{fr}(z_0,z_0)C_s^{-1}(z_0)\vec{S}(z_0), \qquad \text{(XI-20a)}$$

see also Figure XI-6. In this expression the direct source wave is ignored. The source-decomposition operator $C_s^{-1}(z_0)$ is defined by equation (XI-12c) or (XI-15b); the receiver-composition $C_r(z_0)$ operator is defined by equation (XI-17c) or (XI-19b).

In equation (XI-20a), the free surface one-way response matrix $\mathbf{X}_{fr}(z_0,z_0)$ is related to the primary one-way response matrix $\mathbf{X}(z_0,z_0)$ according to

$$\mathbf{X}_{fr}(z_0,z_0) = [\mathbf{I} - \mathbf{X}(z_0,z_0)\mathbf{R}_{fr}^-(z_0)]^{-1}\mathbf{X}(z_0,z_0), \qquad \text{(XI-20b)}$$

see also Figure XI-5. $\mathbf{R}_{fr}^-(z_0)$ represents the free surface reflectivity for upgoing waves (ideally, $\mathbf{R}_{fr}^-(z_0)$=-$\mathbf{I}$). The primary one-way response matrix $\mathbf{X}(z_0,z_0)$ at the surface z_0 is related to the primary one-way response matrix $\mathbf{X}(z_t,z_t)$ at the target depth level z_t according to

$$\mathbf{X}(z_0,z_0) = \mathbf{W}^-(z_0,z_t)\mathbf{X}(z_t,z_t)\mathbf{W}^+(z_t,z_0) + \text{"overburden response"}, \qquad \text{(XI-20c)}$$

see also Figure XI-4. In this expression the internal multiple reflections are ignored. Operators $\mathbf{W}^+$ and $\mathbf{W}^-$ are forward wave field extrapolation operators for downgoing and upgoing waves, respectively, see also chapter V. The total forward model, as described by equations (XI-20a), (XI-20b) and (XI-20c), is visualized in Figure XI-9. Note that this forward model consists of three "layers" which have each their own specific character (Berkhout and Wapenaar, 1989). Layer I is fully determined by the acquisition and the near-surface properties. Layer II is fully determined by the propagation properties of the overburden. Finally, layer III is fully determined by the propagation and reflection properties of the target zone.

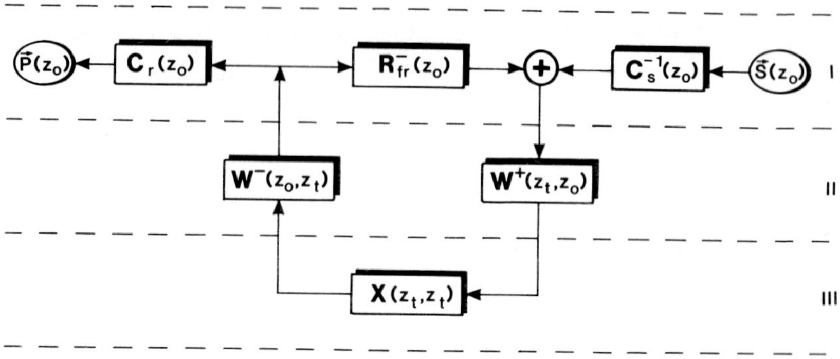

Figure XI-9: Forward model for single-component seismic data.
(The direct source wave, the response of the overburden and the internal multiple reflections are ignored).
Layer I : Acquisition and near-surface properties.
Layer II : Overburden propagation properties.
Layer III: Target propagation and reflection properties.

XI.3 SURFACE RELATED ACOUSTIC PRE-PROCESSING

XI.3.1 Introduction

Before redatuming can be applied the recorded two-way seismic wave fields must be decomposed into one-way wave fields and the surface related multiple reflections must be removed. In the frequency domain this problem can be formulated as follows: given a data vector $\vec{P}(z_0)$, determine the primary one-way response matrix $\mathbf{X}(z_0,z_0)$. Obviously problem is underdetermined: the number of unknowns (i.e., the elements of matrix $\mathbf{X}(z_0,z_0)$) largely exceeds the number of knowns (i.e., the elements of vector $\vec{P}(z_0)$). In practice many seismic experiments are carried out for different lateral positions of the source. Hence, in practice many *independent* data vectors are available. Accordingly, the forward model (XI-20a) for one seismic experiment (i.e., one shot record) can be extended to a forward model for many seismic experiments, yielding

$$\mathbf{P}(z_0) = \mathbf{C}_r(z_0)\mathbf{X}_{fr}(z_0,z_0)\mathbf{C}_s^{-1}(z_0)\mathbf{S}(z_0). \qquad \text{(XI-21a)}$$

Here the columns of the data matrix $\mathbf{P}(z_0)$ contain the different data vectors $\vec{P}(z_0)$ (see also Appendix A, section A.2); the columns of the source matrix $\mathbf{S}(z_0)$ contain the corresponding source vectors $\vec{S}(z_0)$. For point sources, each source vector $\vec{S}(z_0)$ contains only one non-zero element, its value representing the source signature $S(\omega)$. Hence, for identical point sources the columns of matrix $\mathbf{S}(z_0)$ can be ordered in such a way that $\mathbf{S}(z_0)$ represents a scaled identity matrix, according to

$$\mathbf{S}(z_0) = S(\omega)\mathbf{I}. \tag{XI-21b}$$

Thus equation (XI-21a) may be replaced by

$$\mathbf{P}(z_0) = \mathbf{C}_r(z_0)\mathbf{X}^{(s)}_{fr}(z_0,z_0)\mathbf{C}_s^{-1}(z_0), \tag{XI-21c}$$

with

$$\mathbf{X}^{(s)}_{fr}(z_0,z_0) = S(\omega)\mathbf{X}_{fr}(z_0,z_0). \tag{XI-21d}$$

In principle this formulation also holds for the situation with source and receiver *patterns*. Of course, in this case the definitions of the matrices $\mathbf{C}_s^{-1}$ and $\mathbf{C}_r$ must be modified. A further discussion of source and receiver patterns is beyond the scope of this book.

XI.3.2. Acoustic decomposition

Decomposition of the recorded two-way seismic wave fields into one-way (i.e., downgoing and upgoing) wave fields should be preceded by the removal of the direct (i.e., horizontally propagating) wave fields. We do not discuss this procedure; a good reference is Yilmaz (1987).

Our starting point is equation (XI-21c), which is the forward model of a multi-experiment multi-offset single-component seismic dataset, excluding the direct source waves. Assuming that the source signature $S(\omega)$ is unknown, the scaled free surface one-way response matrix can be obtained from the two-way seismic data matrix $\mathbf{P}(z_0)$ by inverting equation (XI-21c), yielding

$$\mathbf{X}_{fr}^{(s)}(z_o,z_o) = \mathbf{C}_r^{-1}(z_o)\mathbf{P}(z_o)\mathbf{C}_s(z_o), \tag{XI-22}$$

see Figure XI-10. For point pressure sources at the free surface, we obtain from equation (XI-12c)

$$\mathbf{C}_s(z_o) = \mathbf{I}, \tag{XI-23a}$$

which means that for this situation decomposition for the source may simply be omitted. For point sources of volume injection just below the free surface, we obtain from equation (XI-15d)

$$\mathbf{C}_s(z_o) \approx -\frac{1}{(z_s-z_o)\omega^2}\mathbf{M}^{-1}(z_s), \tag{XI-23b}$$

which means that for this situation decomposition for the source may simply be carried out by frequency- and space-dependent scaling of the data. For point sources of volume injection significantly below the free surface, we obtain from equation (XI-15b)

$$\mathbf{C}_s(z_o) = \frac{-2j}{\omega^2}\mathbf{M}^{-1}(z_s)\mathbf{H}_1(z_s)\Big(\big[\mathbf{W}^-(z_o,z_s)\big]^*-\mathbf{W}^-(z_o,z_s)\Big)^{-1}, \tag{XI-23c}$$

which means that for this situation decomposition for the source is actually a "deghosting" process.

For velocity receivers at the free surface we obtain from equation (XI-17c)

$$\mathbf{C}_r^{-1}(z_o) = -\frac{\omega}{2}\mathbf{H}_1^{-1}(z_o)\mathbf{M}(z_o), \tag{XI-24a}$$

whereas for pressure receivers just below the free surface we obtain from equation (XI-19d)

$$\mathbf{C}_r^{-1}(z_o) \approx \frac{-j}{2(z_r-z_o)}\mathbf{H}_1^{-1}(z_o), \tag{XI-24b}$$

hence, for both these situations decomposition of the received waves may be carried out by removing the scaled vertical derivative from the data.

For pressure receivers significantly below the free surface we obtain from equation (XI-19b)

$$C_r^{-1}(z_0) = \left([W^+(z_r,z_0)]^* - W^+(z_r,z_0) \right)^{-1} , \qquad \text{(XI-24c)}$$

which means that for this situation decomposition of the received waves is actually a "deghosting" process.

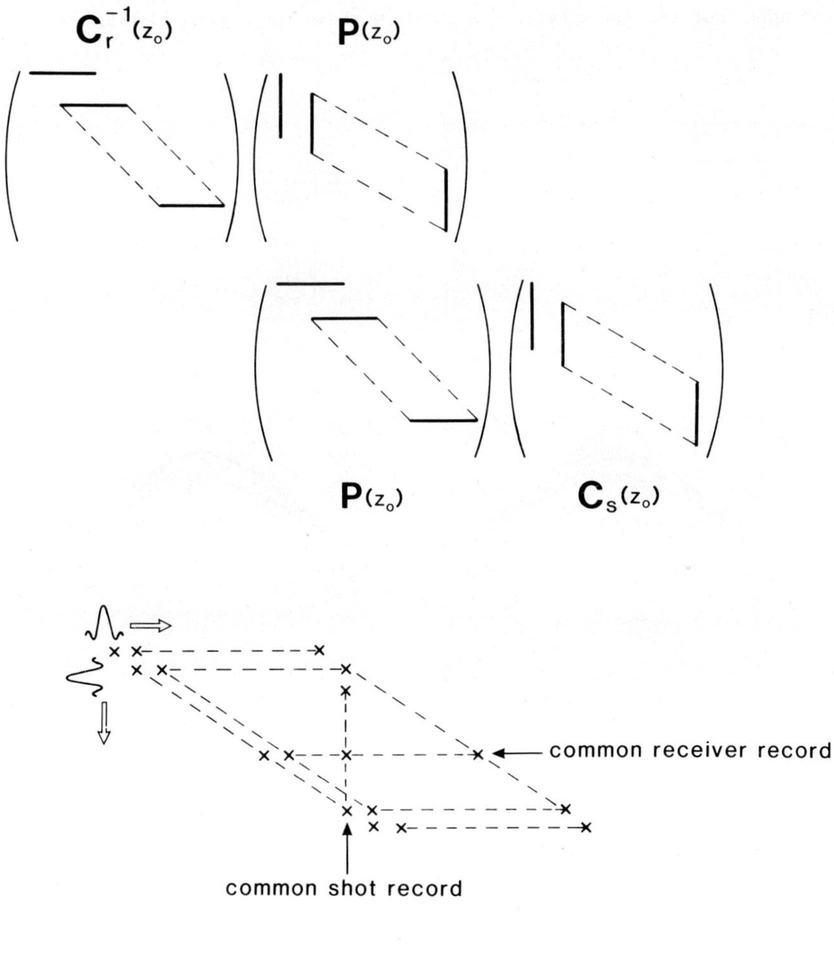

Figure XI-10: According to equation (XI-22), acoustic decomposition involves lateral deconvolution processes along the receivers in each common shot record and along the sources in each common receiver record. The same principle holds for acoustic redatuming, as described by equations (XI-30a) and (XI-30b).(After Berkhout, 1985).

Summarizing, decomposition of the recorded two-way seismic wave fields into one-way wave fields may be carried out by applying the matrix operators $\mathbf{C}_r^{-1}(z_0)$ and $\mathbf{C}_s(z_0)$ to the data matrix $\mathbf{P}(z_0)$, see equation (XI-22) and Figure XI-10. Note that $\mathbf{C}_r^{-1}(z_0)\mathbf{P}(z_0)$ describes a lateral deconvolution process along the columns (i.e., the common shot records) of matrix $\mathbf{P}(z_0)$, whereas $\mathbf{P}(z_0)\mathbf{C}_s(z_0)$ describes a lateral deconvolution process along the rows (i.e., the common receiver records) of matrix $\mathbf{P}(z_0)$.

We illustrate the acoustic decomposition procedure with a simple 2-D example. For the subsurface configuration shown in Figure XI-11a, we

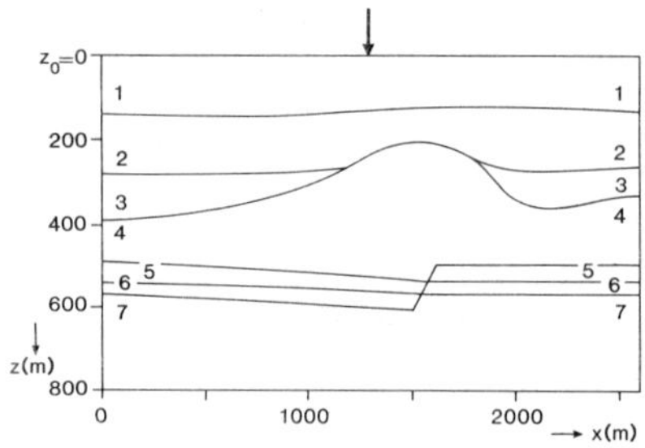

Layer	C(m/s)
1	2400
2	3000
3	3400
4	4100
5	3700
6	4200
7	3500

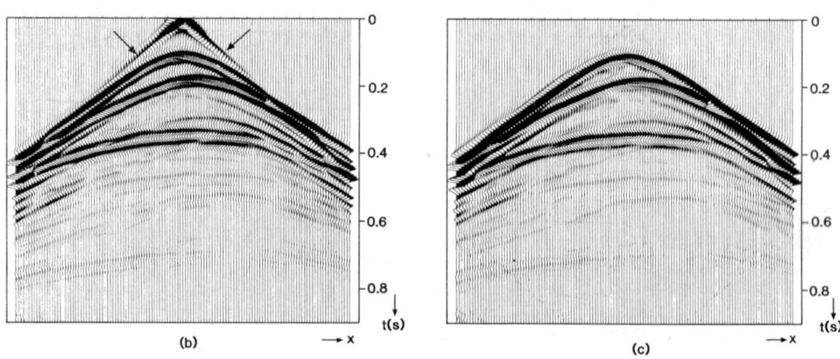

Figure XI-11: Acoustic decomposition.

a. 2-D inhomogeneous subsurface. The pressure receivers are burried 4 m below the free surface.

b. One shot record. The source position is indicated by the arrow in Figure a.

c. The same shot record, after decomposition.

generated a number of seismic shot records by finite difference modeling (Kelly et al., 1976). We used pressure sources at the free surface and pressure receivers at a depth of $z_r - z_o = 4$ m below the free surface. One shot record is shown in the space-time domain in Figure XI-11b. Note that the direct waves decay rapidly with the offset due to the dipole-like behaviour of the source and receivers in combination with the free surface. Therefore they can be easily removed from the data by muting. All shot records are transformed from the time domain to the frequency domain, yielding a data matrix $P(z_0)$ for each frequency in the seismic band (5 Hz $< f = \frac{\omega}{2\pi} <$ 80 Hz). Next, decomposition is carried out by applying equation (XI-22) for each frequency in the seismic band, with $\mathbf{C}_s(z_0) = I$ and $\mathbf{C}_r^{-1}(z_0)$ given by equation (XI-24b)[1)]. Finally, the results are transformed back from the frequency domain to the time domain. Figure XI-11c shows the shot record of Figure XI-11b after decomposition. Note that the main effect of the decomposition is an offset-dependent change of the wave form.

XI.3.3 Acoustic multiple elimination

After the decomposition has been carried out, the scaled free surface one-way response matrix $\mathbf{X}_{fr}^{(s)}(z_0,z_0)$ is available for all frequencies in the seismic band. This response matrix contains significant multiple reflections related to the free surface (see Figure XI-11c). They can be removed by inverting equation (XI-20b), yielding (Berkhout, 1985)

$$\mathbf{X}(z_0,z_0) = \mathbf{X}_{fr}(z_0,z_0)\left[I + \mathbf{R}_{fr}^{-}(z_0)\mathbf{X}_{fr}(z_0,z_0)\right]^{-1}, \tag{XI-25a}$$

or

$$\mathbf{X}(z_0,z_0) = \mathbf{X}_{fr}(z_0,z_0)\left[I + \sum_{m=1}^{\infty}\left(-\mathbf{R}_{fr}^{-}(z_0)\mathbf{X}_{fr}(z_0,z_0)\right)^m\right], \tag{XI-25b}$$

with

$$\mathbf{X}_{fr}(z_0,z_0) = \frac{1}{S(\omega)}\mathbf{X}_{fr}^{(s)}(z_0,z_0), \tag{XI-25c}$$

---

1) For the maximum frequency the wavelength is equal to $\lambda_{min} = c_1/f_{max} = 2400/80$ m= 30 m. Compared to $\frac{1}{4}\lambda_{min}$, the receiver depth $z_r - z_0 = 4$ m is sufficiently small to justify the use of equation (XI-24b).

see equation (XI-21d). Hence, acoustic surface related multiple elimination involves

- *source signature deconvolution*, i.e., removal of $S(\omega)$ from the data, according to equation (XI-25c),
- *multiple prediction and substraction*, according to equation (XI-25b).

Note that the *multiple predictor*

$$\mathbf{X}_{fr}(z_0,z_0) \sum_{m=1}^{\infty} \left[-\mathbf{R}^{-}_{fr}(z_0)\mathbf{X}_{fr}(z_0,z_0)\right]^m$$

is fully determined by the free surface response matrix $\mathbf{X}_{fr}(z_0,z_0)$ and the free surface reflection matrix $\mathbf{R}^{-}_{fr}(z_0)=-\mathbf{I}$. Hence, no knowledge of the subsurface is required for surface related multiple elimination. Only knowledge of the source signature $S(\omega)$ is required for the deconvolution. However, when the source signature is not known, it can be estimated by applying *adaptive* multiple elimination. This can be seen as a standard minimization problem: the multiple reflections are optimumly removed (i.e., the energy in the broad-band data is minimized) when the correct source signature is used for the deconvolution. Hence, using an adaptive procedure, the source deconvolution and the multiple elimination are carried out simultaneously. For an extensive discussion on adaptive elimination of surface related multiple reflections we refer to Verschuur et al. (1989).

We illustrate the acoustic multiple elimination procedure with a 2-D example. We consider the decomposed data of the example in section XI.3.2. One decomposed shot record is shown in Figure XI-12a.
The same shot record after adaptive multiple elimination is shown in Figure XI-12b.[1)] Note that this result clearly shows the primary one-way response (including minor *internal* multiple reflections) of the subsurface

[1)] Although only one shot record is shown, bear in mind that all shot records (the columns of $\mathbf{X}_{fr}(z_0,z_0)$) were involved in the multiple elimination process.

configuration of Figure XI-11a. For comparison, Figure XI-12c shows the primary one-way response, obtained by finite difference modeling.

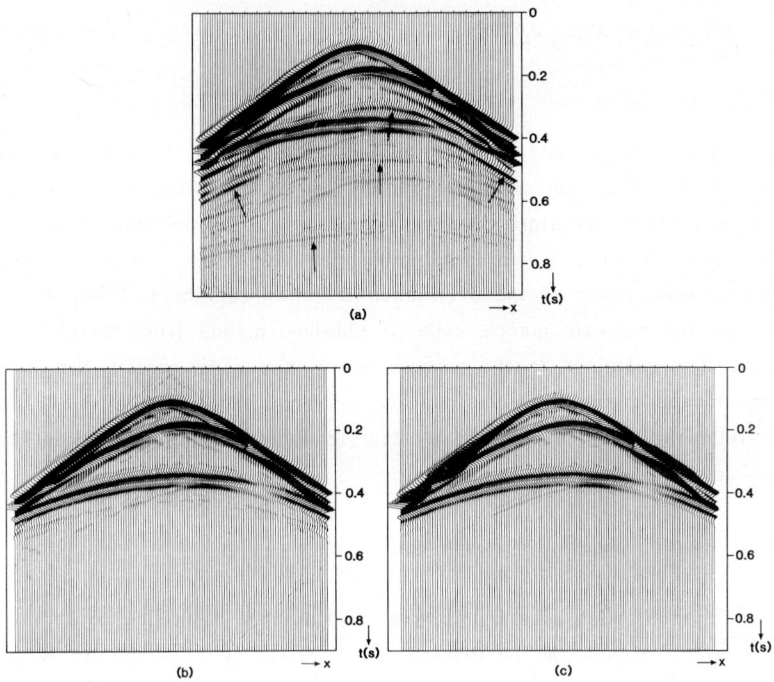

Figure XI-12: Acoustic multiple elimination.

a. The decomposed shot record of Figure XI-11c.

b. The same shot record, after adaptive multiple elimination.

c. For comparison, the same shot record obtained by forward modeling in the subsurface configuration of Figure XI-11a, without the free surface.

XI.4 ACOUSTIC REDATUMING

XI.4.1 Introduction

After surface related acoustic pre-processing, the primary one-way response matrix $X(z_0,z_0)$ is available for all frequencies within the seismic band. The aim of acoustic redatuming is to find the primary one-way response matrix $X(z_t,z_t)$ that would be measured at the target depth level z_t for all frequencies within the seismic band.

According to equation (XI-20c),

$$\mathbf{X}(z_o,z_o) = \mathbf{W}^-(z_o,z_t)\mathbf{X}(z_t,z_t)\mathbf{W}^+(z_t,z_o)^{1)}, \quad \text{(XI-26)}$$

the response matrix $\mathbf{X}(z_o,z_o)$ is a distorted version of the response matrix $\mathbf{X}(z_t,z_t)$, the distortion being determined by the operators $\mathbf{W}^+(z_t,z_o)$ and $\mathbf{W}^-(z_o,z_t)$. These operators describe the propagation properties of the overburden. Hence, acoustic redatuming can only be carried out when the acoustic *macro model* of the overburden is known (Berkhout, 1986). In practice the acoustic macro model is obtained mainly from travel time information contained in the primary one-way response matrix $\mathbf{X}(z_o,z_o)$. A discussion on macro model estimation is beyond the scope of this book. In the following we assume that an accurate acoustic macro model of the overburden is available.

XI.4.2 Acoustic redatuming of a multi-shot record

Acoustic redatuming involves compensation for the distortion caused by propagation through the overburden. By inverting equation (XI-26) we obtain the following expression for the redatumed response matrix (Berkhout, 1985):

$$\mathbf{X}(z_t,z_t) = \mathbf{F}^-(z_t,z_o)\mathbf{X}(z_o,z_o)\mathbf{F}^+(z_o,z_t),^{2)} \quad \text{(XI-27a)}$$

where

$$\mathbf{F}^+(z_o,z_t) \triangleq [\mathbf{W}^+(z_t,z_o)]^{-1} \quad \text{(XI-27b)}$$

and

$$\mathbf{F}^-(z_t,z_o) \triangleq [\mathbf{W}^-(z_o,z_t)]^{-1}. \quad \text{(XI-27c)}$$

---

1) For simplicity the overburden response is ignored.

2) Again, for simplicity the overburden response is ignored. In reality $\mathbf{X}(z_t,z_t)$, as defined by (XI-27), consists of a causal term representing the target response and a non-causal term related to the overburden response. The latter can be easily removed after the redatumed data have been transformed back to the time domain.

In practice, direct inversion of the extrapolation matrices $\mathbf{W}^+$ and $\mathbf{W}^-$ should be avoided. When the contrasts in the overburden are weak to moderate, we may apply the modified matched inverse operators

$$<\mathbf{F}^+(z_o,z_t)> = \left[\mathbf{W}^-(z_o,z_t)\right]^* \tag{XI-28a}$$

and

$$<\mathbf{F}^-(z_t,z_o)> = \left[\mathbf{W}^+(z_t,z_o)\right]^*. \tag{XI-28b}$$

For an extensive discussion of these operators we refer to chapter VII. When the contrasts in the overburden are significant we should apply higher order approximations of the inverse operators. For the ℓ'th order estimates of $\mathbf{F}^+$ and $\mathbf{F}^-$ we may write

$$\mathbf{F}_\ell^+(z_o,z_t) = \sum_{m=o}^{\ell} \left(\Xi(z_o)\right)^m <\mathbf{F}^+(z_o,z_t)> \tag{XI-29a}$$

and

$$\mathbf{F}_\ell^-(z_t,z_o) = \sum_{m=o}^{\ell} \left(\Xi(z_t)\right)^m <\mathbf{F}^-(z_t,z_o)>, \tag{XI-29b}$$

where operators $\Xi(z_o)$ and $\Xi(z_t)$ take care of the scattering effects related to the contrasts in the overburden. For an extensive discussion of these operators we refer to ‘ apter IX.

Consider again equation (XI-27a), where $\mathbf{F}^+$ and $\mathbf{F}^-$ may represent either the modified matched inverse operators $<\mathbf{F}^+>$ and $<\mathbf{F}^->$, respectively, or the ℓ'th order approximations $\mathbf{F}_\ell^+$ and $\mathbf{F}_\ell^-$, respectively. Note that redatuming, as described by equation (XI-27a), could be carried out as a two-step procedure, according to

$$\mathbf{X}(z_t,z_o) = \mathbf{F}^-(z_t,z_o)\mathbf{X}(z_o,z_o), \tag{XI-30a}$$

followed by

$$\mathbf{X}(z_t,z_t) = \mathbf{X}(z_t,z_0)\mathbf{F}^+(z_0,z_t), \tag{XI-30b}$$

see also Figure XI-10. Equation (XI-30a) describes a lateral deconvolution process along the receivers in each common shot record (i.e., along the columns of $\mathbf{X}(z_0,z_0)$). Physically it means that the *receivers* are downward extrapolated from the surface z_0 to the target depth level z_t. Hence, $\mathbf{X}(z_t,z_0)$ represents the one-way target response at z_t, related to sources at z_0. Equation (XI-30b) describes a lateral deconvolution process along the sources in each common receiver record (i.e., along the rows of $\mathbf{X}(z_t,z_0)$). Physically it means that the *sources* are downward extrapolated from the surface z_0 to the target depth level z_t. The redatuming scheme described by Berryhill (1984) is based on a similar principle.

XI.4.3 Acoustic redatuming of single-shot records and stacking

For practical applications, redatuming according to (XI-30a) and (XI-30b) may not be the most efficient solution. Particularly for 3-D applications it involves a cumbersome data reordering process (from common shot records to common receiver records) in between the two steps. Berkhout (1985) and Wapenaar and Berkhout (1987) show that equations (XI-30a) and (XI-30b) can be rewritten as extrapolation *per shot record*, followed by *'stacking'*, without loss of accuracy. Therefore, define vectors $\vec{X}_i(z_0,z_0)$ for $i=1,....,I$, which represent the different common shot records (i.e., the columns of matrix $\mathbf{X}(z_0,z_0)$). Downward extrapolation of the receivers in the i'th shot record from z_0 to z_t is then described by

$$\vec{X}_i(z_t,z_0) = \mathbf{F}^-(z_t,z_0)\vec{X}_i(z_0,z_0), \tag{XI-31a}$$

where $\vec{X}_i(z_t,z_0)$ denotes the one-way target response at z_t, related to the i'th source at z_0. Note that $\vec{X}_i(z_t,z_0)$ represents the i'th column of matrix $\mathbf{X}(z_t,z_0)$, as defined by equation (XI-30a). Define *row* vectors $[\vec{F}_i^+(z_0,z_t)]^T$ for $i=1,...,I$, which contain the rows of $\mathbf{F}^+(z_0,z_t)$.

With simple matrix calculus it can be verified that equation (XI-30b) may thus be replaced by[1)]

$$X(z_t,z_t) = \sum_{i=1}^{I} \vec{X}_i(z_t,z_0)\,[\vec{F}_i^{+}(z_0,z_t)]^T, \tag{XI-31b}$$

see Figure XI-13. Equations (XI-31a) and (XI-31b) confirm that redatuming may be carried out per shot record, followed by 'stacking'. For a discussion on the implementation aspects of 3-D redatuming per shot record we refer to Wapenaar (1986) and Kinneging (1989).

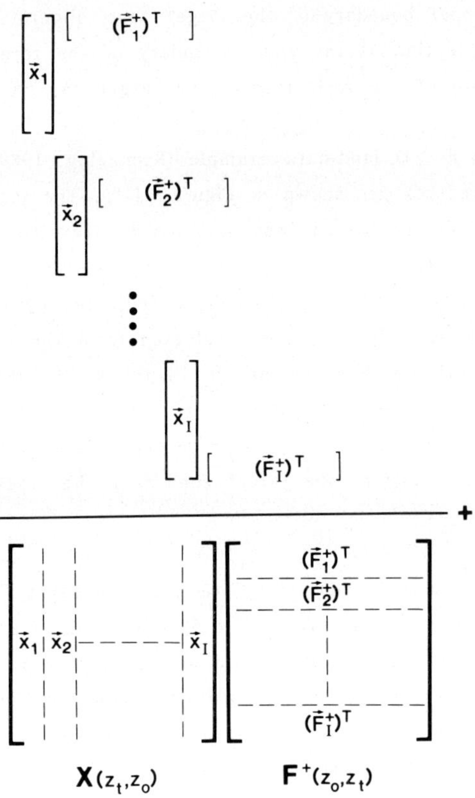

Figure XI-13: Mathematical representation of acoustic redatuming of single-shot records and stacking (equation (XI-31)). The result is identical to acoustic redatuming of a multi-shot record (equation XI-30)).

---

1) The summation convention does not apply to the shot index i.

XI.4.4 Examples of acoustic redatuming

We illustrate the acoustic redatuming procedure with some 2-D and 3-D examples. First we consider the data of the examples in sections XI.3.2 and XI.3.3. The subsurface configuration is shown in Figure XI-14a. The upper boundary of the target zone at z_t= 450 m is indicated by a dashed line. One shot record at the surface z_0= 0 m after decomposition and multiple elimination is shown in Figure XI-14b. Note that the response of the target zone is distorted due to the propagation effects of the overburden. The result of acoustic redatuming per shot record and stacking is shown in Figures XI-14c and XI-14d. Figure XI-14c represents one shot record at the upper boundary of the target zone. Figure XI-14d represents the zero offset section at the upper boundary of the target zone. Note that the structure of the reflectors in the target can be clearly recognized.

Next we discuss a 2-D land-data example (Kinneging, 1989). Two of in total 342 shot records are shown in Figure XI-15. The macro subsurface model, shown in Figure XI-16a, was obtained by traveltime inversion (Van der Made et al., 1984). The shot records were redatumed to the upper boundary of the target zone at z_t=3250 m, indicated by the dashed line in Figure XI-16a. A zero offset section, selected from the redatumed data, is shown in Figure XI-16b. For comparison, Figure XI-16c shows data related to the same target zone, selected from a conventional common midpoint stack at the surface[1]. Note that the redatumed data in Figure XI-16b show much more details of the target zone than the stacked data in Figure XI-16c.

Finally we discuss a 3-D redatuming example (Kinneging, 1989). Cross-sections of two 3-D shot records (50 x 50 traces each) are shown in Figure XI-17. These are two of in total 600 shot records that were modeled for the 3-D subsurface configuration shown in Figure XI-18a (only the responses of the five point diffractors at z_t=1000 m were modeled; the acquisition surface was non-reflecting).

---

[1] For an introduction to conventional seismic processing we refer to Yilmaz (1987).

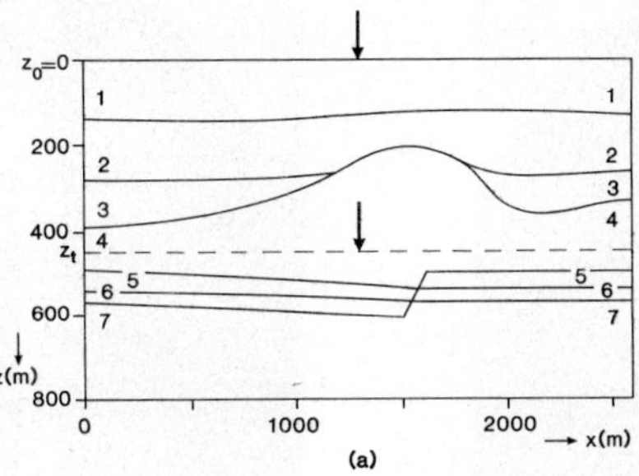

Layer	C(m/s)
1	2400
2	3000
3	3400
4	4100
5	3700
6	4200
7	3500

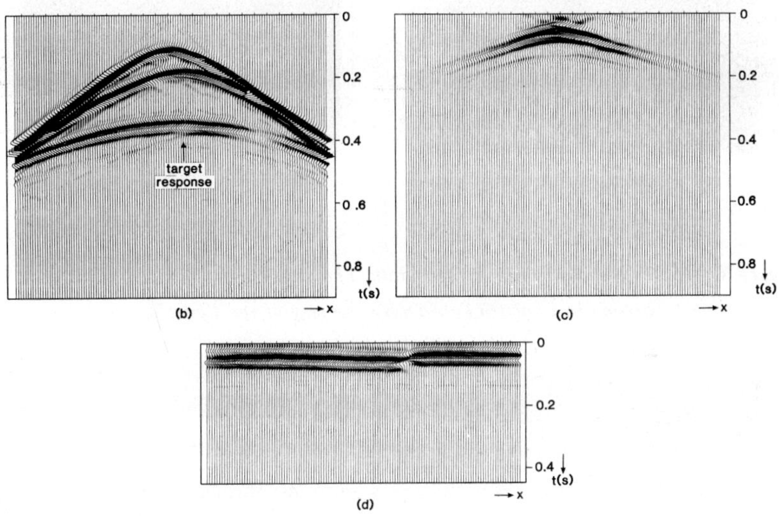

Figure XI-14: Acoustic redatuming.

a. 2-D inhomogeneous subsurface. The dashed line indicates the target upper boundary.

b. The pre-processed shot record of Figure XI-12b. The source position is indicated by the arrow at z_o in Figure a.

c. One shot record, selected from the redatumed data. The source position is indicated by the arrow at z_t in Figure a.

d. Zero offset section, selected from the redatumed data at z_t.

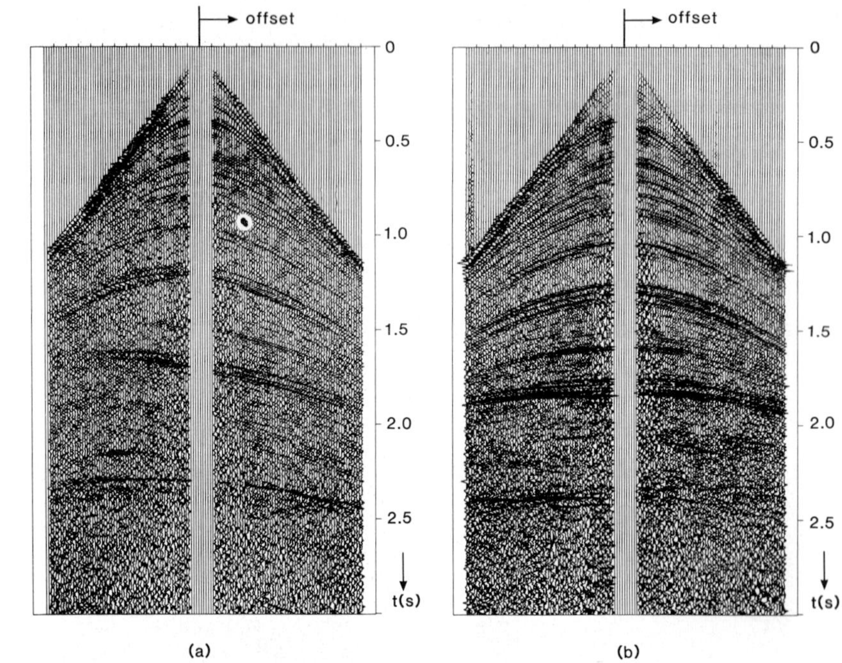

Figure XI-15: Two shot records of land-data. The source positions are indicated by arrows in Figure XI-16a. (By courtesy of the Nederlandse Aardolie Maatschappij, Assen, The Netherlands)

Figure XI-18b shows two vertical cross-sections through the subsurface. The shot records were redatumed to the target depth z_t=1000 m. Two zero offset cross-sections, selected from the 3-D redatumed data, are shown in Figure XI-18c. Note that these results clearly show the point diffractors, indicated by the dots in Figure XI-18b, at their correct lateral positions at t=o.

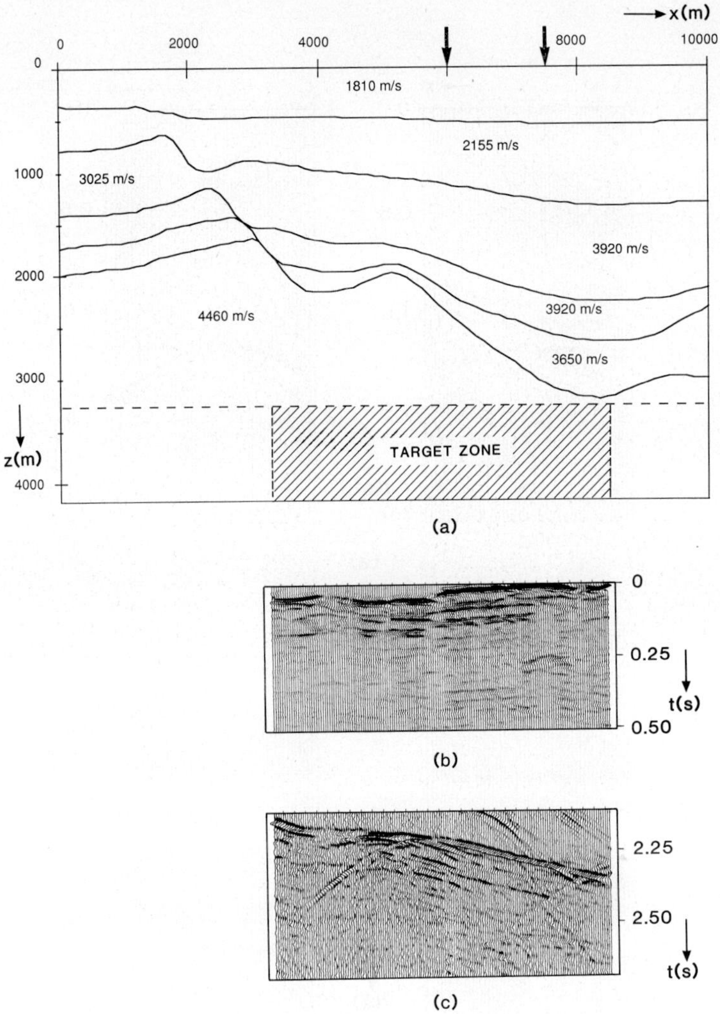

Figure XI-16: Acoustic redatuming of land-data.

a. Macro subsurface model. The dashed line indicates the target upper boundary.

b. Zero offset section, selected from the redatumed shot records at z_t=3250 m.

c. The same target zone, selected from the conventional common midpoint stack at the surface.

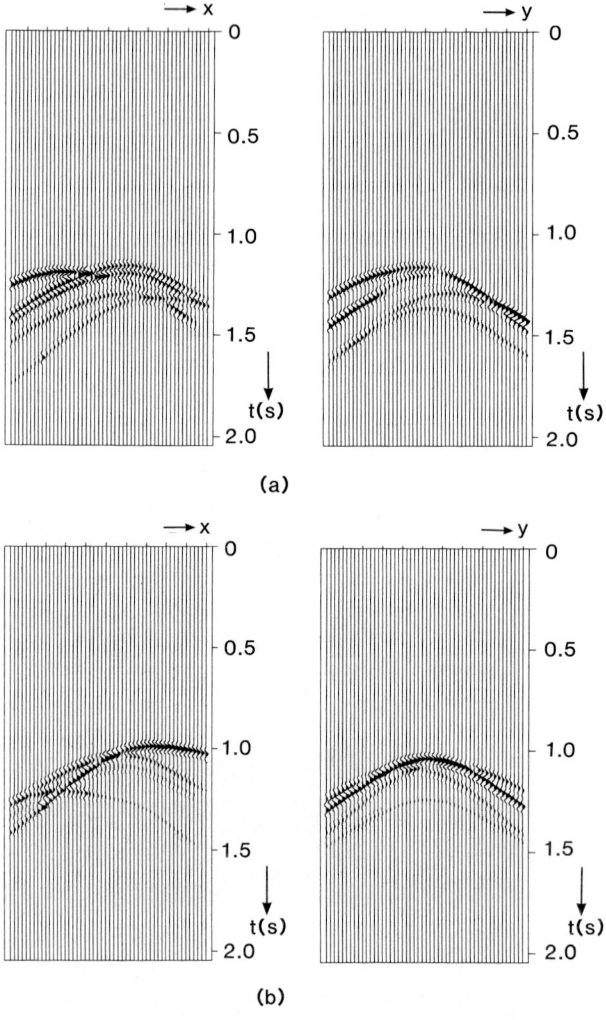

Figure XI-17: Cross-sections of two 3-D shot records. The source positions are indicated by the arrows in Figure XI-18a.

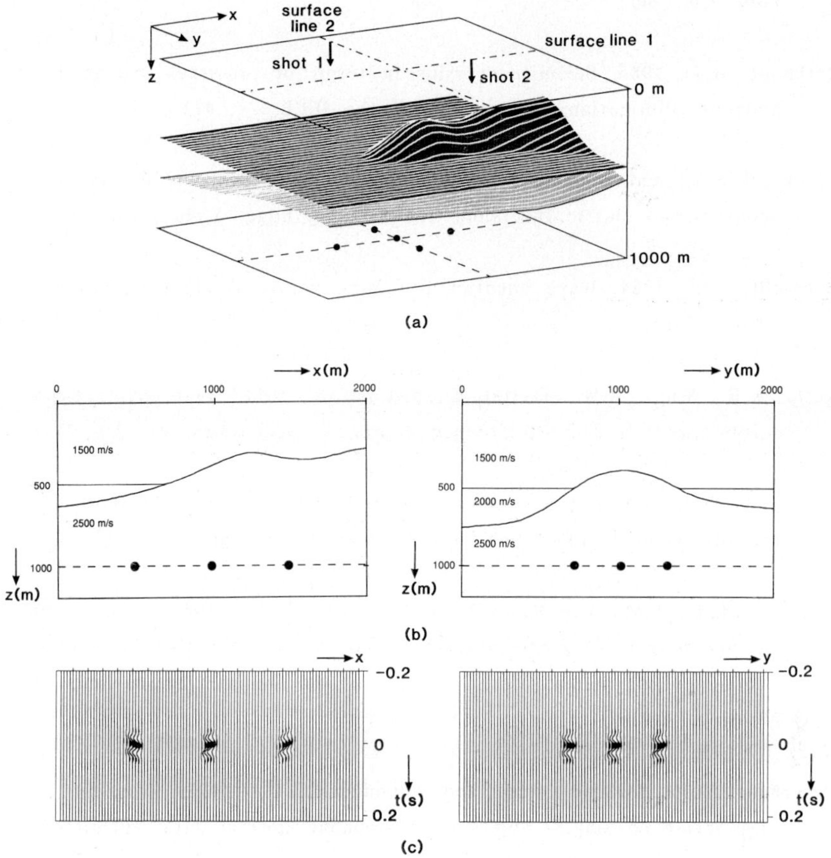

Figure XI-18: 3-D acoustic redatuming.

a. 3-D subsurface model with five point diffractors at the target depth z_t=1000 m.

b. Two vertical cross-sections through the subsurface model.

c. Two zero offset cross-sections, selected from the redatumed shot records at z_t=1000 m.

XI.5 REFERENCES

Berkhout, A.J., 1985, Seismic migration: imaging of acoustic energy by wave field extrapolation. A. Theoretical aspects: Elsevier Science Publ. Co., Inc.

Berkhout, A.J., 1986, Seismic inversion in terms of pre-stack migration and multiple elimination: Proceedings of the IEEE, *74*, 415 - 427.

Berkhout, A.J., and Wapenaar, C.P.A., 1989, Delphi: Delft Philosophy on acoustic and elastic inversion: The Leading Edge, *8*, in press.

Berryhill, J.R., 1984, Wave equation datuming before stack: Geophysics, *49*, 2064 - 2066.

Kelly, K.R., Ward, R.W., Treitel, S., and Alford, R.M., 1976, Synthetic seismograms: a finite difference approach: Geophysics, *41*, 2 - 27.

Kinneging, N.A., 1989, Three-dimensional redatuming of seismic shot records: D.Sc thesis, Delft University of Technology.

Van der Made, P.M., Van Riel, P., and Berkhout, A.J., 1984, Velocity and subsurface geometry inversion by a parameter estimation in complex inhomogeneous media: presented at the 54th Annual International SEG Meeting, Atlanta.

Verschuur, D.J., Berkhout, A.J., and Wapenaar, C.P.A., 1989, Multiple elimination on single- and multi-component seismic data: submitted for publication in Geophysics.

Wapenaar, C.P.A., 1986, Pre-stack migration in two and three dimensions: D.Sc thesis, Delft University of Technology.

Wapenaar, C.P.A., and Berkhout, A.J., 1987, Full pre-stack versus shot record migration: presented at the 57th Annual International SEG Meeting, New Orleans.

Yilmaz, O., 1987, Seismic data processing: Society of Exploration Geophysicists.

XII

ELASTIC REDATUMING OF MULTI-COMPONENT SEISMIC DATA

XII.1 INTRODUCTION

In agreement with the acoustic situation, elastic redatuming of multi-component seismic data is preferably preceded by pre-processing at the acquisition surface. By surface related elastic pre-processing, the recorded *two-way* wave fields are decomposed into *one-way* P- and S-wave fields and the multiple reflections and conversions related to the earth's surface are removed. An important consequence is that the decomposed primary one-way P- and S-wave responses can be redatumed *independently* in a manner that is very similar to acoustic redatuming.

Before we discuss the actual elastic processing scheme, we present a forward model for multi-component seismic data. It is important to bear in mind that the description of this forward model is not a proposal for a numerical modeling scheme for multi-component seismic data. It only serves as an introduction to the elastic processing scheme which is discussed in sections XII.3 and XII.4.

XII.2 FORWARD MODEL FOR MULTI-COMPONENT SEISMIC DATA

First we derive an expression for the one-way seismic response at z_0 of a single interface at $z_1>z_0$ between two homogeneous isotropic elastic half-spaces (see Figure XI-1 for the acoustic equivalence). A downgoing elastic wave field is incident from the upper half-space $z<z_0$. In the space-frequency domain, the P- and S-wave potentials associated to this downgoing wave field at z_0 read $\Phi^+(x,y,z_0;\omega)$ and $\vec{\Psi}^+(x,y,z_0;\omega)$, respectively. The components of the S-wave potential at z_0 read $\Psi_x^+(x,y,z_0;\omega)$, $\Psi_y^+(x,y,z_0;\omega)$ and $\Psi_z^+(x,y,z_0;\omega)$. Only two of these components are independent, see section II.2.5. In the following we only consider the x- and y-components. In the matrix notation of Appendix A, the wave fields $\Phi^+(x,y,z_0;\omega)$, $\Psi_x^+(x,y,z_0;\omega)$ and $\Psi_y^+(x,y,z_0;\omega)$ are replaced by the vectors $\vec{\Phi}^+(z_0)$, $\vec{\Psi}_x^+(z_0)$ and $\vec{\Psi}_y^+(z_0)$, respectively. In analogy with equation (IV-62d),

these three vectors are combined in one *three-component* data vector $\vec{\underset{\sim}{D}}^{+}(z_o)$:

$$\vec{\underset{\sim}{D}}^{+}(z_o) \triangleq \begin{bmatrix} \vec{\Phi}^{+}(z_o) \\ \vec{\Psi}_x^{+}(z_o) \\ \vec{\Psi}_y^{+}(z_o) \end{bmatrix} . \tag{XII-1}$$

The relationship between the downgoing wave fields at z_o and z_1 is described by the forward extrapolation matrix $\underset{\sim}{\mathbf{W}}^{+}(z_1,z_o)$, according to

$$\vec{\underset{\sim}{D}}^{+}(z_1) = \underset{\sim}{\mathbf{W}}^{+}(z_1,z_o)\vec{\underset{\sim}{D}}^{+}(z_o), \tag{XII-2a}$$

or

$$\begin{bmatrix} \vec{\Phi}^{+}(z_1) \\ \vec{\Psi}_x^{+}(z_1) \\ \vec{\Psi}_y^{+}(z_1) \end{bmatrix} = \begin{bmatrix} \mathbf{W}^{+}_{\phi,\phi}(z_1,z_o) & 0 & 0 \\ 0 & \mathbf{W}^{+}_{\psi_x,\psi_x}(z_1,z_o) & 0 \\ 0 & 0 & \mathbf{W}^{+}_{\psi_y,\psi_y}(z_1,z_o) \end{bmatrix} \begin{bmatrix} \vec{\Phi}^{+}(z_o) \\ \vec{\Psi}_x^{+}(z_o) \\ \vec{\Psi}_y^{+}(z_o) \end{bmatrix}, \tag{XII-2b}$$

see section IV.3.2. The relationship between the downgoing and upgoing wave fields at z_1 is described by the reflection matrix $\underset{\sim}{\mathbf{R}}^{+}(z_1)$, according to

$$\vec{\underset{\sim}{D}}^{-}(z_1) = \underset{\sim}{\mathbf{R}}^{+}(z_1)\vec{\underset{\sim}{D}}^{+}(z_1), \tag{XII-3a}$$

or

$$\begin{bmatrix} \vec{\Phi}^{-}(z_1) \\ \vec{\Psi}_x^{-}(z_1) \\ \vec{\Psi}_y^{-}(z_1) \end{bmatrix} = \begin{bmatrix} \mathbf{R}^{+}_{\phi,\phi}(z_1) & \mathbf{R}^{+}_{\phi,\psi_x}(z_1) & \mathbf{R}^{+}_{\phi,\psi_y}(z_1) \\ \mathbf{R}^{+}_{\psi_x,\phi}(z_1) & \mathbf{R}^{+}_{\psi_x,\psi_x}(z_1) & \mathbf{R}^{+}_{\psi_x,\psi_y}(z_1) \\ \mathbf{R}^{+}_{\psi_y,\phi}(z_1) & \mathbf{R}^{+}_{\psi_y,\psi_x}(z_1) & \mathbf{R}^{+}_{\psi_y,\psi_y}(z_1) \end{bmatrix} \begin{bmatrix} \vec{\Phi}^{+}(z_1) \\ \vec{\Psi}_x^{+}(z_1) \\ \vec{\Psi}_y^{+}(z_1) \end{bmatrix}, \tag{XII-3b}$$

see section IV.3.3. Finally, the relationship between the upgoing wave fields at z_1 and z_0 is described by the forward extrapolation matrix $\underset{\sim}{W}^-(z_0,z_1)$, according to

$$\vec{\underset{\sim}{D}}^-(z_0) = \underset{\sim}{W}^-(z_0,z_1)\vec{\underset{\sim}{D}}^-(z_1), \tag{XII-4a}$$

or

$$\begin{bmatrix} \vec{\Phi}^-(z_0) \\ \vec{\Psi}_x^-(z_0) \\ \vec{\Psi}_y^-(z_0) \end{bmatrix} = \begin{bmatrix} W^-_{\phi,\phi}(z_0,z_1) & 0 & 0 \\ 0 & W^-_{\psi_x,\psi_x}(z_0,z_1) & 0 \\ 0 & 0 & W^-_{\psi_y,\psi_y}(z_0,z_1) \end{bmatrix} \begin{bmatrix} \vec{\Phi}^-(z_1) \\ \vec{\Psi}_x^-(z_1) \\ \vec{\Psi}_y^-(z_1) \end{bmatrix}, \tag{XII-4b}$$

see again section IV.3.2. The combination of equations (XII-2), (XII-3) and (XII-4) yields the following relationship between the downgoing and upgoing wave fields at z_0:

$$\vec{\underset{\sim}{D}}^-(z_0) = \underset{\sim}{X}(z_0,z_0)\vec{\underset{\sim}{D}}^+(z_0), \tag{XII-5a}$$

where

$$\underset{\sim}{X}(z_0,z_0) = \underset{\sim}{W}^-(z_0,z_1)\underset{\sim}{R}^+(z_1)\underset{\sim}{W}^+(z_1,z_0), \tag{XII-5b}$$

(see Figure XI-2 for the acoustic equivalence). Matrix $\underset{\sim}{X}(z_0,z_0)$, as defined by equation (XII-5b), is the multi-component one-way response matrix. Note that we may also write

$$\begin{bmatrix} \vec{\Phi}^-(z_0) \\ \vec{\Psi}_x^-(z_0) \\ \vec{\Psi}_y^-(z_0) \end{bmatrix} = \begin{bmatrix} X_{\phi,\phi}(z_0,z_0) & X_{\phi,\psi_x}(z_0,z_0) & X_{\phi,\psi_y}(z_0,z_0) \\ X_{\psi_x,\phi}(z_0,z_0) & X_{\psi_x,\psi_x}(z_0,z_0) & X_{\psi_x,\psi_y}(z_0,z_0) \\ X_{\psi_y,\phi}(z_0,z_0) & X_{\psi_y,\psi_x}(z_0,z_0) & X_{\psi_y,\psi_y}(z_0,z_0) \end{bmatrix} \begin{bmatrix} \vec{\Phi}^+(z_0) \\ \vec{\Psi}_x^+(z_0) \\ \vec{\Psi}_y^+(z_0) \end{bmatrix}, \tag{XII-6a}$$

where the single-component one-way response matrices are given by

$$\mathbf{X}_{\Omega_2,\Omega_1}(z_o,z_o) = \mathbf{W}^-_{\Omega_2,\Omega_2}(z_o,z_1)\mathbf{R}^+_{\Omega_2,\Omega_1}(z_1)\mathbf{W}^+_{\Omega_1,\Omega_1}(z_1,z_o), \qquad \text{(XII-6b)}$$

where Ω_1 and Ω_2 each may stand for ϕ or ψ_x or ψ_y. Note the high degree of similarity with the acoustic one-way response matrix $\mathbf{X}(z_o,z_o)$, as defined by equation (XI-3b):

$$\mathbf{X}(z_o,z_o) = \mathbf{W}^-(z_o,z_1)\mathbf{R}^+(z_1)\mathbf{W}^+(z_1,z_o). \qquad \text{(XII-7)}$$

Let us now consider an arbitrarily inhomogeneous anisotropic elastic half-space $z>z_o$ below a reflection free surface z_o. Furthermore, let us distinguish between an *overburden* $z_o<z\le z_t$ and a *target zone* $z>z_t$ (see Figure XI-3 for the acoustic equivalence). In analogy with equation (XI-8) we obtain the following relationship between the multi-component one-way response matrices at the surface z_o and at the upper boundary z_t of the target zone:

$$\underset{\sim}{\mathbf{X}}(z_o,z_o) = \underset{\sim}{\mathbf{W}}^-(z_o,z_t)\underset{\sim}{\mathbf{X}}(z_t,z_t)\underset{\sim}{\mathbf{W}}^+(z_t,z_o)$$

$$+ \text{"overburden response"}, \qquad \text{(XII-8a)}$$

(see Figure XI-4 for the acoustic equivalence), where the forward extrapolation matrices read

$$\underset{\sim}{\mathbf{W}}^+(z_t,z_o) = \begin{bmatrix} \mathbf{W}^+_{\phi,\phi}(z_t,z_o) & \mathbf{W}^+_{\phi,\psi_x}(z_t,z_o) & \mathbf{W}^+_{\phi,\psi_y}(z_t,z_o) \\ \mathbf{W}^+_{\psi_x,\phi}(z_t,z_o) & \mathbf{W}^+_{\psi_x,\psi_x}(z_t,z_o) & \mathbf{W}^+_{\psi_x,\psi_y}(z_t,z_o) \\ \mathbf{W}^+_{\psi_y,\phi}(z_t,z_o) & \mathbf{W}^+_{\psi_y,\psi_x}(z_t,z_o) & \mathbf{W}^+_{\psi_y,\psi_y}(z_t,z_o) \end{bmatrix} \qquad \text{(XII-8b)}$$

and

$$\underset{\sim}{\mathbf{W}}^{-}(z_0,z_t) = \begin{bmatrix} W^{-}_{\phi,\phi}(z_0,z_t) & W^{-}_{\phi,\psi_x}(z_0,z_t) & W^{-}_{\phi,\psi_y}(z_0,z_t) \\ W^{-}_{\psi_x,\phi}(z_0,z_t) & W^{-}_{\psi_x,\psi_x}(z_0,z_t) & W^{-}_{\psi_x,\psi_y}(z_0,z_t) \\ W^{-}_{\psi_y,\phi}(z_0,z_t) & W^{-}_{\psi_y,\psi_x}(z_0,z_t) & W^{-}_{\psi_y,\psi_y}(z_0,z_t) \end{bmatrix}, \quad \text{(XII-8c)}$$

see section VI.5.2. For the general inhomogeneous anisotropic situation, equation (XII-8a) is not exact. The main approximation is the negligence of internal multiple reflections. In section XII.4, equation (XII-8) will be used as the starting point for the derivation of an elastic redatuming scheme.

Sofar we assumed that the surface z_0 is reflection-free. In practical seismic situations, however, surface z_0 represents the earth's *free surface* which is a perfect reflector for the upcoming waves $\vec{\underset{\sim}{D}}^{-}(z_0)$. The free surface reflection matrix $\underset{\sim}{R}^{-}_{fr}(z_0)$ is defined by equation (IV-73b). In analogy with equation (XI-11) we obtain for this situation

$$\vec{\underset{\sim}{D}}^{-}(z_0) = \underset{\sim}{X}_{fr}(z_0,z_0)\vec{\underset{\sim}{D}}^{+}_{s}(z_0), \quad \text{(XII-9a)}$$

where $\vec{\underset{\sim}{D}}^{+}_{s}(z_0)$ represents the downgoing source wave fields at z_0 and where the free surface multi-component one-way response matrix $\underset{\sim}{X}_{fr}(z_0,z_0)$ is defined as

$$\underset{\sim}{X}_{fr}(z_0,z_0) = [I - \underset{\sim}{X}(z_0,z_0)\underset{\sim}{R}^{-}_{fr}(z_0)]^{-1}\underset{\sim}{X}(z_0,z_0), \quad \text{(XII-9b)}$$

or

$$\underset{\sim}{X}_{fr}(z_0,z_0) = \left[I + \sum_{m=1}^{\infty} \left(\underset{\sim}{X}(z_0,z_0)\underset{\sim}{R}^{-}_{fr}(z_0)\right)^m\right]\underset{\sim}{X}(z_0,z_0). \quad \text{(XII-9c)}$$

The latter expression clearly shows that the free surface generates an infinite number of multiple reflections and conversions (see Figure XI-5 for the acoustic equivalence).

Next we discuss the relationship between the one-way wave fields in the forward model (XII-9) and the two-way seismic data (see Figure XI-6 for the acoustic equivalence). According to equation (IV-62a), the relationship between two-way and one-way elastic wave fields reads

$$\begin{bmatrix} \vec{\underset{\sim}{V}}(z) \\ \vec{\underset{\sim}{\tau}}_z(z) \end{bmatrix} = \begin{bmatrix} \underset{\sim}{L}_1^+(z) & \underset{\sim}{L}_1^-(z) \\ \underset{\sim}{L}_2^+(z) & \underset{\sim}{L}_2^-(z) \end{bmatrix} \begin{bmatrix} \vec{\underset{\sim}{D}}^+(z) \\ \vec{\underset{\sim}{D}}^-(z) \end{bmatrix}, \tag{XII-10a}$$

or

$$\vec{\underset{\sim}{V}}(z) = \underset{\sim}{L}_1^+(z)\vec{\underset{\sim}{D}}^+(z) + \underset{\sim}{L}_1^-(z)\vec{\underset{\sim}{D}}^-(z) \tag{XII-10b}$$

and

$$\vec{\underset{\sim}{\tau}}_z(z) = \underset{\sim}{L}_2^+(z)\vec{\underset{\sim}{D}}^+(z) + \underset{\sim}{L}_2^-(z)\vec{\underset{\sim}{D}}^-(z), \tag{XII-10c}$$

where the *three-component* velocity and traction vectors $\vec{\underset{\sim}{V}}(z)$ and $\vec{\underset{\sim}{\tau}}_z(z)$ are defined according to

$$\vec{\underset{\sim}{V}}(z) = \begin{bmatrix} \vec{V}_x(z) \\ \vec{V}_y(z) \\ \vec{V}_z(z) \end{bmatrix} \tag{XII-10d}$$

and

$$\vec{\underset{\sim}{\tau}}_z(z) = \begin{bmatrix} \vec{\tau}_{xz}(z) \\ \vec{\tau}_{yz}(z) \\ \vec{\tau}_{zz}(z) \end{bmatrix}, \tag{XII-10e}$$

respectively, and where matrix $\underset{\sim}{L}_\alpha^\pm(z)$ for α=1, 2 is defined by equation (IV-62e).

Unlike in section XI.2, in the following we restrict ourselves to the situation where the sources and the receivers are *at* the free surface (Figure XII-1). In analogy with equation (XI-12b) we define a *source-decomposition* operator $\underset{\sim}{C}_s^{-1}(z_0)$ which describes the relationship between the traction source vector $\vec{\underset{\sim}{\tau}}_{z,s}(z_0)$[1)] at the free surface and the downgoing source wave fields $\vec{\underset{\sim}{D}}_s^{+}(z_0)$, according to

$$\vec{\underset{\sim}{D}}_s^{+}(z_0) \triangleq \underset{\sim}{C}_s^{-1}(z_0)\vec{\underset{\sim}{\tau}}_{z,s}(z_0). \qquad \text{(XII-11a)}$$

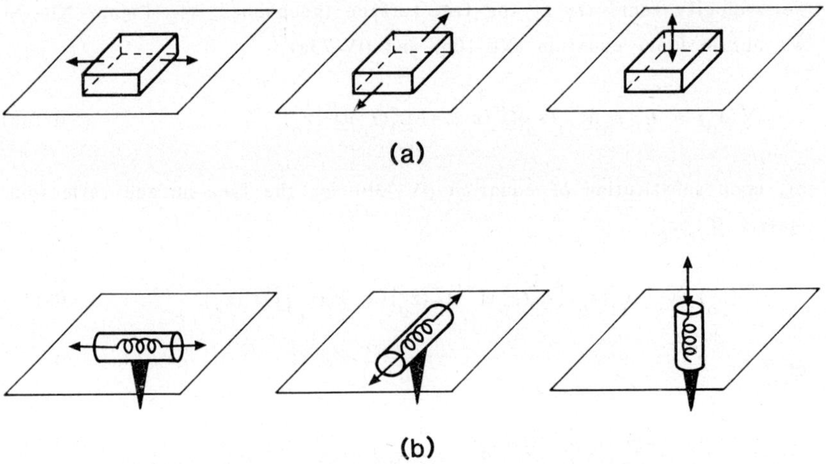

Figure XII-1: Multi-component data acquisition.

a. Three differently oriented seismic vibrators, imposing stresses in the x-, y- and z-direction to the earth's surface.

b. Three differently oriented geophones, measuring the x-, y- and z-components of the particle velocity at the earth's surface.

---

1) The multi-component vector $\vec{\underset{\sim}{\tau}}_{z,s}(z_0)$ contains the vectors $\vec{\tau}_{xz,s}(z_0)$, $\vec{\tau}_{yz,s}(z_0)$ and $\vec{\tau}_{zz,s}(z_0)$. For a point source of tensile stress (a vertical vibrator), vector $\vec{\tau}_{zz,s}(z_0)$ contains only one non-zero element, its value representing the source signature $S(\omega)$. Similarly, for a point source of shearing stress (a horizontal vibrator), one of the vectors $\vec{\tau}_{xz,s}(z_0)$ or $\vec{\tau}_{yz,s}(z_0)$ contains only one non-zero element, its value representing $S(\omega)$. When the vibrators are not ideal point sources, as in Figure XII-1a, then the source vectors contain the stress *distribution* at z_0.

Since for this situation the upgoing source wave fields at z_0 must be zero, we obtain from equation (XII-10c)

$$\vec{\underset{\sim}{\tau}}_{z,s}(z_0) = \underset{\sim}{L}_2^+(z_0)\vec{\underset{\sim}{D}}_s^+(z_0). \tag{XII-11b}$$

Hence, the source-decomposition operator reads

$$\underset{\sim}{C}_s^{-1}(z_0) = [\underset{\sim}{L}_2^+(z_0)]^{-1}. \tag{XII-12}$$

Next we derive an expression for the *receiver-composition* operator $\underset{\sim}{C}_r(z_0)$. For velocity receivers at the free surface (geophones, see Figure XII-1b) we obtain from equation (XII-10b) and (IV-73a)[1)]

$$\vec{\underset{\sim}{V}}(z_0) = \underset{\sim}{L}_1^+(z_0)\underset{\sim}{R}_{fr}^-(z_0)\vec{\underset{\sim}{D}}^-(z_0) + \underset{\sim}{L}_1^-(z_0)\vec{\underset{\sim}{D}}^-(z_0), \tag{XII-13a}$$

or, upon substitution of equation (IV-73b) for the free surface reflection matrix $\underset{\sim}{R}_{fr}^-(z_0)$,

$$\vec{\underset{\sim}{V}}(z_0) = [-\underset{\sim}{L}_1^+(z_0)\left(\underset{\sim}{L}_2^+(z_0)\right)^{-1}\underset{\sim}{L}_2^-(z_0) + \underset{\sim}{L}_1^-(z_0)]\vec{\underset{\sim}{D}}^-(z_0), \tag{XII-13b}$$

or

$$\vec{\underset{\sim}{V}}(z_0) = [\underset{\sim}{N}_1^-(z_0)]^{-1}\vec{\underset{\sim}{D}}^-(z_0), \tag{XII-13c}$$

with $\underset{\sim}{N}_1^-(z_0)$ defined in equation (IV-64). Hence, if we define the receiver-composition process in analogy with equation (XI-17b) as

$$\vec{\underset{\sim}{V}}(z_0) \triangleq \underset{\sim}{C}_r(z_0)\vec{\underset{\sim}{D}}^-(z_0), \tag{XII-13d}$$

then the receiver-composition operator $\underset{\sim}{C}_r(z_0)$ reads

$$\underset{\sim}{C}_r(z_0) = [\underset{\sim}{N}_1^-(z_0)]^{-1}. \tag{XII-13e}$$

---

1) We ignore the horizontally propagating surface waves ("ground-roll").

We summarize our expressions for the forward model for multi-component seismic data. According to equations (XII-9a), (XII-11a) and (XII-13d), the three-component velocity vector $\vec{\underset{\sim}{V}}(z_0)$ is related to the traction source vector $\vec{\underset{\sim}{\tau}}_{z,s}(z_0)$ according to

$$\vec{\underset{\sim}{V}}(z_0) = \underset{\sim}{C}_r(z_0)\underset{\sim}{X}_{fr}(z_0,z_0)\underset{\sim}{C}_s^{-1}(z_0)\vec{\underset{\sim}{\tau}}_{z,s}(z_0). \tag{XII-14a}$$

In this expression the surface waves are ignored. The free surface one-way response matrix $\underset{\sim}{X}_{fr}(z_0,z_0)$ is related to the primary one-way response matrix $\underset{\sim}{X}(z_0,z_0)$, according to

$$\underset{\sim}{X}_{fr}(z_0,z_0) = [I - \underset{\sim}{X}(z_0,z_0)\underset{\sim}{R}^-_{fr}(z_0)]^{-1}\underset{\sim}{X}(z_0,z_0), \tag{XII-14b}$$

where

$$\underset{\sim}{X}(z_0,z_0) = \begin{bmatrix} X_{\phi,\phi}(z_0,z_0) & X_{\phi,\psi_x}(z_0,z_0) & X_{\phi,\psi_y}(z_0,z_0) \\ X_{\psi_x,\phi}(z_0,z_0) & X_{\psi_x,\psi_x}(z_0,z_0) & X_{\psi_x,\psi_y}(z_0,z_0) \\ X_{\psi_y,\phi}(z_0,z_0) & X_{\psi_y,\psi_x}(z_0,z_0) & X_{\psi_y,\psi_y}(z_0,z_0) \end{bmatrix}. \tag{XII-14c}$$

The multi-component one-way response matrix $\underset{\sim}{X}(z_0,z_0)$ at the surface z_0 is related to the multi-component one-way response matrix $\underset{\sim}{X}(z_t,z_t)$ at the target depth level z_t according to

$$\underset{\sim}{X}(z_0,z_0) = \underset{\sim}{W}^-(z_0,z_t)\underset{\sim}{X}(z_t,z_t)\underset{\sim}{W}^+(z_t,z_0)$$

$$+ \text{"overburden response"}. \tag{XII-14d}$$

In this expression the internal multiple reflections are ignored. The forward model, as described by equations (XII-14a), (XII-14b) and (XII-14d), is

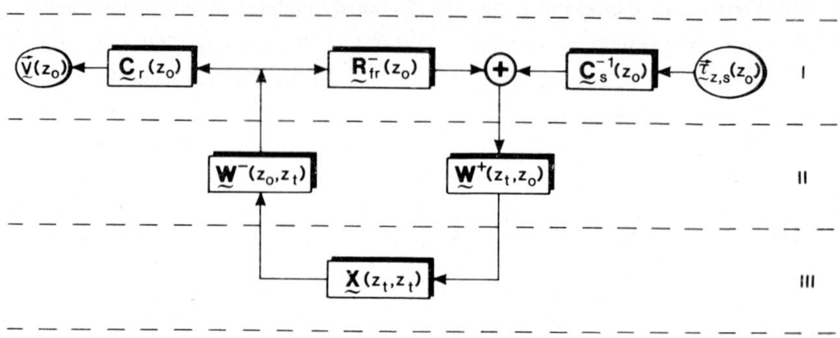

Figure XII-2: Forward model for multi-component seismic data. (The surface waves, the response of the overburden and the internal multiple reflections are ignored).
Layer I : Acquisition and near-surface properties
Layer II : Overburden propagation properties
Layer III: Target propagation and reflection properties.

visualized in Figure XII-2. Finally, note that if wave conversion during propagation may be neglected, we obtain from (XII-14c) and (XII-14d) *independent* expressions for the single-component one-way response matrices:

$$X_{\Omega_2,\Omega_1}(z_0,z_0) = W^-_{\Omega_2,\Omega_2}(z_0,z_t)X_{\Omega_2,\Omega_1}(z_t,z_t)W^+_{\Omega_1,\Omega_1}(z_t,z_0)$$

$$+ \text{"overburden response".} \qquad \text{(XII-14e)}$$

Here Ω_1 and Ω_2 may stand for ϕ or ψ_x or ψ_y.

XII.3 SURFACE RELATED ELASTIC PRE-PROCESSING

XII.3.1 Introduction

The aim of surface related elastic pre-processing is to decompose the recorded two-way seismic wave fields into one-way P- and S-wave fields and to remove the surface related multiple reflections and conversions.

Similar as in the acoustic situation, this can only be accomplished if many seismic experiments are carried out for different lateral positions of the source. Moreover, in the elastic situation ideally three independent seismic experiments should be carried out for each source position by applying three differently oriented seismic vibrators. The resulting measurements can be described by an extended version of the forward model (XII-14a), according to

$$\underset{\sim}{\mathbf{V}}(z_0) = \underset{\sim}{\mathbf{C}}_r(z_0)\underset{\sim}{\mathbf{X}}_{fr}(z_0,z_0)\underset{\sim}{\mathbf{C}}_s^{-1}(z_0)\underset{\sim}{\tau}_{z,s}(z_0). \tag{XII-15a}$$

Here the columns of the data matrix $\underset{\sim}{\mathbf{V}}(z_0)$ contain the different data vectors $\vec{\underset{\sim}{V}}(z_0)$; the columns of the source matrix $\underset{\sim}{\tau}_{z,s}(z_0)$ contain the corresponding source vectors $\vec{\underset{\sim}{\tau}}_{z,s}(z_0)$. When use is made of independent horizontal vibrators, imposing shearing stresses in the x- and y-direction, and vertical vibrators, imposing tensile stresses in the z-direction (Figure XII-1a), then the source vectors can be ordered in such a way that the source matrix may be written as[1)]

$$\underset{\sim}{\tau}_{z,s}(z_0) = \begin{bmatrix} \tau_{xz,s}(z_0) & 0 & 0 \\ 0 & \tau_{yz,s}(z_0) & 0 \\ 0 & 0 & \tau_{zz,s}(z_0) \end{bmatrix}. \tag{XII-15b}$$

Furthermore, for identical point sources this expression may be further simplified to

$$\underset{\sim}{\tau}_{z,s}(z_0) = S(\omega)I, \tag{XII-15c}$$

$S(\omega)$ representing the source signature. Now equation (XII-15a) may be replaced by

$$\underset{\sim}{\mathbf{V}}(z_0) = \underset{\sim}{\mathbf{C}}_r(z_0)\underset{\sim}{\mathbf{X}}_{fr}^{(s)}(z_0,z_0)\underset{\sim}{\mathbf{C}}_s^{-1}(z_0), \tag{XII-15d}$$

---

1) In practice the different vibrators may be oriented in arbitrary directions. In this case, mutually perpendicular vibrators can be simulated by applying a weighted summation to the different responses. A similar remark can be made for the geophones (Cliet and Dubesset, 1987).

with

$$\underset{\sim}{\mathbf{X}}^{(s)}_{fr}(z_o,z_o) = S(\omega)\underset{\sim}{\mathbf{X}}_{fr}(z_o,z_o) \qquad \text{(XII-15e)}$$

and

$$\underset{\sim}{\mathbf{V}}(z_o) = \begin{bmatrix} \mathbf{V}_{x,x}(z_o) & \mathbf{V}_{x,y}(z_o) & \mathbf{V}_{x,z}(z_o) \\ \mathbf{V}_{y,x}(z_o) & \mathbf{V}_{y,y}(z_o) & \mathbf{V}_{y,z}(z_o) \\ \mathbf{V}_{z,x}(z_o) & \mathbf{V}_{z,y}(z_o) & \mathbf{V}_{z,z}(z_o) \end{bmatrix} . \qquad \text{(XII-15f)}$$

Here any of the sub-matrices $\mathbf{V}_{i,j}(z_o)$ for i=x,y,z and j=x,y,z represents a (monochromatic) single-component seismic survey, carried out with geophones oriented in the i-direction and vibrators oriented in the j-direction. In Figure XII-3a the 2-D situation is shown for one element of matrix $\mathbf{V}_{z,z}(z_o)$.
Similarly, in Figures XII-3b, c and d the 2-D situation is shown for the corresponding elements in matrices $\mathbf{V}_{x,z}(z_o)$, $\mathbf{V}_{z,x}(z_o)$ and $\mathbf{V}_{x,x}(z_o)$, respectively.

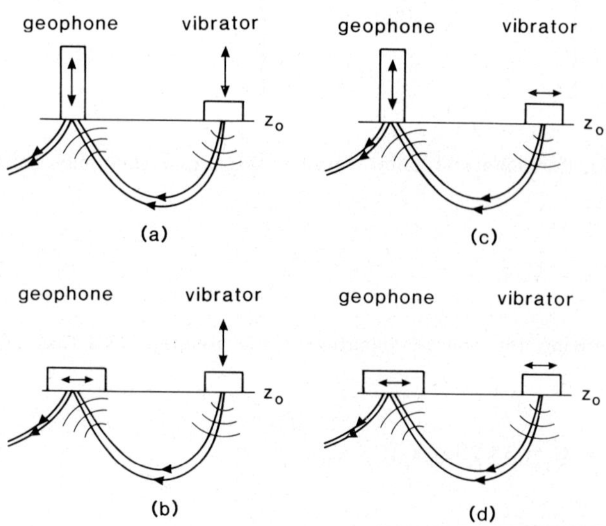

Figure XII-3: 2-D visualization of multi-component data acquisition at a free surface z_o. The double raypaths symbolically represent P- and S-waves.

XII.3.2 Elastic decomposition

Elastic decomposition should be preceded by the removal of the surface waves. We do not discuss this procedure; a good reference is Beresford-Smith and Rango (1989).
Our starting point is equation (XII-15d), which is the forward model of a multi-experiment multi-offset multi-component seismic dataset, excluding the surface waves. Assuming that the source signature $S(\omega)$ is unknown, the scaled free surface one-way response matrix can be obtained from the two-way seismic data matrix $\underset{\sim}{\mathbf{V}}(z_0)$ by inverting equation (XII-15d), yielding (Wapenaar et al., 1989),

$$\underset{\sim}{\mathbf{X}}_{fr}^{(s)}(z_0,z_0) = \underset{\sim}{\mathbf{C}}_r^{-1}(z_0)\underset{\sim}{\mathbf{V}}(z_0)\underset{\sim}{\mathbf{C}}_s(z_0), \qquad \text{(XII-16a)}$$

where

$$\underset{\sim}{\mathbf{C}}_s(z_0) = \underset{\sim}{\mathbf{L}}_2^+(z_0) \qquad \text{(XII-16b)}$$

and

$$\underset{\sim}{\mathbf{C}}_r^{-1}(z_0) = \underset{\sim}{\mathbf{N}}_1^-(z_0), \qquad \text{(XII-16c)}$$

matrices $\underset{\sim}{\mathbf{L}}_2^+(z_0)$ and $\underset{\sim}{\mathbf{N}}_1^-(z_0)$ being defined by equations (IV-62e) and (IV-64b), respectively. Note that $\underset{\sim}{\mathbf{C}}_r^{-1}(z_0)\underset{\sim}{\mathbf{V}}(z_0)$ describes a lateral deconvolution process along the columns (i.e., the common shot records) of matrix $\underset{\sim}{\mathbf{V}}(z_0)$, whereas $\underset{\sim}{\mathbf{V}}(z_0)\underset{\sim}{\mathbf{C}}_s(z_0)$ describes a lateral deconvolution process along the rows (i.e., the common receiver records) of matrix $\underset{\sim}{\mathbf{V}}(z_0)$, (see Figure XI-10 for the acoustic equivalence).

In analogy with equation (XII-14c), the decomposed data matrix $\underset{\sim}{\mathbf{X}}_{fr}^{(s)}(z_0,z_0)$ may be written as

$$\underset{\sim}{\mathbf{X}}_{fr}^{(s)}(z_0,z_0) = \begin{bmatrix} X_{\phi,\phi}(z_0,z_0) & X_{\phi,\psi_x}(z_0,z_0) & X_{\phi,\psi_y}(z_0,z_0) \\ X_{\psi_x,\phi}(z_0,z_0) & X_{\psi_x,\psi_x}(z_0,z_0) & X_{\psi_x,\psi_y}(z_0,z_0) \\ X_{\psi_y,\phi}(z_0,z_0) & X_{\psi_y,\psi_x}(z_0,z_0) & X_{\psi_y,\psi_y}(z_0,z_0) \end{bmatrix}_{fr}^{(s)} . \qquad \text{(XII-17)}$$

Any of the sub-matrices simulates a (monochromatic) single-component one-way seismic survey at the free surface. Matrices $[\mathbf{X}_{\phi,\phi}(z_0,z_0)]^{(s)}_{fr}$ and $[\mathbf{X}_{\phi,\psi_\alpha}(z_0,z_0)]^{(s)}_{fr}$ for α=x,y represent seismic surveys in terms of received upgoing P-waves related to sources in terms of downgoing P-waves or downgoing S_x- or S_y-waves. Similarly, matrices $[\mathbf{X}_{\psi_\beta,\phi}(z_0,z_0)]^{(s)}_{fr}$ and $[\mathbf{X}_{\psi_\beta,\psi_\alpha}(z_0,z_0)]^{(s)}_{fr}$ for β=x,y and α=x,y represent seismic surveys in terms of received upgoing S_x- or S_y-waves related to sources in terms of downgoing P-waves or downgoing S_x- or S_y-waves. In Figure XII-4a the 2-D situation is shown for one element of matrix $[\mathbf{X}_{\phi,\phi}(z_0,z_0)]^{(s)}_{fr}$. Similarly, in Figures XII-4b,c and d the 2-D situation is shown for the corresponding elements in matrices $[\mathbf{X}_{\psi_y,\phi}(z_0,z_0)]^{(s)}_{fr}$, $[\mathbf{X}_{\phi,\psi_y}(z_0,z_0)]^{(s)}_{fr}$ and $[\mathbf{X}_{\psi_y,\psi_y}(z_0,z_0)]^{(s)}_{fr}$, respectively. Note that in the 2-D situation ψ_y refers to SV-waves.

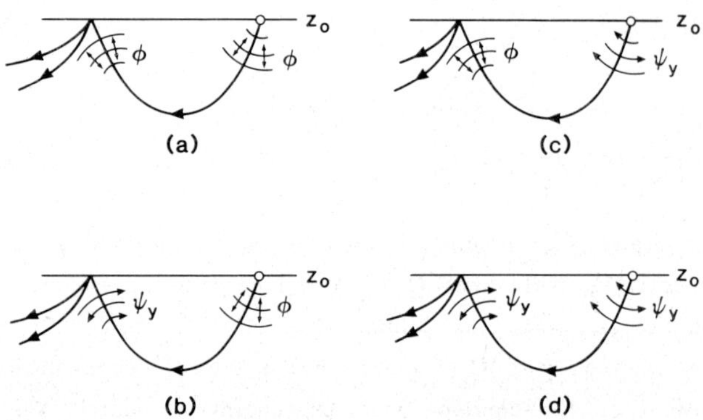

Figure XII-4: 2-D visualization of decomposed data at a free surface z_0.

We illustrate the elastic decomposition procedure with a simple 2-D example. For the subsurface configuration shown in Figure XII-5, we generated 128 multi-component seismic shot records by finite difference modeling (Kelly et al., 1976; Haimé, 1987). We used vertical and horizontal vibrators as well as vertical and horizontal geophones at the free surface z_0. One multi-component shot record is shown in the space-time domain in Figure XII-6.

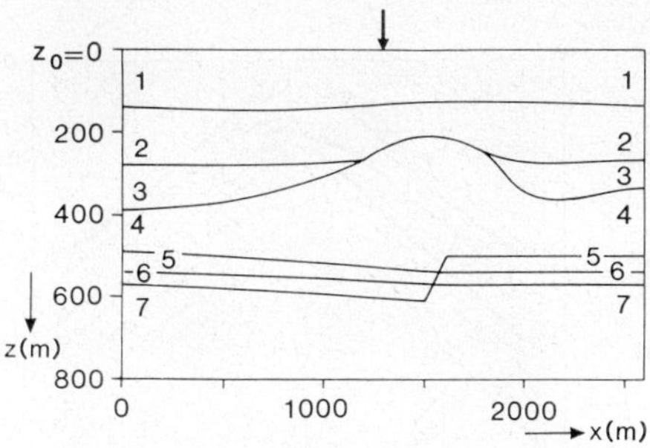

Layer	Cp(m/s)	Cs(m/s)	ρ(kg/m³)
1	2400	1400	1000
2	3000	2000	1600
3	3000	2000	2100
4	4100	2200	2200
5	3700	2000	2300
6	4200	2400	2000
7	3500	2100	1800

Figure XII-5: 2-D inhomogeneous elastic subsurface. The multi-component vibrators and geophones are situated at the free surface z_o=0 m.

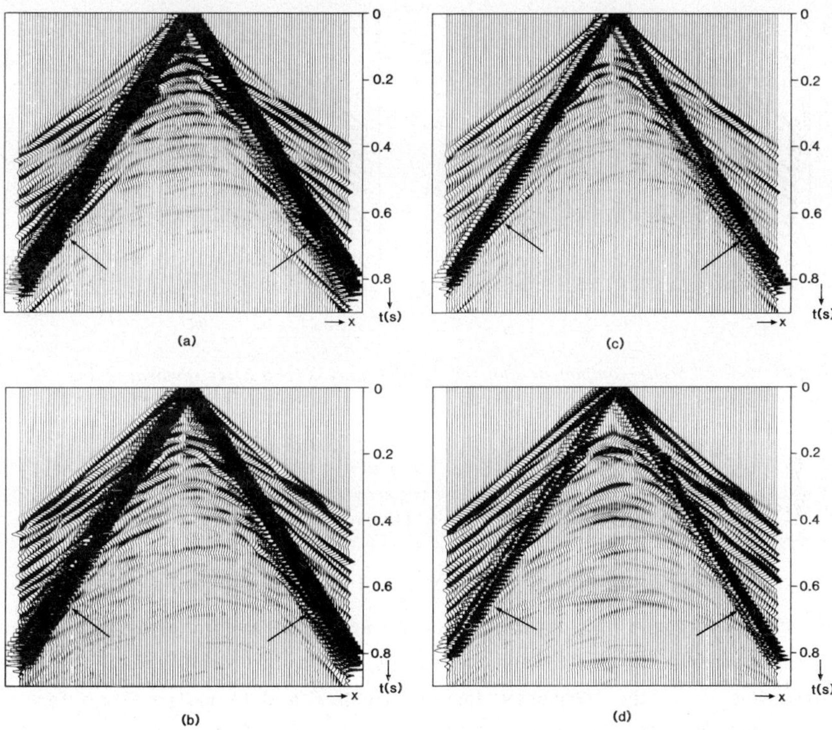

Figure XII-6: Multi-component shot record. The source position is indicated by the arrow in Figure XII-5.

a. Pseudo P-P data ($V_{z,z}$, see Figure XII-3a)

b. Pseudo SV-P data ($V_{x,z}$, see Figure XII-3b)

c. Pseudo P-SV data ($V_{z,x}$, see Figure XII-3c)

d. Pseudo SV-SV data ($V_{x,x}$, see Figure XII-3d)

The arrows indicate the ground-roll.

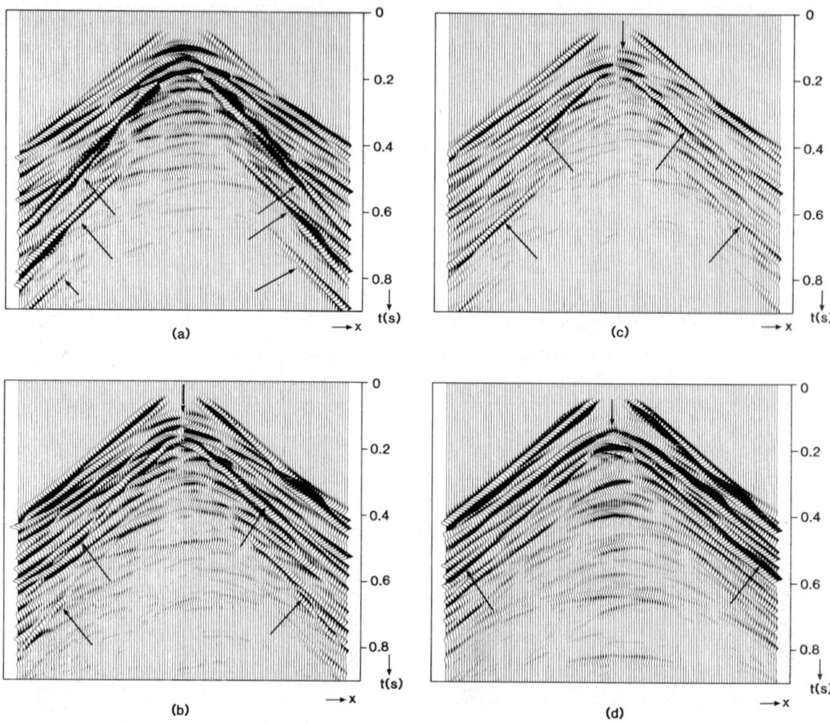

Figure XII-7: Multi-component shot record of Figure XII-6 after removal of the ground-roll.

a. Pseudo P-P data | *c. Pseudo P-SV data*

b. Pseudo SV-P data | *d. Pseudo SV-SV data*

The arrows indicate spurious events.

Figure XII-7 shows the same multi-component shot record after removal of the ground-roll. All multi-component shot records are transformed from the time domain to the frequency domain, yielding a data matrix $\underset{\sim}{\mathbf{V}}(z_0)$ for each frequency in the seismic band (5 Hz<f= $\frac{\omega}{2\pi}$ < 80 Hz). Next, decomposition is carried out by applying equation (XII-16a) for each frequency in the seismic band. Finally, the results are transformed back from the frequency domain to the time domain. Figure XII-8 shows one multi-component shot record after decomposition (Herrmann et al., 1989). Note that the spurious events, indicated by the arrows in Figure XII-7, vanished completely.

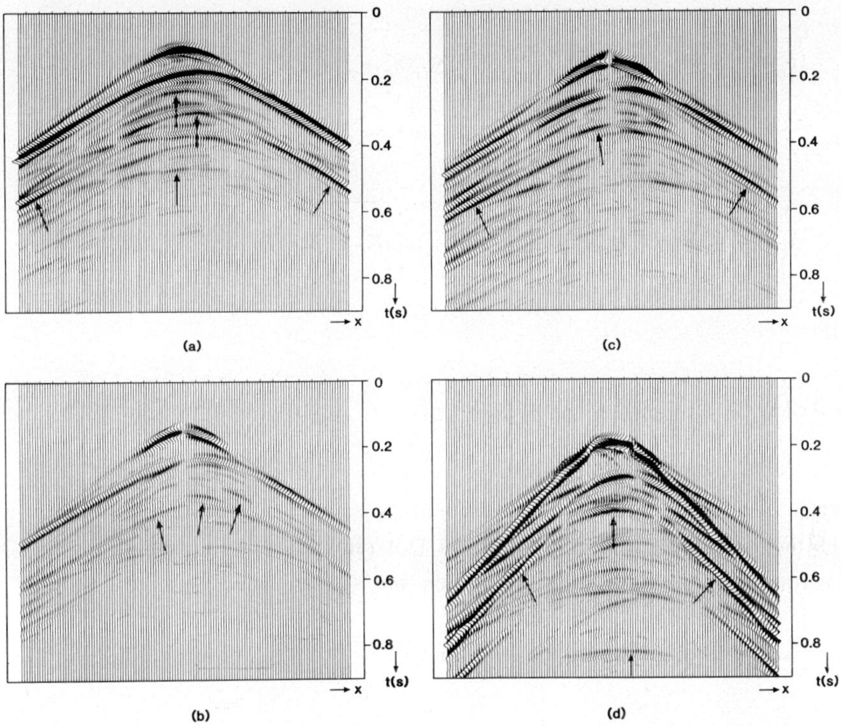

Figure XII-8: Multi-component shot record, after decomposition into one-way P- and SV-wave responses. The source position is indicated by the arrow in Figure XII-5.

a. True P-P data (see Figure XII-4a)

b. True SV-P data (see Figure XII-4b)

c. True P-SV data (see Figure XII-4c)

d. True SV-SV data (see Figure XII-4d)

The arrows indicate surface related multiple reflections and conversions.

XII.3.3 Elastic multiple elimination

After the decomposition has been carried out, the scaled multi-component free surface one-way response matrix $\underset{\sim}{\mathbf{X}}^{(s)}_{fr}(z_0,z_0)$ is available for all frequencies in the seismic band. This response matrix contains significant multiple reflections and conversions related to the free surface (see Figure XII-8). They can be removed by inverting equation (XII-14b), yielding

$$\underset{\sim}{X}(z_0,z_0) = \underset{\sim}{X}_{fr}(z_0,z_0)[I + \underset{\sim}{R}^-_{fr}(z_0)\underset{\sim}{X}_{fr}(z_0,z_0)]^{-1}, \tag{XII-18a}$$

or

$$\underset{\sim}{X}(z_0,z_0) = \underset{\sim}{X}_{fr}(z_0,z_0)\left[I + \sum_{m=1}^{\infty}\left(-\underset{\sim}{R}^-_{fr}(z_0)\underset{\sim}{X}_{fr}(z_0,z_0)\right)^m\right], \tag{XII-18b}$$

with

$$\underset{\sim}{X}_{fr}(z_0,z_0) = \frac{1}{S(\omega)}\underset{\sim}{X}^{(s)}_{fr}(z_0,z_0). \tag{XII-18c}$$

Using an adaptive procedure, the source deconvolution (equation (XII-18c)) and the multiple elimination (equation (XII-18b)) may be carried out simultaneously. For a further discussion see section XI.3.3.

Note that the final result may be written as

$$\underset{\sim}{X}(z_0,z_0) = \begin{bmatrix} X_{\phi,\phi}(z_0,z_0) & X_{\phi,\psi_x}(z_0,z_0) & X_{\phi,\psi_y}(z_0,z_0) \\ X_{\psi_x,\phi}(z_0,z_0) & X_{\psi_x,\psi_x}(z_0,z_0) & X_{\psi_x,\psi_y}(z_0,z_0) \\ X_{\psi_y,\phi}(z_0,z_0) & X_{\psi_y,\psi_x}(z_0,z_0) & X_{\psi_y,\psi_y}(z_0,z_0) \end{bmatrix}. \tag{XII-19}$$

Any of the sub-matrices simulates a (monochromatic) single-component one-way seismic survey at a reflection-free surface. In Figure XII-9a the 2-D situation is shown for one element of matrix $\mathbf{X}_{\phi,\phi}(z_0,z_0)$. Similarly, in Figures XII-9b, c and d the 2-D situation is shown for the corresponding elements in matrices $\mathbf{X}_{\psi_y,\phi}(z_0,z_0)$, $\mathbf{X}_{\phi,\psi_y}(z_0,z_0)$ and $\mathbf{X}_{\psi_y,\psi_y}(z_0,z_0)$, respectively.

We illustrate the elastic multiple elimination procedure with a 2-D example (Verschuur et al., 1989). We consider the decomposed data of the

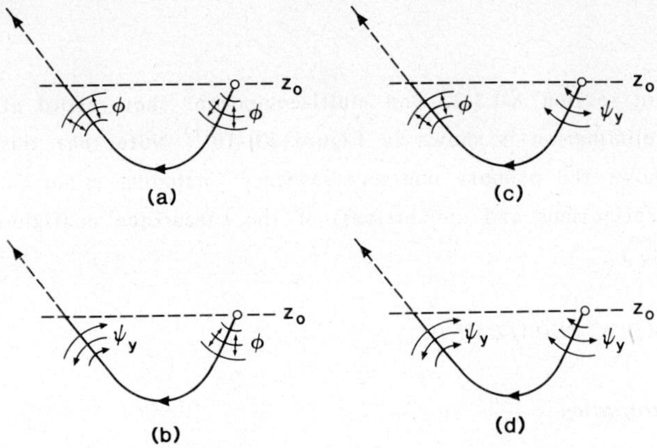

Figure XII-9: 2-D visualization of decomposed data at a reflection-free surface z_o (after surface related elastic multiple elimination).

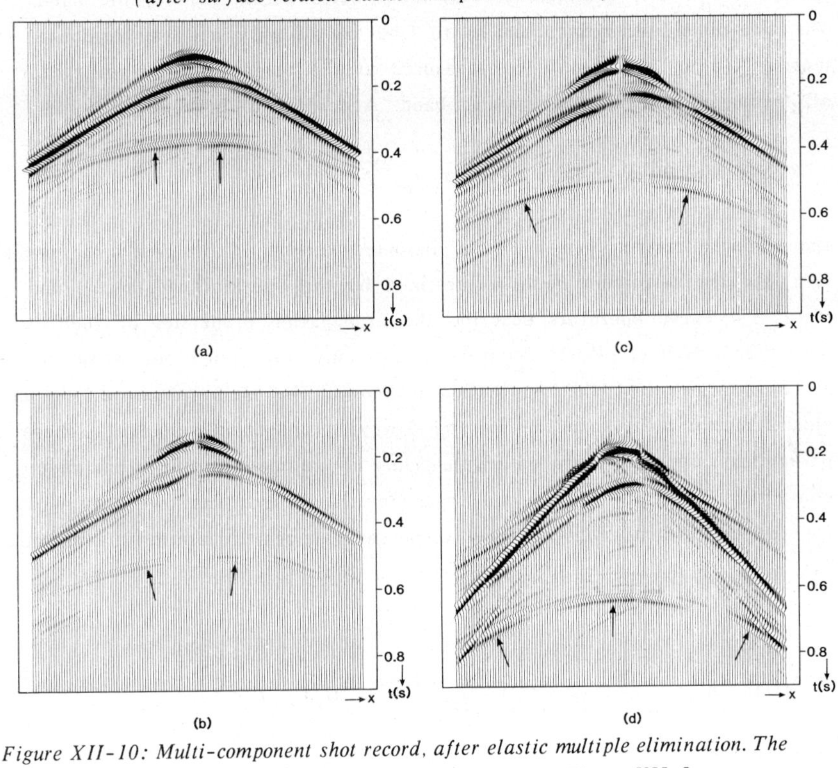

Figure XII-10: Multi-component shot record, after elastic multiple elimination. The source position is indicated by the arrow in Figure XII-5.

a. True P-P data (see Figure XII-9a)

b. True SV-P data (see Figure XII-9b)

c. True P-SV data (see Figure XII-9c)

d. True SV-SV data (see Figure XII-9d)

The arrows indicate the response of the target reflectors below z_t=450 m.

example in section XII.3.2. One multi-component shot record after adaptive multiple elimination is shown in Figure XII-10.[1] Note that this result clearly shows the primary one-way response (including minor *internal* multiple reflections and conversions) of the subsurface configuration of Figure XII-5.

XII.4 ELASTIC REDATUMING

XII.4.1 Introduction

After surface related elastic pre-processing, the primary one-way response matrix $\underset{\sim}{X}(z_0,z_0)$ is available for all frequencies within the seismic band. The aim of elastic redatuming is to find the primary one-way response matrix $\underset{\sim}{X}(z_t,z_t)$ that would be measured at the target depth level z_t for all frequencies within the seismic band. According to equation (XII-14d),

$$\underset{\sim}{X}(z_0,z_0) = \underset{\sim}{W}^-(z_0,z_t)\underset{\sim}{X}(z_t,z_t)\underset{\sim}{W}^+(z_t,z_0)^{2)}, \qquad \text{(XII-20a)}$$

the response matrix $\underset{\sim}{X}(z_0,z_0)$ is a distorted version of the response matrix $\underset{\sim}{X}(z_t,z_t)$, the distortion being determined by the operators $\underset{\sim}{W}^+(z_t,z_0)$ and $\underset{\sim}{W}^-(z_0,z_t)$. These operators describe the propagation properties of the overburden. Hence, elastic redatuming can only be carried out when the *elastic macro model* of the overburden is known. The main parameters in an elastic macro model are the P- and S-wave propagation velocities. These parameters can be estimated *independently* if we may assume that wave conversion *during propagation* can be neglected. In that case we may write for the single-component one-way response matrices, according to equation (XII-14e),

$$\mathbf{X}_{\Omega_2,\Omega_1}(z_0,z_0) = \mathbf{W}^-_{\Omega_2,\Omega_2}(z_0,z_t)\mathbf{X}_{\Omega_2,\Omega_1}(z_t,z_t)\mathbf{W}^+_{\Omega_1,\Omega_1}(z_t,z_0),^{2)} \qquad \text{(XII-20b)}$$

---

1) Although only one multi-component shot record is shown, bear in mind that all multi-component shot records (the columns of $\underset{\sim}{X}^{(s)}_{fr}(z_0,z_0)$) were involved in the multiple elimination process.

2) For simplicity the overburden response is ignored.

where Ω_1 and Ω_2 may stand for ϕ or ψ_x or ψ_y. Note that the expression for $\mathbf{X}_{\phi,\phi}(z_o,z_o)$ contains the P-wave extrapolation operators $\mathbf{W}^+_{\phi,\phi}(z_t,z_o)$ and $\mathbf{W}^-_{\phi,\phi}(z_o,z_t)$, hence the P-wave macro model may be obtained from the travel time information contained in the P-wave response matrix $\mathbf{X}_{\phi,\phi}(z_o,z_o)$. Similar arguments lead to the conclusion that the S-wave macro model may be obtained from the travel time information contained in the S-wave response matrices $\mathbf{X}_{\psi_x,\psi_x}(z_o,z_o)$ and $\mathbf{X}_{\psi_y,\psi_y}(z_o,z_o)$. Differences between the S-wave macro models obtained from $\mathbf{X}_{\psi_x,\psi_x}(z_o,z_o)$ and $\mathbf{X}_{\psi_y,\psi_y}(z_o,z_o)$ may be an indication for *azimuthal anisotropy* (see section II.4.1 and Martin and Davis, 1987). A further discussion on elastic macro model estimation is beyond the scope of this book. In the following we assume that an accurate macro model of the overburden is available.

XII.4.2 Principle of elastic redatuming

Elastic redatuming involves compensation for the distortion caused by propagation through the overburden. By inverting equation (XII-20a) we obtain the following expression for the redatumed response matrix:

$$\underset{\sim}{\mathbf{X}}(z_t,z_t) = \underset{\sim}{\mathbf{F}}^-(z_t,z_o)\underset{\sim}{\mathbf{X}}(z_o,z_o)\underset{\sim}{\mathbf{F}}^+(z_o,z_t),^{1)} \qquad \text{(XII-21a)}$$

where

$$\underset{\sim}{\mathbf{F}}^+(z_o,z_t) \triangleq [\underset{\sim}{\mathbf{W}}^+(z_t,z_o)]^{-1} \qquad \text{(XII-21b)}$$

and

$$\underset{\sim}{\mathbf{F}}^-(z_t,z_o) \triangleq [\underset{\sim}{\mathbf{W}}^-(z_o,z_t)]^{-1}. \qquad \text{(XII-21c)}$$

---

1) Again, for simplicity the overburden response is ignored . In reality $\underset{\sim}{\mathbf{X}}(z_t,z_t)$, as defined by (XII-21), consists of a causal term representing the target response and a non-causal term related to the overburden response. The latter can be easily removed after the redatumed data have been transformed back to the time domain.

In practice, direct inversion of the extrapolation matrices $\underset{\sim}{\mathbf{W}}^+$ and $\underset{\sim}{\mathbf{W}}^-$ should be avoided. Assuming that the contrasts in the overburden are weak to moderate, we may apply the modified matched inverse operators

$$<\underset{\sim}{\mathbf{F}}^+(z_0,z_t)> = [\underset{\sim}{\mathbf{W}}^-(z_0,z_t)]^* \tag{XII-22a}$$

and

$$<\underset{\sim}{\mathbf{F}}^-(z_t,z_0)> = [\underset{\sim}{\mathbf{W}}^+(z_t,z_0)]^*. \tag{XII-22b}$$

For an extensive discussion of these operators we refer to chapter VIII.

Elastic redatuming of multi-component data by means of equation (XII-21a) is useful only when the P- and S-wave macro models are fully consistent. In practice, however, these macro models contain (small) errors. In section VIII.3.5 we argued that, as long as the contrasts in the medium are weak to moderate, wave conversion during propagation is preferably ignored. Therefore redatuming should preferably be based on inverting equation (XII-20b), yielding

$$\mathbf{X}_{\Omega_2,\Omega_1}(z_t,z_t) = \mathbf{F}^-_{\Omega_2,\Omega_2}(z_t,z_0)\mathbf{X}_{\Omega_2,\Omega_1}(z_0,z_0)\mathbf{F}^+_{\Omega_1,\Omega_1}(z_0,z_t), \tag{XII-23a}$$

where Ω_1 and Ω_2 may stand for ϕ or ψ_x or ψ_y. Since we assumed that the contrasts in the overburden are weak to moderate we may apply the modified matched inverse operators

$$<\mathbf{F}^+_{\Omega_1,\Omega_1}(z_0,z_t)> = [\mathbf{W}^-_{\Omega_1,\Omega_1}(z_0,z_t)]^* \tag{XII-23b}$$

and

$$<\mathbf{F}^-_{\Omega_2,\Omega_2}(z_t,z_0)> = [\mathbf{W}^+_{\Omega_2,\Omega_2}(z_t,z_0)]^*. \tag{XII-23c}$$

When the contrasts in the overburden are significant we should apply higher order approximations of the inverse operators $\mathbf{F}^{+}_{\Omega_1,\Omega_1}(z_0,z_t)$ and $\mathbf{F}^{-}_{\Omega_2,\Omega_2}(z_t,z_0)$ (see section X.3) rather than returning to equation (XII-21a).

Finally, note that equation (XII-23a), which describes redatuming of *elastic* data, is fully equivalent with equation (XI-27a), which describes redatuming of acoustic data. Hence, redatuming of any of the one-way response matrices $\mathbf{X}_{\Omega_2,\Omega_1}(z_0,z_0)$ may also be carried out *per shot record*, followed by *stacking*, just as we explained in section XI.4.3 for the acoustic situation.

XII.4.3 Example of elastic redatuming

We illustrate the elastic redatuming procedure with a 2-D example. We consider the data of the examples in sections XII.3.2 and XII.3.3. The subsurface configuration is again shown in Figure XII-11. The upper boundary of the target zone at z_t= 450 m is indicated by a dashed line. One multi-component shot record at the surface z_0=o m after elastic decomposition and multiple elimination was shown in Figure XII-10. Note that the response of the target zone is distorted due to the propagation effects of the overburden. The four data types (P-P, SV-P, P-SV and SV-SV) were redatumed fully independently from z_0 to z_t. Figure XII-12 shows one multi-component shot record at the upper boundary of the target zone. Note that this result clearly shows the *angle-dependent* reflectivity properties of the reflectors in the target. Finally, Figure XII-13 shows the P-P and SV-SV zero offset sections, selected from the redatumed data at the upper boundary of the target zone. Note that the structure of the reflectors in the target can be clearly recognized.

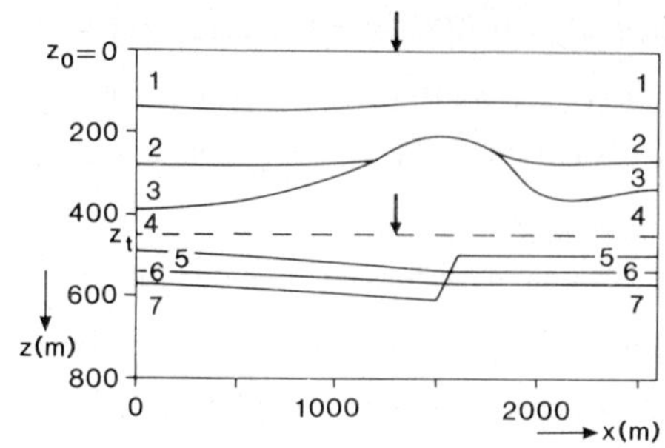

Layer	Cp(m/s)	Cs(m/s)	ρ(kg/m³)
1	2400	1400	1000
2	3000	2000	1600
3	3000	2000	2100
4	4100	2200	2200
5	3700	2000	2300
6	4200	2400	2000
7	3500	2100	1800

Figure XII-11: 2-D inhomogeneous elastic subsurface. The dashed line indicates the target upper boundary.

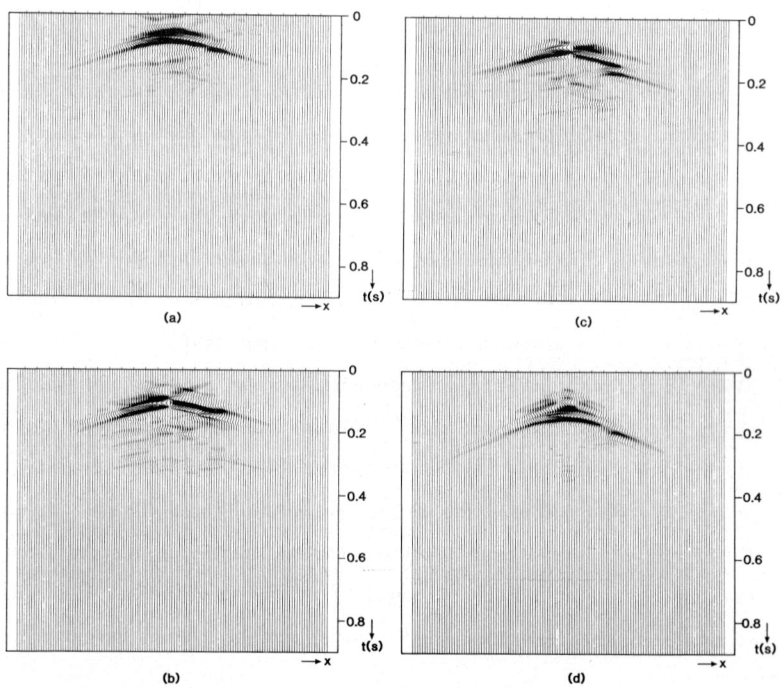

Figure XII-12: Multi-component shot record, selected from the redatumed data at z_t. The source position is indicated by the arrow at z_t=450 m in Figure XII-11.

a. True P-P data *c. True P-SV data*

b. True SV-P data *d. True SV-SV data*

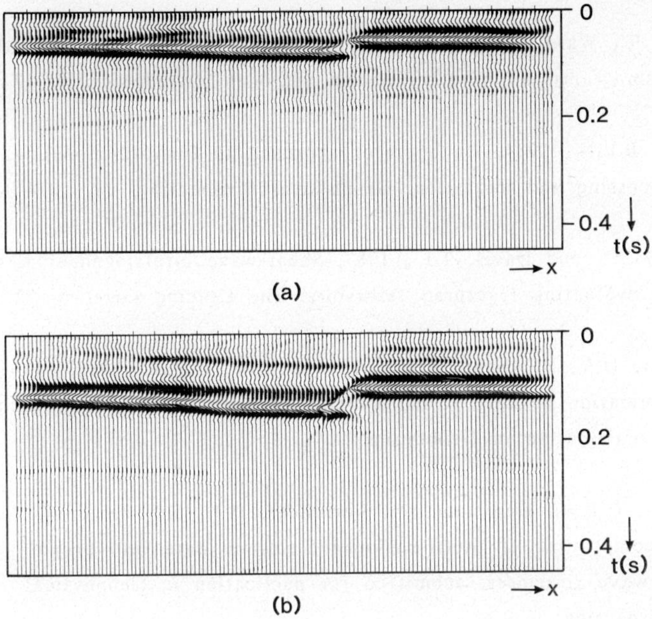

Figure XII-13: Zero offset sections, selected from the redatumed data at z_t.
a. True P-P data
b. True SV-SV data.

XII.5 REFERENCES

Beresford-Smith, G., and Rango, R., 1989, Suppression of ground roll by windowing in two domains: First Break, *7*, 55 - 63.

Cliet, C., and Dubesset, M., 1987, Three-component recordings: Interest for land seismic source study: Geophysics, *52*, 1048 - 1059.

Dankbaar, J.W.M., 1985, Separation of P- and S-waves: Geophysical Prospecting, *33*, 970 - 986.

Haimé, G.C., 1987, Full elastic inverse wave field extrapolation through an inhomogeneous medium: M.Sc. thesis, Delft University of Technology.

Herrmann, P., Verschuur, D.J., Wapenaar, C.P.A., and Berkhout, A.J., 1989, Decomposition of multi-component seismic data into primary P- and S-wave responses: presented at the 51st EAEG Meeting, Berlin.

Kelly, K.R., Ward, R.W., Treitel, S., and Alford, R.M., 1976, Synthetic seismograms: a finite difference approach: Geophysics, *41*, 2 - 27.

Kennett, B.L.N., 1984, An operator approach to forward modeling, data processing and migration: Geophysical Prospecting, *32*, 1074 - 1090.

Martin, M.A., and Davis, T.L., 1987, Shear-wave birefringence: A new tool for evaluating fractured reservoirs: The Leading Edge, *6*, 22 - 28.

Verschuur, D.J., Berkhout, A.J., and Wapenaar, C.P.A., 1989, Multiple elimination on single- and multi-component seismic data: submitted for publication in Geophysics.

Wapenaar, C.P.A., Herrmann, P., Verschuur, D.J., and Berkhout, A.J., 1989, Decomposition of multi-component seismic data into primary P- and S- wave responses: submitted for publication in Geophysical Prospecting.

APPENDIX A

MATRIX NOTATION

A.1 INTRODUCTION

In this book we often encounter operations that are written as a generalized spatial convolution integral, according to

$$Q(x,y,z;\omega) = \iint_{-\infty}^{\infty} H(x,y,z;x',y',z';\omega)P(x',y',z';\omega)dx'dy'. \tag{A-1}$$

Here P represents a quantity related to an acoustic or elastic wave field (for instance the acoustic pressure or one component of the particle velocity), H represents an operator (for instance for downward wave field extrapolation) and Q represents again a quantity related to an acoustic or elastic wave field (for instance the downward extrapolated wave field). In this Appendix we discuss a matrix notation for wave fields and operators, as introduced by Berkhout (1985), generalized for 2-D and 3-D applications. This matrix notation has the important advantage that operations of the form (A-1) are replaced by simple matrix products, like

$$\vec{Q}(z) = \mathbf{H}(z,z')\vec{P}(z'), \tag{A-2}$$

where vectors $\vec{P}$ and $\vec{Q}$ contain discretized versions of the wave fields P and Q and where matrix H represents the discretized operator. This notation suits very well with the seismic situation, where we always deal with sampled wave fields. Furthermore, with this notation we can easily describe more complicated operations. For instance, if (A-2) represents *forward* wave field extrapolation, then *inverse* wave field extrapolation is simply described by

$$\vec{P}(z') = [\mathbf{H}(z,z')]^{-1}\vec{Q}(z). \tag{A-3}$$

In the following we discuss the matrix notation in more detail. We separately treat wave fields and operators.

A.2 MATRIX NOTATION FOR WAVE FIELDS

Consider a 2-D wave field, measured at a constant depth level as a function of lateral position and time, described by

$$p(x,z_0;t), \tag{A-4a}$$

where

- p : wave field (for instance the acoustic pressure),
- x : lateral coordinate of the receivers,
- z_0 : depth level of the acquisition surface,
- t : time.

After a Fourier transformation from time to frequency, as defined by (III-1a), this wave field is described by

$$P(x,z_0;\omega), \tag{A-4b}$$

where

- P : Fourier transformed wave field,
- ω : circular frequency.

In the following we only consider the frequency domain representation, that is, we assume that monochromatic wave fields $P(x,z_0;\omega_i)$ are available for a range of ω_i values. All these monochromatic wave fields can be treated independently. If we consider one frequency component ω_i only, then the discretized version of the wave field can be represented by a *vector*, according to

$$\vec{P}(z_0) = \begin{bmatrix} P(-K\Delta x,z_0;\omega_i) \\ \vdots \\ P(k\Delta x,z_0;\omega_i) \\ \vdots \\ P(K\Delta x,z_0;\omega_i) \end{bmatrix}, \tag{A-4c}$$

where Δx is the distance between the receivers.

For the seismic situation this vector may represent the (monochromatic) data in one common shot record. Let us now write this vector symbolically as

$$\vec{P}(z_o) = \begin{bmatrix} P_{-K} \\ \vdots \\ P_k \\ \vdots \\ P_K \end{bmatrix} \downarrow x_r \quad , \tag{A-5a}$$

where x_r denotes that the different elements in this vector correspond to the different lateral positions of the receivers. With this notation we can write the (monochromatic) data $P(x_r,z_o;x_s,z_o;\omega_i)$ in a 2-D seismic survey symbolically as a *matrix*, according to

$$\mathbf{P}(z_o) = \overset{\rightarrow x_s}{\begin{bmatrix} P_{-K,-M} \cdots\cdots P_{-K,m} \cdots\cdots P_{-K,M} \\ \vdots \quad\quad \vdots \quad\quad \vdots \\ P_{k,-M} \cdots\cdots P_{k,m} \cdots\cdots P_{k,M} \\ \vdots \quad\quad \vdots \quad\quad \vdots \\ P_{K,-M} \cdots\cdots P_{K,m} \cdots\cdots P_{K,M} \end{bmatrix}} \downarrow x_r \quad , \tag{A-5b}$$

where x_s denotes the different lateral positions of the sources.
Each element $P_{k,m}$ corresponds to a fixed lateral receiver coordinate $x_{r,k}$ and a fixed lateral source coordinate $x_{s,m}$. Each column (fixed x_s) in this data matrix represents one (monochromatic) common shot record; each row (fixed x_r) represents one common receiver record; the diagonal ($x_s=x_r$) represents zero offset data and the anti-diagonal ($x_s=-x_r$) represents common midpoint data.

The (monochromatic) data $P(x_r,y_r,z_o;x_s,y_s,z_o;\omega_i)$ in a 3-D seismic areal survey can also be represented by a matrix (Kinneging et al, 1989), according to

$$
\mathbf{P}(z_0) = \begin{bmatrix} \mathbf{P}_{-L,-N} & \cdots\cdots & \mathbf{P}_{-L,n} & \cdots\cdots & \mathbf{P}_{-L,N} \\ \vdots & & \vdots & & \vdots \\ \mathbf{P}_{\ell,-N} & \cdots\cdots & \mathbf{P}_{\ell,n} & \cdots\cdots & \mathbf{P}_{\ell,N} \\ \vdots & & \vdots & & \vdots \\ \mathbf{P}_{L,-N} & \cdots\cdots & \mathbf{P}_{L,n} & \cdots\cdots & \mathbf{P}_{L,N} \end{bmatrix} \begin{matrix} \rightarrow y_s \\ \\ \\ \\ \downarrow y_r \end{matrix} , \qquad \text{(A-5c)}
$$

where y_r denotes the different cross-line positions of the receivers and where y_s denotes the different cross-line positions of the sources. Each sub-matrix $\mathbf{P}_{\ell,n}$ corresponds to a fixed cross-line receiver coordinate $y_{r,\ell}$ and a fixed cross-line source coordinate $y_{s,n}$. The elements in the sub-matrix itself are defined as in (A-5b). Note that each column (fixed x_s, y_s) of the total matrix $\mathbf{P}(z_0)$ represents one (monochromatic) common shot record and each row (fixed x_r, y_r) represents one common receiver record. In this book a data matrix $\mathbf{P}(z_0)$ may represent either a 2-D seismic survey, as in (A-5b), or a 3-D seismic areal survey, as in (A-5c). Hence, a data vector $\vec{P}(z_0)$ (one column of $\mathbf{P}(z_0)$) may represent either a 2-D or a 3-D seismic shot record.

A.3 MATRIX NOTATION FOR OPERATORS

Consider again the operation described by (A-1). If we replace the wave fields P and Q as well as the operator H by their discretized versions, then the integrals are replaced by summations, according to

$$
Q(k\Delta x,\ell\Delta y,z;\omega_i) = \sum_{m=-M}^{M} \sum_{n=-N}^{N} H(k\Delta x,\ell\Delta y,z;m\Delta x,n\Delta y,z';\omega_i)P(m\Delta x,n\Delta y,z';\omega_i)\Delta x\Delta y, \qquad \text{(A-6)}
$$

for $k=-K, \ldots, K$ and $\ell=-L, \ldots, L$.

Here it is assumed that M and N are "sufficiently large" and that Δx and Δy are "sufficiently small". The latter condition can always be satisfied as we deal with *band-limited* seismic data. For an extensive discussion on various aspects of discretization the reader is referred to Berkhout (1985).

In analogy with section A.2, we define data vectors $\vec{P}(z')$ and $\vec{Q}(z)$ which contain the discretized wave fields P and Q. Next we replace equation (A-6) by the matrix equation (A-2)

$$\vec{Q}(z) = \mathbf{H}(z,z')\vec{P}(z'). \tag{A-7}$$

This implies that we define the operator matrix $\mathbf{H}(z,z')$ according to

$$\mathbf{H}(z,z') = \begin{bmatrix} \mathbf{H}_{-L,-N} & \cdots\cdots & \mathbf{H}_{-L,n} & \cdots\cdots & \mathbf{H}_{-L,N} \\ \vdots & & \vdots & & \vdots \\ \mathbf{H}_{\ell,-N} & \cdots\cdots & \mathbf{H}_{\ell,n} & \cdots\cdots & \mathbf{H}_{\ell,N} \\ \vdots & & \vdots & & \vdots \\ \mathbf{H}_{L,-N} & \cdots\cdots & \mathbf{H}_{L,n} & \cdots\cdots & \mathbf{H}_{L,N} \end{bmatrix}, \tag{A-8a}$$

where the elements of the sub-matrices $\mathbf{H}_{\ell,n}$ read

$$(\mathbf{H}_{\ell,n})_{k,m} = \Delta x \Delta y \; H(k\Delta x,\ell\Delta y,z;m\Delta x,n\Delta y,z';\omega_i). \tag{A-8b}$$

Note the high degree of similarity of this operator matrix $\mathbf{H}(z,z')$ with the data matrix $\mathbf{P}(z_0)$ defined in (A-5c).
One column in the data matrix $\mathbf{P}(z_0)$ represents the (monochromatic) data $P(x_r,y_r,z_0;x_s,y_s,z_0;\omega_i)$ as a function of (x_r,y_r) for a source at $(x_s=m\Delta x,y_s=n\Delta y,z_0)$. Similarly, one column in the operator matrix $\mathbf{H}(z,z')$ represents the (monochromatic) "spatial impulse response" of the operator $H(x,y,z;x',y',z';\omega_i)$ as a function of (x,y) for an "impulse" at $(x'=m\Delta x,y'=n\Delta y,z')$.

Finally, let us consider again operation (A-1) and choose a very special operator, given by

$$H(x,y,z;x',y',z'=z;\omega) = h(x,y,z)\delta(x-x')\delta(y-y'). \tag{A-9a}$$

Substitution into (A-1), taking z'=z, yields

$$Q(x,y,z;\omega) = h(x,y,z) \int\int_{-\infty}^{\infty} \delta(x-x')\delta(y-y')P(x',y',z;\omega)dx'dy', \tag{A-9b}$$

or

$$Q(x,y,z;\omega) = h(x,y,z)P(x,y,z;\omega), \tag{A-9c}$$

which is a simple product of two scalar functions. The discretized version of the "operator" (A-9a) reads

$$H(k\Delta x,\ell\Delta y,z;m\Delta x,n\Delta y,z'=z;\omega_i) = h(k\Delta x,\ell\Delta y,z) \frac{\delta_{km}}{\Delta x} \frac{\delta_{\ell n}}{\Delta y} . \tag{A-9d}$$

Hence, the matrix representation of the product of two scalar functions, as described by (A-9c), becomes

$$\vec{Q}(z) = \mathbf{H}(z)\vec{P}(z), \tag{A-10a}$$

where the elements of the sub-matrices $\mathbf{H}_{\ell,n}$ of $\mathbf{H}(z)$ read

$$(\mathbf{H}_{\ell,n})_{k,m} = h(k\Delta x,\ell\Delta y,z)\delta_{km}\delta_{\ell n}. \tag{A-10b}$$

Hence, for this situation matrix $\mathbf{H}(z)$ is a *diagonal* matrix, the diagonal elements representing the discretized version of the scalar function h at depth z.

A.4 REFERENCES

Berkhout, A.J., 1985, Seismic migration: Imaging of acoustic energy by wave field extrapolation. A. Theoretical aspects: Elsevier Science Publ. Co., Inc.

Kinneging, N.A., Budejicky, V., Wapenaar, C.P.A., and Berkhout, A.J., 1989, Efficient 2-D and 3-D shot record redatuming: Geophysical Prospecting, *37*, 493 - 530.

APPENDIX B

INTERACTIONS OF ONE-WAY ACOUSTIC WAVE FIELDS

B.1. INTRODUCTION

In chapters V and VII we frequently encounter integrals of the form

$$P(\vec{r}_A,\omega) = \iint_{-\infty}^{\infty} \frac{1}{\rho(\vec{r})}\left[\frac{\partial G(\vec{r},\vec{r}_A,\omega)}{\partial z} P(\vec{r},\omega) - G(\vec{r},\vec{r}_A,\omega)\frac{\partial P(\vec{r},\omega)}{\partial z}\right]_{z_0} dxdy \quad \text{(B-1a)}$$

and

$$P_0(\vec{r}_A,\omega) = \iint_{-\infty}^{\infty} \frac{1}{\rho(\vec{r})}\left[\left(\frac{\partial G(\vec{r},\vec{r}_A,\omega)}{\partial z}\right)^* P(\vec{r},\omega) - \left(G(\vec{r},\vec{r}_A,\omega)\right)^* \frac{\partial P(\vec{r},\omega)}{\partial z}\right]_{z_0} dxdy\,, \quad \text{(B-1b)}$$

where P and G satisfy the acoustic two-way wave equation. Throughout this Appendix, we assume that P and G at z_0 consist of independent downgoing and upgoing waves (see Figure B-1), which we denote by

$$P(\vec{r},\omega) \triangleq P^+(\vec{r},\omega) + P^-(\vec{r},\omega) \text{ at } z=z_0 \quad \text{(B-2a)}$$

and

$$G(\vec{r},\vec{r}_A,\omega) \triangleq G^+(\vec{r},\vec{r}_A,\omega) + G^-(\vec{r},\vec{r}_A,\omega) \text{ at } z=z_0, \quad \text{(B-2b)}$$

where the downgoing waves (P^+, G^+) and the upgoing waves (P^-,G^-) satisfy independent acoustic one-way wave equations at z_0 [1)].

By substituting equations (B-2a) and (B-2b) into the integrals (B-1a) and (B-1b) and by applying the one-way wave equations at $z=z_0$, the "interaction" of downgoing and upgoing waves can be analysed. It will be

---

1) The relationship between the acoustic two-way wave equation and the acoustic one-way wave equations is extensively discussed in chapter III.

shown that the only contribution to P and P_0 comes from wave fields $P^{\pm}$ and $G^{\mp}$ which propagate in *opposite* directions through z_0.

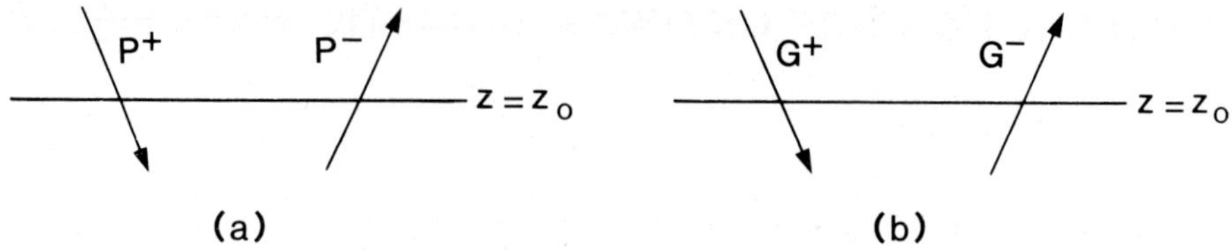

Figure B-1: At $z=z_0$, the wave fields P and G consist of independent downgoing and upgoing waves.

B.2 ANALYSIS FOR FORWARD PROPAGATION

We analyse integral (B-1a), where both P and G represent *causal* or *forward propagating* acoustic wave fields, (Berkhout and Wapenaar, 1989). Upon substitution of (B-2a) and (B-2b) into (B-1a) we obtain

$$P(\vec{r}_A,\omega) = \iint_{-\infty}^{\infty} \frac{1}{\rho} \Big[\Big(\frac{\partial G^+}{\partial z} + \frac{\partial G^-}{\partial z}\Big)\big(P^+ + P^-\big) - \big(G^+ + G^-\big)\Big(\frac{\partial P^+}{\partial z} + \frac{\partial P^-}{\partial z}\Big)\Big]_{z_0} dxdy. \tag{B-3}$$

The downgoing and upgoing waves satisfy *independent* one-way wave equations at $z=z_0$ when the vertical derivatives of the medium parameters vanish at $z=z_0$, which is expressed by

$$\frac{\partial K(x,y,z)}{\partial z} = 0 \qquad \text{at } z=z_0 \tag{B-4a}$$

and

$$\frac{\partial \rho(x,y,z)}{\partial z} = 0 \qquad \text{at } z=z_0, \tag{B-4b}$$

where K and ρ represent the adiabatic compression modulus and the volume density of mass, respectively, of the acoustic medium. To show the principle first, we assume for the moment

$$\nabla K(x,y,z) = \vec{o} \qquad \text{at } z=z_o \tag{B-5a}$$

and

$$\nabla \rho(x,y,z) = \vec{o} \qquad \text{at } z=z_o, \tag{B-5b}$$

meaning that, in addition to (B-4), the medium is also assumed to be *laterally* invariant at z_o. In analogy with equation (III-4a), we define the 2-D spatial Fourier transform $\tilde{A}(k_x,k_y)$ of a space-dependent function $A(x,y)$ by

$$\tilde{A}(k_x,k_y) = \iint_{-\infty}^{\infty} A(x,y)e^{j(k_x x+k_y y)} dxdy. \tag{B-6a}$$

Similarly, we define

$$\tilde{B}(k_x,k_y) = \iint_{-\infty}^{\infty} B(x,y)e^{j(k_x x+k_y y)} dxdy. \tag{B-6b}$$

With these definitions, the generalized version of Parseval's theorem reads

$$\iint_{-\infty}^{\infty} A(x,y)B(x,y)dxdy = \left(\frac{1}{2\pi}\right)^2 \iint_{-\infty}^{\infty} \tilde{A}(-k_x,-k_y)\tilde{B}(k_x,k_y)dk_x dk_y. \tag{B-7}$$

Note that the left-hand side of this equation represents nothing but the *spatial* cross-correlation function of $A(x,y)$ and $B(x,y)$ for zero shift. Applying this theorem to equation (B-3) yields

$$P(\vec{r}_A,\omega) = \left(\frac{1}{2\pi}\right)^2 \iint_{-\infty}^{\infty} \frac{1}{\rho(z_o)} \Bigg[\left(\frac{\partial \tilde{G}'^+}{\partial z} + \frac{\partial \tilde{G}'^-}{\partial z}\right)(\tilde{P}^+ + \tilde{P}^-)$$

$$- (\tilde{G}'^+ + \tilde{G}'^-)\left(\frac{\partial \tilde{P}^+}{\partial z} + \frac{\partial \tilde{P}^-}{\partial z}\right) \Bigg]_{z_o} dk_x dk_y , \tag{B-8a}$$

where

$$\tilde{P}^{\pm} \triangleq \tilde{P}^{\pm}(k_x,k_y,z;\omega) \tag{B-8b}$$

and

$$\tilde{G}'^{\pm} \triangleq \tilde{G}^{\pm}(-k_x,-k_y,z;x_A,y_A,z_A;\omega). \tag{B-8c}$$

The prime (') denotes that k_x and k_y are replaced by $-k_x$ and $-k_y$, respectively. According to equations (III-45a) and (III-45b), $\tilde{P}^{\pm}$ and $\tilde{G}'^{\pm}$ satisfy the following one-way wave equations

$$\frac{\partial \tilde{P}^{\pm}}{\partial z} = \mp jk_z\tilde{P}^{\pm} \qquad \text{at } z=z_o \tag{B-9a}$$

and

$$\frac{\partial \tilde{G}'^{\pm}}{\partial z} = \mp jk_z\tilde{G}'^{\pm} \qquad \text{at } z=z_o, \tag{B-9b}$$

where

$$k_z(z_o) \triangleq +\sqrt{k^2(z_o) - k_x^2 - k_y^2} \qquad \text{for } k_x^2+k_y^2 \le k^2(z_o) \tag{B-10a}$$

and

$$k_z(z_o) \triangleq -j\sqrt{k_x^2 + k_y^2 - k^2(z_o)} \qquad \text{for } k_x^2+k_y^2 > k^2(z_o), \tag{B-10b}$$

(see equations (III-32e) and (III-32f)), with

$$k^2(z_o) = \omega^2\rho(z_o)/K(z_o). \tag{B-10c}$$

Substitution of one-way wave equations (B-9a) and (B-9b) into expression (B-8a) yields

$$P(\vec{r}_A,\omega) = \left(\frac{1}{2\pi}\right)^2 \iint_{-\infty}^{\infty} \frac{-jk_z(z_o)}{\rho(z_o)} \Big[\left(\tilde{G}'^{+} - \tilde{G}'^{-}\right)\left(\tilde{P}^{+} + \tilde{P}^{-}\right)$$

$$- \left(\tilde{G}'^{+} + \tilde{G}'^{-}\right)\left(\tilde{P}^{+} - \tilde{P}^{-}\right)\Big]_{z_o} dk_x dk_y, \tag{B-11a}$$

or

$$P(\vec{r}_A,\omega) = \left(\frac{1}{2\pi}\right)^2 \iint_{-\infty}^{\infty} \frac{-jk_z(z_0)}{\rho(z_0)} \left[2\tilde{G}'^{+}\tilde{P}^{-} - 2\tilde{G}'^{-}\tilde{P}^{+}\right]_{z_0} dk_x dk_y. \quad \text{(B-11b)}$$

Note that this integral does not contain the terms $\tilde{G}'^{+}\tilde{P}^{+}$ and $\tilde{G}'^{-}\tilde{P}^{-}$. Applying the one-way wave equation (B-9a) or (B-9b), we obtain either

$$P(\vec{r}_A,\omega) = -2\left(\frac{1}{2\pi}\right)^2 \iint_{-\infty}^{\infty} \frac{1}{\rho(z_0)} \left[\tilde{G}'^{+} \frac{\partial \tilde{P}^{-}}{\partial z} + \tilde{G}'^{-} \frac{\partial \tilde{P}^{+}}{\partial z}\right]_{z_0} dk_x dk_y, \quad \text{(B-12a)}$$

or, equivalently,

$$P(\vec{r}_A,\omega) = 2\left(\frac{1}{2\pi}\right)^2 \iint_{-\infty}^{\infty} \frac{1}{\rho(z_0)} \left[\frac{\partial \tilde{G}'^{+}}{\partial z}\tilde{P}^{-} + \frac{\partial \tilde{G}'^{-}}{\partial z}\tilde{P}^{+}\right]_{z_0} dk_x dk_y. \quad \text{(B-12b)}$$

Applying Parseval's theorem (B-7) again, yields either

$$P(\vec{r}_A,\omega) = -2\iint_{-\infty}^{\infty} \frac{1}{\rho(z_0)} \left[G^{+} \frac{\partial P^{-}}{\partial z} + G^{-} \frac{\partial P^{+}}{\partial z}\right]_{z_0} dxdy, \quad \text{(B-13a)}$$

or, equivalently,

$$P(\vec{r}_A,\omega) = 2\iint_{-\infty}^{\infty} \frac{1}{\rho(z_0)} \left[\frac{\partial G^{+}}{\partial z}P^{-} + \frac{\partial G^{-}}{\partial z}P^{+}\right]_{z_0} dxdy. \quad \text{(B-13b)}$$

In both integrals, the first term contains the wave fields G^{+} and P^{-}, which propagate in *opposite* directions through z_0 (Figure B-2a). Similarly, the second terms contains the wave fields G^{-} and P^{+}, which also propagate in *opposite* directions through z_0 (Figure B-2b).

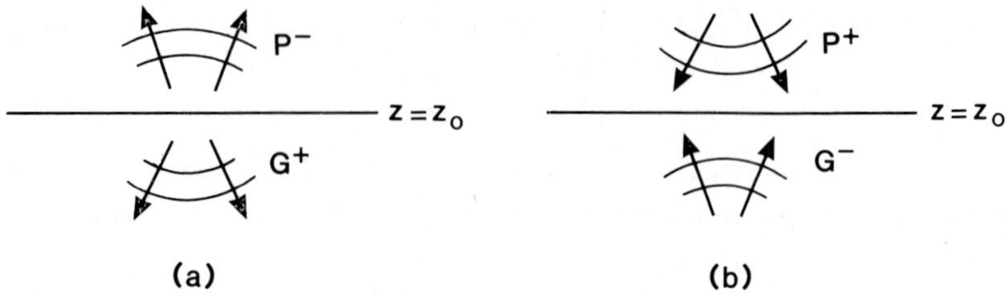

Figure B-2: Integral (B-3) contains all possible products of $G^{\pm}$ *and* $P^{\pm}$*. Equations (B-13a) and (B-13b) show that only the waves propagating in* ***opposite*** *directions have "interaction" at* z_o*.*

We generalize these results for the situation where the medium parameters are laterally variant at z_0. We only assume that the vertical derivatives of the medium parameters vanish at z_0, which is expressed by equation (B-4). We introduce scaled pressure functions P_s and G_s, according to

$$P_s = P/\sqrt{\rho} \qquad \text{at } z=z_0 \tag{B-14a}$$

and

$$G_s = G/\sqrt{\rho} \qquad \text{at } z=z_0, \tag{B-14b}$$

hence, equation (B-3) may be rewritten as

$$P(\vec{r}_A,\omega) = \iint_{-\infty}^{\infty} \left[\left(\frac{\partial G_s^+}{\partial z} + \frac{\partial G_s^-}{\partial z}\right)\left(P_s^+ + P_s^-\right) - \left(G_s^+ + G_s^-\right)\left(\frac{\partial P_s^+}{\partial z} + \frac{\partial P_s^-}{\partial z}\right)\right]_{z_0} dxdy. \tag{B-15}$$

Substitution of $P=\sqrt{\rho}\ P_s$ and $G=\sqrt{\rho}\ G_s$ into the two-way wave equations

$$\nabla.\left(\frac{1}{\rho}\nabla P\right) + \frac{\omega^2}{K} P = o \qquad \text{at } z=z_0 \tag{B-16a}$$

and

$$\nabla.\left(\frac{1}{\rho}\nabla G\right) + \frac{\omega^2}{K}G = o \qquad \text{at } z=z_o, \qquad \text{(B-16b)}$$

yields

$$\nabla^2 P_s + k_s^2 P_s = o \qquad \text{at } z=z_o \qquad \text{(B-17a)}$$

and

$$\nabla^2 G_s + k_s^2 G_s = o \qquad \text{at } z=z_o, \qquad \text{(B-17b)}$$

with

$$k_s^2 = \frac{\omega^2\rho}{K} - \frac{3}{4}\left|\frac{\nabla\rho}{\rho}\right|^2 + \frac{\nabla^2\rho}{2\rho} \qquad \text{at } z=z_o, \qquad \text{(B-17c)}$$

see also Brekhovskikh (1980). In analogy with (III-61), equation (B-17a) can be rewritten as

$$\frac{\partial^2 P_s(\vec{r},\omega)}{\partial z^2} = -H_2(\vec{r},\omega)P_s(\vec{r},\omega) \qquad \text{at } z=z_o, \qquad \text{(B-18a)}$$

where

$$H_2(\vec{r},\omega) = k_s^2(\vec{r},\omega) + \frac{\partial^2}{\partial x^2} + \frac{\partial^2}{\partial y^2} \qquad \text{at } z=z_o. \qquad \text{(B-18b)}$$

Assuming band-limited wave fields, the spatial differentiations $\partial^2/\partial x^2$ and $\partial^2/\partial y^2$ can be written as spatial convolutions along the x- and y-axis, respectively, according to

$$\frac{\partial^2 A(x,y)}{\partial x^2} = \int_{-\infty}^{\infty} d_2(x-x')A(x',y)dx' \qquad \text{(B-19a)}$$

and

$$\frac{\partial^2 A(x,y)}{\partial y^2} = \int_{-\infty}^{\infty} d_2(y-y')A(x,y')dy', \qquad \text{(B-19b)}$$

with $d_2(x)$ and $d_2(y)$ as defined in (III-62). With these definitions, two-way wave equation (B-18) can be written, in analogy with (III-63), as a generalized spatial convolution integral, according to

$$\left.\frac{\partial^2 P_s(\vec{r},\omega)}{\partial z^2}\right|_{z_0} = -\iint_{-\infty}^{\infty} \left[H_2(\vec{r},\vec{r}',\omega)P_s(\vec{r}',\omega)\right]_{z_0} dx'dy', \tag{B-20a}$$

where

$$H_2(\vec{r},\vec{r}',\omega) = k_s^2(\vec{r},\omega)\delta(x-x')\delta(y-y') + d_2(x-x')\delta(y-y') + \delta(x-x')d_2(y-y'), \tag{B-20b}$$

with

$$\vec{r} = (x,y,z) \tag{B-20c}$$

and

$$\vec{r}' = (x',y',z'=z). \tag{B-20d}$$

Note that operator H_2 is symmetric in $\vec{r}$ and $\vec{r}'$:

$$H_2(\vec{r}',\vec{r},\omega) = H_2(\vec{r},\vec{r}',\omega). \tag{B-20e}$$

Let us now, in analogy with (III-67), implicitly define an operator H_1 according to

$$H_2(\vec{r},\vec{r}'',\omega) = \iint_{-\infty}^{\infty} H_1(\vec{r},\vec{r}',\omega)H_1(\vec{r}',\vec{r}'',\omega)dx'dy', \tag{B-21a}$$

with

$$\vec{r}'' = (x'',y'',z''=z). \tag{B-21b}$$

Due to the symmetry property of operator H_2, operator H_1 is also symmetric in $\vec{r}$ and $\vec{r}'$:

$$H_1(\vec{r}',\vec{r},\omega) = H_1(\vec{r},\vec{r}',\omega). \tag{B-21c}$$

From (B-20) and (B-21) we obtain, in analogy with (III-72), the following *one-way wave equations* for the scaled pressure functions:

$$\left.\frac{\partial P_s^{\pm}(\vec{r},\omega)}{\partial z}\right|_{z_0} = \mp j \iint_{-\infty}^{\infty} \left[H_1(\vec{r},\vec{r}',\omega)P_s^{\pm}(\vec{r}',\omega)\right]_{z_0} dx'dy'. \tag{B-22a}$$

Similarly,

$$\left.\frac{\partial G_s^{\pm}(\vec{r}',\vec{r}_A,\omega)}{\partial z}\right|_{z_0} = \mp j \iint_{-\infty}^{\infty} \left[H_1(\vec{r}',\vec{r},\omega)G_s^{\pm}(\vec{r},\vec{r}_A,\omega)\right]_{z_0} dxdy. \tag{B-22b}$$

Substitution of (B-22) into (B-15) yields

$$P(\vec{r}_A,\omega) = \iint_{-\infty}^{\infty} \Big[\Big\{\iint_{-\infty}^{\infty} -jH_1(\vec{r},\vec{r}',\omega)\big(G_s^{+}(\vec{r},\vec{r}_A,\omega)-G_s^{-}(\vec{r},\vec{r}_A,\omega)\big)\, dxdy\Big\}\times$$

$$\big(P_s^{+}(\vec{r}',\omega)+P_s^{-}(\vec{r}',\omega)\big)\Big]_{z_0} dx'dy'$$

$$- \iint_{-\infty}^{\infty} \Big[\big(G_s^{+}(\vec{r},\vec{r}_A,\omega)+G_s^{-}(\vec{r},\vec{r}_A,\omega)\big)\times$$

$$\Big\{\iint_{-\infty}^{\infty} -jH_1(\vec{r},\vec{r}',\omega)\big(P_s^{+}(\vec{r}',\omega)-P_s^{-}(\vec{r}',\omega)\big)\, dx'dy'\Big\}\Big]_{z_0} dxdy, \tag{B-23a}$$

or

$$P(\vec{r}_A,\omega) = 2\iint_{-\infty}^{\infty}\iint_{-\infty}^{\infty} \Big[-jH_1(\vec{r},\vec{r}',\omega)\Big\{G_s^{+}(\vec{r},\vec{r}_A,\omega)P_s^{-}(\vec{r}',\omega)$$

$$- G_s^{-}(\vec{r},\vec{r}_A,\omega)P_s^{+}(\vec{r}',\omega)\Big\}\Big]_{z_0} dxdy\ dx'dy'. \tag{B-23b}$$

Applying the one-way wave equation (B-22a) or (B-22b), we obtain

$$P(\vec{r}_A,\omega) = -2\int\!\!\int_{-\infty}^{\infty} \left[G_s^{+}(\vec{r},\vec{r}_A,\omega)\frac{\partial P_s^{-}(\vec{r},\omega)}{\partial z} + G_s^{-}(\vec{r},\vec{r}_A,\omega)\frac{\partial P_s^{+}(\vec{r},\omega)}{\partial z}\right]_{z_0} dxdy, \qquad \text{(B-24a)}$$

or, equivalently

$$P(\vec{r}_A,\omega) = 2\int\!\!\int_{-\infty}^{\infty} \left[\frac{\partial G_s^{+}(\vec{r}\,',\vec{r}_A,\omega)}{\partial z} P_s^{-}(\vec{r}\,',\omega) + \frac{\partial G_s^{-}(\vec{r}\,',\vec{r}_A,\omega)}{\partial z} P_s^{+}(\vec{r}\,',\omega)\right]_{z_0} dx'dy'. \qquad \text{(B-24b)}$$

Expressions (B-24a) and (B-24b) are the generalized versions of expressions (B-13a) and (B-13b), respectively. Again, note that only the waves propagating in opposite directions have interaction at z_0 (see Figure B-2).

Substituting (B-14) into (B-24a) and choosing G^{+}=o yields

$$P(\vec{r}_A,\omega) = -2\int\!\!\int_{-\infty}^{\infty} \left[\frac{1}{\rho(\vec{r})} G^{-}(\vec{r},\vec{r}_A,\omega) \frac{\partial P^{+}(\vec{r},\omega)}{\partial z}\right]_{z_0} dxdy, \qquad \text{(B-25a)}$$

which is the one-way version of the Rayleigh I integral (V-44a).

Substituting (B-14) into (B-24b), choosing G^{+}=o and omitting the primes, yields

$$P(\vec{r}_A,\omega) = 2\int\!\!\int_{-\infty}^{\infty} \left[\frac{1}{\rho(\vec{r})} \frac{\partial G^{-}(\vec{r},\vec{r}_A,\omega)}{\partial z} P^{+}(\vec{r},\omega)\right]_{z_0} dxdy, \qquad \text{(B-25b)}$$

which is the one-way version of the Rayleigh II integral (V-44b).

B. 3 ANALYSIS FOR BACKWARD PROPAGATION

We analyse integral (B-1b), where P represents a causal or forward propagating acoustic wave field, whereas G^{*} represents an *anti-causal* or *backward propagating* acoustic wave field (Wapenaar et al., 1989). Upon substitution of (B-2a) and (B-2b) into (B-1b) we obtain

$$P_0(\vec{r}_A,\omega) = \iint_{-\infty}^{\infty} \frac{1}{\rho} \left[\left(\frac{\partial G^+}{\partial z} + \frac{\partial G^-}{\partial z}\right)^* (P^+ + P^-) - (G^+ + G^-)^* \left(\frac{\partial P^+}{\partial z} + \frac{\partial P^-}{\partial z}\right) \right]_{z_0} dxdy. \tag{B-26}$$

The downgoing and upgoing waves satisfy independent one-way wave equations at $z=z_0$ when the vertical derivatives of the medium parameters vanish at $z=z_0$, which is expressed by equation (B-4). In this section we only show the principle, assuming for simplicity that the medium parameters are also laterally invariant at $z=z_0$, which is expressed by equation (B-5).

Consider the following version of Parseval's theorem

$$\iint_{-\infty}^{\infty} A^*(x,y)B(x,y)dxdy = \left(\frac{1}{2\pi}\right)^2 \iint_{-\infty}^{\infty} \tilde{A}^*(k_x,k_y)\tilde{B}(k_x,k_y)dk_x dk_y. \tag{B-27}$$

Applying this theorem to equation (B-26) yields

$$P_0(\vec{r}_A,\omega) = \left(\frac{1}{2\pi}\right)^2 \iint_{-\infty}^{\infty} \frac{1}{\rho(z_0)} \left[\left(\frac{\partial \tilde{G}^+}{\partial z} + \frac{\partial \tilde{G}^-}{\partial z}\right)^* (\tilde{P}^+ + \tilde{P}^-) - (\tilde{G}^+ + \tilde{G}^-)^* \left(\frac{\partial \tilde{P}^+}{\partial z} + \frac{\partial \tilde{P}^-}{\partial z}\right) \right]_{z_0} dk_x dk_y, \tag{B-28a}$$

where

$$\tilde{P}^{\pm} \triangleq \tilde{P}^{\pm}(k_x,k_y,z;\omega) \tag{B-28b}$$

and

$$\tilde{G}^{\pm} \triangleq \tilde{G}^{\pm}(k_x,k_y,z;x_A,y_A,z_A;\omega), \tag{B-28c}$$

where, in analogy with (B-9), $\tilde{P}^{\pm}$ and $\tilde{G}^{\pm}$ satisfy the following one-way wave equations

$$\frac{\partial \tilde{P}^{\pm}}{\partial z} = \mp jk_z \tilde{P}^{\pm} \qquad \text{at } z=z_o \tag{B-29a}$$

and

$$\frac{\partial \tilde{G}^{\pm}}{\partial z} = \mp jk_z \tilde{G}^{\pm} \qquad \text{at } z=z_o, \tag{B-29b}$$

with k_z defined by (B-10). Consequently, the *backward propagating* wave field $(\tilde{G}^{\pm})^*$ satisfies the following one-way wave equation

$$\left(\frac{\partial \tilde{G}^{\pm}}{\partial z}\right)^* = \pm jk_z^*(\tilde{G}^{\pm})^* \qquad \text{at } z=z_o. \tag{B-29c}$$

Note that, according to equations (B-10a) and (B-10b),

$$k_z^* = k_z \qquad \text{for } k_x^2+k_y^2 \le k^2(z_o) \tag{B-30a}$$

and

$$k_z^* = -k_z \qquad \text{for } k_x^2+k_y^2 > k^2(z_o). \tag{B-30b}$$

Hence, unlike in section B.2, it appears to be necessary to analyse equation (B-28a) separately for the propagating wavenumber area ($k_x^2+k_y^2 \le k^2(z_o)$) and the evanescent wavenumber area ($k_x^2+k_y^2 > k^2(z_o)$). Substitution of one-way wave equations (B-29a) and (B-29c) into expression (B-28a) yields

$$P_0(\vec{r}_A,\omega) = \left(\frac{1}{2\pi}\right)^2 \iint\limits_{k_x^2+k_y^2\leq k^2(z_0)} \frac{-jk_z(z_0)}{\rho(z_0)} \Big[(-\tilde{G}^+ + \tilde{G}^-)^* (\tilde{P}^+ + \tilde{P}^-)$$

$$- (\tilde{G}^+ + \tilde{G}^-)^* (\tilde{P}^+ - \tilde{P}^-)\Big]_{z_0} dk_x dk_y$$

$$+ \left(\frac{1}{2\pi}\right)^2 \iint\limits_{k_x^2+k_y^2> k^2(z_0)} \frac{-jk_z(z_0)}{\rho(z_0)} \Big[(\tilde{G}^+ - \tilde{G}^-)^* (\tilde{P}^+ + \tilde{P}^-)$$

$$- (\tilde{G}^+ + \tilde{G}^-)^* (\tilde{P}^+ - \tilde{P}^-)\Big]_{z_0} dk_x dk_y, \qquad \text{(B-31a)}$$

or

$$P_0(\vec{r}_A,\omega) = \left(\frac{1}{2\pi}\right)^2 \iint\limits_{k_x^2+k_y^2\leq k^2(z_0)} \frac{-jk_z(z_0)}{\rho(z_0)} \Big[-2(\tilde{G}^+)^* \tilde{P}^+ + 2(\tilde{G}^-)^* \tilde{P}^-\Big]_{z_0} dk_x dk_y$$

$$+ \left(\frac{1}{2\pi}\right)^2 \iint\limits_{k_x^2+k_y^2> k^2(z_0)} \frac{-jk_z(z_0)}{\rho(z_0)} \Big[2(\tilde{G}^+)^* \tilde{P}^- - 2(\tilde{G}^-)^* \tilde{P}^+\Big]_{z_0} dk_x dk_y. \qquad \text{(B-31b)}$$

In equations (B-31a) and (B-31b), the second integral over the evanescent wavenumber area ($k_x^2+k_y^2>k^2(z_0)$) is negligible when the source of $\tilde{P}^{\pm}$ and the source of $\tilde{G}^{\pm}$ are not both in the direct vicinity of z_0. Hence, for this situation we may approximate (B-31b) by

$$P_0(\vec{r}_A,\omega) \approx \left(\frac{1}{2\pi}\right)^2 \iint\limits_{k_x^2+k_y^2\leq k^2(z_0)} \frac{-jk_z(z_0)}{\rho(z_0)} \Big[-2(\tilde{G}^+)^* \tilde{P}^+ + 2(\tilde{G}^-)^* \tilde{P}^-\Big]_{z_0} dk_x dk_y. \qquad \text{(B-32)}$$

Note that this integral does not contain the terms $(\tilde{G}^+)^* \tilde{P}^-$ and $(\tilde{G}^-)^* \tilde{P}^+$. Applying the one-way wave equation (B-29a) or (B-29c), and adding a negligible integral over the evanescent wavenumber area, we obtain

$$P_0(\vec{r}_A,\omega) \approx -2\left(\frac{1}{2\pi}\right)^2 \iint_{-\infty}^{\infty} \frac{1}{\rho(z_0)} \left[(\tilde{G}^+)^* \frac{\partial \tilde{P}^+}{\partial z} + (\tilde{G}^-)^* \frac{\partial \tilde{P}^-}{\partial z}\right]_{z_0} dk_x dk_y, \tag{B-33a}$$

or, equivalently,

$$P_0(\vec{r}_A,\omega) \approx 2\left(\frac{1}{2\pi}\right)^2 \iint_{-\infty}^{\infty} \frac{1}{\rho(z_0)} \left[\left(\frac{\partial \tilde{G}^+}{\partial z}\right)^* \tilde{P}^+ + \left(\frac{\partial \tilde{G}^-}{\partial z}\right)^* \tilde{P}^-\right]_{z_0} dk_x dk_y. \tag{B-33b}$$

Applying Parseval's theorem (B-27) again, yields

$$P_0(\vec{r}_A,\omega) \approx -2\iint_{-\infty}^{\infty} \frac{1}{\rho(z_0)} \left[(G^+)^* \frac{\partial P^+}{\partial z} + (G^-)^* \frac{\partial P^-}{\partial z}\right]_{z_0} dxdy, \tag{B-34a}$$

or, equivalently,

$$P_0(\vec{r}_A,\omega) \approx 2\iint_{-\infty}^{\infty} \frac{1}{\rho(z_0)} \left[\left(\frac{\partial G^+}{\partial z}\right)^* P^+ + \left(\frac{\partial G^-}{\partial z}\right)^* P^-\right]_{z_0} dxdy. \tag{B-34b}$$

In both integrals, the first term contains the wave fields $(G^+)^*$ and P^+, which propagate in *opposite* directions through z_0 (Figure B-3a). Similarly, the second term contains the wave fields $(G^-)^*$ and P^-, which also propagate in *opposite* directions through z_0 (Figure B-3b). Note that, unlike expressions (B-13a) and (B-13b) in section B.2, expressions (B-34a) and (B-34b) are not exact. The approximation concerns the negligence of evanescent waves at $z=z_0$.

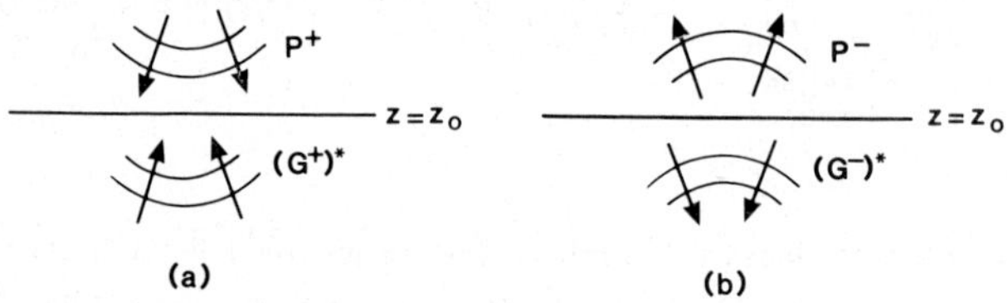

Figure B-3: Integral (B-26) contains all possible products of $(G^{\pm})^*$ and $P^{\pm}$. Equations (B-34a) and (B-34b) show that only the waves propagating in **opposite** *directions have "interaction" at z_0.*

Expressions (B-34a) and (B-34b) may be generalized for laterally varying medium parameters at $z=z_0$ in a similar way as described in section B.2. A further discussion is beyond the scope of this Appendix.

Choosing $G^+=0$ and generalizing (B-34a) and (B-34b) for laterally varying medium parameters at $z=z_0$, yields

$$P_0(\vec{r}_A,\omega) \approx -2\int\int_{-\infty}^{\infty}\left[\frac{1}{\rho(\vec{r})}\left(G^-(\vec{r},\vec{r}_A,\omega)\right)^* \frac{\partial P^-(\vec{r},\omega)}{\partial z}\right]_{z_0} dxdy \tag{B-35a}$$

and

$$P_0(\vec{r}_A,\omega) \approx 2\int\int_{-\infty}^{\infty}\left[\frac{1}{\rho(\vec{r})}\left(\frac{\partial G^-(\vec{r},\vec{r}_A,\omega)}{\partial z}\right)^* P^-(\vec{r},\omega)\right]_{z_0} dxdy, \tag{B-35b}$$

respectively. These are the one-way versions of the Rayleigh I and Rayleigh II integrals (VII-65a) and (VII-65b), respectively, for inverse wave field extrapolation.

B.4. REFERENCES

Berkhout, A.J., and Wapenaar C.P.A., 1989, One-way versions of the Kirchhoff integral: Geophysics, *54*, 460 - 467.

Brekhovskikh, L.M., 1980, Waves in layered media: Academic Press, Inc.

Wapenaar, C.P.A., Peels, G.L., Budejicky, V., and Berkhout, A.J., 1989, Inverse extrapolation of primary seismic waves: Geophysics, *54*, 853 - 863.

APPENDIX C

INTERACTIONS OF ONE-WAY ELASTIC WAVE FIELDS

C.1. INTRODUCTION

In chapters VI and VIII we frequently encounter integrals of the form

$$\Omega(\vec{r}_A,\omega) = \iint_{-\infty}^{\infty} [\vec{\Theta}_{z,\Omega}(\vec{r},\vec{r}_A,\omega).\vec{V}(\vec{r},\omega) - \vec{G}_\Omega(\vec{r},\vec{r}_A,\omega).\vec{\tau}_z(\vec{r},\omega)]_{z_0} dxdy \qquad \text{(C-1a)}$$

and

$$\Omega_0(\vec{r}_A,\omega) = \iint_{-\infty}^{\infty} [(\vec{\Theta}_{z,\Omega}(\vec{r},\vec{r}_A,\omega))^*.\vec{V}(\vec{r},\omega)+(\vec{G}_\Omega(\vec{r},\vec{r}_A,\omega))^*.\vec{\tau}_z(\vec{r},\omega)]_{z_0} dxdy, \qquad \text{(C-1b)}$$

where $(\vec{V},\vec{\tau}_z)$ and $(\vec{G}_\Omega,\vec{\Theta}_{z,\Omega})$ represent two elastic wave fields in terms of the particle velocity vectors $\vec{V}$ and $\vec{G}_\Omega$ and the traction vectors $\vec{\tau}_z$ and $\vec{\Theta}_{z,\Omega}$. Throughout this Appendix, we assume that at z_0 the elastic wave fields consist of independent downgoing and upgoing waves, which we denote by

$$\vec{V}(\vec{r},\omega) \triangleq \vec{V}^+(\vec{r},\omega) + \vec{V}^-(\vec{r},\omega) \qquad \text{at } z=z_0, \qquad \text{(C-2a)}$$

$$\vec{\tau}_z(\vec{r},\omega) \triangleq \vec{\tau}_z^+(\vec{r},\omega) + \vec{\tau}_z^-(\vec{r},\omega) \qquad \text{at } z=z_0, \qquad \text{(C-2b)}$$

$$\vec{G}_\Omega(\vec{r},\vec{r}_A,\omega) \triangleq \vec{G}_\Omega^+(\vec{r},\vec{r}_A,\omega) + \vec{G}_\Omega^-(\vec{r},\vec{r}_A,\omega) \qquad \text{at } z=z_0 \qquad \text{(C-2c)}$$

and

$$\vec{\Theta}_{z,\Omega}(\vec{r},\vec{r}_A,\omega) \triangleq \vec{\Theta}^+_{z,\Omega}(\vec{r},\vec{r}_A,\omega) + \vec{\Theta}^-_{z,\Omega}(\vec{r},\vec{r}_A,\omega) \text{ at } z=z_0. \qquad \text{(C-2d)}$$

Furthermore, we assume that at z_0 the downgoing and upgoing waves consist of independent P- and S-waves (see also Figure C-1), which we denote by

$$[\vec{V}^{\pm}(\vec{r},\omega)]_{z_0} \triangleq \frac{-1}{j\omega\rho}[\nabla\Phi^{\pm}(\vec{r},\omega) + \nabla\times\vec{\Psi}^{\pm}(\vec{r},\omega)]_{z_0}, \tag{C-3a}$$

with (for the source-free situation at z_0)

$$[\nabla.\vec{\Psi}^{\pm}(\vec{r},\omega)]_{z_0} \triangleq 0 \tag{C-3b}$$

and

$$[\vec{G}^{\pm}_{\Omega}(\vec{r},\vec{r}_A,\omega)]_{z_0} \triangleq \frac{-1}{j\omega\rho}[\nabla\Gamma^{\pm}_{\phi,\Omega}(\vec{r},\vec{r}_A,\omega) + \nabla\times\vec{\Gamma}^{\pm}_{\psi,\Omega}(\vec{r},\vec{r}_A,\omega)]_{z_0}, \tag{C-3c}$$

with (for the source-free situation at z_0)

$$[\nabla.\vec{\Gamma}^{\pm}_{\psi,\Omega}(\vec{r},\vec{r}_A,\omega)]_{z_0} \triangleq 0, \tag{C-3d}$$

where the downgoing P-waves potentials $(\Phi^+,\Gamma^+_{\phi,\Omega})$, the downgoing S-wave potentials $(\vec{\Psi}^+, \vec{\Gamma}^+_{\psi,\Omega})$, the upgoing P-wave potentials $(\Phi^-,\Gamma^-_{\phi,\Omega})$ and the upgoing S-wave potentials $(\vec{\Psi}^-,\vec{\Gamma}^-_{\psi,\Omega})$ satisfy independent elastic one-way wave equations at z_0 [1)].

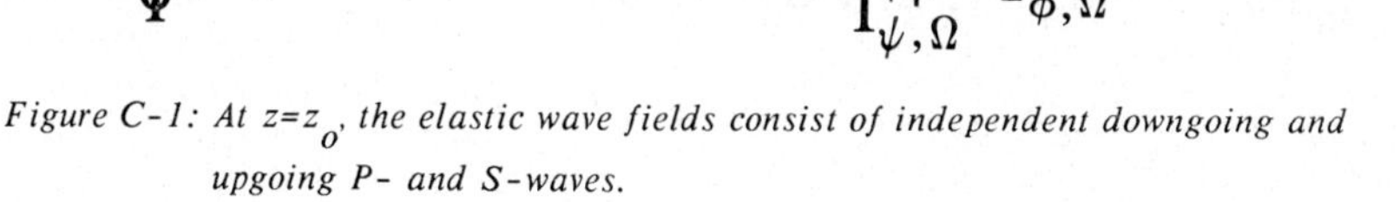

Figure C-1: At $z=z_0$, the elastic wave fields consist of independent downgoing and upgoing P- and S-waves.

---

1) The relationship between the elastic two-way wave equation and the elastic one-way wave equations is extensively discussed in chapter IV.

By substituting equations (C-2) and (C-3) into the integrals (C-1a) and (C-1b) and by applying the one-way wave equations at $z=z_0$, the "interaction" of downgoing and upgoing waves can be analysed. We will show that the only contribution to Ω and Ω_0 comes from P-waves ($\Phi^{\pm}$, $\Gamma^{\mp}_{\phi,\Omega}$) and from S-waves ($\vec{\Psi}^{\pm}$, $\vec{\Gamma}^{\mp}_{\psi,\Omega}$) which propagate in *opposite* directions through z_0.

C.2 ANALYSIS FOR FORWARD PROPAGATION

We analyse integral (C-1a), where both the elastic wave fields ($\vec{V},\vec{\tau}_z$) and ($\vec{G}_\Omega$, $\vec{\Theta}_{z,\Omega}$) are *causal* or *forward propagating*, (Wapenaar and Haimé, 1989).

Upon substitution of (C-2) into (C-1a) we obtain

$$\Omega(\vec{r}_A,\omega) = \int\!\!\int_{-\infty}^{\infty} \left[(\vec{\Theta}^{+}_{z,\Omega}+\vec{\Theta}^{-}_{z,\Omega}) \cdot (\vec{V}^{+}+\vec{V}^{-}) - (\vec{G}^{+}_{\Omega}+\vec{G}^{-}_{\Omega}) \cdot (\vec{\tau}^{+}_{z}+\vec{\tau}^{-}_{z}) \right]_{z_0} dxdy. \tag{C-4}$$

The downgoing and upgoing waves satisfy *independent* one-way wave equations at $z=z_0$ when the vertical derivatives of the medium parameters vanish at $z=z_0$, which is expressed by

$$\frac{\partial c_{ijk\ell}(x,y,z)}{\partial z} = 0 \qquad \text{at } z=z_0 \tag{C-5a}$$

and

$$\frac{\partial \rho(x,y,z)}{\partial z} = 0 \qquad \text{at } z=z_0, \tag{C-5b}$$

where $c_{ijk\ell}$ and ρ represent the stiffness tensor and the volume density of mass, respectively, of the elastic medium. For simplicity, however, we assume in addition to (C-5) that the medium is isotropic and laterally invariant at $z=z_0$, which is expressed by

$$c_{ijk\ell} = \lambda\delta_{ij}\delta_{k\ell} + \mu\left[\delta_{ik}\delta_{j\ell} + \delta_{i\ell}\delta_{jk}\right] \qquad \text{at } z=z_0, \tag{C-6a}$$

$$\nabla\lambda(x,y,z) = \vec{o} \qquad \text{at } z=z_o, \qquad \text{(C-6b)}$$

$$\nabla\mu(x,y,z) = \vec{o} \qquad \text{at } z=z_o \qquad \text{(C-6c)}$$

and

$$\nabla\rho(x,y,z) = \vec{o} \qquad \text{at } z=z_o, \qquad \text{(C-6d)}$$

where λ and μ are the Lamé coefficients. Applying Parseval's theorem (B-7) to equation (C-4) yields

$$\Omega(\vec{r}_A,\omega) = \left(\frac{1}{2\pi}\right)^2 \iint_{-\infty}^{\infty} \Big[\left(\tilde{\vec{\Theta}}{}'^{,+}_{z,\Omega} + \tilde{\vec{\Theta}}{}'^{,-}_{z,\Omega}\right).\left(\tilde{\vec{V}}^+ + \tilde{\vec{V}}^-\right)$$

$$-\left(\tilde{\vec{G}}{}'^{,+}_{\Omega} + \tilde{\vec{G}}{}'^{,-}_{\Omega}\right).\left(\tilde{\vec{\tau}}^+_z + \tilde{\vec{\tau}}^-_z\right)\Big]_{z_o} dk_x dk_y, \qquad \text{(C-7a)}$$

where

$$\tilde{\vec{V}}^{\pm} \triangleq \tilde{\vec{V}}^{\pm}(k_x,k_y,z;\omega) \qquad \text{(C-7b)}$$

$$\tilde{\vec{\tau}}^{\pm}_z \triangleq \tilde{\vec{\tau}}^{\pm}_z(k_x,k_y,z;\omega) \qquad \text{(C-7c)}$$

$$\tilde{\vec{G}}{}'^{,\pm}_{\Omega} \triangleq \tilde{\vec{G}}^{\pm}_{\Omega}(-k_x,-k_y,z;x_A,y_A,z_A;\omega) \qquad \text{(C-7d)}$$

and

$$\tilde{\vec{\Theta}}{}'^{,\pm}_{z,\Omega} \triangleq \tilde{\vec{\Theta}}^{\pm}_{z,\Omega}(-k_x,-k_y,z;x_A,y_A,z_A;\omega). \qquad \text{(C-7e)}$$

The prime (') denotes that k_x and k_y are replaced by $-k_x$ and $-k_y$, respectively.

In analogy with (C-3a) and (C-3b) we express $\tilde{\vec{V}}^{\pm}$ in terms of P- and S-waves potentials, according to

$$\tilde{\vec{V}}^{\pm} \triangleq \tilde{\vec{D}}_p^{\pm} \tilde{\Phi}^{\pm} + \tilde{D}_s^{\pm} \tilde{\vec{\Psi}}^{\pm} \qquad \text{at } z=z_o, \tag{C-8a}$$

with

$$\tilde{\vec{D}}_s^{\pm} \cdot \tilde{\vec{\Psi}}^{\pm} \triangleq o \qquad \text{at } z=z_o, \tag{C-8b}$$

where

$$\tilde{\vec{D}}_p^{\pm} = \frac{-1}{j\omega\rho} \begin{bmatrix} -jk_x \\ -jk_y \\ \mp jk_{z,p} \end{bmatrix}, \tag{C-8c}$$

$$\tilde{\vec{D}}_s^{\pm} = \frac{-1}{j\omega\rho} \begin{bmatrix} -jk_x \\ -jk_y \\ \mp jk_{z,s} \end{bmatrix} \tag{C-8d}$$

and

$$\tilde{D}_s^{\pm} = \frac{-1}{j\omega\rho} \begin{bmatrix} 0 & \pm jk_{z,s} & -jk_y \\ \mp jk_{z,s} & 0 & jk_x \\ jk_y & -jk_x & 0 \end{bmatrix}. \tag{C-8e}$$

Note that we made use of the one-way wave equations (IV-29c) and (IV-29d) for the P- and S-wave potentials, respectively:

$$\frac{\partial \tilde{\Phi}^{\pm}}{\partial z} = \mp jk_{z,p} \tilde{\Phi}^{\pm} \qquad \text{at } z=z_o \tag{C-9a}$$

and

$$\frac{\partial \tilde{\vec{\Psi}}^{\pm}}{\partial z} = \mp jk_{z,s} \tilde{\vec{\Psi}}^{\pm} \qquad \text{at } z=z_o, \tag{C-9b}$$

where

$$k_{z,p}(z_o) \triangleq +\sqrt{k_p^2(z_o) - k_x^2 - k_y^2} \qquad \text{for } k_x^2+k_y^2 \le k_p^2(z_o), \tag{C-10a}$$

$$k_{z,p}(z_o) \triangleq -j\sqrt{k_x^2 + k_y^2 - k_p^2(z_o)} \qquad \text{for } k_x^2+k_y^2 > k_p^2(z_o), \tag{C-10b}$$

$$k_{z,s}(z_o) \triangleq +\sqrt{k_s^2(z_o) - k_x^2 - k_y^2} \qquad \text{for } k_x^2+k_y^2 \le k_s^2(z_o) \tag{C-10c}$$

and

$$k_{z,s}(z_o) \triangleq -j\sqrt{k_x^2 + k_y^2 - k_s^2(z_o)} \qquad \text{for } k_x^2+k_y^2 > k_s^2(z_o), \tag{C-10d}$$

with

$$k_p^2(z_o) = \frac{\omega^2\rho(z_o)}{\lambda(z_o)+2\mu(z_o)} \tag{C-10e}$$

and

$$k_s^2(z_o) = \frac{\omega^2\rho(z_o)}{\mu(z_o)} \,. \tag{C-10f}$$

Similarly, in analogy with (C-3c) and (C-3d) we express $\tilde{\vec{G}}_{\Omega}^{,\pm}$ in terms of P- and S-wave potentials, according to

$$\tilde{\vec{G}}_{\Omega}^{,\pm} \triangleq \tilde{\vec{D}}_p^{,\pm} \, \tilde{\Gamma}_{\phi,\Omega}^{,\pm} + \tilde{\mathbf{D}}_s^{,\pm} \, \tilde{\vec{\Gamma}}_{\psi,\Omega}^{,\pm} \qquad \text{at } z=z_o, \tag{C-11a}$$

with

$$\tilde{\vec{D}}_s^{,\pm} \cdot \tilde{\vec{\Gamma}}_{\psi,\Omega}^{,\pm} \triangleq o \qquad \text{at } z=z_o. \tag{C-11b}$$

Here we made use of the one-way wave equations

$$\frac{\partial \tilde{\Gamma}_{\phi,\Omega}^{,\pm}}{\partial z} = \mp jk_{z,p} \tilde{\Gamma}_{\phi,\Omega}^{,\pm} \tag{C-12a}$$

and

$$\frac{\partial \tilde{\vec{\Gamma}}'^{\pm}_{\psi,\Omega}}{\partial z} = \mp jk_{z,s} \tilde{\vec{\Gamma}}'^{\pm}_{\psi,\Omega} . \tag{C-12b}$$

Note that

$$\tilde{\vec{D}}'^{\pm}_{p} = -\tilde{\vec{D}}^{\mp}_{p} , \tag{C-13a}$$

$$\tilde{\vec{D}}'^{\pm}_{s} = -\tilde{\vec{D}}^{\mp}_{s} \tag{C-13b}$$

and

$$\tilde{D}'^{\pm}_{s} = -\tilde{D}^{\mp}_{s} . \tag{C-13c}$$

Next, we express $\tilde{\vec{\tau}}^{\pm}_{z}$ in terms of $\tilde{\vec{V}}^{\pm}$. In analogy with equation (IV-3b) we write

$$j\omega\tilde{\vec{\tau}}^{\pm}_{z} = -jk_{\beta}\mathbf{C}_{3\beta}\tilde{\vec{V}}^{\pm} + \mathbf{C}_{33}\frac{\partial\tilde{\vec{V}}^{\pm}}{\partial z} , \tag{C-14a}$$

with $\mathbf{C}_{3\beta}$ and $\mathbf{C}_{33}$ defined by (II-23b). At $z=z_0$, where the medium is assumed to be homogeneous and isotropic (see also equation (C-6)), we obtain

$$j\omega\tilde{\vec{\tau}}^{\pm}_{z} \doteq \begin{bmatrix} \mu\frac{\partial}{\partial z} & 0 & -\mu jk_x \\ 0 & \mu\frac{\partial}{\partial z} & -\mu jk_y \\ -\lambda jk_x & -\lambda jk_y & (\lambda+2\mu)\frac{\partial}{\partial z} \end{bmatrix} \tilde{\vec{V}}^{\pm}. \tag{C-14b}$$

Upon substitution of (C-8) and one-way wave equations (C-9a) and (C-9b) we may express $\tilde{\vec{\tau}}^{\pm}_{z}$ in terms of P- and S-wave potentials, according to

$$\tilde{\vec{\tau}}^{\pm}_{z} = \tilde{\mathbf{E}}^{\pm}_{p}\,\tilde{\vec{D}}^{\pm}_{p}\,\tilde{\Phi}^{\pm} + \tilde{\mathbf{E}}^{\pm}_{s}\,\tilde{\mathbf{D}}^{\pm}_{s}\,\tilde{\vec{\Psi}}^{\pm} \qquad \text{at } z=z_0, \tag{C-15a}$$

where

$$\tilde{\mathbf{E}}_{p}^{\pm} = \frac{1}{j\omega} \begin{bmatrix} \mp\mu jk_{z,p} & 0 & -\mu jk_{x} \\ 0 & \mp\mu jk_{z,p} & -\mu jk_{y} \\ -\lambda jk_{x} & -\lambda jk_{y} & \mp(\lambda+2\mu)jk_{z,p} \end{bmatrix} \tag{C-15b}$$

and

$$\tilde{\mathbf{E}}_{s}^{\pm} = \frac{1}{j\omega} \begin{bmatrix} \mp\mu jk_{z,s} & 0 & -\mu jk_{x} \\ 0 & \mp\mu jk_{z,s} & -\mu jk_{y} \\ -\lambda jk_{x} & -\lambda jk_{y} & \mp(\lambda+2\mu)jk_{z,s} \end{bmatrix}. \tag{C-15c}$$

Similarly, we may express $\tilde{\vec{\Theta}}'^{\pm}_{z,\Omega}$ in terms of P- and S-wave potentials, according to

$$\tilde{\vec{\Theta}}'^{\pm}_{z,\Omega} = \tilde{\mathbf{E}}'^{\pm}_{p} \; \tilde{\vec{\mathbf{D}}}'^{\pm}_{p} \tilde{\Gamma}'^{\pm}_{\phi,\Omega} + \tilde{\mathbf{E}}'^{\pm}_{s} \; \tilde{\mathbf{D}}'^{\pm}_{s} \; \tilde{\vec{\Gamma}}'^{\pm}_{\psi,\Omega} \qquad \text{at } z=z_{o}. \tag{C-16a}$$

Note that

$$\tilde{\mathbf{E}}'^{\pm}_{p} = -\tilde{\mathbf{E}}^{\mp}_{p} \tag{C-16b}$$

and

$$\tilde{\mathbf{E}}'^{\pm}_{s} = -\tilde{\mathbf{E}}^{\mp}_{s} \quad . \tag{C-16c}$$

From equations (C-8), (C-10), (C-11), (C-13), (C-15) and (C-16) we may derive

$$\left[\tilde{\vec{\Theta}}'^{\pm}_{z,\Omega} \cdot \tilde{\vec{V}}^{\pm} - \tilde{\vec{G}}'^{\pm}_{\Omega} \cdot \tilde{\vec{\tau}}^{\pm}_{z}\right]_{z_{o}} = o. \tag{C-17}$$

This equation expresses that no interaction occurs between elastic wave fields which propagate in the same direction through $z=z_0$.

From the same equations we may also derive

$$\left[\tilde{\vec{\Theta}}^{\pm}_{z,\Omega} \cdot \tilde{\vec{V}}^{\mp} - \tilde{\vec{G}}^{\pm}_{\Omega} \cdot \tilde{\vec{\tau}}^{\mp}_{z}\right]_{z_0} = \pm \frac{2}{j\omega\rho(z_0)} \left[jk_{z,p}\tilde{\Gamma}^{\pm}_{\phi,\Omega}\tilde{\Phi}^{\mp} + jk_{z,s}\tilde{\vec{\Gamma}}^{\pm}_{\psi,\Omega} \cdot \tilde{\vec{\Psi}}^{\mp}\right]_{z_0}. \tag{C-18}$$

This equation expresses that interaction occurs between P-waves ($\Phi^{\pm}$, $\Gamma^{\mp}_{\phi,\Omega}$) which propagate in *opposite* directions through $z=z_0$ and between S-waves ($\vec{\Psi}^{\pm}$, $\vec{\Gamma}^{\mp}_{\psi,\Omega}$) which propagate in *opposite* directions through $z=z_0$. *No interaction* occurs between P- and S-waves.

Substitution of (C-17), (C-18) and the one-way wave equations (C-9) or (C-12) into equation (C-7a) and applying Parseval's theorem (B-7) again, yields

$$\Omega(\vec{r}_A,\omega) = \frac{2}{j\omega}\iint_{-\infty}^{\infty} \frac{1}{\rho(z_0)}\left[\Gamma^{+}_{\phi,\Omega}\frac{\partial\Phi^{-}}{\partial z} + \vec{\Gamma}^{+}_{\psi,\Omega} \cdot \frac{\partial\vec{\Psi}^{-}}{\partial z} + \Gamma^{-}_{\phi,\Omega}\frac{\partial\Phi^{+}}{\partial z} + \vec{\Gamma}^{-}_{\psi,\Omega} \cdot \frac{\partial\vec{\Psi}^{+}}{\partial z}\right]_{z_0} dxdy. \tag{C-19a}$$

or, equivalently,

$$\Omega(\vec{r}_A,\omega) = \frac{-2}{j\omega}\iint_{-\infty}^{\infty} \frac{1}{\rho(z_0)}\left[\frac{\partial\Gamma^{+}_{\phi,\Omega}}{\partial z}\Phi^{-} + \frac{\partial\vec{\Gamma}^{+}_{\psi,\Omega}}{\partial z} \cdot \vec{\Psi}^{-} + \frac{\partial\Gamma^{-}_{\phi,\Omega}}{\partial z}\Phi^{+} + \frac{\partial\vec{\Gamma}^{-}_{\psi,\Omega}}{\partial z} \cdot \vec{\Psi}^{+}\right]_{z_0} dxdy. \tag{C-19b}$$

Choosing $\Gamma^{+}_{\phi,\Omega}=o$ and $\vec{\Gamma}^{+}_{\psi,\Omega}=\vec{o}$ in (C-19a) and (C-19b) yields the elastic one-way Rayleigh I and Rayleigh II integrals (VI-66a) and (VI-66b), respectively.

Finally we derive an alternative representation of equation (C-19b) by eliminating the z-component of the S-wave potential $\vec{\Psi}^{\pm}$. From equations (C-8b) and (C-8d) we obtain

$$\pm jk_{z,s}\tilde{\bar{\Psi}}_z^{\pm} = +jk_x\tilde{\bar{\Psi}}_x^{\pm} + jk_y\tilde{\bar{\Psi}}_y^{\pm} \qquad \text{at } z=z_0.$$

Substitution of this expression into equation (C-18) yields

$$[\tilde{\vec{\Theta}}'^{\pm}_{z,\Omega}\cdot\tilde{\vec{V}}^{\bar{\mp}} - \tilde{\vec{G}}'^{\pm}_{\Omega}\cdot\tilde{\vec{\tau}}_z^{\bar{\mp}}]_{z_0} =$$

$$\frac{2}{j\omega\rho(z_0)}\Big[\pm jk_{z,p}\tilde{\Gamma}'^{\pm}_{\phi,\Omega}\tilde{\bar{\Phi}}^{\mp} + \big(\pm jk_{z,s}\tilde{\Gamma}'^{\pm}_{\psi_x,\Omega} + jk_x\tilde{\Gamma}'^{\pm}_{\psi_z,\Omega}\big)\tilde{\bar{\Psi}}_x^{\mp} + \big(\pm jk_{z,s}\tilde{\Gamma}'^{\pm}_{\psi_y,\Omega} + jk_y\tilde{\Gamma}'^{\pm}_{\psi_z,\Omega}\big)\tilde{\bar{\Psi}}_y^{\mp}\Big]_{z_0}. \tag{C-20}$$

Substitution of (C-17), (C-20) and the one-way wave equations (C-12) into equation (C-7a) and applying Parseval's theorem (B-7) again, yields

$$\Omega(\vec{r}_A,\omega) =$$

$$\frac{-2}{j\omega}\iint_{-\infty}^{\infty}\frac{1}{\rho(z_0)}\Bigg[\frac{\partial\Gamma^+_{\phi,\Omega}}{\partial z}\Phi^- + \Big(\frac{\partial\Gamma^+_{\psi_x,\Omega}}{\partial z} - \frac{\partial\Gamma^+_{\psi_z,\Omega}}{\partial x}\Big)\Psi_x^- + \Big(\frac{\partial\Gamma^+_{\psi_y,\Omega}}{\partial z} - \frac{\partial\Gamma^+_{\psi_z,\Omega}}{\partial y}\Big)\Psi_y^-$$

$$+ \frac{\partial\Gamma^-_{\phi,\Omega}}{\partial z}\Phi^+ + \Big(\frac{\partial\Gamma^-_{\psi_x,\Omega}}{\partial z} - \frac{\partial\Gamma^-_{\psi_z,\Omega}}{\partial x}\Big)\Psi_x^+ + \Big(\frac{\partial\Gamma^-_{\psi_y,\Omega}}{\partial z} - \frac{\partial\Gamma^-_{\psi_z,\Omega}}{\partial y}\Big)\Psi_y^+\Bigg]_{z_0} dxdy, \tag{C-21a}$$

or, using a more compact notation,

$$\Omega(\vec{r}_A,\omega) = \frac{-2}{j\omega}\iint_{-\infty}^{\infty}\frac{1}{\rho(z_0)}\Bigg[\frac{\partial\Gamma^+_{\phi,\Omega}}{\partial z}\Phi^- + \big(\epsilon_{3\alpha j}\epsilon_{jik}\partial_i\Gamma^+_{\psi_k,\Omega}\big)\Psi_\alpha^-$$

$$+ \frac{\partial\Gamma^-_{\phi,\Omega}}{\partial z}\Phi^+ + \big(\epsilon_{3\alpha j}\epsilon_{jik}\partial_i\Gamma^-_{\psi_k,\Omega}\big)\Psi_\alpha^+\Bigg]_{z_0} dxdy, \tag{C-21b}$$

where ϵ_{jik} is the alternating tensor as defined by equation (VI-27).

Choosing $\Gamma^{+}_{\phi,\Omega}$=o and $\Gamma^{+}_{\psi_k,\Omega}$=o in (C-21b) yields the modified elastic one-way Rayleigh II integral (VI-71).

C.3 ANALYSIS FOR BACKWARD PROPAGATION

We analyse integral (C-1b), where $(\vec{V}, \vec{\tau}_z)$ represents a causal or forward propagating elastic wave field, whereas $(\vec{G}^*_{\Omega}, \vec{\Theta}^*_{z,\Omega})$ represents an *anti-causal* or *backward propagating* elastic wave field. Upon substitution of (C-2) into (C-1b) we obtain

$$\Omega_o(\vec{r}_A,\omega) = \iint_{-\infty}^{\infty} \left[\left(\vec{\Theta}^{+}_{z,\Omega}+\vec{\Theta}^{-}_{z,\Omega}\right)^{*}.(\vec{V}^{+}+\vec{V}^{-}) + \left(\vec{G}^{+}_{\Omega}+\vec{G}^{-}_{\Omega}\right)^{*}.\left(\vec{\tau}^{+}_{z}+\vec{\tau}^{-}_{z}\right)\right]_{z_o} dxdy. \tag{C-22}$$

The downgoing and upgoing waves satisfy *independent* one-way wave equations at $z=z_o$ when the vertical derivatives of the medium parameters vanish at $z=z_o$, which is expressed by equation (C-5). For simplicity, however, we assume in addition to (C-5) that the medium is isotropic and laterally invariant at $z=z_o$, which is expressed by equation (C-6).

Applying Parseval's theorem (B-27) to equation (C-22) yields

$$\Omega_o(\vec{r}_A,\omega) = \left(\frac{1}{2\pi}\right)^2 \iint_{-\infty}^{\infty} \left[\left(\tilde{\vec{\Theta}}^{+}_{z,\Omega}+\tilde{\vec{\Theta}}^{-}_{z,\Omega}\right)^{*} . (\tilde{\vec{V}}^{+}+\tilde{\vec{V}}^{-}) + \left(\tilde{\vec{G}}^{+}_{\Omega}+\tilde{\vec{G}}^{-}_{\Omega}\right)^{*} . \left(\tilde{\vec{\tau}}^{+}_{z} + \tilde{\vec{\tau}}^{-}_{z}\right)\right]_{z_o} dk_x dk_y. \tag{C-23}$$

In analogy with section C.2, we express the downgoing and upgoing wave fields in terms of P- and S-wave potentials, according to

$$\tilde{\vec{V}}^{\pm} \triangleq \tilde{\vec{D}}^{\pm}_{p} \tilde{\Phi}^{\pm} + \tilde{D}^{\pm}_{s} \tilde{\vec{\Psi}}^{\pm} \qquad \text{at } z=z_o, \tag{C-24a}$$

with

$$\vec{\tilde{D}}_s^{\pm} \cdot \vec{\tilde{\Psi}}^{\pm} \triangleq o \qquad \text{at } z=z_o \tag{C-24b}$$

and

$$\left(\vec{\tilde{G}}_{\Omega}^{\pm}\right)^* \triangleq \left(\vec{\tilde{D}}_p^{\pm}\right)^* \left(\tilde{\Gamma}_{\phi,\Omega}^{\pm}\right)^* + \left(\tilde{\mathbf{D}}_s^{\pm}\right)^* \left(\vec{\tilde{\Gamma}}_{\psi,\Omega}^{\pm}\right)^* \qquad \text{at } z=z_o, \tag{C-24c}$$

with

$$\left(\vec{\tilde{D}}_s^{\pm}\right)^* \cdot \left(\vec{\tilde{\Gamma}}_{\psi,\Omega}^{\pm}\right)^* \triangleq o \qquad \text{at } z=z_o. \tag{C-24d}$$

Similarly,

$$\vec{\tilde{\tau}}_z^{\pm} = \tilde{\mathbf{E}}_p^{\pm} \vec{\tilde{D}}_p^{\pm} \tilde{\Phi}^{\pm} + \tilde{\mathbf{E}}_s^{\pm} \tilde{\mathbf{D}}_s^{\pm} \vec{\tilde{\Psi}}^{\pm} \qquad \text{at } z=z_o \tag{C-24e}$$

and

$$\left(\vec{\tilde{\Theta}}_{z,\Omega}^{\pm}\right)^* \triangleq \left(\tilde{E}_p^{\pm}\right)^* \left(\vec{\tilde{D}}_p^{\pm}\right)^* \left(\tilde{\Gamma}_{\phi,\Omega}^{\pm}\right)^* + \left(\tilde{E}_s^{\pm}\right)^* \left(\tilde{\mathbf{D}}_s^{\pm}\right)^* \left(\vec{\tilde{\Gamma}}_{\psi,\Omega}^{\pm}\right)^*$$

$$\text{at } z=z_o. \tag{C-24f}$$

Here $\vec{\tilde{D}}_p^{\pm}$, $\vec{\tilde{D}}_s^{\pm}$, $\tilde{\mathbf{D}}_s^{\pm}$, $\tilde{\mathbf{E}}_p^{\pm}$ and $\tilde{\mathbf{E}}_s^{\pm}$ are defined as in section C.2. Note that

$$\left(\vec{\tilde{D}}_p^{\pm}\right)^* = \vec{\tilde{D}}_p^{\pm} \qquad \text{for } k_x^2+k_y^2 \le k_p^2(z_o), \tag{C-25a}$$

$$\left(\vec{\tilde{D}}_s^{\pm}\right)^* = \vec{\tilde{D}}_s^{\pm} \qquad \text{for } k_x^2+k_y^2 \le k_s^2(z_o), \tag{C-25b}$$

$$\left(\tilde{\mathbf{D}}_s^{\pm}\right)^* = \tilde{\mathbf{D}}_s^{\pm} \qquad \text{for } k_x^2+k_y^2 \le k_s^2(z_o), \tag{C-25c}$$

$$\left(\tilde{\mathbf{E}}_p^{\pm}\right)^* = \tilde{\mathbf{E}}_p^{\pm} \qquad \text{for } k_x^2+k_y^2 \le k_p^2(z_o) \tag{C-25d}$$

and

$$\left(\tilde{\mathbf{E}}_s^{\pm}\right)^* = \tilde{\mathbf{E}}_s^{\pm} \qquad \text{for } k_x^2+k_y^2 \le k_s^2(z_o). \tag{C-25e}$$

These relations are valid only for the propagating wavenumber areas $k_x^2+k_y^2 \le k_p^2(z_o)$ and $k_x^2+k_y^2 \le k_s^2(z_o)$, because only in those areas the wavenumbers $k_{z,p}$ and $k_{z,s}$, respectively, are real (see equation (C-10)).

Hence, assuming *evanescent waves* are negligible at $z=z_0$, we may derive in analogy with (C-17),

$$[(\tilde{\vec{\Theta}}^{\pm}_{z,\Omega})^{*}.\tilde{\vec{V}}^{\mp} + (\tilde{\vec{G}}^{\pm}_{\Omega})^{*}.\tilde{\vec{\tau}}^{\mp}_{z}]_{z_0} \approx 0 \qquad \text{for all } (k_x,k_y). \tag{C-26}$$

This equation expresses that no interaction occurs between elastic wave fields which propagate in the same direction through $z=z_0$ (keep in mind that the asterisk (*) denotes back-propagation).

Similarly, we may derive in analogy with (C-18),

$$[(\tilde{\vec{\Theta}}^{\pm}_{z,\Omega})^{*}.\tilde{\vec{V}}^{\pm} + (\tilde{\vec{G}}^{\pm}_{\Omega})^{*}.\tilde{\vec{\tau}}^{\pm}_{z}]_{z_0} \approx \tag{C-27}$$

$$\mp \frac{2}{j\omega\rho(z_0)}[jk_{z,p}(\tilde{\Gamma}^{\pm}_{\phi,\Omega})^{*}\tilde{\Phi}^{\pm} + jk_{z,s}(\tilde{\vec{\Gamma}}^{\pm}_{\psi,\Omega})^{*}.\tilde{\vec{\Psi}}^{\pm}]_{z_0} \qquad \text{for all } (k_x,k_y).$$

This equation expresses that interaction occurs between P-waves which propagate in *opposite* directions through $z=z_0$ and between S-waves which propagate in *opposite* directions through $z=z_0$. *No interaction* occurs between P- and S-waves.

Substitution of (C-26), (C-27) and the one-way wave equations (C-9) or (C-12) into equation (C-23) and applying Parseval's theorem (B-27) again, yields

$$\Omega_0(\vec{r}_A,\omega) \approx \frac{2}{j\omega}\int\int_{-\infty}^{\infty}\frac{1}{\rho(z_0)}\Bigg[(\Gamma^{+}_{\phi,\Omega})^{*}\frac{\partial\Phi^{+}}{\partial z} + (\vec{\Gamma}^{+}_{\psi,\Omega})^{*}.\frac{\partial\vec{\Psi}^{+}}{\partial z}$$

$$+ (\Gamma^{-}_{\phi,\Omega})^{*}\frac{\partial\Phi^{-}}{\partial z} + (\vec{\Gamma}^{-}_{\psi,\Omega})^{*}.\frac{\partial\vec{\Psi}^{-}}{\partial z}\Bigg]_{z_0}dxdy, \tag{C-28a}$$

or, equivalently,

$$\Omega_o(\vec{r}_A,\omega) \approx \frac{-2}{j\omega}\int\int_{-\infty}^{\infty}\frac{1}{\rho(z_o)}\Bigg[\left(\frac{\partial\Gamma^+_{\phi,\Omega}}{\partial z}\right)^*\Phi^+ + \left(\frac{\partial\vec{\Gamma}^+_{\psi,\Omega}}{\partial z}\right)^*\cdot\vec{\Psi}^+$$

$$+\left(\frac{\partial\Gamma^-_{\phi,\Omega}}{\partial z}\right)^*\Phi^- + \left(\frac{\partial\vec{\Gamma}^-_{\psi,\Omega}}{\partial z}\right)^*\cdot\vec{\Psi}^-\Bigg]_{z_o}dxdy. \qquad \text{(C-28b)}$$

Note that, unlike expressions (C-19a) and (C-19b) in section C.2, expressions (C-28a) and (C-28b) are not exact. The approximation concerns the negligence of evanescent waves at $z=z_o$.

Choosing $\Gamma^+_{\phi,\Omega}=o$ and $\vec{\Gamma}^+_{\psi,\Omega}=\vec{o}$ in (C-28a) and (C-28b) yields the elastic one-way Rayleigh I and Rayleigh II integrals (VIII-31a) and (VIII-31b), respectively, for inverse wave field extrapolation.

Finally, we may derive an alternative representation of equation (C-28b) by eliminating the z-component of the S-wave potential $\vec{\Psi}^{\pm}$. In analogy with equation (C-21b) we obtain

$$\Omega_o(\vec{r}_A,\omega) \approx \frac{-2}{j\omega}\int\int_{-\infty}^{\infty}\frac{1}{\rho(z_o)}\Bigg[\left(\frac{\partial\Gamma^+_{\phi,\Omega}}{\partial z}\right)^*\Phi^+ + \left(\epsilon_{3\alpha j}\epsilon_{jik}\partial_i\Gamma^+_{\psi_k,\Omega}\right)^*\Psi^+_\alpha$$

$$+\left(\frac{\partial\Gamma^-_{\phi,\Omega}}{\partial z}\right)^*\Phi^- + \left(\epsilon_{3\alpha j}\epsilon_{jik}\partial_i\Gamma^-_{\psi_k,\Omega}\right)^*\Psi^-_\alpha\Bigg]_{z_o}dxdy. \qquad \text{(C-29)}$$

C.4 REFERENCE

Wapenaar, C.P.A., and Haimé, G.C., 1989, Elastic extrapolation of primary seismic P- and S-waves: Geophysical Prospecting, *37*, in press.

INDEX

A

Acoustic waves .... 7 - 33
Airgun 11
Alternating tensor 214, 458
Anisotropic solid 54
Anti-causal 16, 442, 459
Azimuth angle 19, 65, 78, 97
Azimuthal anisotropy 65, 419

B

Backward propagating 16, 443, 459
Bessel function 270
Betti's reciprocity theorem 205
Birefringence (shear wave) 65
Boundary conditions 83, 94, 138, 338, 359

C

Cauchy's formula 40, 228
Causal 16, 434, 451
Composition
  acoustic 91, 377, 378
  elastic 120, 133, 151, 406
Compression modulus 11, 46
Compressional waves 7, 35, 47, 59
Conservation
  angular momentum 42
  mass 8
  momentum 8, 40
Constitutive relation 9, 42
Continuity (equation of) 8, 11, 78, 99
Conversion 134, 327, 415
Convolution (spatial) 101, 148, 259, 271, 305, 425
Cross-correlation (spatial) 435

D

Decomposition
  acoustic .... 92, 374, 376, 381 - 385
  elastic .... 120, 133, 151, 405, 411 - 414
Deconvolution (spatial) 260, 307, 384, 390, 411
Deghosting 382, 383
Dipole
  source 14, 187, 238
  wave field 18
Direct waves 381
Dirichlet boundary condition 175, 177
Downgoing waves 29, 89, 91, 433, 449

E

Eigenvalue decomposition 88, 120
Einstein's summation convention 6
Elastic waves .... 35 - 74
Evanescent waves 30, 71, 78, 262, 446, 462

F
Finite difference modeling 251, 296
Forward model 2, 3
acoustic .... 368 - 380
elastic .... 399 - 408
Forward propagating 16, 434, 451
Fourier transformations .... 76 - 78
properties 79, 102, 147, 435, 443
spatial 76
temporal 76
Free space Green's wave field
acoustic 169, 174, 192
elastic 222, 235, 242
Free surface 69, 141, 156, 176, 229, 372, 403

G
Gauss' theorem 8, 162, 205
Gaussian beam modeling 197
Geophones 377, 406
Ghost 375, 377
Green's functions/wave fields
acoustic .... 166 - 170
backward propagating 167, 171, 208, 279, 312
boundary conditions 175, 181, 228, 232
elastic .... 206 - 210
forward propagating 167, 171, 208, 279, 312
one-way 185
P- and S-waves .... 213 - 224
two-way 178
Green's matrix
acoustic 195
elastic 248
Ground-roll 406

H
Homogeneous plane waves 19
Hooke's law 43
Huygen's principle 187, 239
Hydrophones 377

I
Inhomogeneous plane waves 24
Interactions of wave fields
one-way acoustic .... 282, 433 - 447
one-way elastic .... 316, 449 - 462
Interface 83, 163, 274
Isotropic solid 46, 56
Iterative extrapolation .... 345 - 356, 361 - 366

K
Kirchhoff-Helmholtz integral
acoustic .... 171 - 174, 279 - 281
elastic .... 210 - 213, 312 - 315
P- and S-waves .... 224 - 228
Kronecker symbol 42

L
Lamé coefficients 46, 56
Lamé potentials 47
Longitudinal 22, 59

M
Macro subsurface model 174, 213, 258, 388, 418
Matched filter 264, 273, 308, 341
Matrix notation .... 425 - 431
Migration 2
Monochromatic
  plane waves 22
  wave fields 426
Monopole
  source 13, 187, 238
  wave field 18
Motion (equation of) 9, 11, 41, 78, 99, 116, 144
Multi-component data 3, 247, 310, 325
Multi-component operators 247, 311, 325
Multi-component shot record 413, 414, 415, 417, 422
Multiple elimination
  acoustic .... 385 - 387
  elastic .... 415 - 418
Multiple prediction 386
Multiple reflections 3, 93, 134, 180, 291, 318, 354, 374, 403
Multi-shot record 388
Muting 385

N
Neumann boundary condition 175, 177
Newton's law 8, 40

O
One-way response matrix
  acoustic 369
  elastic 401
Overburden 4, 370, 402

P
Parseval's theorem 435, 443
Pascal's law 37
Phase
  slowness 20, 54
  velocity 20
Phase-shift operator 193, 245
Plane waves
  acoustic .... 18 - 33
  elastic .... 54 - 72
  evanescent 30, 71, 78
  homogeneous 19, 54
  inhomogeneous 24, 67
  monochromatic 22, 31, 70, 77
  propagating 30, 71, 78
Polarization 53, 63

Potential
  Green's P-waves ... 218
  Green's S-waves ... 218
  P-waves ... 47, 132, 233
  S-waves ... 48, 132, 233
Pre-processing (surface related)
  acoustic ... 380 - 387
  elastic ... 408 - 418
Pressure ... 9, 11
Primary waves ... 93, 108, 291, 318, 338
Propagating waves ... 30, 71, 78, 262
Propagation velocity ... 16, 19, 52, 53
Propagator ... 83
Pseudo P- and S-waves ... 328
P-waves ... 47
  qP ... 60, 62

Q
Quasi-longitudinal ... 62
Quasi-transversal ... 63

R
Rayleigh integral
  acoustic one-way (forward) ... 180 - 185
  acoustic one-way (inverse) ... 291 - 294
  acoustic two-way ... 176 - 180
  elastic one-way (forward) ... 231 - 236
  elastic one-way (inverse) ... 319 - 323
  elastic two-way ... 230 - 231
Rayleigh's reciprocity theorem ... 163
Rayleigh wave ... 28, 70, 143
Reciprocity
  acoustic ... 159 - 166, 167, 168, 195
  elastic ... 201 - 206, 207, 209, 221, 241
Recursive extrapolation ... 337 - 345, 358 - 361
Redatuming ... 5
  acoustic ... 387 - 397
  elastic ... 418 - 423
Reference medium ... 171, 181, 210, 258
Reflection operators
  acoustic ... 96, 111
  elastic ... 139, 154
Representation theorem
  acoustic ... 166 - 174
  elastic ... 206 - 228
Resolution (spatial) ... 267, 309
Rigid surface ... 175, 229

S
Seismic inversion ... 1, 4
Shear modulus ... 46
Shear waves ... 7, 35, 48, 59
Shear wave birefringence ... 65
Single-component data ... 2, 367
Single-shot record ... 390
Slowness
  phase ... 20, 54
  surface ... 57, 137
  vector ... 20, 54, 57

Snell's law ... 97
Sommerfeld's radiation condition ... 166, 176, 230, 279, 314
Sources
  airgun ... 11
  dipole ... 14, 187, 238
  force ... 9, 213
  mass injection ... 8
  moment ... 50, 214
  monopole ... 13, 187, 238
  one-way ... 92, 108
  pressure ... 374
  P-waves ... 50
  strain ... 43
  stress ... 44
  S-waves ... 50
  traction ... 405
  vibrator ... 405
  volume injection ... 11, 50, 213, 374
Space-frequency domain ... 76
Space-time domain ... 76
Spatial wavelet ... 267
Spherical waves
  acoustic ... 13 - 18
  elastic ... 49 - 53
Square-root operator ... 103, 110
State (equation of) ... 9, 11
Stiffness coefficients ... 43, 56
Strain ... 37
Stress ... 36
Stress-displacement relation ... 44
Stress-strain relation ... 43
Stress-velocity relation ... 44, 116, 144
Summation convention ... 6, 41
Surface waves ... 70
S-waves ... 48
  SH, qSH ... 58, 60, 63, 64
  SV, qSV ... 58, 60, 63, 64
  S_x, S_y, S_z ... 53, 412

T
Target zone ... 4, 370, 402
Taylor series ... 82, 90
Three-component data ... 149, 400, 404
Tilt angle ... 19, 62, 78, 97
Traction ... 36, 405
Transmission operators
  acoustic ... 96, 111
  elastic ... 140, 154
Transversal ... 59
Transverse isotropy ... 55, 56, 121, 136
True amplitude ... 291, 318

U
Upgoing waves ... 29, 89, 91, 433, 449

V

Velocity
- acoustic propagation 16
- particle 10
- phase 20
- P-wave propagation 52, 59, 127
- S-wave propagation 53, 59, 127

Vibrator 405
Voigt notation 56

W

Wave equation
- acoustic 12, 32
- elastic 45, 72, 134, 144, 151
- one-way 3, 32, 72, 92, 105, 108, 121, 134, 151
- P-waves 48, 51
- S-waves 48, 51
- two-way 2, 12, 45, 81, 87, 99, 117, 144

Wave field extrapolation
- acoustic .... 83, 101, 185 - 198
- elastic .... 119, 146, 236 - 254
- forward .... 159 - 255
- integral formulation .... 185 - 193, 236 - 245
- inverse .... 257 - 366
- matrix formulation .... 194 - 198, 245 - 254, 294 - 296, 323 - 327
- multi-component 247
- one-way .... 32, 159 - 366
- P-waves 243
- S-waves 243
- two-way 83, 101, 119, 146

Wave front
- plane 19, 25
- spherical 16

Wave length 22, 268
Wave-number frequency domain 76
WKB-solutions 93, 193, 272

Z

Zero offset data
- acoustic 392, 394
- elastic 421